Study and Solutions Guide

for

Calculus

Early Transcendendal Functions

by

Larson, Hostetler, and Edwards

DAVID E. HEYD

The Pennsylvania State University

The Behrend College

D. C. Heath and Company

Lexington, Massachusetts   Toronto

International Standard Book Number: 0-669-39351-7

9 8 7 6 5 4 3 2

Preface

This *Guide* is designed as a supplement to *Calculus, Early Transcendental Functions* by Roland E. Larson, Robert P. Hostetler, and Bruce H. Edwards. All references to chapters, theorems, and exercises relate to the main text. Although this supplement is not a substitute for good study habits, it can be of great value when incorporated into a well-planned course of study. The following suggestions are given to assist you in the use of the text, your lecture notes, and this *Guide*.

- *Read the section in the text for general content before class.* You will be surprised how much more you will acquire from the lecture if you are aware of the objectives of the section and the types of problems that will be solved. If you are familiar with the topic, you will understand more of the lecture and you will be able to take fewer (and better) notes.

- *As soon after class as possible, work problems from the exercise set.* The exercise sets in the text are divided into groups of similar problems and are presented in approximately the same order as the section topics. Try to get an overall picture of the various types of problems in the set. As you work your way through the exercise set, reread your class notes and the portion of the section that covers each type of problem. Pay particular attention to the solved examples.

- *Learning calculus takes much practice.* You cannot learn calculus merely by reading any more than you can learn to play the piano or to bowl merely by reading. Only after you have practiced the techniques of a section and have discovered your weak points can you make good use of the supplementary solutions in this guide.

- *Technology.* Graphing utilities and symbolic algebra systems are now readily available. A computer icon is shown next to the exercises in this *Guide* where a computer/calculator is required or simply helpful. The computer and calculator are merely tools. Your ability to effectively use these tools requires that you continually sharpen your problem solving skills and your understanding of fundamental mathematical principles.

During many years of teaching I have found that good study habits are essential for success in mathematics. My students have found the following additional suggestions to be helpful in making the best use of their time.

- *Write neatly in pencil.* A notebook filled with unorganized scribbling is of little value.

- *Work at a deliberate and methodical pace without skipping steps.* When you hurry through a problem you are more apt to make careless arithmetic or algebraic errors that, in the long run, waste time.

- *Keep up with the work.* This suggestion is crucial because calculus is a very structured topic. If you cannot do the problems in one section, you are not likely to be able to do the problems in the next. The night before a quiz or test is not the time to start working problems. In some instances cramming may help you pass an examination, but it is an inferior way to learn and retain essential concepts.

- After working some of the assigned exercises with access to the examples and answers, *try at least one of each type of exercise with the book closed.* This will increase your confidence on quizzes and tests.

- Do not be overly concerned with finding the most efficient way to solve a problem. *Your first goal is to find one way that works.* Short cuts and clever methods come later.

- If you have trouble with the algebra of calculus, refer to the algebra review at the beginning of this guide.

I wish to acknowledge several people whose help and encouragement were invaluable in the production of this *Guide*. First, I am grateful to Roland E. Larson, Robert P. Hostetler, and Bruce H. Edwards for the privilege of working with them on the main text. I also wish to thank Jessica L. Pflueger for the computer graphics, and the staff at D. C. Heath and Company. I am grateful to my wife, Jean, for her love and support and to our children, Ed, Ruth and Andy, for their assistance during the months I've worked on this project.

David E. Heyd

Contents

0 ALGEBRA REVIEW

0.1 Monomial Factors

Factor as indicated:

(a) $3x^4 + 4x^3 - x^2 = x^2($ $)$

(b) $2\sqrt{x} + 6x^{3/2} = 2\sqrt{x}($ $)$

(c) $e^{-x} - xe^{-x} + 2x^2e^{-x} = e^{-x}($ $)$

(d) $x^{-1} - 2 + x = x^{-1} = x^{-1}($ $)$

(e) $\dfrac{x}{2} - 6x^2 = \dfrac{x}{2}($ $)$

(f) $\sin x + \tan x = \sin x($ $)$

(g) $\dfrac{1}{2x^2 + 4x} = \dfrac{1}{2x}\left($ $\right)$

Solution:

(a) $3x^4 + 4x^3 - x^2 = x^2(3x^2 + 4x - 1)$

(b) $2\sqrt{x} + 6x^{3/2} = 2\sqrt{x}(1 + 3x)$

(c) $e^{-x} - xe^{-x} + 2x^2e^{-x} = e^{-x}(1 - x + 2x^2)$

(d) $x^{-1} - 2 + x = x^{-1} = x^{-1}(1 - 2x + x^2)$

(e) $\dfrac{x}{2} - 6x^2 = \dfrac{x}{2}(1 - 12x)$

(f) $\sin x + \tan x = \sin x + \dfrac{\sin x}{\cos x} = \sin x\left(1 + \dfrac{1}{\cos x}\right)$

$= \sin x(1 + \sec x)$

(g) $\dfrac{1}{2x^2 + 4x} = \dfrac{1}{2x}\left(\dfrac{1}{x+2}\right)$

0.2   Binomial Factors

Factor as indicated:

(a) $(x-1)^2(x) - (x-1) = (x-1)(\quad\quad)$

(b) $3(x^2+4)(x^2+1) + 6(x^2+4)^2 = 3(x^2+4)(\quad\quad)$

(c) $\sqrt{x^2+1} - \dfrac{x^2}{\sqrt{x^2+1}} = \dfrac{1}{\sqrt{x^2+1}}(\quad\quad)$

(d) $(x-3)^3(x+2) - 2(x-3)^2(x+2)^2$
$$= (x-3)^2(x+2)(\quad\quad)$$

(e) $(2x+1)^{3/2}(x^{1/2}) + (2x+1)^{5/2}(x^{-1/2})$
$$= (2x+1)^{3/2}(x^{-1/2})(\quad\quad)$$

Solution:

(a) $(x-1)^2(x) - (x-1) = (x-1)[(x-1)x - 1]$
$$= (x-1)(x^2 - x - 1)$$

(b) $3(x^2+4)(x^2+1) + 6(x^2+4)^2$
$$= 3(x^2+4)[(x^2+1) + 2(x^2+4)]$$
$$= 3(x^2+4)(3x^2+9)$$

(c) $\sqrt{x^2+1} - \dfrac{x^2}{\sqrt{x^2+1}} = (x^2+1)^{1/2} - x^2(x^2+1)^{-1/2}$
$$= (x^2+1)^{-1/2}[(x^2+1) - x^2]$$
$$= \dfrac{1}{\sqrt{x^2+1}}$$

(d) $(x-3)^3(x+2) - 2(x-3)^2(x+2)^2$
$$= (x-3)^2(x+2)[(x-3) - 2(x+2)]$$
$$= (x-3)^2(x+2)(-x-7)$$

(e) $(2x+1)^{3/2}(x^{1/2}) + (2x+1)^{5/2}(x^{-1/2})$
$$= (2x+1)^{3/2}(x^{-1/2})[x + (2x+1)]$$
$$= (2x+1)^{3/2}(x^{-1/2})(3x+1)$$

0.3   Factoring Quadratic Expressions

Factor as indicated:

(a) $x^2 - 3x + 2 = ($ $)($ $)$

(b) $x^2 - 9 = ($ $)($ $)$

(c) $x^2 + 5x - 6 = ($ $)($ $)$

(d) $x^2 + 5x + 6 = ($ $)($ $)$

(e) $2x^2 + 5x - 3 = ($ $)($ $)$

(f) $e^{2x} + 2 + e^{-2x} = ($ $)^2$

(g) $x^4 - 7x^2 + 12 = ($ $)($ $)($ $)$

(h) $1 - \sin^2 x = ($ $)($ $)$

Solution:

(a) $x^2 - 3x + 2 = (x - 2)(x - 1)$

(b) $x^2 - 9 = (x + 3)(x - 3)$

(c) $x^2 + 5x - 6 = (x + 6)(x - 1)$

(d) $x^2 + 5x + 6 = (x + 2)(x + 3)$

(e) $2x^2 + 5x - 3 = (2x - 1)(x + 3)$

(f) $e^{2x} + 2 + e^{-2x} = (e^x + e^{-x})^2$

(g) $x^4 - 7x^2 + 12 = (x^2 - 3)(x^2 - 4) = (x^2 - 3)(x + 2)(x - 2)$

(h) $1 - \sin^2 x = (1 + \sin x)(1 - \sin x)$

0.4   Cancellation

Reduce each expression to lowest terms:

(a) $\dfrac{3x + 9}{6x}$

(b) $\dfrac{x^2}{x^{1/2}}$

(c) $\dfrac{(x + 1)^3(x - 2) + 3(x + 1)^2}{(x + 1)^4}$

(d) $\dfrac{x^{1/2} - x^{1/3}}{x^{1/6}}$

(e) $\dfrac{\sqrt{x-1}+(x-1)^{3/2}}{\sqrt{x-1}}$ (f) $\dfrac{1-(\sin x + \cos x)^2}{2\sin x}$

Solution:

(a) $\dfrac{3x+9}{6x} = \dfrac{3(x+3)}{3(2x)} = \dfrac{x+3}{2x}$

(b) $\dfrac{x^2}{x^{1/2}} = \dfrac{(x^{1/2})(x^{3/2})}{x^{1/2}} = x^{3/2}$

(c) $\dfrac{(x+1)^3(x-2)+3(x+1)^2}{(x+1)^4} = \dfrac{(x+1)^2[(x+1)(x-2)+3]}{x+1)^4}$

$\qquad = \dfrac{x^2-x+1}{(x+1)^2}$

(d) $\dfrac{x^{1/2}-x^{1/3}}{x^{1/6}} = \dfrac{x^{1/6}(x^{2/6}-x^{1/6})}{x^{1/6}} = x^{1/3}-x^{1/6}$

(e) $\dfrac{\sqrt{x-1}+(x-1)^{3/2}}{\sqrt{x-1}} = \dfrac{\sqrt{x-1}[1+(x-1)]}{\sqrt{x-1}} = x$

(f) $\dfrac{1-(\sin x + \cos x)^2}{2\sin x} = \dfrac{1-(\sin^2 x + 2\sin x \cos x + \cos^2 x)}{2\sin x}$

$\qquad = \dfrac{1-(\sin^2 x + \cos^2 x) - 2\sin x \cos x}{2\sin x}$

$\qquad = \dfrac{1-1-2\sin x \cos x}{2\sin x} = -\cos x$

0.5   Quadratic Formula

	Equation	*Solve for*
(a)	$x^2-4x-1=0$	x
(b)	$2x^2+x-3=0$	x
(c)	$\cos^2 x + 3\cos x + 2 = 0$	$\cos x$
(d)	$x^2-xy-(1+y^2)=0$	x
(e)	$x^4-4x^2+2=0$	x^2

Solution:

(a) $x = \dfrac{4 \pm \sqrt{16+4}}{2} = \dfrac{4 \pm \sqrt{20}}{2} = \dfrac{4 \pm 2\sqrt{5}}{2} = 2 \pm \sqrt{5}$

(b) $x = \dfrac{-1 \pm \sqrt{1+24}}{4} = \dfrac{-1 \pm 5}{4}$

$x = \dfrac{4}{4} = 1$ or $x = -\dfrac{6}{4} = -\dfrac{3}{2}$

(c) $\cos x = \dfrac{-3 \pm \sqrt{9-8}}{2} = \dfrac{-3 \pm 1}{2}$

$\cos x = -\dfrac{2}{2} = -1$ or $\cos x = -\dfrac{4}{2} = -2$

(d) $x = \dfrac{y \pm \sqrt{y^2 + 4(1+y^2)}}{2} = \dfrac{y \pm \sqrt{y^2 + 4 + 4y^2}}{2}$

$= \dfrac{y \pm \sqrt{5y^2 + 4}}{2}$

(e) $x^2 = \dfrac{4 \pm \sqrt{16-8}}{2} = \dfrac{4 \pm \sqrt{8}}{2} = \dfrac{4 \pm 2\sqrt{2}}{2} = 2 \pm \sqrt{2}$

0.6 Synthetic Division

Use synthetic division to factor as indicated:

(a) $x^3 - 4x^2 + 2x + 1 = (x-1)(\quad)$

(b) $2x^3 + 5x + 7 = (x+1)(\quad)$

(c) $x^4 - 3x^3 + x^2 + x + 2 = (x-2)(\quad)$

(d) $4x^4 + 3x^2 - 1 = (2x-1)(\quad)$

Solution:

(a) $\qquad x^3 - 4x^2 + 2x + 1$

$$
\begin{array}{r|rrrr}
1 & 1 & -4 & 2 & 1 \\
  &   & 1 & -3 & -1 \\
\hline
  & 1 & -3 & -1 & 0
\end{array}
$$

$x^3 - 4x^2 + 2x + 1 = (x-1)(x^2 - 3x - 1)$

(b) $2x^3 + 5x + 7$

$$
\begin{array}{r|rrrr}
-1 & 2 & 0 & 5 & 7 \\
   &   & -2 & 2 & -7 \\
\hline
   & 2 & -2 & 7 & 0
\end{array}
$$

$$2x^3 + 5x + 7 = (x+1)(2x^2 - 2x + 7)$$

(c) $x^4 - 3x^3 + x^2 + x + 2$

$$
\begin{array}{r|rrrrr}
2 & 1 & -3 & 1 & 1 & 2 \\
  &   & 2 & -2 & -2 & -2 \\
\hline
  & 1 & -1 & -1 & -1 & 0
\end{array}
$$

$$x^4 - 3x^3 + x^2 + x + 2 = (x-2)(x^3 - x^2 - x - 1)$$

(d) $4x^4 + 3x^2 - 1$

$$
\begin{array}{r|rrrrr}
\frac{1}{2} & 4 & 0 & 3 & 0 & -1 \\
            &   & 2 & 1 & 2 & 1 \\
\hline
            & 4 & 2 & 4 & 2 & 0
\end{array}
$$

$$
\begin{aligned}
4x^4 + 3x^2 - 1 &= \left(x - \frac{1}{2}\right)(4x^3 + 2x^2 + 4x + 2) \\
&= (2x - 1)(2x^3 + x^2 + 2x + 1)
\end{aligned}
$$

0.7   Special Products

Factor completely (into linear or irreducible quadratic factors):

(a) $x^3 - 27$ (b) $x^3 - 3x^2 + 3x - 1$

(c) $x^3 + 6x^2 + 12x + 8$ (d) $x^4 - 25$

(e) $x^4 - 8x^3 + 24x^2 - 32x + 16$

Solution:

(a) $x^3 - 27 = (x-3)(x^2 + 3x + 9)$

(b) $x^3 - 3x^2 + 3x - 1 = (x-1)^3$

(c) $x^3 + 6x^2 + 12x + 8 = x^3 + 3(2)x^2 + 3(2^2)x + 2^3 = (x+2)^3$

(d) $x^4 - 25 = (x^2 + 5)(x^2 - 5) = (x^2 + 5)(x + \sqrt{5})(x - \sqrt{5})$

(e) $x^4 - 8x^3 + 24x^2 - 32x + 16$
$$= x^4 - 4(2)x^3 + 6(2^2)x^2 - 4(2^3)x + 2^4 = (x-2)^4$$

0.8   Factoring by Grouping

Factor completely (into linear or irreducible quadratic factors):

(a) $x^3 + 4x^2 - 2x - 8$

(b) $x^3 + 2x^2 + 3x + 6$

(c) $5\cos^2 x - 5\sin^2 x + \sin x + \cos x$

(d) $\cos^2 x + 4\cos x + 4 - \tan^2 x$

Solution:

(a) $x^3 + 4x^2 - 2x - 8 = x^2(x+4) - 2(x+4)$
$$= (x^2 - 2)(x+4)$$
$$= (x + \sqrt{2})(x - \sqrt{2})(x+4)$$

(b) $x^3 + 2x^2 + 3x + 6 = x^2(x+2) + 3(x+2)$
$$= (x^2 + 3)(x+2)$$

(c) $5\cos^2 x - 5\sin^2 x + \sin x + \cos x$
$$= 5(\cos^2 x - \sin^2 x) + (\sin x + \cos x)$$
$$= 5(\cos x + \sin x)(\cos x - \sin x) + (\cos x + \sin x)$$
$$= (\cos x + \sin x)[5(\cos x - \sin x) + 1]$$

(d) $\cos^2 x + 4\cos x + 4 - \tan^2 x$
$$= (\cos x + 2)^2 - \tan^2 x$$
$$= (\cos x + 2 + \tan x)(\cos x + 2 - \tan x)$$

0.9 Simplifying

Rewrite each of the following in simplest form:

(a) $\dfrac{(x-1)(x+3)-(x+1)^2}{x+1}$

(b) $\dfrac{\sqrt{x^2+1}-\dfrac{1}{\sqrt{x^2+1}}}{x^2+1}$

(c) $\dfrac{x^2-5x+6}{x^2-4x+4}$

(d) $\dfrac{1}{x+1}-\dfrac{1}{x-1}-\dfrac{1}{x^2-1}$

(e) $\dfrac{x(-2x)}{2\sqrt{1-x^2}}+\sqrt{1-x^2}+\dfrac{1}{\sqrt{1-x^2}}$

Solution:

(a) $\dfrac{(x-1)(x+3)-(x+1)^2}{x+1}=\dfrac{(x^2+2x-3)-(x^2+2x+1)}{x+1}$

$$=\dfrac{-4}{x+1}$$

(b) $\dfrac{\sqrt{x^2+1}-\dfrac{1}{\sqrt{x^2+1}}}{x^2+1}=\dfrac{\dfrac{1}{\sqrt{x^2+1}}(x^2+1-1)}{x^2+1}$

$$=\dfrac{x^2+1-1}{\sqrt{x^2+1}(x^2+1)}=\dfrac{x^2}{(x^2+1)^{3/2}}$$

(c) $\dfrac{x^2-5x+6}{x^2-4x+4}=\dfrac{(x-2)(x-3)}{(x-2)^2}=\dfrac{x-3}{x-2}$

(d) $\dfrac{1}{x+1}-\dfrac{1}{x-1}-\dfrac{2}{x^2-1}=\dfrac{(x-1)-(x+1)-2}{x^2-1}=\dfrac{-4}{x^2-1}$

(e) $\dfrac{x(-2x)}{2\sqrt{1-x^2}}+\sqrt{1-x^2}+\dfrac{1}{\sqrt{1-x^2}}$

$$=\dfrac{-x^2}{\sqrt{1-x^2}}+\dfrac{1-x^2}{\sqrt{1-x^2}}+\dfrac{1}{\sqrt{1-x^2}}$$

$$=\dfrac{2-2x^2}{\sqrt{1-x^2}}$$

$$=\dfrac{2(1-x^2)}{\sqrt{1-x^2}}=2\sqrt{1-x^2}$$

0.10   Rationalizing

Remove the sum or difference from the denominator by multiplying the numerator and denominator by the conjugate of the denominator.

(a) $\dfrac{1}{1 - \cos x}$ (b) $\dfrac{x}{1 - \sqrt{x^2 + 1}}$ (c) $\dfrac{2}{x + \sqrt{x^2 + 1}}$

Solution:

(a) $\dfrac{1}{1 - \cos x} = \left(\dfrac{1}{1 - \cos x}\right)\left(\dfrac{1 + \cos x}{1 + \cos x}\right)$

$\qquad\qquad = \dfrac{1 + \cos x}{1 - \cos^2 x} = \dfrac{1 + \cos x}{\sin^2 x}$

(b) $\left(\dfrac{x}{1 - \sqrt{x^2 + 1}}\right)\left(\dfrac{1 + \sqrt{x^2 + 1}}{1 + \sqrt{x^2 + 1}}\right)$

$\qquad = \dfrac{x(1 + \sqrt{x^2 + 1})}{1 - (x^2 + 1)}$

$\qquad = \dfrac{x(1 + \sqrt{x^2 + 1})}{-x^2} = \dfrac{1 + \sqrt{x^2 + 1}}{-x}$

(c) $\left(\dfrac{2}{x + \sqrt{x^2 + 1}}\right)\left(\dfrac{x - \sqrt{x^2 + 1}}{x - \sqrt{x^2 + 1}}\right)$

$\qquad = \dfrac{2(x - \sqrt{x^2 + 1})}{x^2 - (x^2 + 1)} = -2(x - \sqrt{x^2 + 1})$

0.11   Algebraic Errors to Avoid

Error	*Correct form*	*Comment*
$a - (x - b) \neq a - x - b$	$a - (x - b) = a - x + b$	Change all signs when distributing negative through parentheses.
$(a + b)^2 \neq a^2 + b^2$	$(a + b)^2 = a^2 + 2ab + b^2$	Don't forget middle term when squaring binomials.
$\left(\dfrac{1}{2}a\right)\left(\dfrac{1}{2}b\right) \neq \dfrac{1}{2}ab$	$\left(\dfrac{1}{2}a\right)\left(\dfrac{1}{2}b\right) = \dfrac{1}{4}(ab)$	1/2 occurs twice as a factor.
$\dfrac{a}{x + b} \neq \dfrac{a}{x} + \dfrac{a}{b}$	Leave as $\dfrac{a}{x + b}$	Don't add denominators when adding fractions.
$\dfrac{1}{a} + \dfrac{1}{b} \neq \dfrac{1}{a + b}$	$\dfrac{1}{a} + \dfrac{1}{b} = \dfrac{a + b}{ab}$	Use definition for adding fractions.
$\dfrac{\frac{x}{a}}{b} \neq \dfrac{bx}{a}$	$\dfrac{\frac{x}{a}}{b} = \left(\dfrac{x}{a}\right)\left(\dfrac{1}{b}\right) = \dfrac{x}{ab}$	Multiply by reciprocal of the denominator.
$\dfrac{1}{3x} \neq \dfrac{1}{3}x$	$\dfrac{1}{3x} = \dfrac{1}{3} \cdot \dfrac{1}{x}$	Use definition for multiplying fractions.
$1/x + 2 \neq \dfrac{1}{x + 2}$	$1/x + 2 = \dfrac{1}{x} + 2$	Be careful when using a slash to denote division.
$(x^2)^3 \neq x^5$	$(x^2)^3 = x^{2 \cdot 3} = x^6$	Multiply exponents when an exponential form is raised to a power.
$2x^3 \neq (2x)^3$	$2x^3 = 2(x^3)$	Exponents have priority over coeffiecients.
$\dfrac{1}{x^2 + x^3} \neq x^{-2} + x^{-3}$	Leave as $\dfrac{1}{x^2 + x^3}$	Don't shift term-by-term from denominator to numerator.
$\sqrt{5x} \neq 5\sqrt{x}$	$\sqrt{5x} = \sqrt{5}\sqrt{x}$	Radicals apply to every factor inside radical.
$\sqrt{x^2 + a^2} \neq x + a$	Leave as $\sqrt{x^2 + a^2}$	Don't apply radicals term by term.
$\dfrac{a + bx}{a} \neq 1 + bx$	$\dfrac{a + bx}{a} = 1 + \dfrac{b}{a}x$	Cancel common factors, *not* common terms.
$\dfrac{a + ax}{a} \neq a + x$	$\dfrac{a + ax}{a} = 1 + x$	Factor *before* canceling.

PREREQUISITES

THE CARTESIAN PLANE AND FUNCTIONS

1 The Cartesian plane

1. Plot the points $(2, 1)$ and $(4, 5)$. Find the distance between the points and find the midpoint of the line segment joining the points.

Solution:

Let $(2, 1) = (x_1, y_1)$ and $(4, 5) = (x_2, y_2)$. Then

$$d = \sqrt{(x_2 - x_1)^2 + (y_2 - y_1)^2} = \sqrt{(4 - 2)^2 + (5 - 1)^2}$$
$$= \sqrt{2^2 + 4^2} = \sqrt{20} = 2\sqrt{5}$$

$$\text{midpoint} = \left(\frac{x_1 + x_2}{2}, \frac{y_1 + y_2}{2} \right) = \left(\frac{2 + 4}{2}, \frac{1 + 5}{2} \right) = (3, 3)$$

7. Show that the points $(4, 0), (2, 1)$, and $(-1, -5)$ are vertices of a right triangle.

Solution:

Let $d_1 =$ distance between $(4, 0)$ and $(2, 1)$.

$$d_1{}^2 = (2 - 4)^2 + (1 - 0)^2 = (-2)^2 + 1^2 = 5$$

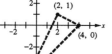

Let d_2 = distance between $(2, 1)$ and $(-1, -5)$.

$$d_2{}^2 = (-1 - 2)^2 + (-5 - 1)^2 = (-3)^2 + (-6)^2 = 45$$

Let d_3 = distance between $(4, 0)$ and $(-1, -5)$.

$$d_3{}^2 = (-1 - 4)^2 + (-5 - 0)^2 = (-5)^2 + (-5)^2 = 50$$

Since $d_1{}^2 + d_2{}^2 = 5 + 45 = 50 = d_3{}^2$, the triangle must be a right triangle.

15. Use the distance formula to determine if the points $(-2, 1), (-1, 0)$, and $(2, -2)$ lie on a straight line.

Solution:

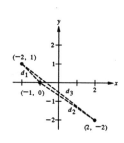

Let d_1 = distance between $(-2, 1)$ and $(-1, 0)$. Then

$$d_1 = \sqrt{[-1 - (-2)]^2 + (0 - 1)^2} = \sqrt{1^2 + (-1)^2} = \sqrt{2}$$

Let d_2 = distance between $(-1, 0)$ and $(2, -2)$. Then

$$d_2 = \sqrt{[2 - (-1)]^2 + (-2 - 0)^2} = \sqrt{3^2 + (-2)^2} = \sqrt{13}$$

Let d_3 = distance between $(-2, 1)$ and $(2, -2)$. Then

$$d_3 = \sqrt{[2 - (-2)]^2 + (-2 - 1)^2} = \sqrt{4^2 + (-3)^2} = \sqrt{25} = 5$$

The points $(-2, 1), (-1, 0)$, and $(2, -2)$ lie on a line only if $d_1 + d_2 = d_3$. Since $\sqrt{2} + \sqrt{13} \approx 5.02 \neq 5$, the points are *not* collinear.

21. Find the relationship between x and y so that (x, y) is equidistant from $(4, -1)$ and $(-2, 3)$.

Solution:

Let d_1 = distance between $(4, -1)$ and (x, y). Then

$$\begin{aligned} d_1 &= \sqrt{(x - 4)^2 + [y - (-1)]^2} \\ &= \sqrt{x^2 - 8x + 16 + y^2 + 2y + 1} \\ &= \sqrt{x^2 - 8x + y^2 + 2y + 17} \end{aligned}$$

Let d_2 = distance between $(-2, 3)$ and (x, y). Then

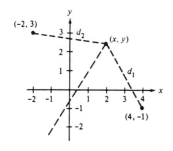

$$d_2 = \sqrt{[x - (-2)]^2 + (y - 3)^2}$$
$$= \sqrt{x^2 + 4x + 4 + y^2 - 6y + 9}$$
$$= \sqrt{x^2 + 4x + y^2 - 6y + 13}$$

Setting d_1 equal to d_2, you have

$$\sqrt{x^2 - 8x + y^2 + 2y + 17} = \sqrt{x^2 + 4x + y^2 - 6y + 13}$$
$$x^2 - 8x + y^2 + 2y + 17 = x^2 + 4x + y^2 - 6y + 13$$
$$-12x + 8y = -4$$
$$3x - 2y = 1$$

23. Use the Midpoint Rule successively to find the three points that divide the line segment joining (x_1, y_1) and (x_2, y_2) into four equal parts.

Solution:

The midpoint of the given line segment is $\left(\dfrac{x_1 + x_2}{2}, \dfrac{y_1 + y_2}{2} \right)$.

The midpoint between (x_1, y_1) and $\left(\dfrac{x_1 + x_2}{2}, \dfrac{y_1 + y_2}{2} \right)$ is

$$\left(\frac{x_1 + \dfrac{x_1 + x_2}{2}}{2}, \frac{y_1 + \dfrac{y_1 + y_2}{2}}{2} \right)$$
$$= \left(\frac{1}{2} \left(\frac{2x_1 + x_1 + x_2}{2} \right), \frac{1}{2} \left(\frac{2y_1 + y_1 + y_2}{2} \right) \right)$$
$$= \left(\frac{3x_1 + x_2}{4}, \frac{3y_1 + y_2}{4} \right)$$

The midpoint between $\left(\dfrac{x_1 + x_2}{2}, \dfrac{y_1 + y_2}{2} \right)$ and (x_2, y_2) is

$$\left(\frac{\dfrac{x_1 + x_2}{2} + x_2}{2}, \frac{\dfrac{y_1 + y_2}{2} + y_2}{2} \right)$$
$$= \left(\frac{1}{2} \left(\frac{x_1 + x_2 + 2x_2}{2} \right), \frac{1}{2} \left(\frac{y_1 + y_2 + 2y_2}{2} \right) \right)$$
$$= \left(\frac{x_1 + 3x_2}{4}, \frac{y_1 + 3y_2}{4} \right)$$

Thus the three points are

$$\left(\frac{3x_1 + x_2}{4}, \frac{3y_1 + y_2}{4}\right), \left(\frac{x_1 + x_2}{2}, \frac{y_1 + y_2}{2}\right)$$

$$\left(\frac{x_1 + 3x_2}{4}, \frac{y_1 + 3y_2}{4}\right)$$

33. Write the general equation of the circle with center at $(2, -1)$ and radius 4.

Solution:

Let $(2, -1) = (h, k)$ and $r = 4$. Then using the standard form of the equation of a circle, you have

$$(x - h)^2 + (y - k)^2 = r^2$$
$$(x - 2)^2 + [y - (-1)]^2 = 4^2$$
$$x^2 - 4x + 4 + y^2 + 2y + 1 = 16$$
$$x^2 + y^2 - 4x + 2y - 11 = 0$$

39. Write the general equation of the circle passing through the points $(0, 0)$, $(0, 8)$, and $(6, 0)$.

Solution:

Since the general form of the equation of the circle is

$$x^2 + y^2 + Dx + Ey + F = 0$$

you must find the coefficients D, E, and F such that the given points are solution points.

Solution Point	Resulting equation	Coefficient
$(0,0)$	$(0)^2 + (0)^2 + D(0) + E(0) + F = 0$	$F = 0$
$(0,8)$	$(0)^2 + (8)^2 + D(0) + E(8) + F = 0$	$E = -8$
$(6,0)$	$(6)^2 + (0)^2 + D(6) + E(0) + F = 0$	$D = -6$

Therefore, the general equation is $x^2 + y^2 - 6x - 8y = 0$.

53. Sketch the inequality $x^2 + y^2 - 4x + 2y + 1 \leq 0$.

Solution:

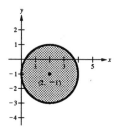

$$x^2 + y^2 - 4x + 2y + 1 \leq 0$$
$$(x^2 - 4x + 4) + (y^2 + 2y + 1) \leq -1 + 4 + 1$$
$$(x - 2)^2 + (y + 1)^2 \leq 4$$

Therefore, the inequality is satisfied by the set of all points lying on the boundary and in the interior of the circle with center $(2, -1)$ and radius 2.

59. Prove that an angle inscribed in a semicircle is a right angle.

Solution:

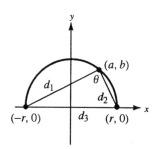

For simplicity, assume the semicircle is centered at the origin with a radius r (see the accompanying figure). If (a, b) is a point on the semicircle, then it must satisfy the equation $a^2 + b^2 = r^2$. To verify that the angle at (a, b) is a right angle, it is sufficient to show that $d_1{}^2 + d_2{}^2 = d_3{}^2$.

$$d_1{}^2 = [a - (-r)]^2 + (b - 0)^2$$
$$d_2{}^2 = (a - r)^2 + (b - 0)^2$$
$$d_1{}^2 + d_2{}^2 = (a^2 + 2ar + r^2 + b^2) + (a^2 - 2ar + r^2 + b^2)$$
$$= 2a^2 + 2b^2 + 2r^2$$
$$= 2(a^2 + b^2) + 2r^2$$
$$= 2r^2 + 2r^2 = 4r^2 = (2r)^2 = d_3{}^2$$

2   Graphs of Equations

9. Find the intercepts of the graph of $y = x^2 + x - 2$.

Solution:

To find the x-intercepts, let $y = 0$. Then

$$x^2 + x - 2 = 0$$

and by factoring (or by the quadratic formula),

$$(x + 2)(x - 1) = 0.$$

Therefore, $y = 0$ when $x = -2$ or $x = 1$ and the x-intercepts are $(-2, 0)$ and $(1, 0)$. To find the y-intercepts, let $x = 0$. Then $y = 0^2 + 0 - 2 = -2$, and the y-intercept is $(0, -2)$.

15. Find the intercepts of the graph of $x^2 y - x^2 + 4y = 0$.

Solution:

To find the x-intercepts you let $y = 0$. Then $x^2(0) - x^2 + 4(0) = -x^2 = 0$, which implies that $x = 0$. Letting $y = 0$ yields the same result and the only intercept is $(0, 0)$.

21. Check $y^2 = x^3 - 4x$ for symmetry about both axes and the origin.

Solution:

There is *no symmetry* about the y-axis since replacing x with $-x$ in the equation yields

$$y^2 = (-x)^3 - 4(-x) = -x^3 + 4x$$

which is *not* equivalent to the original equation.

There is *symmetry* about the x-axis since replacing y with $-y$ in the equation yields

$$(-y)^2 = x^3 - 4x \qquad \text{or} \qquad y^2 = x^3 - 4x$$

which *is* equivalent to the original equation.

There is *no symmetry* about the origin since replacing x with $-x$ and y with $-y$ in the equation yields

$$(-y)^2 = (-x)^3 - 4(-x) \qquad \text{or} \qquad y^2 = -x^3 + 4x$$

which is *not* equivalent to the original equation.

25. Check $y = x/(x^2 + 1)$ for symmetry about both axes and the origin.

Solution:

There is *no symmetry* about the y-axis since replacing x with $-x$ in the equation yields

$$y = \frac{-x}{(-x)^2 + 1} = \frac{-x}{x^2 + 1}$$

which is *not* equivalent to the original equation.

There is *no symmetry* about the x-axis since replacing y with $-y$ in the equation yields

$$-y = \frac{x}{x^2 + 1}$$

which is *not* equivalent to the original equation.

There is *symmetry* with respect to the origin since replacing x with $-x$ and y with $-y$ in the equation yields

$$-y = \frac{-x}{(-x)^2 + 1} \qquad \text{or} \qquad y = \frac{x}{x^2 + 1}$$

which *is* equivalent to the original equation.

39. Sketch the graph of the equation $y = 1 - x^2$. Identify any intercepts and test for symmetry.

Solution:

To find the x-intercepts, let $y = 0$. Then

$$0 = 1 - x^2$$

and by factoring
$$0 = (1 + x)(1 - x).$$

Therefore, the x-intercepts are $(-1, 0)$ and $(1, 0)$. To find any the y-intercept, let $x = 0$. Then $y = 1$ and the y-intercept is $(0, 1)$.

There is symmetry with respect to the y-axis since replacing x with $-x$ in the equation yields

$$y = 1 - (-x)^2 \qquad \text{or} \qquad y = 1 - x^2$$

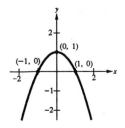

which is equivalent to the original equation.

Some solution points are:

x	0	± 1	± 2
y	1	0	-3

59. Find the points of intersection of the graphs of $x + y = 7$ and $3x - 2y = 11$.

Solution:

To solve the two equations

$$x + y = \ 7$$
$$3x - 2y = 11$$

simultaneously, multiply the first equation by 2 and add. Thus

$$2x + 2y = 14$$
$$(+) \quad 3x - 2y = 11$$
$$\overline{\hspace{1.2cm} 5x \hspace{1.2cm} = 25}$$
$$x = \ 5$$

Substituting $x = 5$ into the first equation, you have

$$5 + y = 7 \qquad \text{or} \qquad y = 2$$

Thus, the point of intersection is $(5, 2)$.

61. Find the points of intersection of the graphs of $x^2 + y^2 = 5$ and $x - y = 1$.

Solution:

To solve the two equations

$$x^2 + y^2 = 5$$
$$x - y = 1$$

simultaneously, solve the second equation for x and obtain

$$x = y + 1.$$

Substituting into the first equation yields

$$(y + 1)^2 + y^2 = 5$$
$$(y^2 + 2y + 1) + y^2 = 5$$
$$2y^2 + 2y - 4 = 0$$
$$2(y + 2)(y - 1) = 0$$

which implies that $y = -2$ or $y = 1$. Therefore,

$$x = (-2) + 1 = -1 \qquad \text{or} \qquad x = 1 + 1 = 2$$

and the points of intersection are $(-1, -2)$ and $(2, 1)$.

 65. Use a graphing utility to graph the equations $y = x^3 - 2x^2 + x - 1$ and $y = -x^2 + 3x - 1$. Then find the points of intersection of the graphs.

Solution:

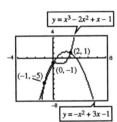

The graphs are shown in the accompanying figure. Equating the two expressions for y yields

$$x^3 - 2x^2 + x - 1 = -x^2 + 3x - 1$$
$$x^3 - 2x^2 + x^2 + x - 3x - 1 + 1 = 0$$
$$x^3 - x^2 - 2x = 0$$
$$x(x - 2)(x + 1) = 0$$

Thus, $x = 0$, 2, or -1, and substituting these values into the second (or first) equation, you have

$$y = -(0)^2 + 3(0) - 1 = -1$$
$$y = -(2)^2 + 3(2) - 1 = 1$$
$$y = -(-1)^2 + 3(-1) - 1 = -5$$

Therefore, the points of intersection are $(0, -1), (2, 1)$, and $(-1, -5)$.

69. For what values of k does the graph of $y = kx^3$ pass through the points (a) $(1, 4)$, (b) $(-2, 1)$, (c) $(0, 0)$, and (d) $(-1, -1)$?

Solution:

(a) If $x = 1$ and $y = 4$, then $4 = k(1)^3$ and you obtain $k = 4$.

(b) If $x = -2$ and $y = 1$, then $1 = k(-2)^3$ and you obtain $k = -\frac{1}{8}$.

(c) If $x = y = 0$, then $0 = k(0)^3$ and k may have *any value*.

(d) If $x = y = -1$, then $-1 = k(-1)^3$ and you obtain $k = 1$.

3   Lines in the Plane

9. Plot the points $(3, -4)$ and $(5, 2)$ and find the slope of the line passing through them.

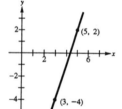

Solution:

Let $(3, -4) = (x_1, y_1)$ and $(5, 2) = (x_2, y_2)$. The slope of the line passing through (x_1, y_1) and (x_2, y_2) is

$$m = \frac{y_2 - y_1}{x_2 - x_1} = \frac{2 - (-4)}{5 - 3} = \frac{6}{2} = 3$$

23. Find an equation of the line passing through $(2, 1)$ and $(0, -3)$ and sketch its graph.

Solution:

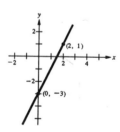

Let $(2, 1) = (x_1, y_1)$ and $(0, -3) = (x_2, y_2)$. The slope of the line passing through (x_1, y_1) and (x_2, y_2) is

$$m = \frac{y_2 - y_1}{x_2 - x_1} = \frac{-3 - 1}{0 - 2} = 2.$$

Using the point-slope form of the equation of a line, you have

$$y - y_1 = m(x - x_1)$$
$$y - 1 = 2(x - 2)$$
$$y - 1 = 2x - 4$$
$$0 = 2x - y - 3 \qquad \text{General form}$$

29. Find an equation of the line passing through $(0, 3)$ with a slope of $m = \frac{3}{4}$ and sketch its graph.

Solution:

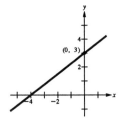

Using the slope-intercept form of the equation of a line, you have

$$y = mx + b$$
$$y = \frac{3}{4}x + 3$$
$$4y = 3x + 12$$
$$0 = 3x - 4y + 12$$

41. Write an equation of the line passing through $(2, 1)$ (a) parallel and (b) perpendicular to the line $4x - 2y = 3$.

Solution:

The line given by $4x - 2y = 3$ has a slope of 2 since

$$4x - 2y = 3$$
$$-2y = -4x + 3$$
$$y = 2x - \frac{3}{2} = mx + b$$

(a) The line through $(2, 1)$ parallel to $4x - 2y = 3$ must also have a slope of 2. Thus its equation must be

$$y - y_1 = m(x - x_1)$$
$$y - 1 = 2(x - 2)$$
$$-2x + y + 3 = 0$$
$$2x - y - 3 = 0$$

(b) The line through $(2, 1)$ perpendicular to $4x - 2y = 3$ must have a slope of $m = -\frac{1}{2}$. Thus its equation must be

$$y - y_1 = m(x - x_1)$$
$$y - 1 = -\frac{1}{2}(x - 2)$$
$$2y - 2 = -x + 2$$
$$x + 2y - 4 = 0$$

55. Use slope to determine if the points $(-2, 1), (-1, 0)$, and $(2, -2)$ are collinear.

Solution:

Let $(-2, 1) = (x_1, y_1)$, $(-1, 0) = (x_2, y_2)$, and $(2, -2) = (x, y)$. The point (x, y) lies on the line passing through (x_1, y_1) and (x_2, y_2) if and only if

$$\frac{y - y_1}{x - x_1} = m = \frac{y - y_2}{x - x_2}.$$

Since

$$\frac{-2 - 1}{2 - (-2)} = -\frac{3}{4} \neq -\frac{2}{3} = \frac{-2 - 0}{2 - (-1)},$$

the three points are *not* collinear.

67. A real estate office handles an apartment complex with 50 units. When the rent is \$380 per month, all 50 units are occupied. However, when the rent is \$425, the average number of occupied units drops to 47. Assume that the relationship between the monthly rent p and the demand x is linear. (Note: Here we use the term *demand* to refer to the number of occupied units.) (a) Write a linear equation giving the quantity demanded x in terms of the rent p.

(b) (Linear extrapolation) Use this equation to predict the number of units occupied if the rent is raised to \$455.

(c) (Linear interpolation) Predict the number of units occupied if the rent is lowered to \$395.

Solution:

(a) Two solution points to the linear equation are $(x_1, p_1) = (50, 380)$ and $(x_2, p_2) = (47, 425)$. Therefore, the slope is

$$m = \frac{425 - 380}{47 - 50} = -15$$

and the equation of the line is

$$p - 380 = -15(x - 50)$$

$$p = -15x + 1130 \quad \text{or} \quad x = \frac{1}{15}(1130 - p)$$

(b) When $p = \$455$, $x = \dfrac{1}{15}(1130 - 455) = 45$ units.

(c) When $p = \$395$, $x = \dfrac{1}{15}(1130 - 395) = 49$ units.

73. Find the distance between the parallel lines $x + y = 1$ and $x + y = 5$.

Solution:

A point on the line $x + y = 1$ is $(2, -1)$. The distance between the given parallel lines is equal to the distance from $(2, -1)$ to the line $x + y = 5$. Letting $(2, -1) = (x_1, y_1)$ and $x + y - 5 = Ax + By + C = 0$, you have

$$d = \frac{|Ax_1 + By_1 + C|}{\sqrt{A^2 + B^2}}$$
$$= \frac{|1(2) + 1(-1) - 5|}{\sqrt{1^2 + 1^2}}$$
$$= \frac{|-4|}{\sqrt{2}} = \frac{4}{\sqrt{2}} = 2\sqrt{2}$$

4   Functions

3. Given $f(x) = \sqrt{x + 3}$, find:

   (a) $f(-2)$ (b) $f(6)$ (c) $f(c)$ (d) $f(x + \Delta x)$.

Solution:

   (a) $f(-2) = \sqrt{-2 + 3} = \sqrt{1} = 1$ (b) $f(6) = \sqrt{6 + 3} = \sqrt{9} = 3$

   (c) $f(c) = \sqrt{c + 3}$ (d) $f(x + \Delta x) = \sqrt{x + \Delta x + 3}$

9. Given $f(x) = x^3$, find $\dfrac{f(x + \Delta x) - f(x)}{\Delta x}$.

Solution:

$$
\begin{aligned}
\frac{f(x + \Delta x) - f(x)}{\Delta x} &= \frac{(x + \Delta x)^3 - x^3}{\Delta x} \\
&= \frac{x^3 + 3x^2\,\Delta x + 3x(\Delta x)^2 + (\Delta x)^3 - x^3}{\Delta x} \\
&= \frac{\Delta x[3x^2 + 3x\,\Delta x + (\Delta x)^2]}{\Delta x} \\
&= 3x^2 + 3x\,\Delta x + (\Delta x)^2
\end{aligned}
$$

17. Find the domain and range of the function $f(x) = \sqrt{9 - x^2}$ and sketch its graph.

Solution:

Since $9 - x^2$ must be nonnegative ($9 - x^2 \geq 0$), the domain is $[-3, 3]$. The range is $[0, 3]$. There is symmetry with respect to the y-axis since

$$ y = \sqrt{9 - (-x)^2} = \sqrt{9 - x^2} $$

is equivalent to the original equation. Squaring both members of the equation, you have

$$ y^2 = 9 - x^2 \qquad \text{or} \qquad x^2 + y^2 = 3^2. $$

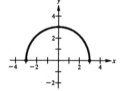

This is the standard form of an equation of a circle with center $(0,0)$ and radius 3. Therefore, the graph of $f(x) = \sqrt{9 - x^2}$ is a semicircle in the first and second quadrants with center $(0,0)$ and radius 3.

35. Determine if y is a function of x for the equation $y^2 = x^2 - 1$.

Solution:

Solving the equation for y, yields

$$
\begin{aligned}
y^2 &= x^2 - 1 \\
y &= \pm\sqrt{x^2 - 1}, \quad |x| \geq 1
\end{aligned}
$$

For each value of the independent variable x such that $|x| > 1$, there corresponds two values of the dependent variable y. Therefore, y is not a function of x.

44. Given $f(x) = 1/x$ and $g(x) = x^2 - 1$, find:

   (a) $f[g(2)]$ (b) $g[f(2)]$ (c) $f[g(1/\sqrt{2})]$
   (d) $g[f(1/\sqrt{2})]$ (e) $g[f(x)]$ (f) $f[g(x)]$

Solution:

   (a) $f[g(2)] = f(2^2 - 1) = f(3) = \dfrac{1}{3}$

   (b) $g[f(2)] = g\left(\dfrac{1}{2}\right) = \left(\dfrac{1}{2}\right)^2 - 1 = \dfrac{1}{4} - 1 = -\dfrac{3}{4}$

   (c) $f\left[g\left(\dfrac{1}{\sqrt{2}}\right)\right] = f\left[\left(\dfrac{1}{\sqrt{2}}\right)^2 - 1\right] = f\left(-\dfrac{1}{2}\right) = \dfrac{1}{-\frac{1}{2}} = -2$

   (d) $g\left[f\left(\dfrac{1}{\sqrt{2}}\right)\right] = g\left(\dfrac{1}{1/\sqrt{2}}\right) = g(\sqrt{2}) = (\sqrt{2})^2 - 1 = 1$

   (e) $g[f(x)] = g\left(\dfrac{1}{x}\right) = \left(\dfrac{1}{x}\right)^2 - 1 = \dfrac{1}{x^2} - 1 = \dfrac{1 - x^2}{x^2}$

   (f) $f[g(x)] = f(x^2 - 1) = \dfrac{1}{x^2 - 1}$

55. Determine if the function $f(x) = x(4 - x^2)$ is even, odd, or neither.

Solution:

The function is odd since

$$f(-x) = (-x)[4 - (-x)^2] = -x(4 - x^2) = -f(x)$$

73. You are in a boat 2 miles from the nearest point on the coast. You are to go to a point Q, 3 miles down the coast and 1 mile inland (see figure). You can row at 2 miles per hour and walk at 4 miles per hour. Express the total time T of the trip as a function of x.

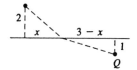

Solution:

S = distance on water = $\sqrt{x^2 + 4}$

L = distance on land = $\sqrt{1 + (3 - x)^2} = \sqrt{x^2 - 6x + 10}$

$$\text{Total time} = (\text{time on water}) + (\text{time on land})$$

$$T = \frac{S}{(\text{rate on water})} + \frac{L}{(\text{rate on land})}$$

$$= \frac{\sqrt{x^2 + 4}}{2} + \frac{\sqrt{x^2 - 6x + 10}}{4}$$

5   Review of Trigonometric Functions

7. Express the following angles in degree measure:

(a) $3\pi/2$ (b) $7\pi/6$

(c) $-7\pi/12$ (d) -2.367

Solution:

Since $180° = \pi$ radians, it follows that that 1 radian $= 180°/\pi$.

(a) $\dfrac{3\pi}{2}$ radians $= \left(\dfrac{3\pi}{2}\right)\left(\dfrac{180°}{\pi}\right) = 270°$

(b) $\dfrac{7\pi}{6}$ radians $= \left(\dfrac{7\pi}{6}\right)\left(\dfrac{180°}{\pi}\right) = 210°$

(c) $-\dfrac{7\pi}{12}$ radians $= \left(-\dfrac{7\pi}{12}\right)\left(\dfrac{180°}{\pi}\right) = -105°$

(d) -2.367 radians $= (-2.367)\left(\dfrac{180°}{\pi}\right) = -135.619°$

19. Find $\cot\theta$ given $\cos\theta = \frac{4}{5}$.

Solution:

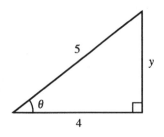

Using the fact that

$$\cos\theta = \frac{4}{5},$$

construct the accompanying figure and obtain

$$\cot\theta = \frac{4}{y} = \frac{4}{\sqrt{25 - 16}} = \frac{4}{3}.$$

25. Evaluate the sine, cosine, and tangent of the following angles *without* using a calculator:

(a)   225°   (b)   −225°   (c) $\dfrac{5\pi}{3}$ (d) $\dfrac{11\pi}{6}$

Solution:

(a) The angle is in quadrant III and the reference angle is $225° - 180° = 45°$. Therefore,

$$\sin 225° = -\sin 45° = -\frac{\sqrt{2}}{2}$$

$$\cos 225° = -\cos 45° = -\frac{\sqrt{2}}{2}$$

$$\tan 225° = \tan 45° = 1$$

(b) The angle is in quadrant II and the reference angle is $225° - 180° = 45°$. Therefore,

$$\sin(-225°) = \sin 45° = \frac{\sqrt{2}}{2}$$

$$\cos(-225°) = -\cos 45° = -\frac{\sqrt{2}}{2}$$

$$\tan(-225°) = -\tan 45° = -1$$

(c) The angle is in quadrant IV and the reference angle is $2\pi - \frac{5\pi}{3} = \frac{\pi}{3}$. Therefore,

$$\sin \frac{5\pi}{3} = -\sin \frac{\pi}{3} = -\frac{\sqrt{3}}{2}$$

$$\cos \frac{5\pi}{3} = \cos \frac{\pi}{3} = \frac{1}{2}$$

$$\tan \frac{5\pi}{3} = -\tan \frac{\pi}{3} = -\sqrt{3}$$

(d) The angle is in quadrant IV and the reference angle is $2\pi - \frac{11\pi}{6} = \frac{\pi}{6}$. Therefore,

$$\sin \frac{11\pi}{6} = -\sin \frac{\pi}{6} = -\frac{1}{2}$$

$$\cos \frac{11\pi}{6} = \cos \frac{\pi}{6} = \frac{\sqrt{3}}{2}$$

$$\tan \frac{11\pi}{6} = -\tan \frac{\pi}{6} = -\frac{\sqrt{3}}{3}$$

37. Solve the equation $\tan^2 \theta - \tan \theta = 0$ for θ where $0 \le \theta < 2\pi$.

Solution:

$$\tan^2 \theta - \tan \theta = 0$$
$$\tan \theta (\tan \theta - 1) = 0$$

If $\tan \theta = 0$, then $\theta = 0$ or $\theta = \pi$. If $\tan \theta - 1 = 0$, you have $\tan \theta = 1$ and $\theta = \pi/4$ or $\theta = 5\pi/4$. Thus for $0 \le \theta < 2\pi$, there are four solutions:

$$\theta = 0, \ \frac{\pi}{4}, \ \pi, \ \frac{5\pi}{4}.$$

43. Solve for y and r in the accompanying triangle.

Solution:

$$\tan 30^\circ = \frac{\sqrt{3}}{3} = \frac{y}{100}$$

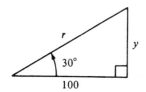

Therefore, $y = 100\sqrt{3}/3$.

$$\sec 30^\circ = \frac{2\sqrt{3}}{3} = \frac{r}{100}$$

Therefore, $r = 200\sqrt{3}/3$.

47. An airplane leaves the runway climbing at 18° with a speed of 275 feet per second (see figure). Find the altitude of the plane after 1 minute.

Solution:

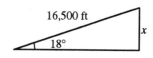

In 1 minute the plane travels

$$(60 \text{ sec})(275 \text{ ft/sec}) = 16,500 \text{ ft}.$$

This distance is the approximate length of the hypotenuse of a right triangle whose side opposite the angle of magnitude 18° is the altitude h of the plane. Therefore,

$$\sin 18^\circ \approx \frac{h}{16,500}$$
$$16,500 \sin 18^\circ \approx h$$
$$h \approx 5100 \text{ ft}$$

65. Sketch the graph of the function $\csc\left(x/2\right)$.

Solution:

The graph of $y = \csc\left(x/2\right)$ has the following characteristics:

$$\text{period: } \frac{2\pi}{\frac{1}{2}} = 4\pi$$

vertical asymptotes: $x = 2n\pi, n$ an integer

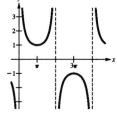

Using the basic shape of the graph of the cosecant function, you sketch one period of the function on the interval $[0, 4\pi]$, following the pattern

minimum: $(\pi, 1)$ maximum: $(3\pi, -1)$.

6   Inverse Functions

5. Show that $f(x) = \sqrt{x-4}$ and $g(x) = x^2 + 4\ (x \geq 0)$ are inverses of each other by showing that $f(g(x)) = g(f(x)) = x$. Then use a graphing utility to graph f and g in the same viewing rectangle. Observe that the graph of g is a reflection of the graph of f in the line $y = x$.

Solution:

$$f(x) = \sqrt{x-4} \quad \text{and} \quad g(x) = x^2 + 4 \quad (x \geq 0)$$

The composite of f with g is given by

$$f(g(x)) = f(x^2 + 4) = \sqrt{(x^2 + 4) - 4} = \sqrt{x^2} = x.$$

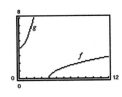

The composite of g with f is given by

$$g(f(x)) = g(\sqrt{x-4}) = (\sqrt{x-4})^2 + 4 = x - 4 + 4 = x.$$

Since $f(g(x)) = g(f(x)) = x$, you can conclude that f and g are inverses of each other. The graphs of f and g are shown in the accompanying figure.

9. Find the inverse of $f(x) = 2x - 3$ and then graph both f and f^{-1}.

Solution:

The function has an inverse because it is increasing on its entire domain. To find an equation for the inverse, let $y = f(x)$ and solve for x in terms of y.

$$2x - 3 = y$$
$$x = \frac{y + 3}{2}$$
$$f^{-1}(y) = \frac{y + 3}{2}$$

Interchanging x and y yields

$$f^{-1}(x) = \frac{x + 3}{2}.$$

Remember that any letter can be used to represent the independent variable. Thus,

$$f^{-1}(y) = \frac{y + 3}{2}, \quad f^{-1}(x) = \frac{x + 3}{2}, \quad \text{and} f^{-1}(t) = \frac{t + 3}{2}$$

represent the same function.

19. Find the inverse of $f(x) = x^{2/3}$ $(x \geq 0)$ and then graph both f and f^{-1}.

Solution:

The function has an inverse because it is increasing on $[0, \infty)$. To find an equation for the inverse, let $y = f(x)$ and solve for x in terms of y.

$$x^{2/3} = y \qquad x \geq 0$$
$$x = y^{3/2} \qquad y \geq 0$$
$$f^{-1}(y) = y^{3/2} \qquad y \geq 0$$

Finally, using x as the independent variable yields

$$f^{-1}(x) = x^{3/2} \qquad x \geq 0$$

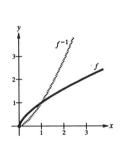

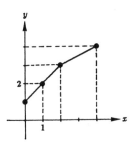

25. Use the graph of the function f in the accompanying figure to complete the table and sketch the graph of f^{-1}.

x	1	2	3	4
$f^{-1}(x)$				

Solution:

Complete the following table by using the accompanying graph for the function f.

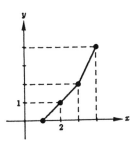

x	0	1	2	4
$f(x)$	1	2	3	4

From the reflective property of inverses, you know the graph of f contains the point (a, b) if and only if the graph of f^{-1} contains the point (b, a). Therefore, you obtain the required table by interchanging the x and y coordinates of the preceding table.

x	1	2	3	4
$f^{-1}(x)$	0	1	2	4

45. Show that $f(x) = 4/x^2$ is one-to-one on the interval $(0, \infty)$ and therefore has an inverse on that interval.

Solution:

A function f is one-to-one if, for a and b in its domain, $f(a) = f(b)$ implies that $a = b$. Let a and b be real numbers in the domain of f such that $f(a) = f(b)$. Then

$$\frac{4}{a^2} = \frac{4}{b^2}$$
$$a^2 = b^2.$$

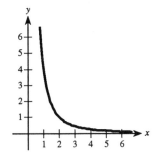

Since $a > 0$ and $b > 0$, it follows that $a = b$ and the function is one-to-one on the specified domain. Therefore, the f has an inverse on the the indicated interval.

(*Note:* From the accompanying graph it follows that the graph of f passes the horizontal line test and therefore f has an inverse on the indicated interval.)

73. Evaluate $\arccos \frac{1}{2}$

Solution:

Since $y = \arccos \frac{1}{2}$ if and only if $\cos y = \frac{1}{2}$, then $y = \pi/3$ in the interval $[0, \pi]$. Thus

$$\arccos \frac{1}{2} = \frac{\pi}{3}.$$

91. Evaluate (a) $\sin(\arctan \frac{3}{4})$ and (b) $\sec(\arcsin \frac{4}{5})$ without the use of a calculator.

Solution:

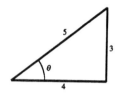

(a) Begin by sketching a triangle to represent θ, "the angle whose tangent is $\frac{3}{4}$". Then

$$\theta = \arctan \frac{3}{4}$$

and

$$\sin \left(\arctan \frac{3}{4} \right) = \sin \theta = \frac{3}{5}.$$

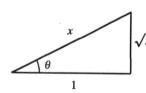

(b) Begin by sketching a triangle to represent θ, "the angle whose sine is $\frac{4}{5}$". Then

$$\theta = \arcsin \frac{4}{5}$$

and

$$\sec \left(\arcsin \frac{4}{5} \right) = \sec \theta = \frac{5}{3}.$$

99. Write an algebraic expression that is equivalent to the expression $\sin(\text{arcsec } x)$.

Solution:

Begin by sketching a triangle to represent θ, "the angle whose secant is x". Then

$$\theta = \text{arcsec } x$$

and

$$\sin(\text{arcsec } x) = \sin \theta = \frac{\sqrt{x^2 - 1}}{x}.$$

103. Write an algebraic expression equivalent to the expression $csc \left(\arctan \dfrac{x}{\sqrt{2}} \right)$.

Solution:

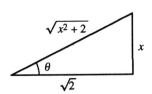

Begin by sketching a triangle to represent θ, "the angle whose tangent is $x/\sqrt{2}$." Then

$$\theta = \arctan \frac{x}{\sqrt{2}}$$

and

$$csc \left(\arctan \frac{x}{\sqrt{2}} \right) = csc \, \theta = \frac{\sqrt{x^2 + 2}}{x}.$$

115. Solve $\arcsin \sqrt{2x} = \arccos \sqrt{x}$ for x.

Solution:

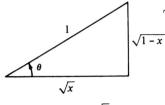

$\theta = \text{Arccos } \sqrt{x}$

$\sin \theta = \sqrt{1-x}$

Taking the sine of each member of the equation, yields

$$\sin(\arcsin \sqrt{2x}) = \sin(\arccos \sqrt{x})$$
$$\sqrt{2x} = \sqrt{1 - x}$$
$$2x = 1 - x$$
$$3x = 1 \quad \Longrightarrow \quad x = \frac{1}{3}.$$

7   Exponential and Logarithmic Functions

9. Solve for x if $\left(\dfrac{1}{3}\right)^{x-1} = 27$.

Solution:

Using the properties of exponents you have

$$\left(\frac{1}{3}\right)^{x-1} = 27$$
$$(3^{-1})^{x-1} = 3^3$$
$$3^{-(x-1)} = 3^3$$
$$-(x-1) = 3$$
$$x - 1 = -3 \quad \Longrightarrow \quad x = -2$$

27. Sketch a graph of $y = e^{-x^2}$.

Solution:

(a) The graph is symmetric with respect to the y-axis since

$$y = e^{-(-x)^2} = e^{-x^2}.$$

(b) The y-intercept is $(0, 1)$.

(c) The x-axis is a horizontal asymptote since $e^{-x^2} \longrightarrow 0$ as $x \longrightarrow \infty$.

(d) The graph lies entirely above the x-axis, since for all x

$$0 < e^{-x^2}.$$

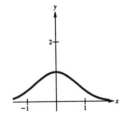

The following table shows solution points of the equation. By plotting these points and using the information above, you have the graph shown in the accompanying figure.

x	0	0.5	1	2
y	1	0.607	0.368	0.135

33. Write the logarithmic equations as exponential equations:

(a) $\ln 2 = 0.6931\ldots$ (b) $\ln 8.4 = 2.128\ldots$

Solution:

By definition,

$$\ln x = b \quad \text{if and only if} \quad e^b = x.$$

(a) Letting $x = 2$, and $b = 0.6931\ldots$, you have
$$\ln 2 = 0.6931\ldots \text{ if and only if } e^{0.6931\ldots} = 2.$$

(b) Letting $x = 8.4$, and $b = 2.128\ldots$, you have
$$\ln 8.4 = 2.128\ldots \text{ if and only if } e^{2.128\ldots} = 8.4.$$

51. Using properties of logarithms and the fact that $\ln 2 \approx 0.6931$ and $\ln 3 \approx 1.0986$, find the following:

(a) $\ln 6$ (b) $\ln \frac{2}{3}$

(c) $\ln 81$ (d) $\ln \sqrt{3}$

Solution:

(a) $\ln 6 = \ln (2 \cdot 3)$
$$= \ln 2 + \ln 3 \approx 0.6931 + 1.0986 = 1.7917$$

(b) $\ln \frac{2}{3} = \ln 2 - \ln 3$
$$\approx 0.6931 - 1.0986 = -0.4055$$

(c) $\ln 81 = \ln (3^4) = 4\ln 3 \approx 4(1.0986) = 4.3944$

(d) $\ln \sqrt{3} = \ln (3^{1/2}) = \frac{1}{2}\ln 3 \approx \frac{1}{2}(1.0986) = 0.5493$

59. Use the properties of logarithms to write
$$\ln \left(\frac{x^2 - 1}{x^3} \right)^3$$
as a sum, difference, or multiple of logarithms.

Solution:

$$\ln \left(\frac{x^2 - 1}{x^3} \right)^3 = 3[\ln (x^2 - 1) - \ln x^3]$$
$$= 3[\ln [(x + 1)(x - 1)] - 3\ln x]$$
$$= 3[\ln (x + 1) + \ln (x - 1) - 3\ln x]$$

65. Use the properties of logarithms to write

$$\frac{1}{3}[2 \ln (x + 3) + \ln x - \ln (x^2 - 1)]$$

as a single quantity.

Solution:

$$\frac{1}{3}[2 \ln (x + 3) + \ln x - \ln (x^2 - 1)]$$

$$= \frac{1}{3}[\ln (x + 3)^2 + \ln x - \ln (x^2 - 1)]$$

$$= \frac{1}{3}[\ln x(x + 3)^2 - \ln (x^2 - 1)]$$

$$= \frac{1}{3} \ln \frac{x(x + 3)^2}{x^2 - 1}$$

$$= \ln \left(\frac{x(x + 3)^2}{x^2 - 1} \right)^{1/3} = \ln \sqrt[3]{\frac{x(x + 3)^2}{x^2 - 1}}$$

1 LIMITS AND THEIR PROPERTIES

1.2 An Introduction to Limits

5. Complete the following table and use the result to estimate the limit

$$\lim_{x \to 3} \frac{\dfrac{1}{x+1} - \dfrac{1}{4}}{x-3}$$

x	2.9	2.99	2.999	3.001	3.01	3.1
$f(x)$						

Solution:

The table lists values of $f(x)$ at several x-values near 3.

x	2.9	2.99	2.999	3.001	3.01	3.1
$f(x)$	-0.0641	-0.0627	-0.0625	-0.0625	-0.0623	-0.0610

As x approaches 3 from the left and from the right $f(x)$ approaches -0.0625. Therefore, we estimate the limit to be $-\frac{1}{16}$.

15. Use the accompanying graph to find the limit (if it exists).

$$\lim_{x \to \pi/2} \tan x$$

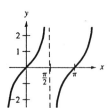

Solution:

From the graph we observe that as x approaches $\pi/2$ from the left, $f(x)$ increases without bound, and as x approaches $\pi/2$ from the right, $f(x)$ decreases without bound. Since $f(x)$ is not approaching a real number L as x approaches $\pi/2$, we say that the limit does *not* exist.

19. Find L such that $\lim_{x \to 2}(3x + 2) = L$ and then find $\delta > 0$ such that $|f(x) - L| < 0.01$ whenever $0 < |x - c| < \delta$.

Solution:

We use the definition of limit to verify that

$$\lim_{x \to 2}(3x + 2) = 8.$$

We are required to show that there exists a δ such that $|f(x) - L| < 0.01$ whenever $0 < |x - 2| < \delta$.

$$|f(x) - L| < 0.01$$
$$|(3x + 2) - 8| < 0.01$$
$$|3x - 6| < 0.01$$
$$3|x - 2| < 0.01$$
$$0 < |x - 2| < \frac{0.01}{3}$$

Therefore, $\delta = \dfrac{0.01}{3}$.

23. Find L such that $\lim_{x \to 2}(x + 3) = L$. Then for $\epsilon > 0$, find $\delta > 0$ such that $|f(x) - L| < \epsilon$ whenever $0 < |x - c| < \delta$.

Solution:

We use the definition of limit to verify that

$$\lim_{x \to 2}(x + 3) = 5.$$

Given $\epsilon > 0$,

$$|(x + 3) - 5| < \epsilon$$
$$|x - 2| < \epsilon = \delta$$

27. Find L such that $\lim\limits_{x \to 0} \sqrt[3]{x} = L$. Then for $\epsilon > 0$, find $\delta > 0$ such that $|f(x) - L| < \epsilon$ whenever $0 < |x - c| < \delta$.

Solution:

We use the definition of limit to verify that
$$\lim_{x \to 0} \sqrt[3]{x} = 0.$$

Given $\epsilon > 0$,
$$|\sqrt[3]{x} - 0| < \epsilon$$
$$(|\sqrt[3]{x}|)^3 < \epsilon^3$$
$$|x| < \epsilon^3 = \delta$$

29. Find L such that $\lim\limits_{x \to 1}(x^2 + 1) = L$. Then for $\epsilon > 0$, find $\delta > 0$, such that $|f(x) - L| < \epsilon$ whenever $0 < |x - c| < \delta$.

Solution:

We use the definition of limit to verify that
$$\lim_{x \to 1}(x^2 + 1) = 2.$$

Given $\epsilon > 0$, you have
$$|(x^2 + 1) - 2| < \epsilon$$
$$|x^2 - 1| < \epsilon$$
$$|(x + 1)(x - 1)| < \epsilon$$
$$|x + 1|\,|x - 1| < \epsilon$$

Without loss of generality, assume that x is in the interval $(0, 2)$. Therefore, $|x + 1| < 3$ and you have
$$|x + 1|\,|x - 1| < 3|x - 1| < \epsilon$$
$$|x - 1| < \frac{\epsilon}{3} = \delta$$

1.3   Properties of Limits

11. Find $\lim\limits_{x \to 3} \sqrt{x + 1}$.

Solution:
$$\lim_{x \to 3} \sqrt{x + 1} = \sqrt{3 + 1} = \sqrt{4} = 2$$

27. Find $\lim\limits_{x \to 1}(\ln 3x + e^x)$.

Solution:

$$\lim_{x \to 1}(\ln 3x + e^x) = \lim_{x \to 1}\ln 3x + \lim_{x \to 1}e^x$$
$$= \ln 3(1) + e^1 = \ln 3 + e \approx 3.817$$

31. If $\lim\limits_{x \to c} f(x) = 4$, find:

(a) $\lim\limits_{x \to c}[f(x)]^3$ (b) $\lim\limits_{x \to c}\sqrt{f(x)}$

(c) $\lim\limits_{x \to c}[3f(x)]$ (d) $\lim\limits_{x \to c}[f(x)]^{3/2}$

Solution:

(a) $\lim\limits_{x \to c}[f(x)]^3 = \left[\lim\limits_{x \to c}f(x)\right]^3 = 4^3 = 64$

(b) $\lim\limits_{x \to c}\sqrt{f(x)} = \sqrt{\lim\limits_{x \to c}f(x)} = \sqrt{4} = 2$

(c) $\lim\limits_{x \to c}[3f(x)] = 3\lim\limits_{x \to c}f(x) = 3(4) = 12$

(d) $\lim\limits_{x \to c}[f(x)]^{3/2} = \left[\lim\limits_{x \to c}f(x)\right]^{3/2} = 4^{3/2} = 8$

1.4   Techniques for Evaluating Limits

5. Find (if it exists) $\lim\limits_{x \to -1}\dfrac{x^2 - 1}{x + 1}$.

Solution:

$$\lim_{x \to -1}\frac{x^2 - 1}{x + 1} = \lim_{x \to -1}\frac{(x + 1)(x - 1)}{x + 1} = \lim_{x \to -1}(x - 1) = -2$$

13. Find (if it exists) $\displaystyle\lim_{x\to 0}\frac{\sqrt{3+x}-\sqrt{3}}{x}$

Solution:

$$\lim_{x\to 0}\frac{\sqrt{3+x}-\sqrt{3}}{x}=\lim_{x\to 0}\left(\frac{\sqrt{3+x}-\sqrt{3}}{x}\cdot\frac{\sqrt{3+x}+\sqrt{3}}{\sqrt{3+x}+\sqrt{3}}\right)$$

$$=\lim_{x\to 0}\frac{3+x-3}{x(\sqrt{3+x}+\sqrt{3})}$$

$$=\lim_{x\to 0}\frac{1}{\sqrt{3+x}+\sqrt{3}}=\frac{1}{2\sqrt{3}}=\frac{\sqrt{3}}{6}$$

15. Find (if it exists) $\displaystyle\lim_{x\to 0}\frac{\dfrac{1}{2+x}-\dfrac{1}{2}}{x}$

Solution:

$$\lim_{x\to 0}\frac{\dfrac{1}{2+x}-\dfrac{1}{2}}{x}=\lim_{x\to 0}\frac{\dfrac{2-(2+x)}{2(2+x)}}{x}$$

$$=\lim_{x\to 0}\frac{-x}{x(2)(2+x)}$$

$$=\lim_{x\to 0}\frac{-1}{2(2+x)}=-\frac{1}{4}$$

19. Find (if it exists)

$$\lim_{\Delta x\to 0}\frac{(x+\Delta x)^2-2(x+\Delta x)+1-(x^2-2x+1)}{\Delta x}.$$

Solution:

$$\lim_{\Delta x\to 0}\frac{(x+\Delta x)^2-2(x+\Delta x)+1-(x^2-2x+1)}{\Delta x}$$

$$=\lim_{\Delta x\to 0}\frac{x^2+2x\Delta x+(\Delta x)^2-2x-2\Delta x+1-x^2+2x-1}{\Delta x}$$

$$=\lim_{\Delta x\to 0}\frac{2x\Delta x+(\Delta x)^2-2\Delta x}{\Delta x}$$

$$=\lim_{\Delta x\to 0}\frac{\Delta x(2x+\Delta x-2)}{\Delta x}$$

$$=\lim_{\Delta x\to 0}(2x+\Delta x-2)=2x-2$$

33. Find (if it exists) $\displaystyle\lim_{x \to \pi/2} \frac{\cos x}{\cot x}$.

Solution:

$$\lim_{x \to \pi/2} \frac{\cos x}{\cot x} = \lim_{x \to \pi/2} \frac{\cos x}{(\cos x)/(\sin x)}$$

$$= \lim_{x \to \pi/2} \sin x$$

$$= 1$$

39. Find (if it exists) $\displaystyle\lim_{x \to 0} \frac{1 - e^{-x}}{e^x - 1}$.

Solution:

$$\lim_{x \to 0} \frac{1 - e^{-x}}{e^x - 1} = \lim_{x \to 0} \left(\frac{1 - e^{-x}}{e^x - 1} \cdot \frac{e^x}{e^x} \right)$$

$$= \lim_{x \to 0} \frac{e^x - 1}{e^x (e^x - 1)} = \lim_{x \to 0} \frac{1}{e^x} = 1$$

47. Use a graphing utility to graph the given function and the equations $y = x$ and $y = -x$ in the same viewing rectangle. Using the graphs to visually observe the Squeeze Theorem, find

$$\lim_{x \to 0} x \sin \frac{1}{x}.$$

Solution:

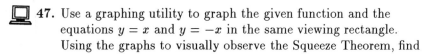

The required graphs are shown in the accompanying figure. What follows is a direct application of the Squeeze Theorem to find the required limit. Since

$$\lim_{x \to 0}(-|x|) = 0 \quad \text{and} \quad \lim_{x \to 0} |x| = 0$$

and

$$-|x| \le x \sin \frac{1}{x} \le |x|,$$

you have

$$0 = \lim_{x \to 0}(-|x|) \le \lim_{x \to 0} x \sin \frac{1}{x} \le \lim_{x \to 0} |x| = 0.$$

Therefore, by the Squeeze Theorem

$$\lim_{x \to 0} x \sin \frac{1}{x} = 0.$$

53. Complete the following table and use the result to estimate the limit

$$\lim_{x \to 0} \frac{\ln(x+1)}{x}.$$

x	−0.1	−0.01	−0.001	0.001	0.01	0.1
$f(x)$						

Use a graphing utility to graph the function and estimate the limit. What is the domain of the function?

Solution:

The table lists values of the function at several x-values near 0.

x	−0.1	−0.01	−0.001	0.001	0.01	0.1
$f(x)$	1.0536	1.0050	1.0005	0.9995	0.9950	0.9531

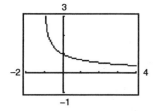

As x approaches 0 from the left and from the right the function approaches 1. Therefore, we estimate the limit to be 1. This agrees with the estimate of the limit obtained from the graph in the accompanying figure. The domain of the function is the set of all real numbers except $x = 0$. You cannot discern from the graph generated by the calculator that $x = 0$ is not in the domain of the function.

1.5   Continuity and One-sided Limits

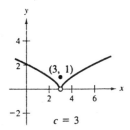

3. Use the graph to visually determine:

(a) $\displaystyle\lim_{x \to c^+} f(x)$ (b) $\displaystyle\lim_{x \to c^-} f(x)$ (c) $\displaystyle\lim_{x \to c} f(x)$

Solution:

(a) $\displaystyle\lim_{x \to c^+} f(x) = 0$ (b) $\displaystyle\lim_{x \to c^-} f(x) = 0$

(c) Since $\displaystyle\lim_{x \to c^+} f(x) = 0 = \lim_{x \to c^-} f(x)$, you have $\displaystyle\lim_{x \to c} f(x) = 0$

[Note that $f(c) \neq \displaystyle\lim_{x \to c} f(x)$.]

12. Find (if it exists) $\lim\limits_{x \to 2} \dfrac{|x-2|}{x-2}$.

Solution:

$$\lim_{x \to 2+} \frac{|x-2|}{x-2} = \lim_{x \to 2+} \frac{x-2}{x-2} = 1$$

$$\lim_{x \to 2-} \frac{|x-2|}{x-2} = \lim_{x \to 2-} \frac{-(x-2)}{x-2} = -1$$

Since the limit from the left is *not equal* to the limit from the right, the limit does *not* exist.

15. Find (if it exists) $\lim\limits_{x \to 3} f(x)$, where

$$f(x) = \begin{cases} \dfrac{x+2}{2}, & x \le 3 \\ \dfrac{12-2x}{2}, & x > 3 \end{cases}$$

Solution:

$$\lim_{x \to 3-} f(x) = \lim_{x \to 3-} \frac{x+2}{5} = \frac{5}{2}$$

$$\lim_{x \to 3+} f(x) = \lim_{x \to 3+} \frac{12-2x}{3} = 2$$

Since the limit from the left is *not equal* to the limit from the right, the limit does *not* exist.

17. Find (if it exists) $\lim\limits_{x \to 1} f(x)$, where

$$f(x) = \begin{cases} x^3 + 1, & x < 1 \\ x+1, & x \ge 1 \end{cases}$$

Solution:

$$\lim_{x \to 1-} f(x) = \lim_{x \to 1-} (x^3 + 1) = 2$$

$$\lim_{x \to 1+} f(x) = \lim_{x \to 1+} (x+1) = 2$$

Since the limit from the left is *equal* to the limit from the right, you have $\lim_{x \to 1} f(x) = 2$.

31. Find the discontinuities (if any) for $f(x) = 1/(x-1)$. Which of the discontinuities are removable?

Solution:

From Theorem 1.11 you know that f is continuous for all x other than $x = 1$. At $x = 1$ the function is discontinuous and the discontinuity is nonremovable since

$$\lim_{x \to 1^-} \frac{1}{x-1} = -\infty \quad \text{and} \quad \lim_{x \to 1^+} \frac{1}{x-1} = \infty.$$

36. Find the discontinuities (if any) for $f(x) = (x+2)/(x^2 - 3x - 10)$ Which of the discontinuities are removable?

Solution:

Since $x^2 - 3x - 10 = (x-5)(x+2)$, $x = 5$ and $x = -2$ are not in the domain of f. By Theorem 1.11, f is continuous for all x other than $x = 5$ or $x = -2$. At $x = 5$ the function is discontinuous and the discontinuity is nonremovable since

$$\lim_{x \to 5^-} \frac{x+2}{(x-5)(x+2)} = \lim_{x \to 5^-} \frac{1}{x-5} = -\infty$$

and

$$\lim_{x \to 5^+} \frac{x+2}{(x-5)(x+2)} = \lim_{x \to 5^+} \frac{1}{x-5} = \infty.$$

At $x = -2$ the function is discontinuous but it is removable since

$$\lim_{x \to -2} \frac{x+2}{x^2 - 3x - 10} = \lim_{x \to -2} \frac{1}{x-5} = -\frac{1}{7}.$$

41. Find the discontinuities (if any) for

$$f(x) = \begin{cases} \dfrac{x}{2} + 1, & x \le 2 \\ 3 - x, & x > 2 \end{cases}$$

Which of the discontinuities are removable?

Solution:

Since f is linear to the right and left of $x = 2$, it is continuous for all x other than possibly at $x = 2$. At $x = 2$,

$$\lim_{x \to 2^-} f(x) = \lim_{x \to 2^-} \left(\frac{x}{2} + 1 \right) = 2$$

$$\lim_{x \to 2^+} f(x) = \lim_{x \to 2^+} (3 - x) = 1$$

Thus f is discontinuous at $x = 2$ and this discontinuity is nonremovable since the limit from the left is *not equal* to the limit from the right.

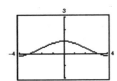

65. Use a graphing utility to obtain a graph of the function $f(x) = \frac{\sin x}{x}$ on the interval $[-4, \, 4]$. Does the graph of the function appear continuous on this interval? Is the function continuous on $[-4, \, 4]$? Write a short paragraph about the importance of examining a function analytically as well as graphically.

Solution:

The graph of the function

$$f(x) = \frac{\sin x}{x}$$

over the interval $[-4, \, 4]$ is shown in the accompanying figure. The graph was produced by a graphing utility and appears continuous. Since $f(0)$ is not defined, the function is not continuous at $x = 0$. This example shows that a removable continuity usually cannot be observed on a graph produced by a graphing utility. Therefore, it is important to examine a function analytically as well as graphically.

76. Verify the applicability of the Intermediate Value Theorem and find the value of c in $[\frac{5}{2}, 4]$ guaranteed by the theorem if $f(x) = (x^2 + x)/(x - 1)$ and $f(c) = 6$.

Solution:

By Theorem 1.11, $f(x) = (x^2 + x)/(x - 1)$ is continuous for every real number except $x = 1$. Therefore, it is continuous on $[\frac{5}{2}, 4]$. Also,

$$f\left(\frac{5}{2}\right) = \frac{35}{6} < 6 < \frac{20}{3} = f(4)$$

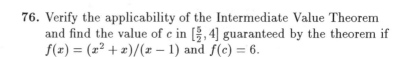

Hence, the Intermediate Value Theorem applies. To find c, solve the equation

$$\frac{x^2 + x}{x - 1} = 6$$

$$\left(\frac{x^2 + x}{x - 1}\right)(x - 1) = 6(x - 1)$$

$$x^2 - 5x + 6 = 0$$

$$(x - 3)(x - 2) = 0$$

The solution of this equation is $x = 2$ or $x = 3$. Since 2 is not in $[\frac{5}{2}, 4]$, it follows that $c = 3$.

79. The number of units in inventory in a small company is given by

$$N(t) = 25\left(2\left[\!\!\left[\frac{t + 2}{2}\right]\!\!\right] - t\right)$$

where t is the time in months. Sketch the graph of this function and discuss its continuity. How often must this company replenish its inventory?

Solution:

Since the number of units in inventory is given by

$$N(t) = 25\left(2\left[\!\!\left[\frac{t + 2}{2}\right]\!\!\right] - t\right) = 25\left(2\left[\!\!\left[\frac{t}{2} + 1\right]\!\!\right] - t\right),$$

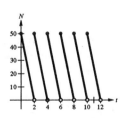

we observe that the greatest integer function is discontinuous at every positive even integer. Therefore, $N(t)$ is discontinuous at every positive even integer. We demonstrate this by evaluating the function for several values of t and plotting the points to generate the accompanying figure.

$$N(0) = 25(2[\![1]\!] - 0) = 25[2(1) - 0] = 50$$
$$N(1) = 25(2[\![1.5]\!] - 1) = 25[2(1) - 1] = 25$$
$$N(1.8) = 25(2[\![1.9]\!] - 1.8) = 25[2(1) - 1.8] = 5$$
$$N(2) = 25(2[\![2]\!] - 2) = 25[2(2) - 2] = 50$$
$$N(3) = 25(2[\![2.5]\!] - 3) = 25[2(2) - 3] = 25$$
$$N(3.8) = 25(2[\![2.9]\!] - 3.8) = 25[2(2) - 3.8] = 5$$
$$N(4) = 25(2[\![3]\!] - 4) = 25[2(3) - 4] = 50$$

The company must replenish its inventory every two months.

1.6   Infinite Limits

5. Determine the limits as x approached -3 from the left and from the right for the function $f(x) = 1/(x^2 - 9)$.

Solution:

When $x = 3$, the values of the numerator and denominator of the function

$$f(x) = \frac{1}{x^2 - 9}$$

are 1 and 0, respectively. Therefore, $x = -3$ is a vertical asymptote. The behavior of f as x approaches -3 from the left and from the right can be seen from the following table:

x	-3.5	-3.1	-3.01	-3.001	-2.999	-2.99	-2.9	-2.5
$f(x)$	0.31	1.64	16.64	166.64	-166.69	-16.69	-1.69	-0.36

Therefore,

$$\lim_{x \to -3^-} \frac{1}{x^2 - 9} = \infty \quad \text{and} \quad \lim_{x \to -3^+} \frac{1}{x^2 - 9} = -\infty$$

15. Find the vertical asymptotes (if any) for $f(x) = \dfrac{1}{e^x - 1}$.

Solution:

Using Theorem 1.14 observe that the graph of f has a vertical asymptote at $x = 0$. This is true because the denominator is zero (and the numerator is not zero) when $x = 0$.

17. Find the vertical asymptotes (if any) for $f(x) = x/(x^2 + x - 2)$.

Solution:

By factoring the denominator you have

$$f(x) = \frac{x}{x^2 + x - 2} = \frac{x}{(x - 1)(x + 2)}.$$

Therefore, the function has two vertical asymptotes, and they are $x = 1$ and $x = -2$.

21. Find the vertical asymptotes (if any) for $f(x) = (x^3+1)/(x+1)$.

Solution:

A superficial investigation of the function may lead to the incorrect conclusion that $x = -1$ is a vertical asymptote. However, notice that the numerator as well as the denominator of the rational function is zero. Hence, Theorem 1.14 does not apply.

$$f(x) = \frac{x^3+1}{x+1} = \frac{(x+1)(x^2-x+1)}{x+1} = x^2 - x + 1, \quad (x \neq -1).$$

Therefore, $x = -1$ is a removable disconitinuity.

28. Determine whether the function $f(x) = \sin(x+1)/(x+1)$ has a vertical asymptote or a removable discontinuity at $x = -1$.

Solution:

By Theorem 1.9

$$\lim_{x \to -1} \frac{\sin(x+1)}{x+1} = 1.$$

Therefore, $x = -1$ is a removable discontinuity.

37. Find (if it exists) $\displaystyle\lim_{x \to 0^+} \frac{2}{\sin x}$.

Solution:

Since

$$\lim_{x \to 0^+} 2 = 2 \qquad \text{and} \qquad \lim_{x \to 0^+} \sin x = 0,$$

if follows that

$$\lim_{x \to 0^+} \frac{2}{\sin x} = \infty.$$

47. A 25-foot ladder is leaning against a house. If the base of the ladder is pulled away from the house at a rate of 2 feet per second, the top will move down the wall at a rate of

$$r = \frac{2x}{\sqrt{625 - x^2}} \text{ feet per second.}$$

(a) Find the rate when x is 7 feet. (b) Find the rate when x is 15 feet. (c) Find the limit of r as $x \to 25^-$.

Solution:

(a) When $x = 7$, you have

$$r = \frac{2(7)}{\sqrt{625 - 7^2}} = \frac{14}{\sqrt{576}} = \frac{7}{12} \text{ feet per second.}$$

(b) When $x = 15$, you have

$$r = \frac{2(15)}{\sqrt{625 - 15^2}} = \frac{30}{\sqrt{400}} = \frac{3}{2} \text{ feet per second.}$$

(c) As x approaches 25 from the left, you have

$$\lim_{x \to 25^-} \frac{2x}{\sqrt{625 - x^2}} = \infty.$$

Review Exercises for Chapter 1

9. Evaluate $\lim\limits_{x \to 0} \dfrac{\dfrac{1}{x+1} - 1}{x}$.

Solution:

$$\lim_{x \to 0} \frac{\dfrac{1}{x+1} - 1}{x} = \lim_{x \to 0} \frac{\dfrac{1 - (x+1)}{x+1}}{x}$$

$$= \lim_{x \to 0} \frac{1 - x - 1}{x(x+1)}$$

$$= \lim_{x \to 0} \frac{-1}{x+1} = -1$$

15. Evaluate $\lim\limits_{\Delta x \to 0} \dfrac{\sin\left(\frac{\pi}{6} + \Delta x\right) - \frac{1}{2}}{\Delta x}$.

Solution:

Since
$$\sin\left(\frac{\pi}{6} + \Delta x\right) = \sin \frac{\pi}{6} \cos \Delta x + \cos \frac{\pi}{6} \sin \Delta x,$$

you have

$$\lim_{\Delta x \to 0} \frac{\sin\left(\frac{\pi}{6} + \Delta x\right) - \frac{1}{2}}{\Delta x} = \lim_{\Delta x \to 0} \frac{\frac{1}{2}\cos \Delta x + \frac{\sqrt{3}}{2}\sin \Delta x - \frac{1}{2}}{\Delta x}$$

$$= \lim_{\Delta x \to 0}\left[\frac{\sqrt{3}\sin \Delta x}{2\,\Delta x} + \frac{1}{2}\left(\frac{\cos \Delta x - 1}{\Delta x}\right)\right]$$

$$= \frac{\sqrt{3}}{2}$$

19. Evaluate $\lim\limits_{x \to -1+} \dfrac{x+1}{x^3+1}$.

Solution:

$$\lim_{x \to -1+} \frac{x+1}{x^3+1} = \lim_{x \to -1+} \frac{x+1}{(x+1)(x^2 - x + 1)}$$

$$= \lim_{x \to -1+} \frac{1}{x^2 - x + 1} = \frac{1}{3}$$

39. Determine the value of c so that the function

$$f(x) = \begin{cases} x + 3, & x \le 2 \\ cx + 6, & x > 2 \end{cases}$$

is continuous for all x.

Solution:

Since f is linear to the right and left of 2, it is continuous for all values of x other than possibly at $x = 2$. Furthermore, since $f(2) = 2 + 3 = 5$, you can make f continuous at $x = 2$ by finding c such that

$$c(2) + 6 = 5$$
$$2c = -1$$
$$c = -\frac{1}{2}$$

Now you have

$$\lim_{x \to 2^-} f(x) = \lim_{x \to 2^-} (x + 3) = 2 + 3 = 5$$

and

$$\lim_{x \to 2^+} f(x) = \lim_{x \to 2^+} \left(-\frac{x}{2} + 6 \right) = -1 + 6 = 5.$$

Thus for $c = -1/2$, f is continuous at $x = 2$ and consequently f is continuous for all x.

49. Determine if $\lim\limits_{x \to 0} \dfrac{|x|}{x} = 1$ is true or false.

Solution:

$$\lim_{x \to 0^+} \frac{|x|}{x} = \lim_{x \to 0^+} \frac{x}{x} = \lim_{x \to 0^+} 1 = 1$$

$$\lim_{x \to 0^-} \frac{|x|}{x} = \lim_{x \to 0^-} \frac{-x}{x} = \lim_{x \to 0^-} (-1) = -1$$

Since the limit from the right and the limit from the left are *not* equal, the limit does *not* exist and the given statement is false.

2 DIFFERENTIATION

2.1 The Derivative and the Tangent Line Problem

9. Use the definition of the derivative to find $f'(x)$ if
$f(x) = 2x^2 + x - 1$.

Solution:

$$f'(x) = \lim_{\Delta x \to 0} \frac{f(x + \Delta x) - f(x)}{\Delta x}$$

$$= \lim_{\Delta x \to 0} \frac{2(x + \Delta x)^2 + (x + \Delta x) - 1 - (2x^2 + x - 1)}{\Delta x}$$

$$= \lim_{\Delta x \to 0} \frac{2x^2 + 4x\Delta x + 2(\Delta x)^2 + x + \Delta x - 1 - 2x^2 - x + 1}{\Delta x}$$

$$= \lim_{\Delta x \to 0} \frac{\Delta x(4x + 2\Delta x + 1)}{\Delta x}$$

$$= \lim_{\Delta x \to 0} (4x + 2\Delta x + 1)$$

$$= 4x + 1$$

13. Use the definition of the derivative to find $f'(x)$ if
$f(x) = 1/(x - 1)$.

Solution:

$$f'(x) = \lim_{\Delta x \to 0} \frac{f(x + \Delta x) - f(x)}{\Delta x}$$

$$= \lim_{\Delta x \to 0} \frac{\dfrac{1}{(x + \Delta x - 1)} - \dfrac{1}{x - 1}}{\Delta x}$$

$$= \lim_{\Delta x \to 0} \frac{x - 1 - (x + \Delta x - 1)}{\Delta x (x + \Delta x - 1)(x - 1)}$$

$$= \lim_{\Delta x \to 0} \frac{-\Delta x}{\Delta x (x + \Delta x - 1)(x - 1)}$$

$$= \lim_{\Delta x \to 0} \frac{-1}{(x + \Delta x - 1)(x - 1)}$$

$$= \frac{-1}{(x - 1)(x - 1)}$$

$$= \frac{-1}{(x - 1)^2}$$

19. Use the definition to find the derivative of $f(x) = x^3$. Sketch the graph of f and find an equation of the tangent line to the graph at $(2, 8)$.

Solution:

$$f'(x) = \lim_{\Delta x \to 0} \frac{f(x + \Delta x) - f(x)}{\Delta x}$$

$$= \lim_{\Delta x \to 0} \frac{(x + \Delta x)^3 - x^3}{\Delta x}$$

$$= \lim_{\Delta x \to 0} \frac{x^3 + 3x^2 \Delta x + 3x(\Delta x)^2 + (\Delta x)^3 - x^3}{\Delta x}$$

$$= \lim_{\Delta x \to 0} \frac{\Delta x [3x^2 + 3x \Delta x + (\Delta x)^2]}{\Delta x}$$

$$= \lim_{\Delta x \to 0} (3x^2 + 3x \Delta x + (\Delta x)^2)$$

$$= 3x^2$$

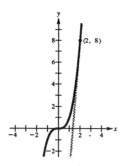

Thus, $f'(x) = 3x^2$ and the slope of the tangent line at $(2, 8)$ is $f'(2) = 3(2)^2 = 12$. Finally, by the point-slope form of the equation of a line, you have

$$y - y_1 = m(x - x_1)$$

$$y - 8 = 12(x - 2)$$

$$y = 12x - 16$$

23. Find an equation of a line that is tangent to the curve $y = x^3$ and is parallel to the line $3x - y + 1 = 0$.

Solution:

The slope of the line given by $3x - y + 1 = 0$ is 3 since

$$3x - y + 1 = 0$$
$$y = 3x + 1 = mx + b$$

Thus, it is necessary to find a point (or points) on the graph of $y = x^3$ such that the tangent line at that point has a slope of 3. From Exercise 19, you know that $dy/dx = 3x^2$ is the slope at any point on the graph of $y = x^3$. Therefore,

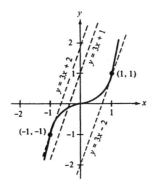

$$\frac{dy}{dx} = 3x^2 = 3$$
$$x^2 = 1 \quad \text{or} \quad x = \pm 1$$

Finally, we conclude that the slope of the graph of $y = x^3$ is 3 at the points $(1, 1)$ and $(-1, -1)$. The tangent lines at these two points are

$$y - 1 = 3(x - 1) \qquad y - (-1) = 3[x - (-1)]$$
$$y = 3x - 2 \qquad\qquad y = 3x + 2$$

25. Find the equations of the two tangent lines to the graph of $f(x) = 4x - x^2$ that pass through the point $(2, 5)$.

Solution:

To begin, find $f'(x)$ as follows:

$$f'(x) = \lim_{\Delta x \to 0} \frac{f(x + \Delta x) - f(x)}{\Delta x}$$
$$= \lim_{\Delta x \to 0} \frac{4(x + \Delta x) - (x + \Delta x)^2 - (4x - x^2)}{\Delta x}$$
$$= \lim_{\Delta x \to 0} \frac{4x + 4\Delta x - x^2 - 2x\Delta x - (\Delta x)^2 - 4x + x^2}{\Delta x}$$
$$= \lim_{\Delta x \to 0} \frac{\Delta x(4 - 2x - \Delta x)}{\Delta x}$$
$$= \lim_{\Delta x \to 0} (4 - 2x - \Delta x)$$
$$= 4 - 2x$$

Now let (x, y) be a point on the graph of $y = 4x - x^2$. Since $dy/dx = 4 - 2x$, the slope of the tangent line at (x, y) is $m = 4 - 2x$. On the other hand, if the tangent line at (x, y) passes through the point $(2, 5)$, then its slope must be

$$m = \frac{y - 5}{x - 2} = \frac{4x - x^2 - 5}{x - 2}$$

Equating these two values of m, you have

$$\frac{4x - x^2 - 5}{x - 2} = 4 - 2x$$
$$4x - x^2 - 5 = (4 - 2x)(x - 2)$$
$$4x - x^2 - 5 = 4x - 2x^2 - 8 + 4x$$
$$x^2 - 4x + 3 = 0$$
$$(x - 3)(x - 1) = 0$$
$$x = 3 \quad \text{or} \quad x = 1$$

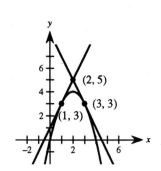

If $x = 3$, then $y = 4(3) - 3^2 = 3$ and $m = 4 - 2(3) = -2$. Therefore,

$$y - 3 = -2(x - 3) \quad \text{or} \quad y = -2x + 9.$$

If $x = 1$, then $y = 4(1) - 1^2 = 3$ and $m = 4 - 2(1) = 2$. Therefore,

$$y - 3 = 2(x - 1) \quad \text{or} \quad y = 2x + 1.$$

39. Use the alternative limit form to find the derivative (if it exists) of $f(x) = x^3 + 2x^2 + 1$ at $x = -2$.

Solution:

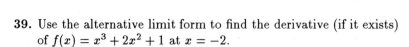

$$f'(-2) = \lim_{x \to -2} \frac{f(x) - f(-2)}{x - (-2)}$$
$$= \lim_{x \to -2} \frac{(x^3 + 2x^2 + 1) - 1}{x + 2}$$
$$= \lim_{x \to -2} \frac{x^2(x + 2)}{x + 2}$$
$$= \lim_{x \to -2} x^2 = 4$$

47. Find every point at which the function $f(x) = (x - 3)^{2/3}$ is differentiable.

Solution:

The graph of $f(x) = (x - 3)^{2/3}$ is continuous at $x = 3$. However, the one-sided limits

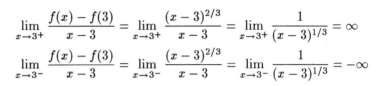

$$\lim_{x \to 3+} \frac{f(x) - f(3)}{x - 3} = \lim_{x \to 3+} \frac{(x-3)^{2/3}}{x - 3} = \lim_{x \to 3+} \frac{1}{(x-3)^{1/3}} = \infty$$

$$\lim_{x \to 3-} \frac{f(x) - f(3)}{x - 3} = \lim_{x \to 3-} \frac{(x-3)^{2/3}}{x - 3} = \lim_{x \to 3-} \frac{1}{(x-3)^{1/3}} = -\infty$$

are not equal. Therefore, f is not differentiable at $x = 3$ and the point $(3, 0)$ is called a **cusp**. Hence, f is differentiable in the intervals $(-\infty, 3)$ and $(3, \infty)$.

2.2   Differentiation Rules and Rates of Change

11. Differentiate $s(t) = t^3 - 2t + 4$.

Solution:

$$s(t) = t^3 - 2t + 4$$
$$s'(t) = 3t^2 - 2$$

15. Differentiate $y = \dfrac{1}{x} - 3 \sin x$.

Solution:

$$y = \frac{1}{x} - 3 \sin x = x^{-1} - 3 \sin x$$
$$y' = -x^{-2} - 3 \cos x$$
$$= -\frac{1}{x^2} - 3 \cos x$$

24. Find $f'(x)$ if $f(x) = \pi/(3x)^2$.

Solution:

$$\text{Function:} \qquad f(x) = \frac{\pi}{(3x)^2}$$

$$\text{Rewrite:} \qquad f(x) = \left(\frac{\pi}{9}\right)x^{-2}$$

$$\text{Derivative:} \qquad f'(x) = \left(\frac{\pi}{9}\right)(-2)x^{-3}$$

$$\text{Simplify:} \qquad f'(x) = \frac{-2\pi}{9x^3}$$

31. Differentiate $y = (2x + 1)^2$ and evaluate the derivative at $(0, 1)$.

Solution:

$$y = (2x + 1)^2 = 4x^2 + 4x + 1$$
$$y' = 8x + 4$$

At the point $(0, 1)$ the derivative is

$$y' = 8(0) + 4 = 4$$

37. Find $f'(t)$ if $f(t) = t^2 - \dfrac{4}{t}$.

Solution:

$$f(t) = t^2 - \frac{4}{t} = t^2 - 4t^{-1}$$

$$f'(t) = 2t - (-1)4t^{-2} = 2t + \frac{4}{t^2}$$

39. Find $f'(x)$ if $f(x) = \dfrac{x^3 - 3x^2 + 4}{x^2}$.

Solution:

$$f(x) = \frac{x^3 - 3x^2 + 4}{x^2} = \frac{x^3}{x^2} - \frac{3x^2}{x^2} + \frac{4}{x^2} = x - 3 + 4x^{-2}$$

$$f'(x) = 1 + 4(-2)x^{-3} = 1 - \frac{8}{x^3} = \frac{x^3 - 8}{x^3}$$

45. Find $f'(x)$ if $f(x) = 4\sqrt{x} + 3\cos x$.

Solution:
$$f(x) = 4\sqrt{x} + 3\cos x = 4x^{1/2} + 3\cos x$$
$$f'(x) = 4\left(\frac{1}{2}\right)x^{-1/2} + 3(-\sin x) = \frac{2}{\sqrt{x}} - 3\sin x$$

47. Find an equation of the line tangent to $y = x^4 - 3x^2 + 2$ at the point $(1, 0)$.

Solution:
$$y = x^4 - 3x^2 + 2$$
$$y' = 4x^3 - 6x$$

Thus the slope of the tangent line at $(1, 0)$ is
$$y' = 4(1)^3 - 6(1) = 4 - 6 = -2$$
The equation of the tangent line at $(1, 0)$ is
$$y - y_1 = m(x - x_1)$$
$$y - 0 = -2(x - 1)$$
$$2x + y - 2 = 0$$

53. At what points, if any, does the graph of $y = x^4 - 3x^2 + 2$ have horizontal tangents?

Solution:

A tangent line is horizontal if the derivative (the slope) at the point of tangency is zero. Since $y' = 4x^3 - 6x = 2x(2x^2 - 3)$, you must find all values of x that satisfy the equation $y' = 2x(2x^2 - 3) = 0$. The solutions to this equation are $x = 0$ and $x = \pm\sqrt{3/2}$. At $x = 0$, you have
$$y = (0)^4 - 3(0)^2 + 2 = 2.$$
At $x = \pm\sqrt{3/2}$, you have
$$y = \left(\pm\sqrt{\frac{3}{2}}\right)^4 - 3\left(\pm\sqrt{\frac{3}{2}}\right)^2 + 2$$
$$= \frac{9}{4} - \frac{9}{2} + 2 = \frac{9 - 18 + 8}{4} = -\frac{1}{4}$$

Thus, the points of horizontal tangency are:

$$(0, 2), \quad \left(\sqrt{\frac{3}{2}}, -\frac{1}{4}\right), \quad \text{and} \quad \left(-\sqrt{\frac{3}{2}}, -\frac{1}{4}\right).$$

61. Sketch the graphs of the two equations $y = x^2$ and $y = -x^2 + 6x - 5$ and sketch the two lines that are tangent to both graphs. Find equations for these lines.

Solution:

Let (x_1, y_1) and (x_2, y_2) be the points of tangency on the graphs of $y = x^2$ and $y = -x^2 + 6x - 5$, respectively. We know that $y_1 = x_1^2$ and $y_2 = -x_2^2 + 6x_2 - 5$. Let m be the slope of the tangent line. Since the line passes through (x_1, y_1) and (x_2, y_2), it follows that

(1) $$m = \frac{y_2 - y_1}{x_2 - x_1} = \frac{-x_2^2 + 6x_2 - 5 - x_1^2}{x_2 - x_1}.$$

Since the line is tangent to $y = x^2$ at (x_1, y_1) and the derivative of this curve is $y' = 2x$, it follows that

$$m = 2x_1$$

Since the line is tangent to $y = -x^2 + 6x - 5$ at (x_2, y_2) and the derivative of this curve is $y' = -2x_2 + 6$, it follows that

$$m = -2x_2 + 6$$

Thus, from the preceding two equations, you have

$$m = 2x_1 = -2x_2 + 6$$

(2) $$x_1 = -x_2 + 3$$

Using equations (1) and (2) yields

$$m = -2x_2 + 6 = \frac{-x_2^2 + 6x_2 - 5 - x_1^2}{x_2 - x_1}$$

$$-2x_2 + 6 = \frac{-x_2^2 + 6x_2 - 5 - (-x_2 + 3)^2}{x_2 - (-x_2 + 3)}$$

$$(2x_2 - 3)(-2x_2 + 6) = -x_2^2 + 6x_2 - 5 - (x_2^2 - 6x_2 + 9)$$

$$-4x_2^2 + 18x_2 - 18 = -2x_2^2 + 12x_2 - 14$$

$$.\quad -2x_2^2 + 6x_2 - 4 = 0$$

$$x_2^2 - 3x_2 + 2 = 0$$

$$(x_2 - 2)(x_2 - 1) = 0$$

$$x_2 = 1 \quad \text{or} \quad x_2 = 2.$$

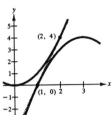

If $x_2 = 1$, then $y_2 = -1^2 + 6(1) - 5 = 0$, $x_1 = -1 + 3 = 2$, and $y_1 = 2^2 = 4$. Thus the line containing $(1, 0)$ and $(2, 4)$ is tangent to both curves. The equation of this line is

$$y - 0 = \left(\frac{4 - 0}{2 - 1}\right)(x - 1) \qquad \text{or} \qquad y = 4x - 4$$

If $x_2 = 2$, then $y_2 = -2^2 + 6(2) - 5 = -4 + 12 - 5 = 3$, $x_1 = -2 + 3 = 1$, and $y_1 = 1^2 = 1$. Thus the line containing $(2, 3)$ and $(1, 1)$ is tangent to both curves. The equation of this line is

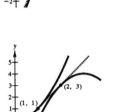

$$y - 1 = \left(\frac{3 - 1}{2 - 1}\right)(x - 1)$$
$$= 2x - 2$$
$$y = 2x - 1$$

75. Find the average rate of change of $f(x) = -1/x$ over the interval $[1, 2]$. Compare this average rate of change to the instantaneous rate of change at the endpoints of the interval.

Solution:

The average rate of change is given by

$$\frac{\Delta y}{\Delta x} = \frac{f(2) - f(1)}{2 - 1} = \frac{-\frac{1}{2} - (-1)}{2 - 1} = \frac{1}{2}.$$

To find the instantaneous rate of change we first find $f'(x)$.

$$f(x) = -\frac{1}{x} = -x^{-1}$$
$$f'(x) = -(-1)x^{-2} = \frac{1}{x^2}$$

The instantaneous rate of change when $x = 1$ is

$$f'(1) = \frac{1}{1^2} = 1$$

and the instantaneous rate of change when $x = 2$ is

$$f'(2) = \frac{1}{2^2} = \frac{1}{4}.$$

77. A silver dollar dropped from the World Trade Center which is 1362 feet high.

(a) Determine the position and velocity functions for the coin.

(b) Find the average velocity on the interval $[1, 2]$.

(c) Find the instantaneous velocity when $t = 1$ and $t = 2$.

(d) Find the time required for the coin to reach ground level.

(e) Find the velocity of the dollar just prior to hitting the ground.

Solution:

(a) Since the initial height of the coin is $s_0 = 1362$ feet and the initial velocity is $v_0 = 0$, you have

$$s(t) = -\frac{1}{2}gt^2 + v_0 t + s_0 = -16t^2 + 1362$$

and

$$v(t) = s'(t) = -32t.$$

(b) The average velocity is given by

$$\frac{s(2) - s(1)}{2 - 1} = 1298 - 1346 = -48 \text{ ft/sec}$$

(c) Since $v(t) = -32t$, the instantaneous velocity at time $t = 1$ is

$$s'(1) = -32(1) = -32 \text{ ft/sec},$$

and at time $t = 2$ is

$$s'(2) = -32(2) = -64 \text{ ft/sec}$$

(d) The dollar will reach ground level when

$$s(t) = -16t^2 + 1362 = 0$$
$$t^2 = \frac{1362}{16}$$
$$t = \frac{1}{4}\sqrt{1362} \approx 9.2 \text{ sec.}$$

(e) The velocity just prior to hitting the ground is

$$v\left(\frac{1}{4}\sqrt{1362}\right) = -32\left(\frac{1}{4}\sqrt{1362}\right) = -8\sqrt{1362}$$

$$\approx -295.2 \text{ ft/sec.}$$

93. Newton's Law of Gravitation states that the gravitational force between two (point) particles of masses m_1 and m_2, respectively, is

$$F = K\frac{m_1 m_2}{r^2}, \ r > 0$$

where r is the distance between the particles. Find the rate of change of force between the particles with respect to r and explain why the rate of change is negative.

Solution:

Since $F = K\dfrac{m_1 m_2}{r^2} = K m_1 m_2 r^{-2}$ for $r > 0$, you have

$$\frac{dF}{dr} = K m_1 m_2 (-2) r^{-3} = -2K\frac{m_1 m_2}{r^3}.$$

Since $r > 0$, it follows that $dF/dr < 0$. This indicates that the gravitational force between the two particles decreases as the distance between them increases.

2.3   The Product and Quotient Rules and Higher-Order Derivatives

3. Differentiate $f(x) = (x^3 - 3x)(2x^2 + 3x + 5)$ and find $f'(0)$.

Solution:

$$f(x) = (x^3 - 3x)(2x^2 + 3x + 5)$$
$$f'(x) = (x^3 - 3x)(4x + 3) + (2x^2 + 3x + 5)(3x^2 - 3)$$
$$= 4x^4 + 3x^3 - 12x^2 - 9x + 6x^4 - 6x^2 + 9x^3 - 9x + 15x^2 - 15$$
$$= 10x^4 + 12x^3 - 3x^2 - 18x - 15$$
$$f'(0) = -15$$

17. Differentiate $f(x) = \dfrac{3 - 2x - x^2}{x^2 - 1}$.

Solution:

$$f(x) = \frac{3 - 2x - x^2}{x^2 - 1}$$

$$= -\frac{x^2 + 2x - 3}{x^2 - 1} = -\frac{(x+3)(x-1)}{(x+1)(x-1)} = -\frac{x+3}{x+1}$$

$$f'(x) = -\frac{(x+1)(1) - (x+3)(1)}{(x+1)^2} = \frac{2}{(x+1)^2}$$

25. Differentiate $f(x) = (3x^3 + 4x)(x - 5)(x + 1)$.

Solution:

We begin by using the Associative Law to group the first two factors and then use the Product Rule twice.

$$f(x) = [(3x^3 + 4x)(x - 5)](x + 1)$$

$$f'(x) = [(3x^3 + 4x)(x - 5)](1)$$

$$+ (x + 1)[(3x^3 + 4x)(1) + (x - 5)(9x^2 + 4)]$$

$$= (3x^3 + 4x)(x - 5) + (x + 1)(3x^3 + 4x)$$

$$+ (x + 1)(x - 5)(9x^2 + 4)$$

$$= 15x^4 - 48x^3 - 33x^2 - 32x - 20$$

27. Differentiate $g(x) = \left(\dfrac{x+1}{x+2}\right)(2x - 5)$.

Solution:

$$g(x) = \frac{(x+1)(2x-5)}{x+2} = \frac{2x^2 - 3x - 5}{x+2}$$

$$g'(x) = \frac{(x+2)(4x-3) - (2x^2 - 3x - 5)(1)}{(x+2)^2}$$

$$= \frac{4x^2 + 5x - 6 - 2x^2 + 3x + 5}{(x+2)^2} = \frac{2x^2 + 8x - 1}{(x+2)^2}$$

33. Differentiate $f(t) = \dfrac{\cos t}{t}$.

Solution:

$$f(t) = \frac{\cos t}{t}$$

$$f'(t) = \frac{t(-\sin t) - (\cos t)(1)}{t^2} = \frac{-(t \sin t + \cos t)}{t^2}$$

41. Differentiate $y = -\csc x - \sin x$.

Solution:

$$y = -\csc x - \sin x$$

$$y' = -(-\csc x \cot x) - \cos x = \frac{1}{\sin x}\left(\frac{\cos x}{\sin x}\right) - \cos x$$

$$= \cos x \left(\frac{1}{\sin^2 x} - 1\right) = \frac{\cos x}{\sin^2 x}(1 - \sin^2 x) = \cos x \cot^2 x$$

45. Differentiate $f(x) = x^2 \tan x$.

Solution:

$$f(x) = x^2 \tan x$$
$$f'(x) = x^2 \sec^2 x + (\tan x)(2x) = x(x \sec^2 x + 2 \tan x)$$

53. Find an equation of the line tangent to the graph of $f(x) = x/(x-1)$ at the point $(2,2)$.

Solution:

$$f(x) = \frac{x}{x-1}$$

$$f'(x) = \frac{(x-1)(1) - (x)(1)}{(x-1)^2} = \frac{x-1-x}{(x-1)^2} = \frac{-1}{(x-1)^2}$$

Therefore, the slope of the tangent line at $(2, 2)$ is $f'(2) = -1/(2-1)^2 = -1$ and an equation of the tangent line at $(2, 2)$ is

$$y - 2 = -1(x - 2)$$
$$= -x + 2$$
$$y = -x + 4$$

69. A population of 500 bacteria is introduced into a culture and grows in number according to the equation

$$P(t) = 500\left(1 + \frac{4t}{50 + t^2}\right)$$

where t is measured in hours. Find the rate at which the population is growing when $t = 2$.

Solution:

$$P'(t) = 500\left[\frac{(50 + t^2)(4) - (4t)(2t)}{(50 + t^2)^2}\right]$$

$$= \frac{500(200 + 4t^2 - 8t^2)}{(50 + t^2)^2} = \frac{2000(50 - t^2)}{(50 + t^2)^2}$$

Therefore, the rate of population growth when $t = 2$ is

$$P'(2) = \frac{2000(50 - 4)}{(50 + 4)^2} \approx 31.55 \text{ bacteria/hr.}$$

75. Find the second derivative of the function $f(x) = \dfrac{x}{x - 1}$.

Solution:

The first derivative of $f(x) = \dfrac{x}{x - 1}$ was found in Exercise 53.

$$f'(x) = \frac{-1}{(x - 1)^2} = \frac{-1}{x^2 - 2x + 1}$$

$$f''(x) = \frac{(x^2 - 2x + 1)(0) - (-1)(2x - 2)}{(x^2 - 2x + 1)^2}$$

$$= \frac{2(x - 1)}{[(x - 1)^2]^2} = \frac{2}{(x - 1)^3}$$

2.4   The Chain Rule

15. Find the first derivative of $y = \sqrt[3]{9x^2 + 4}$.

Solution:

$$y = \sqrt[3]{9x^2 + 4} = (9x^2 + 4)^{1/3}$$

$$\frac{dy}{dx} = \left(\frac{1}{3}\right)(9x^2 + 4)^{-2/3}(18x) = \frac{6x}{(9x^2 + 4)^{2/3}}$$

21. Find the first derivative of $f(t) = \left(\dfrac{1}{t-3}\right)^2$.

Solution:

$$f(t) = \frac{1}{(t-3)^2} = (t-3)^{-2}$$

$$f'(t) = (-2)(t-3)^{-3}(1) = \frac{-2}{(t-3)^3}$$

23. Find the first derivative of $y = \dfrac{1}{\sqrt{x+2}}$.

Solution:

$$y = \frac{1}{\sqrt{x+2}} = (x+2)^{-1/2}$$

$$y' = \left(-\frac{1}{2}\right)(x+2)^{-3/2}(1) = \frac{-1}{2(x+2)^{3/2}}$$

25. Find the first derivative of $f(x) = x^2(x-2)^4$.

Solution:

Applying the Product Rule and the General Power Rule produces

$$f(x) = x^2(x-2)^4$$
$$f'(x) = x^2(4)(x-2)^3(1) + (x-2)^4(2x)$$
$$= 2x(x-2)^3(2x+x-2) = 2x(x-2)^3(3x-2).$$

33. Use a symbolic differentiation utility to find the derivative of the function

$$g(t) = \frac{3t^2}{\sqrt{t^2 + 2t - 1}}.$$

Then use the utility to graph the function and its derivative on the same set of coordinate axes. Describe the behavior of the function the corresponds to the zeros of the graph of the derivative.

Solution:

The first derivative of the function

$$g(t) = \frac{3t^2}{\sqrt{t^2 + 2t - 1}}$$

is

$$g'(t) = \frac{3t(t^2 + 3t - 2)}{(t^2 + 2t - 1)^{3/2}}.$$

(Your differentiation utility may give the derivative in unsimplified form.)

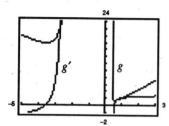

The graphs of the function and its derivative are given in the accompanying figure. The zeros of the derivative are approximately $t = -3.56$ and $t = 0.56$. These are the t-coordinates of the points where the graph of $g(t)$ has horizontal tangents and at these points yield minimum values of the function.

45. Find the derivative of the function $f(\theta) = \dfrac{1}{4}\sin^2 2\theta$.

Solution:

$$f(\theta) = \frac{1}{4}(\sin 2\theta)^2$$

$$f'(\theta) = \frac{1}{4}(2)(\sin 2\theta)(\cos 2\theta)(2)$$

$$= \frac{1}{2}(2\sin 2\theta \cos 2\theta) = \frac{1}{2}\sin 4\theta \qquad \text{Double Angle Identity}$$

53. Find the derivative of the function $g(t) = (e^{-t} + e^t)^3$.

Solution:

$$g(t) = (e^{-t} + e^t)^3$$

$$g'(t) = 3(e^{-t} + e^t)^2[e^{-t}(-1) + e^t(1)] = 3(e^{-t} + e^t)^2(e^t - e^{-t})$$

61. Find the derivative of the function $y = \ln\left(\dfrac{2x}{x-1}\right)$.

Solution:

$$y = \ln\left(\frac{2x}{x-1}\right) = \ln 2 + \ln x - \ln(x-1)$$

$$\frac{dy}{dx} = 0 + \frac{1}{x} - \frac{1}{x-1} = \frac{-1}{x(x-1)}$$

65. Find the derivative of the function $y = \ln(\sin^2 x)$.

Solution:

$$y = \ln(\sin^2 x) = 2\ln(\sin x)$$

$$\frac{dy}{dx} = 2 \cdot \frac{1}{\sin x} \cdot \cos x = 2\cot x$$

87. Find an equation of the tangent line to the graph of $f(x) = \sin 2x$ at the point $(\pi, 0)$.

Solution:

$$f(x) = \sin 2x$$
$$f'(x) = (\cos 2x)(2) = 2 \cos 2x$$

Therefore, the slope of the tangent line at $(\pi, 0)$ is $f'(\pi) = 2 \cos 2\pi = 2$ and an equation of the tangent line at $(\pi, 0)$ is

$$y - 0 = 2(x - \pi)$$
$$y = 2(x - \pi)$$
$$2x - y - 2\pi = 0$$

91. Find $f''(x)$ if $f(x) = \sin x^2$.

Solution:

$$f(x) = \sin x^2$$
$$f'(x) = (\cos x^2)(2x) = 2x \cos x^2$$
$$f''(x) = 2[x(-\sin x^2)(2x) + (\cos x^2)(1)] = 2(\cos x^2 - 2x^2 \sin x^2)$$

103. The speed S of blood that is r centimeters from the center of an artery is given by $S = C(R^2 - r^2)$, where C is a constant, R is the radius of the artery, and S is measured in centimeters per second. Suppose a drug is administered and the artery begins dilating at the rate dR/dt. At a constant distance r, find the rate at which S changes with respect to t for $C = 1.76 \times 10^5$, $R = 1.2 \times 10^{-2}$, and $dR/dt = 10^{-5}$.

Solution:

Using the Chain Rule and the fact that r is constant, you have

$$S = C(R^2 - r^2)$$
$$\frac{dS}{dt} = C\left(2R\frac{dR}{dt} - 0\right)$$

Substituting the given constants yields

$$\frac{dS}{dt} = (1.76 \times 10^5)[2(1.2 \times 10^{-2})(10^{-5})] = 4.224 \times 10^{-2}$$

2.5   Implicit Differentiation

---

5. Use implicit differentiation to find dy/dx if $x^3 - xy + y^2 = 4$.

Solution:

$$x^3 - xy + y^2 = 4$$

$$\frac{d}{dx}[x^3 - xy + y^2] = \frac{d}{dx}[4]$$

$$\frac{d}{dx}[x^3] - \frac{d}{dx}[xy] + \frac{d}{dx}[y^2] = \frac{d}{dx}[4]$$

$$3x^2 - \left[x\frac{dy}{dx} + y(1)\right] + 2y\frac{dy}{dx} = 0$$

$$3x^2 - x\frac{dy}{dx} - y + 2y\frac{dy}{dx} = 0$$

$$(2y - x)\frac{dy}{dx} = y - 3x^2$$

$$\frac{dy}{dx} = \frac{y - 3x^2}{2y - x}$$

11. Use implicit differentiation to find dy/dx if $\sin x + 2\cos 2y = 1$.

Solution:

$$\sin x + 2\cos 2y = 1$$

$$\frac{d}{dx}[\sin x + 2\cos 2y] = \frac{d}{dx}[1]$$

$$\frac{d}{dx}[\sin x] + \frac{d}{dx}[2\cos 2y] = \frac{d}{dx}[1]$$

$$\cos x + 2(-\sin 2y)(2)\frac{dy}{dx} = 0$$

$$\cos x - 4\sin 2y\frac{dy}{dx} = 0$$

$$-4\sin 2y\frac{dy}{dx} = -\cos x$$

$$\frac{dy}{dx} = \frac{\cos x}{4\sin 2y}$$

17. Use implicit differentiation to find dy/dx if $xe^y - 10x + 3y = 0$.

Solution:

$$xe^y - 10x + 3y = 0$$

$$\frac{d}{dx}[xe^y - 10x + 3y] = \frac{d}{dx}[0]$$

$$\frac{d}{dx}[xe^y] - \frac{d}{dx}[10x] + \frac{d}{dx}[3y] = 0$$

$$\left[xe^y \frac{dy}{dx} + e^y\right] - 10 + 3\frac{dy}{dx} = 0$$

$$(xe^y + 3)\frac{dy}{dx} = 10 - e^y$$

$$\frac{dy}{dx} = \frac{10 - e^y}{xe^y + 3}$$

21. Given $xy = 4$, find dy/dx by implicit differentiation and evaluate the derivative at $(-4, -1)$.

Solution:

Implicit differentiation of the equation $xy = 4$ yields

$$x\frac{dy}{dx} + y(1) = 0$$

$$x\frac{dy}{dx} = -y$$

$$\frac{dy}{dx} = -\frac{y}{x}$$

At $(-4, -1)$, $\dfrac{dy}{dx} = -\dfrac{-1}{-4} = -\dfrac{1}{4}$.

27. Given $\tan(x + y) = x$, find dy/dx by implicit differentiation and evaluate the derivative at $(0, 0)$.

Solution:

Implicit differentiation of the equation $\tan(x + y) = x$ yields

$$\sec^2(x + y)\left(1 + \frac{dy}{dx}\right) = 1$$

$$\sec^2(x + y)\frac{dy}{dx} = 1 - \sec^2(x + y)$$

$$\frac{dy}{dx} = \frac{1 - \sec^2(x + y)}{\sec^2(x + y)}$$

$$= \frac{-\tan^2(x + y)}{\tan^2(x + y) + 1} = \frac{-x^2}{x^2 + 1}$$

At $(0,0)$, $\dfrac{dy}{dx} = \dfrac{0}{1} = 0$.

39. Given the equation $9x^2 + 16y^2 = 144$, (a) find two explicit functions by solving the equation for y in terms of x, (b) sketch the graph of the equation and label the parts given by the explicit functions, (c) differentiate the explicit function, and (d) find dy/dx implicitly and show that the result is equivalent to that of part (c).

Solution:

(a) Solving the equation for y, yields

$$9x^2 + 16y^2 = 144$$
$$16y^2 = 144 - 9x^2 = 9(16 - x^2)$$
$$y^2 = \frac{9}{16}(16 - x^2)$$
$$y = \pm\frac{3}{4}\sqrt{16 - x^2}, \qquad -4 \le x \le 4$$

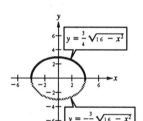

(b) The graph is given in the accompanying figure.

(c) When the explicit functions are differentiated explicitly, you obtain

$$\frac{dy}{dx} = \pm\left(\frac{3}{4}\right)\left(\frac{1}{2}\right)(16 - x^2)^{-1/2}(-2x)$$
$$= \frac{\mp 3x}{4\sqrt{16 - x^2}} = \frac{-3(3x)}{16(\pm\frac{3}{4})\sqrt{16 - x^2}} = \frac{-9x}{16y}$$

(d) Differentiating the equation $9x^2 + 16y^2 = 144$ implicitly yields

$$18x + 32y\,\frac{dy}{dx} = 0$$
$$\frac{dy}{dx} = \frac{-9x}{16y}$$

43. Find d^2y/dx^2 in terms of x and y if $x^2 - y^2 = 16$.

Solution:

$$x^2 - y^2 = 16$$
$$2x - 2yy' = 0$$
$$-2yy' = -2x$$
$$y' = \frac{x}{y}$$

Differentiating implicitly again yields

$$y'' = \frac{y(1) - (x)y'}{y^2} = \frac{y - x(x/y)}{y^2} = \frac{y^2 - x^2}{y^3} = -\frac{16}{y^3}.$$

49. Show that the normal line (the line perpendicular to the tangent line to a curve) at any point on the circle $x^2 + y^2 = r^2$ passes through the origin.

Solution:

By implicit differentiation,

$$x^2 + y^2 = r^2$$
$$2x + 2yy' = 0$$
$$y' = -\frac{x}{y}.$$

Thus, if (x, y) is a point on the circle $x^2 + y^2 = r^2$, the slope of the tangent line at (x, y) is $-x/y$. On the other hand, the slope of the line passing through (x, y) and $(0, 0)$ is

$$m = \frac{y - 0}{x - 0} = \frac{y}{x}$$

Since this slope is the negative reciprocal of y', the line passing through (x, y) and $(0, 0)$ must be perpendicular to the tangent line at (x, y).

53. Show that the graphs of the equations $2x^2 + y^2 = 6$ and $y^2 = 4x$ are orthogonal (the curves intersect at right angles). Sketch the graph of each equation.

Solution:

To find the points of intersection, set $y^2 = 6 - 2x^2$ and $y^2 = 4x$ equal to each other.

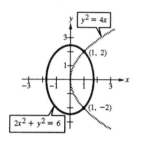

$$4x = 6 - 2x^2$$
$$2x^2 + 4x - 6 = 0$$
$$x^2 + 2x - 3 = 0$$
$$(x + 3)(x - 1) = 0$$
$$x = -3 \quad \text{and} \quad x = 1$$

When $x = 1$, $y = \pm 2$, and when $x = -3$, y is undefined. Thus the two points of intersection are $(1, 2)$ and $(1, -2)$. Differentiating the equation $2x^2 + y^2 = 6$ implicitly yields

$$4x + 2yy' = 0$$
$$y' = -\frac{2x}{y}$$

and differentiating $y^2 = 4x$ implicitly yields

$$2yy' = 4$$
$$y' = \frac{2}{y}.$$

Thus at $(1, 2)$ the slopes of the two curves are

$$\frac{-2(1)}{2} = -1 \quad \text{and} \quad \frac{2}{2} = 1$$

which implies that the tangent lines at this point are perpendicular. Finally, at $(1, -2)$ the slopes of the two curves are

$$\frac{-2(1)}{-2} = 1 \quad \text{and} \quad \frac{2}{-2} = -1$$

and the tangent lines at this point are also perpendicular.

2.6   Related Rates

---

3. Assuming that x and y are differentiable functions of t and $xy = 4$ find:

(a) $\dfrac{dy}{dx}$ when $x = 8$ if $\dfrac{dx}{dt} = 10$

(b) $\dfrac{dx}{dt}$ when $x = 1$ if $\dfrac{dy}{dt} = -6$

Solution:

Since x and y are differentiable functions of t, differentiating the equation $xy = 4$ with respect to t yields

(1)
$$x\frac{dy}{dt} + y\frac{dx}{dt} = 0$$

(a) Solving (1) for $\dfrac{dy}{dt}$, and substituting $y = \dfrac{1}{2}$ and $\dfrac{dx}{dt} = 10$ when $x = 8$, yields

$$\frac{dy}{dt} = -\frac{y}{x}\frac{dx}{dt} = -\frac{\frac{1}{2}}{8}(10) = -\frac{5}{8}.$$

(b) Solving (1) for $\dfrac{dx}{dt}$, and substituting $y = 4$ and $\dfrac{dy}{dt} = -6$ when $x = 1$, yields

$$\frac{dx}{dt} = -\frac{x}{y}\frac{dy}{dt} = -\frac{1}{4}(-6) = \frac{3}{2}.$$

9. A point is moving along the graph of $y = x\ln x$ so that dx/dt is 2 centimeters per second. Find dy/dt for $x = 1$ and $x = e$.

Solution:

Since x and y are differentiable functions of t, differentiating the equation $y = x\ln x$ with respect to t yields

(1) $$\frac{dy}{dt} = x\left(\frac{1}{x}\right)\frac{dx}{dt} + \ln x(1)\frac{dx}{dt} = (1 + \ln x)\frac{dx}{dt}$$

(a) Substituting $x = 1$ and $\dfrac{dx}{dt} = 2$ into (1) yields

$$\frac{dy}{dt} = (1 + \ln 1)2 = 2.$$

(b) Substituting $x = e$ and $\dfrac{dx}{dt} = 2$ in (1) yields

$$\frac{dy}{dt} = (1 + \ln e)2 = 4.$$

21. At a sand and gravel plant, sand is falling off a conveyer and onto a conical pile at the rate of 10 ft³/min. The diameter of the base of the cone is approximately three times the altitude. At what rate is the height of the pile changing when is 15 ft high?

Solution:

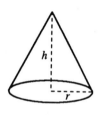

Let

$$V = \text{volume of cone} = \frac{1}{3}\pi r^2 h.$$

Since the diameter of the base is approximately three times the altitude, you have

$$2r = 3h \qquad \text{or} \qquad r = \frac{3}{2}h$$

Therefore,

$$V = \frac{1}{3}\pi \left(\frac{3}{2}h\right)^2 h = \frac{3\pi}{4}h^3$$

Differentiating with respect to t, yields

$$\frac{dV}{dt} = \frac{9\pi}{4}h^2\frac{dh}{dt} \quad \text{or} \quad \frac{4(dV/dt)}{9\pi h^2} = \frac{dh}{dt}.$$

Now, substituting $h = 15$ and $\dfrac{dV}{dt} = 10$, yields the following rate of change of the height of the conical pile.

$$\frac{dh}{dt} = \frac{4(10)}{9\pi(15)^2} = \frac{8}{405\pi} \text{ ft/min.}$$

23. A swimming pool is 40 ft long, 20 ft wide, 4 ft deep at the shallow end, and 9 ft deep at the deep end, the bottom being an inclined plane. Assume that the water is being pumped into the pool at 10 ft³/min and there is 4 ft of water at the deep end.

(a) What percentage of the pool is filled?

(b) At what rate is the water level rising?

Solution:

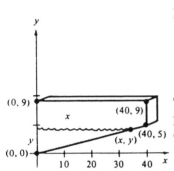

From the figure we see that x and y are related by the equation

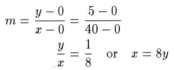

$$m = \frac{y - 0}{x - 0} = \frac{5 - 0}{40 - 0}$$

$$\frac{y}{x} = \frac{1}{8} \quad \text{or} \quad x = 8y$$

The volume of the inclined portion of the pool is given by the product of the width of the pool and the area of the triangular cross section.

$$V_L = 20 \left(\frac{1}{2}\right) xy = 10xy = 10(8y)y = 80y^2$$

When $y = 5$, the inclined portion of the pool has a volume of

$$V_L = 80(5^2) = 2000 \, \text{ft}^3$$

Since the upper rectangular portion of the pool has a volume of

$$V_U = 4(40)(20) = 3200 \, \text{ft}^3$$

the total volume of the pool is

$$V = V_L + V_U = 2000 + 3200 = 5200 \, \text{ft}^3$$

(a) When $y = 4$, the ratio of the filled portion of the pool to the total volume is

$$\frac{80y^2}{V} = \frac{80(4^2)}{5200} = 24.6\%.$$

(b) When $y = 4$, you can find dy/dt by differentiating $V_L = 80y^2$ and substituting $dV_L/dt = 10$.

$$V_L = 80y^2$$

$$\frac{dV_L}{dt} = 160y\frac{dy}{dt}$$

$$10 = 160(4)\frac{dy}{dt}$$

$$\frac{dy}{dt} = \frac{10}{640} = \frac{1}{64} \, \text{ft/min}$$

25. A ladder 25 ft long is leaning against a house. The base of the ladder is pulled away from the house wall at a rate of 2 feet per second.

(a) How fast is the top moving down the wall when the base of the ladder is 7 feet, 15 feet, and 24 feet from the wall?

(b) Consider the triangle formed by the side of the house, the ladder and the ground. Find the rate at which the area of the triangle is changing when $x = 7$.

(c) Find the rate at which the angle between the top of the ladder and the wall of the house is changing when $x = 7$.

Solution:

(a) From the figure it follows that x and y are related by the equation $x^2 + y^2 = (25)^2$. Differentiating this equation with respect to t, yields

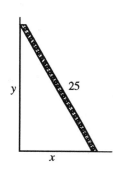

$$2x\frac{dx}{dt} + 2y\frac{dy}{dt} = 0$$

$$y\frac{dy}{dt} = -x\frac{dx}{dt}$$

$$\frac{dy}{dt} = -\frac{x}{y}\frac{dx}{dt}$$

Since $\dfrac{dx}{dt} = 2$, you have $\dfrac{dy}{dt} = -2\dfrac{x}{y}$. When $x = 7$, $y = \sqrt{(25)^2 - 7^2} = \sqrt{576} = 24$.

$$\frac{dy}{dt} = -2\left(\frac{7}{24}\right) = -\frac{7}{12} \approx -0.583 \text{ ft/sec}$$

When $x = 15$, $y = \sqrt{(25)^2 - (15)^2} = \sqrt{400} = 20$.

$$\frac{dy}{dt} = -2\left(\frac{15}{20}\right) = -\frac{3}{2} = -1.5 \text{ ft/sec}$$

When $x = 24$, $y = \sqrt{(25)^2 - (24)^2} = \sqrt{49} = 7$.

$$\frac{dy}{dt} = -2\left(\frac{24}{7}\right) = -\frac{48}{7} \approx -6.857 \text{ ft/sec}$$

(b) The area of the triangle in the accompanying figure is

$$A = \frac{1}{2}xy = \frac{1}{2}x\sqrt{625 - x^2}.$$

Differentiating with respect to t yields

$$\frac{dA}{dt} = \frac{1}{2}\left[x\left(\frac{1}{2}\right)(625 - x^2)^{-1/2}\left(-2x\frac{dx}{dt}\right) + \sqrt{625 - x^2}\,\frac{dx}{dt}\right]$$

$$= \left(\frac{625 - 2x^2}{2\sqrt{625 - x^2}}\right)\frac{dx}{dt}$$

The rate of change of the area when $x = 7$ and $dx/dt = 2$ is

$$\frac{dA}{dt} = \left(\frac{625 - 2(7^2)}{2\sqrt{625 - (7^2)}}\right)(2) = \frac{527}{24}\ \text{ft}^2\text{sec}.$$

(c) If θ is the angle between the top of the ladder and the wall of the house, then

$$\sin\theta = \frac{x}{25}$$

$$\cos\theta\frac{d\theta}{dt} = \frac{1}{25}\frac{dx}{dt}$$

$$\frac{d\theta}{dt} = \frac{\sec\theta}{25}\frac{dx}{dt}.$$

When $x = 7$, $\sin\theta = 7/25$ and $\sec\theta = 25/24$. Since $dx/dt = 2$, the rate of change of the angle θ is

$$\frac{d\theta}{dt} = \frac{25/24}{25}(2) = \frac{1}{12}\ \text{rad/sec}.$$

27. A winch at the top of a 40-foot building pulls a pipe of the same length to a vertical position using the method shown in the accompanying figure. The winch pulls in rope at the rate of -0.5 feet per second. Find the rate of vertical change and the rate of horizontal change at the end of the pipe when $y = 20$.

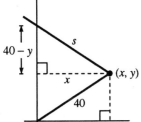

Solution:

The relationship between x and y in the accompanying figure is given by $x = \sqrt{40^2 - y^2}$ and the relationship among the variables x, y, and s is given by $s^2 = x^2 + (40 - y)^2$.

Substituting for x in the second equation and differentiating with respect to t yields

$$s^2 = \left(\sqrt{40^2 - y^2}\right)^2 + (40 - y)^2$$
$$= 40^2 - y^2 + (40 - y)^2$$
$$2s\frac{ds}{dt} = -2y\frac{dy}{dt} + 2(40 - y)(-1)\frac{dy}{dt}$$
$$s\frac{ds}{dt} = -40\frac{dy}{dt}$$
$$-\frac{s}{40}\frac{ds}{dt} = \frac{dy}{dt}$$

When $y = 20$, $s = \sqrt{(40^2 - 20^2) + (40 - 20)^2} = 40$. Substituting this value for s and -0.5 feet per second for ds/dt yields

$$\frac{dy}{dt} = -\frac{40}{40}(-0.5) = \frac{1}{2} \text{ ft/sec.}$$

When $y = 20$, $x = \sqrt{40^2 - 20^2} = 20\sqrt{3}$. Differentiating the equation $s^2 = x^2 + (40 - y)^2$ with respect to t yields

$$2s\frac{ds}{dt} = 2x\frac{dx}{dt} + 2(40 - y)(-1)\frac{dy}{dt}.$$

Solving this equation for dx/dt and substituting the known values for x, y, dy/dt and ds/dt, yields

$$\frac{dx}{dt} = \frac{s(ds/dt) + (40 - y)(dy/dt)}{x}$$
$$= \frac{40(-0.5) + (40 - 20)(1/2)}{20\sqrt{3}}$$
$$= \frac{-20 + 10}{20\sqrt{3}} = -\frac{\sqrt{3}}{6} \text{ ft/sec.}$$

33. A man 6 ft tall walks at a rate of 5 ft/s away from a light that is 15 ft above the ground. When the man is 10 ft from the base of the light:

(a) at what rate is the tip of his shadow moving?

(b) at what rate is the length of his shadow changing?

Solution:

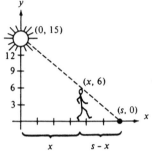

From the figure we see that x and s are related by similar triangles in such a way that

$$\frac{s-x}{6} = \frac{s}{15}$$
$$15s - 15x = 6s$$
$$9s = 15x \quad \Longrightarrow \quad s = \frac{5}{3}x$$

(a) To find ds/dt, given that $dx/dt = 5$, differentiate with respect to t as follows:

$$\frac{ds}{dt} = \frac{5}{3} \cdot \frac{dx}{dt} = \frac{5}{3}(5) = \frac{25}{3} \approx 8.3 \text{ ft/sec}$$

(b) The rate at which the shadow is increasing is

$$\frac{ds}{dt} - \frac{dx}{dt} = \frac{25}{3} - 5 = \frac{10}{3} \approx 3.3 \text{ ft/sec}$$

(Note: The measurement 10 feet given in this problem is a "red herring" since the distance from the base of the light does not affect ds/dt.)

45. A wheel of radius 1 ft revolves at a rate of 10 rev/s. A dot is painted at a point P on the rim of the wheel as shown in the accompanying figure. Find the rate of horizontal movement of the dot for the following angles:
(a) $\theta = 0°$ (b) $\theta = 30°$ (c) $\theta = 60°$

Solution:

From the figure, we have

$$x = \cos \theta$$

We are given

$$\frac{d\theta}{dt} = 10 \text{ rev/sec} = 10(-2\pi) = -20\pi \text{ rad/sec}$$

Therefore,

$$\frac{dx}{dt} = -\sin\theta\frac{d\theta}{dt}$$
$$= -\sin\theta(-20\pi) = 20\pi\sin\theta$$

(a) When $\theta = 0°$, $\dfrac{dx}{dt} = 0$ ft/sec.

(b) When $\theta = 30°$, $\dfrac{dx}{dt} = 20\pi\left(\dfrac{1}{2}\right) = 10\pi$ ft/sec.

(c) When $\theta = 60°$, $\dfrac{dx}{dt} = 20\pi\left(\dfrac{\sqrt{3}}{2}\right) = 10\pi\sqrt{3}$ ft/sec.

Review Exercises for Chapter 2

13. Find the derivative of $f(x) = (3x^2 + 7)(x^2 - 2x + 3)$.

Solution:

$$f(x) = (3x^2 + 7)(x^2 - 2x + 3)$$
$$f'(x) = (3x^2 + 7)(2x - 2) + (x^2 - 2x + 3)(6x)$$
$$= 6x^3 - 6x^2 + 14x - 14 + 6x^3 - 12x^2 + 18x$$
$$= 2(6x^3 - 9x^2 + 16x - 7)$$

17. Find the derivative of $f(x) = \dfrac{x^2 + x - 1}{x^2 - 1}$.

Solution:

$$f(x) = \frac{x^2 + x - 1}{x^2 - 1}$$
$$f'x = \frac{(x^2 - 1)(2x + 1) - (x^2 + x - 1)(2x)}{(x^2 - 1)^2}$$
$$= \frac{2x^3 + x^2 - 2x - 1 - 2x^3 - 2x^2 + 2x}{(x^2 - 1)^2} = -\frac{x^2 + 1}{(x^2 - 1)^2}$$

29. Find the derivative of $f(x) = -x \tan x$.

Solution:

$$f(x) = -x \tan x$$
$$f'(x) = -x \sec^2 x + (\tan x)(-1) = -(x \sec^2 x + \tan x)$$

35. Find the derivative of $g(x) = x^2/e^x$.

Solution:

$$g(x) = \frac{x^2}{e^x}$$
$$g'(x) = \frac{e^x(2x) - x^2(e^x)}{(e^x)^2} = \frac{xe^x(2-x)}{e^{2x}}$$

51. Find the second derivative of $f(x) = \cot x$.

Solution:

$$f(x) = \cot x$$
$$f'x = -\csc^2 x$$
$$f''(x) = -2 \csc x(-\csc x \cot x) = 2 \csc^2 x \cot x$$

57. Use implicit differentiation to find dy/dx for $x^2 + 3xy + y^3 = 10$.

Solution:

$$x^2 + 3xy + y^3 = 10$$
$$\frac{d}{dx}[x^2 + 3xy + y^3] = \frac{d}{dx}[10]$$
$$\frac{d}{dx}[x^2] + \frac{d}{dx}[3xy] + \frac{d}{dx}[y^3] = \frac{d}{dx}[10]$$
$$2x + 3x\frac{dy}{dx} + 3y + 3y^2\frac{dy}{dx} = 0$$
$$(3x + 3y^2)\frac{dy}{dx} = -2x - 3y$$
$$\frac{dy}{dx} = \frac{-(2x + 3y)}{3(x + y^2)}$$

71. Find the equations of the tangent line and the normal line to the graph of $y = \sqrt[3]{(x-2)^2}$ at $(3,1)$.

Solution:

$$y = \sqrt[3]{(x-2)^2} = (x-2)^{2/3}$$
$$y' = \left(\frac{2}{3}\right)(x-2)^{-1/3} = \frac{2}{3(x-2)^{1/3}}$$

Thus the slope of the tangent line at $(3,1)$ is

$$y' = \frac{2}{3(3-2)^{1/3}} = \frac{2}{3}$$

and the equation of the tangent line at $(3,1)$ is

$$y - 1 = \frac{2}{3}(x-3)$$
$$3y - 3 = 2x - 6$$
$$-2x + 3y + 3 = 0$$

Since the slope of the tangent line is $\frac{2}{3}$, the slope of the normal line is $-\frac{3}{2}$ and the equation of the normal line is

$$y - 1 = -\frac{3}{2}(x-3)$$
$$2y - 2 = -3x + 9$$
$$3x + 2y - 11 = 0$$

79. Show that $y = 2\ln x + 3$ is a solution to the differential equation $x(y'') + y' = 0$.

Solution:

$$y = 2(\ln x) + 3$$
$$y' = 2\left(\frac{1}{x}\right)$$
$$y'' = 2\left(\frac{-1}{x^2}\right) = \frac{-2}{x^2}$$
$$x(y'') + y' = x\left(\frac{-2}{x^2}\right) + \left(\frac{2}{x}\right) = \left(\frac{-2}{x}\right) + \left(\frac{2}{x}\right) = 0$$

85. What is the smallest initial velocity that is required to throw a stone to the top of a 49-foot silo?

Solution:

We assume that the stone is thrown from an initial height of $s_0 = 0$. Thus the position equation is

$$s = -16t^2 + v_0 t$$

The maximum value of s occurs when $ds/dt = 0$ and thus you have

$$\frac{ds}{dt} = -32t + v_0 = 0$$

$$-32t = -v_0 \quad \Longrightarrow \quad t = \frac{v_0}{32}$$

This means that the maximum height is

$$s = -16\left(\frac{v_0}{32}\right)^2 + v_0\left(\frac{v_0}{32}\right) = \frac{v_0{}^2}{64}$$

If s is to attain a value of 49, you must have

$$\frac{v_0{}^2}{64} = 49$$

$$v_0{}^2 = 3136 \quad \Longrightarrow \quad v_0 = 56 \text{ ft/sec}$$

93. The cross section of a 5-foot trough is an isosceles trapezoid with lower base 2 feet, upper base 3 feet, and altitude 2 feet. Water is running into the trough at the rate of 1 cubic foot per minute. How fast is the water level rising when the water is 1 foot deep?

Solution:

The figure is a cross section of the trough when the water is at a depth of h feet.

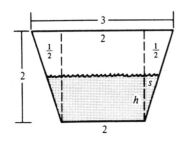

$$\frac{s}{h} = \frac{\frac{1}{2}}{2}$$

$$s = \frac{1}{4}h$$

A = area of cross section of water at depth h

$$= 2h + 2\left(\frac{1}{2}sh\right)$$

$$= 2h + \left(\frac{1}{4}h\right)h = 2h + \frac{1}{4}h^2$$

V = volume of water in trough at depth h

$$= 5A = 5\left(2h + \frac{1}{4}h^2\right)$$

Differentiating with respect to t, yields

$$\frac{dV}{dt} = 5\left(2 + \frac{1}{2}h\right)\frac{dh}{dt} = \frac{5}{2}(4 + h)\frac{dh}{dt}$$

$$\frac{2(dV/dt)}{5(4 + h)} = \frac{dh}{dt}$$

Therefore, when $dV/dt = 1$ and $h = 1$, we have

$$\frac{dh}{dt} = \frac{2(1)}{5(4 + 1)} = \frac{2}{25} \text{ ft/min.}$$

3 APPLICATIONS OF DIFFERENTIATION

3.1 Extrema on an interval

15. Locate the extrema of $f(x) = -x^2 + 3x$ on the interval $[0, 3]$.

Solution:
$$f(x) = -x^2 + 3x$$
$$f'(x) = -2x + 3 = 0$$

Therefore, $x = \dfrac{3}{2}$ is a critical number in $[0, 3]$.

The extrema of f are determined by evaluating f at the critical number and at the endpoints of $[0, 3]$.

$$f(0) = -0^2 + 3(0) = 0 \qquad \text{Minimum}$$
$$f\left(\frac{3}{2}\right) = -\left(\frac{3}{2}\right)^2 + 3\left(\frac{3}{2}\right) = \frac{9}{4} \qquad \text{Maximum}$$
$$f(3) = -3^2 + 3(3) = 0 \qquad \text{Minimum}$$

19. Locate the extrema of $f(x) = 3x^{2/3} - 2x$ on the interval $[-1, 1]$.

Solution:
$$f(x) = 3x^{2/3} - 2x$$
$$f'(x) = 2x^{-1/3} - 2$$
$$= 2\left(\frac{1 - \sqrt[3]{x}}{\sqrt[3]{x}}\right)$$

Therefore, $x = 1$ and $x = 0$ are critical numbers in the interval $[-1, 1]$. [$f'(0)$ is undefined.]

The extrema of f are determined by evaluating f at the critical numbers and at the endpoints of the interval $[-1, 1]$.

$$f(-1) = 3(-1)^{2/3} - 2(-1) = 5 \qquad \text{Maximum}$$
$$f(0) = 3(0)^{2/3} - 2(0) = 0 \qquad \text{Minimum}$$
$$f(1) = 3(1)^{2/3} - 2(1) = 1$$

23. Locate the extrema of $y = e^x \sin x$ on the interval $[0, \pi]$.

Solution:

$$y = e^x \sin x$$
$$\frac{dy}{dx} = e^x \cos x + e^x \sin x = e^x (\cos x + \sin x)$$

Find any critical numbers in the interval $[0, \pi]$, by solving the following equation.

$$\frac{dy}{dx} = 0$$
$$\cos x + \sin x = 0$$
$$1 + \tan x = 0$$
$$\tan x = -1 \quad \Longrightarrow \quad x = \frac{3\pi}{4}$$

The extrema of y are determined by evaluating y at the critical number and at the endpoints of the interval $[0, \pi]$.

$$y(0) = e^0 \sin 0 = 0 \qquad \text{Minimum}$$
$$y(\tfrac{3\pi}{4}) = e^{3\pi/4} \sin \tfrac{3\pi}{4} \approx 7.46 \qquad \text{Maximum}$$
$$y(\pi) = e^\pi \sin \pi = 0 \qquad \text{Minimum}$$

31. Determine from the graph of f if f possesses a relative minimum in the interval (a, b).

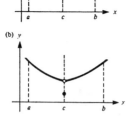

(a) y

(b) y

Solution:

(a) Since f is decreasing in the interval (a, c) and increasing in the interval (c, b), the only possible minimum in the interval (a, b) would occur at $x = c$. However, $f(c)$ is greater than $f(x)$ for x near c. Thus, there is no minimum.

(b) Since $f(c) \leq f(x)$ for all x in (a, b), $f(c)$ is a minimum.

39. Use a symbolic differentiation utility to find the maximum value of $|f''(x)|$ on the interval $[0, 2]$, if $f(x) = \sqrt{1 + x^3}$. (This value is used in the error estimate for the Trapezoid Rule as discussed in Section 5.6.)

Solution:

To find the maximum value of $|f''(x)|$ in $[0, 2]$, select the maximum of $|f''(0)|$, $|f''(2)|$, and $|f''(c)|$, where c is any critical number of $f''(x)$ [i.e., $f'''(c) = 0$ or $f'''(c)$ does not exist.] Using a symbolic differentiation utility yields the following derivatives.

$$f(x) = \sqrt{1 + x^3}$$

$$f'(x) = \frac{3x^2}{2\sqrt{1 + x^3}}$$

$$f''(x) = \frac{3x(x^3 + 4)}{4(1 + x^3)^{3/2}}$$

$$f'''(x) = \frac{-3(x^6 + 20x^3 - 8)}{8(1 + x^3)^{5/2}}$$

Therefore, $f'''(x) = 0$ when $x^6 + 20x^3 - 8 = 0$. Solving this equation using the symbolic differentiation utility yields the critical numbers

$$x = \begin{cases} \sqrt[3]{-10 + 6\sqrt{3}} & \approx 0.732 \\ \sqrt[3]{-10 - 6\sqrt{3}} & \approx -2.732 \end{cases}$$

The critical number in the interval $[0, 2]$ is $\sqrt[3]{-10 + 6\sqrt{3}}$. Evaluating the absolute value of the second derivative at the critical number and the endpoints of the interval, yields

$$|f''(\sqrt[3]{-10 + 6\sqrt{3}}) \approx 1.468$$
$$|f''(0)| = 0$$
$$|f''(2)| = \frac{2}{3}.$$

The maximum value of $|f''(x)|$ in $[0, 2]$ is

$$\left| f''(\sqrt[3]{-10 + 6\sqrt{3}}) \right| \approx |f''(0.732)| \approx 1.468.$$

45. The formula for the power output P of a battery is given by $P = VI - RI^2$ where V is the electromotive force in volts, R is the resistance, and I is the current. Find the current (measured in amperes) that corresponds to a maximum value of P in a battery for which $V = 12$ volts and $R = 0.5$ ohms. (Assume that a 15 amp fuse bounds the output in the interval $0 \le I \le 15$.)

Solution:

Since $V = 12$ and $R = 0.5$, you have

$$P = 12I - \frac{1}{2}I^2$$

$$\frac{dP}{dI} = 12 - I.$$

Therefore, $I = 12$ is a critical number on the interval $[0, 15]$. Determine the maximum of P by evaluating P at the critical number and at the endpoints of the interval $[0, 15]$. Since $P(0) = 0$, $P(12) = 72$, and $P(15) = 67.5$, it follows that the power P is maximum when $I = 12$.

55. Use a graphing utility to graph the function $f(x) = \ln x$. Then graph

$$P_1(x) = f(1) + f'(1)(x - 1)$$

and

$$P_2(x) = f(1) + f'(1)(x - 1) + \tfrac{1}{2}f''(1)(x - 1)^2$$

on the same viewing rectangle. Compare the values of f, P_1, and P_2 and their first derivatives at $x = 1$.

Solution:

$$
\begin{aligned}
f(x) &= \ln x, & f(1) &= 0 \\
f'(x) &= \frac{1}{x}, & f'(1) &= 1 \\
f''(x) &= -\frac{1}{x^2}, & f''(1) &= -1
\end{aligned}
$$

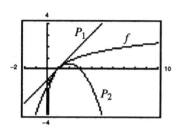

$$P_1(x) = f(1) + f'(1)(x - 1) = x - 1$$
$$P_2(x) = f(1) + f'(1)(x - 1) + \tfrac{1}{2}f''(1)(x - 1)^2$$
$$= (x - 1) - \tfrac{1}{2}(x - 1)^2$$
$$P_1'(x) = 1$$
$$P_2'(x) = 1 - (x - 1) = 2 - x$$
$$f(1) = P_1(1) = P_2(1) = 0$$
$$f'(1) = P_1'(1) = P_2'(1) = 1$$

The values of f, P_1, and P_2, and their first derivatives agree at $x = 1$. The graphs of the functions are shown in the accompanying figure.

3.2   Rolle's Theorem and the Mean Value Theorem

5. Determine whether Rolle's Theorem can be applied to the function $f(x) = (x - 1)(x - 2)(x - 3)$ on the interval $[1, 3]$. If it can be applied, find all values of c in the interval such that $f'(c) = 0$.

Solution:

Since f is a polynomial, it is continuous and differentiable for all x. Also, the zeros of f are $x = 1$, $x = 2$, and $x = 3$. Thus, Rolle's Theorem can be applied on the intervals $[1, 2]$ and $[2, 3]$. Setting $f'(x) = 0$ yields

$$f'(x) = 3x^2 - 12x + 11 = 0$$
$$x = \frac{12 \pm \sqrt{144 - 132}}{6} = \frac{12 \pm 2\sqrt{3}}{6} = \frac{6 \pm \sqrt{3}}{3}.$$

Therefore, in the interval $[1, 2]$,

$$f'\left(\frac{6 - \sqrt{3}}{3}\right) = 0 \quad \text{where} \quad c = \frac{6 - \sqrt{3}}{3} \approx 1.423.$$

and in the interval $[2, 3]$,

$$f'\left(\frac{6 + \sqrt{3}}{3}\right) = 0 \quad \text{where} \quad c = \frac{6 + \sqrt{3}}{3} \approx 2.577.$$

9. Determine whether Rolle's Theorem can be applied to the function $f(x) = x^{2/3} - 1$ on the interval $[-8, 8]$. If it can be applied, find all values of c such that $f'(c) = 0$.

Solution:

We first observe that f is continuous on the specified interval. The zeros of f are $x = \pm 1$. Since

$$f'(x) = \frac{2}{3x^{1/3}},$$

we observe that $f'(0)$ is undefined and therefore, f is not differentiable at $x = 0$. Thus, Rolle's Theorem cannot be applied to this function on the specified interval.

17. Determine whether Rolle's Theorem can be applied to the function

$$f(x) = \frac{x}{2} - \sin\frac{\pi x}{6}$$

on the interval $[-1, 0]$. If it can be applied, find all values of c in the interval such that $f'(c) = 0$.

Solution:

We observe that f is continuous and differentiable for all real x. Also, from the accompanying graph we determine that $f(-1) = f(0) = 0$. Thus, Rolle's Theorem can be applied to the interval $[-1, 0]$. Setting $f'(x) = 0$ yields

$$f'(x) = \frac{1}{2} - \frac{\pi}{6}\cos\frac{\pi x}{6} = 0$$

$$\cos\frac{\pi x}{6} = \frac{3}{\pi}$$

$$x = \pm\frac{6}{\pi}\arccos\frac{3}{\pi}$$

In the interval $[-1, 0]$

$$f'\left(-\frac{6}{\pi}\arccos\frac{3}{\pi}\right) \approx f'(-0.5756) = 0$$

31. Apply the Mean Value Theorem to $f(x) = x/(x+1)$ on the interval $[-\frac{1}{2}, 2]$. Find all values of c in $[-\frac{1}{2}, 2]$ such that

$$f'(c) = \frac{f(2) - f(-\frac{1}{2})}{2 - (-\frac{1}{2})}.$$

Solution:

Since $f(x) = x/(x+1)$ is continuous on $[0, 1]$ and differentiable on $(0, 1)$, you can apply the Mean Value Theorem.

$$f(x) = \frac{x}{x+1}$$

(1) $$f'(x) = \frac{(x+1)(1) - x(1)}{(x+1)^2} = \frac{1}{(x+1)^2}$$

(2) $$\frac{f(2) - f(-\frac{1}{2})}{2 - (-\frac{1}{2})} = \frac{\frac{2}{3} - (-1)}{\frac{5}{2}} = \frac{2}{3}$$

Equating the right-hand members of equations (1) and (2) yields

$$\frac{1}{(x+1)^2} = \frac{2}{3}$$

$$x + 1 = \pm\sqrt{\frac{3}{2}} \quad \Longrightarrow \quad x = -1 \pm \frac{\sqrt{6}}{2}.$$

Therefore, on the interval $[-\frac{1}{2}, 2]$, $c = -1 + \frac{\sqrt{6}}{2}$.

35. Apply the Mean Value Theorem to $f(x) = x - 2\sin x$ on the interval $[-\pi, \pi]$. Find all values of c in $[-\pi, \pi]$ such that

$$f'(c) = \frac{f(\pi) - f(-\pi)}{\pi - (-\pi)}.$$

Solution:

Since f is continuous and differentiable for all real x, you can apply the Mean Value Theorem over the specified interval.

$$f(x) = x - 2\sin x$$
$$f'(x) = 1 - 2\cos x$$

$$1 - 2\cos x = \frac{f(\pi) - f(-\pi)}{\pi - (-\pi)}$$

$$1 - 2\cos x = \frac{\pi - (-\pi)}{2\pi} = 1$$

$$\cos x = 0 \quad \Longrightarrow \quad x = \pm\frac{\pi}{2}$$

Therefore, on the interval $[-\pi, \pi]$, $c = \pm\frac{\pi}{2}$.

53. Show that $x^{2n+1} + ax + b$ cannot have two real roots, where $a > 0$ and n is any positive integer.

Solution:

The polynomial $f(x) = x^{2n+1} + ax + b$ is continuous and differentiable for all x. Therefore, by Rolle's Theorem, if $f(x) = 0$ for two distinct values of x, there must be at least one value of x such that $f'(x) = 0$. However,

$$f(x) = x^{2n+1} + ax + b$$
$$f'(x) = (2n+1)x^{2n} + a = 0$$
$$(x^n)^2 = -\frac{a}{2n+1}$$

$$(\text{positive number}) = (\text{negative number})$$

Therefore, $f'(x) = 0$ has no solution and consequently $f(x) = 0$ cannot have two real zeros.

3.3   Increasing and Decreasing Functions, and The First Derivative Test

11. Find the critical numbers (if any) of $f(x) = 2x^3 + 3x^2 - 12x$, find the open intervals on which f is increasing or decreasing, and locate all relative extrema.

Solution:

$$f(x) = 2x^3 + 3x^2 - 12x$$
$$f'(x) = 6x^2 + 6x - 12 = 6(x+2)(x-1)$$

Therefore, $f'(x) = 0$ when $x = -2$ or $x = 1$. Since f is a polynomial, it is differentiable for all x and the only critical numbers are $x = -2$ and $x = 1$.

Interval	$-\infty < x < -2$	$-2 < x < 1$	$1 < x < \infty$
Test value	$x = -3$	$x = 0$	$x = 2$
Sign of $f'(x)$	$f'(-3) = 24 > 0$	$f'(0) = -12 < 0$	$f'(2) = 24 > 0$
Conclusion	f is increasing	f is decreasing	f is increasing

When $x = -2$, you have $f(-2) = 2(-2)^3 + 3(-2)^2 - 12(-2) = 20$.
When $x = 1$, you have $f(1) = 2 + 3 - 12 = -7$. Therefore, if
follows that $(-2, 20)$ is a relative maximum and $(1, -7)$ is a
relative minimum.

15. Find the critical numbers (if any) of $f(x) = x^{1/3} + 1$, find the
open intervals on which f is increasing or decreasing, and locate
all relative extrema.

Solution:

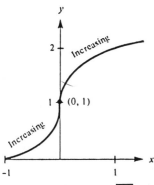

$$f(x) = x^{1/3} + 1$$

$$f'(x) = \left(\frac{1}{3}\right)x^{-2/3} = \frac{1}{3x^{2/3}}$$

Since f is continuous for all x and differentiable for all x
other than $x = 0$, the only critical number is $x = 0$. ($f'(0)$ is
undefined.) We also observe that $f'(x) > 0$ for all x not equal
to zero. Therefore, if follows that f is increasing for all x and
there are **no** relative extrema.

27. Find the critical numbers (if any) of $f(x) = (3 - x)e^{x-3}$, find
the open intervals on which f is increasing or decreasing, and
locate all relative extrema. Use a graphing utility to confirm
your results.

Solution:

$$f(x) = (3 - x)e^{x-3}$$
$$f'(x) = (3 - x)e^{x-3} + e^{x-3}(-1)$$
$$= (2 - x)e^{x-3}$$

Since f is differentiable for all x, the only critical number is
$x = 2$.

Interval	$-\infty < x < 2$	$2 < x < \infty$
Test value	$x = 0$	$x = 3$
Sign of $f'(x)$	$f'(0) = 2e^{-3} > 0$	$f'(3) = -1 < 0$
Conclusion	f increasing	f decreasing

When $x = 2$, you have $f(2) = e^{-1}$. Therefore, $(2, e^{-1})$ is a
relative maximum. The accompanying figure is the graph of f
produced by a graphing utility.

29. Find the critical numbers (if any) of $f(x) = (x/2) + \cos x$, on the interval $0 \le x < 2\pi$, find the intervals where f is increasing or decreasing, and locate all relative extrema. Use a graphing utility to confirm your results.

Solution:

$$f(x) = \frac{x}{2} + \cos x$$

$$f'(x) = \frac{1}{2} - \sin x$$

Hence, $f'(x) = 0$ when $x = \pi/6$ or $x = 5\pi/6$. Since f is continuous and differentiable for all x, the only critical numbers are $x = \pi/6$ and $5\pi/6$.

Interval	$0 < x < \pi/6$	$\pi/6 < x < 5\pi/6$	$5\pi/6 < x < 2\pi)$
Test value	$x = \pi/12$	$x = \pi/2$	$x = \pi$
Sign of $f'(x)$	$f'(\pi/12) > 0$	$f'(\pi/2) < 0$	$f'(\pi) > 0$
Conclusion	f increasing	f decreasing	f increasing

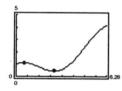

We conclude that a relative maximum occurs at $\left(\dfrac{\pi}{6}, \dfrac{\pi + 6\sqrt{3}}{12}\right)$ and a relative minimum at $\left(\dfrac{5\pi}{6}, \dfrac{5\pi - 6\sqrt{3}}{12}\right)$. The accompanying figure is the graph of f produced by a graphing utility.

49. Coughing forces the trachea (windpipe) to contract, which affects the velocity v of the air passing through the trachea. Suppose the velocity of the air during coughing is given by

$$v = k(R - r)r^2$$

where k is a constant, R is the normal radius of the trachea, and r is the radius during coughing. What radius will produce the maximum air velocity?

Solution:

$$v = k(R - r)r^2 = k(Rr^2 - r^3)$$

$$\frac{dv}{dr} = k(2Rr - 3r^2) = kr(2R - 3r)$$

Therefore, $dv/dr = 0$ when $r = 0$ or $r = \frac{2}{3}R$. Since v is continuous and differentiable for all r, the only critical numbers are $r = 0$ and $r = 2R/3$.

Interval	$-\infty < r < 0$	$0 < r < \frac{2}{3}R$	$\frac{2}{3}R < r < \infty$
Test value	$r = -R$	$r = \frac{R}{3}$	$r = \frac{4R}{3}$
Sign of $v'(r)$	$v'(-R) < 0$	$v'(\frac{R}{3}) > 0$	$v'(\frac{4R}{3}) < 0$
Conclusion	v decreasing	v increasing	v decreasing

We conclude that the velocity v is maximum when $r = \dfrac{2R}{3}$.

55. Find a, b, c, and d so that the function $f(x) = ax^3 + bx^2 + cx + d$ has a relative minimum at $(0,0)$ and a relative maximum at $(2,2)$.

Solution:

In order for $(0, 0)$ and $(2, 2)$ to be solution points to f, you must have

$$f(0) = 0 = a(0^3) + b(0^2) + c(0) + d$$
$$0 = d$$
$$f(2) = 2 = a(2^3) + b(2^2) + c(2) + d$$
$$2 = 8a + 4b + 2c + 0$$
(1) $$1 = 4a + 2b + c$$

Also, for f to have relative extrema at $(0,0)$ and $(2,2)$, $f'(x)$ must be zero at these points. Since $f'(x) = 3ax^2 + 2bx + c$, you have

$$f'(0) = 0 = 3a(0^2) + 2b(0) + c$$
$$0 = c$$
$$f'(2) = 0 = 3a(2^2) + 2b(2) + c$$
$$0 = 12a + 4b + 0$$
(2) $$0 = 3a + b$$

Letting $c = 0$ in equation 1 and multiplying equation 2 by 2, yields

$$
\begin{array}{rcl}
4a + 2b & = & 1 \\
6a + 2b & = & 0 \\
\hline
-2a & = & 1 \\
a & = & -\dfrac{1}{2}
\end{array}
$$

$$4\left(-\frac{1}{2}\right) + 2b = 1$$

$$2b = 3$$

$$b = \frac{3}{2}$$

Finally, we conclude that $a = -\frac{1}{2}, b = \frac{3}{2}, c = 0,$ and $d = 0.$
Thus,

$$f(x) = -\frac{1}{2}x^3 + \frac{3}{2}x^2.$$

3.4   Concavity and the Second Derivative Test

11. Identify all relative extrema for $f(x) = x^3 - 3x^2 + 3$. Use the
Second-Derivative Test when applicable.

Solution:
$$f(x) = x^3 - 3x^2 + 3$$
$$f'(x) = 3x^2 - 6x = 3x(x - 2)$$
$$f'(x) = 0 \quad \text{when} \quad x = 0, \ 2$$
$$f''(x) = 6x - 6$$

At $x = 0$, you have $f(0) = 3$, $f'(0) = 0$, and $f''(0) = -6$.
Therefore, by the Second-Derivative Test, $(0, 3)$ is a relative
maximum.

At $x = 2$, you have $f(2) = -1$, $f'(2) = 0$, and $f''(2) = 6$.
Therefore, by the Second-Derivative Test, $(2, -1)$ is a relative
minimum.

17. Identify all relative extrema for $f(x) = x + (4/x)$. Use the Second-Derivative Test when applicable.

Solution:

$$f(x) = x + \frac{4}{x}$$

$$f'(x) = 1 - \frac{4}{x^2} = \frac{x^2 - 4}{x^2}$$

$$f'(x) = 0 \text{ when } x = \pm 2$$

(Note that f' is undefined when $x = 0$. It is **not** a critical number since it is not in the domain of f.)

$$f''(x) = (-4)(-2)x^{-3} = \frac{8}{x^3}$$

At $x = -2$, you have $f(-2) = -4$, $f'(-2) = 0$, and $f''(-2) = -1$. Therefore, by the Second-Derivative Test, $(-2, -4)$ is a relative maximum.

Since f is symmetrical with respect to the origin, $(2, 4)$ is a relative minimum.

19. Identify all relative extrema for $f(x) = \cos x - x$ on the interval $[0, 4\pi]$. Use the Second-Derivative Test when applicable.

Solution:

$$f(x) = \cos x - x$$

$$f'(x) = -\sin x - 1$$

Therefore, $f'(x) = 0$ when $\sin x = -1$. The critical numbers in the interval $[0, 4\pi]$ are

$$x = \frac{3\pi}{2} \quad \text{and} \quad x = \frac{7\pi}{2}.$$

However, there are no relative extrema since $f'(x) \leq 0$ for all x.

27. Find any relative extrema and points of inflection and sketch the graph of $f(x) = x(x-4)^3$.

Solution:

$$f(x) = x(x-4)^3$$
$$f'(x) = x[3(x-4)^2(1)] + (x-4)^3(1)$$
$$= (x-4)^2[3x + (x-4)] = 4(x-4)^2(x-1)$$
$$f'(x) = 0 \quad \text{when} \quad x = 1,\ 4$$
$$f''(x) = 4[(x-4)^2(1) + (x-1)(2)(x-4)(1)]$$
$$= 4(x-4)[(x-4) + 2(x-1)] = 12(x-4)(x-2)$$

Possible points of inflection occur at $x = 2$ and $x = 4$. By testing the intervals determined by these x-values, you can conclude that the both yield points of inflection.

Interval	$-\infty < x < 2$	$2 < x < 4$	$4 < x < \infty$
Test value	$x = 0$	$x = 3$	$x = 5$
Sign of $f''(x)$	$f''(0) > 0$	$f''(3) < 0$	$f''(5) > 0$
Conclusion	Concave upward	Concave downward	Concave upward

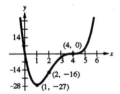

Therefore, $(1, -27)$ is a relative minimum, and $(2, -16)$ and $(4, 0)$ are points of inflection. Note that there is **not** a relative extrema at $(4, 0)$ even though $x = 4$ is a critical number. The first derivative remains nonnegative as x increases through the critical number $x = 4$.

29. Find any relative extrema and points of inflection and sketch the graph of $f(x) = x\sqrt{x+3}$.

Solution:

$$f(x) = x\sqrt{x+3} \qquad \text{domain: } [-3, \infty)$$
$$f'(x) = x\left(\frac{1}{2}\right)(x+3)^{-1/2}(1) + (x+3)^{1/2}(1)$$
$$= \frac{x}{2\sqrt{x+3}} + \sqrt{x+3}$$
$$= \frac{x + 2(x+3)}{2\sqrt{x+3}} = \frac{3(x+2)}{2\sqrt{x+3}}$$
$$f'(x) = 0 \quad \text{when} \quad x = -2$$

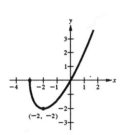

$$f''(x) = \frac{3}{2}\left[\frac{\sqrt{x+3}(1) - (x+2)(\frac{1}{2})(x+3)^{-1/2}(1)}{x+3}\right]$$

$$= \frac{3}{2}\left[\frac{\sqrt{x+3} - (x+2)/2\sqrt{x+3}}{x+3}\right]$$

$$= \frac{3(x+4)}{4(x+3)^{3/2}} > 0 \qquad \text{for all } x \text{ in}(-3, \infty)$$

We conclude that the graph is concave upward for each x in the domain of f and therefore that $(-2, -2)$ is a relative minimum.

35. Find any relative extrema and points of inflection and sketch the graph of $f(x) = 2\sin x + \sin 2x$ in the interval $[0, 2\pi]$.

Solution:

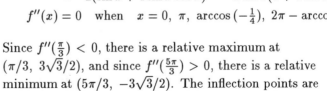
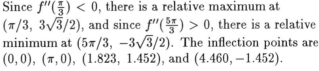

$$f(x) = 2\sin x + \sin 2x$$
$$f'(x) = 2\cos x + 2\cos 2x$$
$$= 2[\cos x + 2\cos^2 x - 1] = 2(2\cos x - 1)(\cos x + 1)$$
$$f'(x) = 0 \quad \text{when} \quad x = \frac{\pi}{3},\ \pi,\ \frac{5\pi}{3}$$
$$f''(x) = -2\sin x - 4\sin 2x$$
$$= -2(\sin x + 4\sin x \cos x) = -2\sin x(1 + 4\cos x)$$
$$f''(x) = 0 \quad \text{when} \quad x = 0,\ \pi,\ \arccos\left(-\tfrac{1}{4}\right),\ 2\pi - \arccos\tfrac{1}{4}$$

Since $f''(\frac{\pi}{3}) < 0$, there is a relative maximum at $(\pi/3,\ 3\sqrt{3}/2)$, and since $f''(\frac{5\pi}{3}) > 0$, there is a relative minimum at $(5\pi/3,\ -3\sqrt{3}/2)$. The inflection points are $(0, 0)$, $(\pi, 0)$, $(1.823,\ 1.452)$, and $(4.460, -1.452)$.

37. Find (if any) the extrema and the points of inflection, and sketch the graph of $f(x) = x^2 e^{-x}$.

Solution:

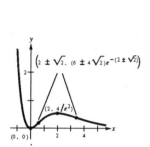

$$f(x) = x^2 e^{-x} \qquad \text{Intercepts:} \quad (0, 0)$$
$$f'(x) = x^2(-e^{-x}) + e^{-x}(2x) = xe^{-x}(-x + 2)$$
$$f''(x) = (xe^{-x})(-1) + (-x + 2)[x(-e^{-x}) + e^{-x}(1)]$$
$$= e^{-x}[-x + (-x + 2)(-x + 1)] = e^{-x}(x^2 - 4x + 2)$$

Solving the equation $f' = 0$ yields the critical numbers $x = 0$, and $x = 2$. Solve the equation $x^2 - 4x + 2 = 0$ to determine where $f''(x) = 0$. The solutions of this equation are $x = 2 \pm \sqrt{2}$.

Interval	$f(x)$	$f'(x)$	$f''(x)$	Shape of graph
$-\infty < x < 0$		$-$	$+$	decreasing, concave up
$x = 0$	0	0	$+$	relative minimum
$0 < x < 2 - \sqrt{2}$		$+$	$+$	increasing, concave up
$x = 2 - \sqrt{2}$	0.191	$+$	0	point of inflection
$2 - \sqrt{2} < x < 2$		$+$	$-$	increasing, concave down
$x = 2$	0.541	0	$-$	relative maximum
$2 < x < 2 + \sqrt{2}$		$-$	$-$	decreasing, concave down
$x = 2 + \sqrt{2}$	0.384	$-$	0	point of inflection
$2 + \sqrt{2} < x < \infty$		$-$	$+$	decreasing, concave up

41. Find any relative extrema and inflection points for
$y = (x^2/2) - \ln x$, and sketch the graph of the function.

Solution:

$$y = \frac{x^2}{2} - \ln x \qquad \text{Domain: } 0 < x$$

$$y' = x - \frac{1}{x}$$

$$y'' = 1 + \frac{1}{x^2} > 0$$

We first observe that $y' = 0$ when $x = \pm 1$. However, $x = -1$ is
not in the domain of the function. Since y'' is positive for all x
in the domain, the graph is concave up and $(1, \frac{1}{2})$ is a relative
minimum point. Furthermore, since y'' is never zero, there are
no points of inflection. By plotting a few points, you can sketch
the graph shown in the accompanying figure.

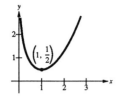

x	0.25	0.5	1	1.5	2	3
y	1.418	0.818	0.5	0.720	1.307	3.401

63. Find a, b, c and d so that the function $f(x) = ax^3 + bx^2 + cx + d$ has a relative maximum at $(3, 3)$, a relative minimum at $(5, 1)$, and a point of inflection at $(4, 2)$.

Solution:

$$f(x) = ax^3 + bx^2 + cx + d$$
$$f'(x) = 3ax^2 + 2bx + c$$
$$f''(x) = 6ax + 2b = 2(3ax + b)$$

$$
\begin{array}{lll}
f(3) = 3 & \implies & 27a + 9b + 3c + d = 3 \quad (1) \\
f(4) = 2 & \implies & 64a + 16b + 4c + d = 2 \quad (2) \\
f(5) = 1 & \implies & 125a + 25b + 5c + d = 1 \quad (3) \\
f'(3) = 0 & \implies & 27a + 6b + c = 0 \quad (4) \\
f'(5) = 0 & \implies & 75a + 10b + c = 0 \quad (5) \\
f''(4) = 0 & \implies & 2(12a + b) = 0 \quad (6)
\end{array}
$$

From (6) you get $b = -12a$. Substituting this into (5), you obtain

$$75a + 10(-12a) + c = 0$$
$$-45a + c = 0$$
$$c = 45a$$

Substitution into (1) and (2) yields

$$27a + 9(-12a) + 3(45a) + d = 3$$
$$54a + d = 3 \quad (7)$$
$$64a + 16(-12a) + 4(45a) + d = 2$$
$$52a + d = 2 \quad (8)$$

To solve simultaneously, subtract (8) from (7) and obtain

$$2a = 1 \qquad a = \frac{1}{2}$$

From (7)

$$d = 3 - 54\left(\frac{1}{2}\right) = -24$$
$$c = 45\left(\frac{1}{2}\right) = \frac{45}{2}$$
$$b = -12\left(\frac{1}{2}\right) = -6$$

Therefore,

$$f(x) = \frac{1}{2}x^3 - 6x^2 + \frac{45}{2}x - 24 = \frac{1}{2}(x^3 - 12x^2 + 45x - 48)$$

67. Prove that a cubic function with three real zeros has a point of inflection whose x-coordinate is the average of the three zeros.

Solution:

We assume the three zeros of the cubic are r_1, r_2, and r_3. Then

$$f(x) = a(x - r_1)(x - r_2)(x - r_3)$$
$$f'(x) = a[(x - r_1)(x - r_2) + (x - r_1)(x - r_3) + (x - r_2)(x - r_3)]$$
$$f''(x) = a[(x - r_1) + (x - r_2) + (x - r_1) + (x - r_3) + (x - r_2)$$
$$+ (x - r_3)]$$
$$= a[6x - 2(r_1 + r_2 + r_3)]$$

Consequently, $f''(x) = 0$ if

$$x = \frac{2(r_1 + r_2 + r_3)}{6} = \frac{r_1 + r_2 + r_3}{3} = \text{average of } r_1, r_2 \text{ and } r_3$$

74. The graph of $y = x \sin 1/x$ is shown in the accompanying figure. Show that the graph is concave downward to the right of $x = 1/\pi$.

Solution:

$$f'(x) = -\frac{1}{x} \cos\left(\frac{1}{x}\right) + \sin\left(\frac{1}{x}\right)$$

$$f''(x) = -\frac{1}{x}\left[-\sin\left(\frac{1}{x}\right)\right]\left(-\frac{1}{x^2}\right) + \cos\left(\frac{1}{x}\right)\left(\frac{1}{x^2}\right)$$

$$+ \cos\left(\frac{1}{x}\right)\left(-\frac{1}{x^2}\right)$$

$$= -\frac{1}{x^3} \sin\left(\frac{1}{x}\right)$$

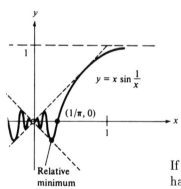

If $x > 1/\pi$, then $1/x < \pi$ and $\sin(1/x) > 0$. Since $x > 0$, you have

$$f''(x) = -\frac{1}{x^3} \sin\left(\frac{1}{x}\right) = (-)(+) < 0$$

Therefore, the graph is concave downward for $x > \dfrac{1}{\pi}$.

🖥 **77.** Use a graphing utility to obtain the graphs of the functions

$$f(x) = 2(\sin x + \cos x)$$

$$P_1(x) = f(a) + f'(a)(x - a), \text{ and}$$

$$P_2(x) = f(a) + f'(a)(x - a) + \tfrac{1}{2}f''(a)(x - a)^2$$

where $a = \pi/4$. Notice that the functions P_1 and P_2 give linear and quadratic approximations to f "near" the point $(a, \ f(a))$.

Solution:

We begin by evaluating the function f and its first and second derivatives at $x = \pi/4$.

$$f(x) = 2(\sin x + \cos x) \qquad\qquad f\left(\frac{\pi}{4}\right) = 2\sqrt{2}$$

$$f'(x) = 2(\cos x - \sin x) \qquad\qquad f'\left(\frac{\pi}{4}\right) = 0$$

$$f''(x) = -2(\sin x + \cos x) \qquad\qquad f''\left(\frac{\pi}{4}\right) = -2\sqrt{2}$$

$$P_1(x) = 2\sqrt{2} + 0\left(x - \frac{\pi}{4}\right) = 2\sqrt{2}$$

$$P_2(x) = 2\sqrt{2} + 0\left(x - \frac{\pi}{4}\right) + \frac{1}{2}(-2\sqrt{2})\left(x - \frac{\pi}{4}\right)^2$$

$$= 2\sqrt{2} - \sqrt{2}\left(x - \frac{\pi}{4}\right)^2.$$

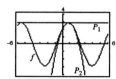

The graphs of f, P_1, and P_2 produced by a graphing utility are shown in the accompanying figure.

3.5   Limits at Infinity

11. Evaluate $\displaystyle\lim_{x\to\infty} \frac{x}{x^2-1}$.

Solution:

Divide both the numerator and denominator by x^2 (the highest power of x in the denominator).

$$\lim_{x\to\infty} \frac{x}{x^2-1} = \lim_{x\to\infty} \frac{x/x^2}{(x^2/x^2)-(1/x^2)} = \lim_{x\to\infty} \frac{1/x}{1-(1/x^2)}$$
$$= \frac{0}{1-0} = 0$$

(Note that the degree of the numerator is less than the degree of the denominator. See Exercise 73 in this section.)

13. Evaluate $\displaystyle\lim_{x\to-\infty} \frac{5x^2}{x+3}$.

Solution:

Divide both the numerator and denominator by x (the highest power of x in the denominator). Since

$$\lim_{x\to-\infty} \frac{5x^2}{x+3} = \lim_{x\to-\infty} \frac{5x^2/x}{(x/x)+(3/x)}$$
$$= \lim_{x\to-\infty} \frac{5x}{1+(3/x)} = -\infty,$$

the limit does not exist. (Note that the degree of the numerator is greater than the degree of the denominator. See Exercise 73 in this section.)

17. Evaluate $\displaystyle \lim_{x \to -\infty} \left(\frac{2x}{x-1} + \frac{3x}{x+1} \right)$.

Solution:

Divide both the numerator and denominator of each fraction by x (the highest power of x in the denominator of each fraction).

$$\lim_{x \to -\infty} \left(\frac{2x}{x-1} + \frac{3x}{x+1} \right) = \lim_{x \to -\infty} \left[\frac{2x/x}{(x/x)-(1/x)} + \frac{3x/x}{(x/x)+(1/x)} \right]$$

$$= \lim_{x \to -\infty} \left[\frac{2}{1-(1/x)} + \frac{3}{1+(1/x)} \right]$$

$$= \frac{2}{1-0} + \frac{3}{1+0} = 5$$

(Note that for each of the rational functions in this exercise, the denominator and numerator are of equal degree. See Exercise 73 in this section.)

19. Evaluate $\displaystyle \lim_{x \to -\infty} \frac{x}{\sqrt{x^2-x}}$.

Solution:

Divide the numerator and denominator of the function x and note that $x = -\sqrt{x^2}$ when $x < 0$.

$$\lim_{x \to -\infty} \frac{x}{\sqrt{x^2-x}} = \lim_{x \to -\infty} \frac{x/x}{\sqrt{x^2-x}/(-\sqrt{x^2})}$$

$$= \lim_{x \to -\infty} \frac{1}{-\sqrt{1-(1/x)}} = -\frac{1}{\sqrt{1+0}} = -1$$

29. Evaluate $\displaystyle \lim_{x \to \infty} (2 - 5e^{-x})$.

Solution:

$$\lim_{x \to \infty} (2 - 5e^{-x}) = \lim_{t \to \infty} 2 - 5 \lim_{x \to \infty} e^{-x}$$

$$= 2 - 5(0) = 2$$

35. Evaluate $\lim\limits_{x \to \infty} (x - \sqrt{x^2 + x})$.

Solution:

$$\lim_{x \to \infty} (x - \sqrt{x^2 + x}) = \lim_{x \to \infty} (x - \sqrt{x^2 + x}) \frac{x + \sqrt{x^2 + x}}{x + \sqrt{x^2 + x}}$$

$$= \lim_{x \to \infty} \frac{x^2 - (x^2 + x)}{x + \sqrt{x^2 + x}}$$

$$= \lim_{x \to \infty} \frac{-x}{x + \sqrt{x^2 + x}}$$

$$= \lim_{x \to \infty} \frac{-x/x}{(x/x) + \sqrt{x^2 + x}/\sqrt{x^2}}$$

$$= \lim_{x \to \infty} \frac{-1}{1 + \sqrt{1 + (1/x)}}$$

$$= \frac{-1}{1 + \sqrt{1 + 0}} = -\frac{1}{2}$$

(Note: For $x > 0$, $x = \sqrt{x^2}$.)

37. Sketch the graph of $y = (2 + x)/(1 - x)$. As a sketching aid examine the equation for intercepts, symmetry, and asymptotes.

Solution:

If $x = 0$, then $y = 2$ and the y-intercept occurs at $(0, 2)$. If $y = 0$, then $2 + x = 0$, $x = -2$, and the x-intercept is $(-2, 0)$. There is no symmetry with respect to either axis or to the origin. There is a vertical asymptote at $x = 1$. Furthermore,

$$\lim_{x \to 1^-} \frac{2 + x}{1 - x} = \infty \quad \text{and} \quad \lim_{x \to 1^+} \frac{2 + x}{1 - x} = -\infty$$

$$\lim_{x \to \pm\infty} \frac{2 + x}{1 - x} = \lim_{x \to \pm\infty} \frac{2/x + x/x}{1/x - x/x}$$

$$= \lim_{x \to \pm\infty} \frac{\dfrac{2}{x} + 1}{\dfrac{1}{x} - 1} = \frac{0 + 1}{0 - 1} = -1$$

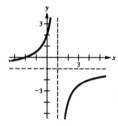

Therefore, there is a horizontal asymptote (to the right and left) at $y = -1$. Start the graph by plotting the following solution points.

x	-3	-1	0.5	2	3	4
y	-0.25	0.5	5	-4	-2.5	-2

47. Sketch the graph of $y = x^3/\sqrt{x^2 - 4}$. As a sketching aid examine the equation for intercepts, symmetry, and asymptotes.

Solution:

Since $x^2 - 4$ must be positive, the domain is $(-\infty, -2)$ and $(2, \infty)$. There is no symmetry with respect to either axis. However, there is symmetry with respect to the origin since

$$(-y) = \frac{(-x)^3}{\sqrt{(-x)^2 - 4}}$$

$$-y = \frac{-x^3}{\sqrt{x^2 - 4}}$$

$$y = \frac{x^3}{\sqrt{x^2 - 4}}$$

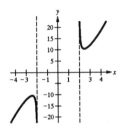

which is equivalent to the original equation. Since the denominator of $x^3/\sqrt{x^2 - 4}$ is zero when $x = 2$ or $x = -2$, there are vertical asymptotes at $x = 2$ and $x = -2$.

x	2.25	2.50	2.75	3.00	4.00
y	11.05	10.45	11.02	12.07	18.48

57. (a) Use a graphing utility to sketch the graphs of

$$f(x) = \frac{x^3 - 3x^2 + 2}{x(x - 3)} \quad \text{and} \quad g(x) = x + \frac{2}{x(x - 3)}$$

on the same coordinate axes. (b) Verify algebraically that f and g represent the same function. (c) Zoom out sufficiently far so that the graph appears as a line. What is this line? (Note that the points of discontinuity are not readily seen when you zoom out.)

Solution:

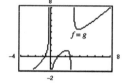

(a) The accompanying figure produced by a graphing utility shows graphically that f and g appear to represent the same function. Note that $x = 0$ and $x = 3$ are vertical asymptotes.

(b)

$$f(x) = \frac{x^3 - 3x^2 + 2}{x(x-3)}$$

$$= \frac{x(x^2 - 3x) + 2}{x^2 - 3x}$$

$$= \frac{x(x^2 - 3x)}{x^2 - 3x} + \frac{2}{x^2 - 3x}$$

$$= x + \frac{2}{x(x-3)} = g(x)$$

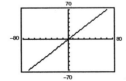

(c) When you zoom out sufficiently far on a graphing utility the graph of the function appears as a line. This line is the slant asymptote $y = x$ and is shown in the accompanying figure.

3.6   A Summary of Curve Sketching

7. Sketch the graph of $y = 3x^4 + 4x^3$, choosing a scale that allows all relative extrema and points of inflection to be identified on the sketch.

Solution:

$$y = 3x^4 + 4x^3 = x^3(3x + 4)$$ Intercepts: $(0,0), (-\frac{4}{3}, 0)$

$$y' = 12x^3 + 12x^2 = 12x^2(x + 1)$$ Critical numbers:

$$x = 0,\ x = -1$$

$$y'' = 36x^2 + 24x = 12x(3x + 2)$$ Possible inflection points:

$$(-\frac{2}{3}, -\frac{16}{27}),\ (0,0)$$

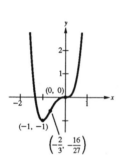

x	y	y'	y''	*Shape of graph*
$-\infty < x < -1$		$-$	$+$	decreasing, concave up
$x = -1$	-1	0	$+$	relative minimum
$-1 < x < -\frac{2}{3}$		$+$	$+$	increasing, concave up
$x = -\frac{2}{3}$	$-\frac{16}{27}$	$+$	0	point of inflection
$-\frac{2}{3} < x < 0$		$+$	$-$	increasing, concave down
$x = 0$	0	0	0	point of inflection
$0 < x < \infty$		$+$	$+$	increasing, concave up

17. Sketch the graph of $y = x\sqrt{4-x}$. Label the intercepts, relative extrema, points of inflection, and the domain.

Solution:

$$y = x\sqrt{4-x} \qquad \text{Domain: } x \leq 4, \quad \text{Intercepts: } (0,0),\ (4,0)$$

$$y' = x\left(\frac{1}{2}\right)(-1)(4-x)^{-1/2} + (4-x)^{1/2}$$

$$= \frac{-x}{2\sqrt{4-x}} + \sqrt{4-x}$$

$$= \frac{-x+8-2x}{2\sqrt{4-x}}$$

$$= \frac{-3x+8}{2\sqrt{4-x}} \qquad \text{Critical numbers: } \frac{8}{3},\, 4$$

$$y'' = \frac{2(4-x)^{1/2}(-3) - (-3x+8)(2)(1/2)(-1)(4-x)^{-1/2}}{4(4-x)}$$

$$= \frac{-6(4-x) + (-3x+8)}{4(4-x)^{3/2}}$$

$$= \frac{-24+6x-3x+8}{4(4-x)^{3/2}} = \frac{3x-16}{4(4-x)^{3/2}}$$

(Note: $x = \frac{16}{3}$ is not in the domain of the function.)

x	y	y'	y''	*Shape of graph*
$-\infty < x < \dfrac{8}{3}$		$+$	$-$	increasing, concave down
$x = \dfrac{8}{3}$	$\dfrac{16\sqrt{3}}{9}$	0	$-$	relative maximum
$\dfrac{8}{3} < x < 4$		$-$	$-$	decreasing, concave down
$x = 4$	0	undefined	undefined	

19. Sketch the graph of $y = 3x^{2/3} - 2x$, choosing a scale that allows all relative extrema and points of inflection to be identified on the sketch.

Solution:

$$y = 3x^{2/3} - 2x = x^{2/3}(3 - 2x^{1/3}) \quad \text{Intercepts: } (0,0), \ \left(\frac{27}{8}, 0\right)$$

$$y' = 2x^{-1/3} - 2 = \frac{2(1 - x^{1/3})}{x^{1/3}} \quad \text{Critical numbers:}$$

$$x = 0, \ x = 1$$

$$y'' = \left(\frac{-1}{3}\right)(2)x^{-4/3} = \frac{-2}{3x^{4/3}}$$

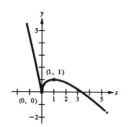

x	y	y'	y''	Shape of graph
$-\infty < x < 0$		$-$	$-$	decreasing, concave down
$x = 0$	0	undefined	undefined	relative minimum
$0 < x < 1$		$+$	$-$	increasing, concave down
$x = 1$	1	0	$-$	relative maximum
$1 < x < \infty$		$-$	$-$	decreasing, concave down

25. Sketch the graph of $y = x \ln x$, choosing a scale that allows any relative extrema and points of inflection to be identified on the sketch.

Solution:

The domain of the function is all real numbers in the interval $(0, \infty)$.

$$y' = x\left(\frac{1}{x}\right) + \ln x = 1 + \ln x$$

$$y'' = \frac{1}{x}$$

Solving the equation $y' = 1 + \ln x = 0$ yields the critical number is $x = e^{-1}$. The second derivative is positive for all real numbers in the domain of the function. Therefore, the graph is concave up and has a minimum at $(e^{-1}, -e^{-1})$.

27. Sketch the graph of $y = \sin x - \frac{1}{18}\sin 3x$. Label any intercepts, relative extrema, points of inflection, or asymptotes.

Solution:

$$y' = \cos x - \frac{1}{6}\cos 3x = \cos x - \frac{1}{6}(4\cos^3 x - 3\cos x)$$

$$= -\frac{1}{6}\cos x(4\cos^2 x - 9) \qquad \text{Critical numbers: } x = \frac{\pi}{2}, \frac{3\pi}{2}$$

$$y'' = -\frac{1}{6}[\cos x(8\cos x)(-\sin x) + (4\cos^2 x - 9)(-\sin x)]$$

$$= \frac{1}{2}\sin x(4\cos^2 x - 3)$$

$$y'' = 0 \quad \text{when} \quad x = 0, \frac{\pi}{6}, \frac{5\pi}{6}, \pi, \frac{7\pi}{6}, \frac{11\pi}{6}$$

x	y	y'	y''	*Shape of graph*
$0 < x < \dfrac{\pi}{6}$		$+$	$+$	increasing, concave up
$x = \dfrac{\pi}{6}$	$\dfrac{4}{9}$	$+$	0	point of inflection
$\dfrac{\pi}{6} < x < \dfrac{\pi}{2}$		$+$	$-$	increasing, concave down
$x = \dfrac{\pi}{2}$	$\dfrac{19}{18}$	0	$-$	relative maximum
$\dfrac{\pi}{2} < x < \dfrac{5\pi}{6}$		$-$	$-$	decreasing, concave down
$x = \dfrac{5\pi}{6}$	$\dfrac{4}{9}$	$-$	0	point of inflection
$\dfrac{5\pi}{6} < x < \pi$		$-$	$+$	decreasing, concave up
$x = \pi$	0	$-$	0	point of inflection
$\pi < x < \dfrac{7\pi}{6}$		$-$	$-$	decreasing, concave down
$x = \dfrac{7\pi}{6}$	$-\dfrac{4}{9}$	$-$	0	point of inflection
$\dfrac{7\pi}{6} < x < \dfrac{3\pi}{2}$		$-$	$+$	decreasing, concave up
$x = \dfrac{3\pi}{2}$	$-\dfrac{19}{18}$	0	$+$	relative minimum
$\dfrac{3\pi}{2} < x < \dfrac{11\pi}{6}$		$+$	$+$	increasing, concave up
$x = \dfrac{11\pi}{6}$	$-\dfrac{4}{9}$	$+$	0	point of inflection
$\dfrac{11\pi}{6} < x < 2\pi$		$+$	$-$	increasing, concave down

31. Sketch the graph of $y = [1/(x-2)] - 3$. Label the intercepts, relative extrema, points of inflection, and the domain.

Solution:

$$y = \frac{1}{x-2} - 3$$

$$= \frac{1 - 3(x-2)}{x-2}$$

$$= \frac{7 - 3x}{x-2} \qquad \text{Domain: all } x \neq 2, \quad \text{Intercepts: } \left(\frac{7}{3}, 0\right), \left(0, -\frac{7}{2}\right)$$

$$y' = \frac{-1}{(x-2)^2} \quad y'' = \frac{2}{(x-2)^3}$$

The graph has a vertical asymptote at $x = 2$. Since

$$\lim_{x \to \pm\infty} \left(\frac{1}{x-2} - 3\right) = -3,$$

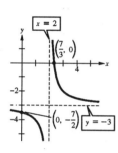

there is a horizontal asymptote at $y = -3$.

x	y'	y''	*Shape of graph*
$-\infty < x < 2$	$-$	$-$	decreasing, concave down
$2 < x < \infty$	$-$	$+$	decreasing, concave up

41. Sketch the graph of $g(t) = 10/(1 + 4e^{-t})$. Label any intercepts, relative extrema, points of inflection, or asymptotes.

Solution:

$$g(t) = \frac{10}{1 + 4e^{-t}} = 10(1 + 4e^{-t})^{-1}$$

$$g'(t) = 10(-1)(1 + 4e^{-t})^{-2}(-4e^{-t}) = \frac{40}{(1 + 4e^{-t})^2}$$

$$g''(t) = 40 \left[\frac{(1 + 4e^{-t})^2(-e^{-t}) - e^{-t}(2)(1 + 4e^{-t})(-4e^{-t})}{(1 + 4e^{-t})^4}\right]$$

$$= 40 \left[\frac{(1 + 4e^{-t})(-e^{-t}) + 8e^{-t}}{(1 + 4e^{-t})^3}\right]$$

$$= \frac{40e^{-t}(4e^{-t} - 1)}{(1 + 4e^{-t})^3}$$

Observe that $g'(t) > 0$ for all real numbers t. Therefore, g is an increasing function and there are no extrema. Solving the equation $g''(t) = 0$ yields

$$4e^{-t} - 1 = 0$$
$$4 - e^t = 0 \quad \Longrightarrow \quad t = \ln 4$$

Since $g''(t) > 0$ for $t < \ln 4$ and $g''(t) < 0$ for $t > \ln 4$, the point $(\ln 4, 5)$ is a point of inflection. Recall that

$$\lim_{t \to \infty} e^{-t} = 0 \quad \text{and} \quad \lim_{t \to -\infty} e^{-t} = \infty.$$

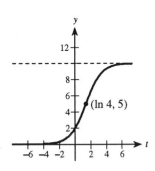

Using these limits, it follows that

$$\lim_{t \to \infty} \frac{10}{1 + 4e^{-t}} = 10$$
$$\lim_{t \to -\infty} \frac{10}{1 + 4e^{-t}} = 0.$$

Hence, $y = 0$ and $y = 10$ are horizontal asymptotes.

53. Create a function f whose graph has a vertical asymptote $x = 5$ and a slant asymptote $y = 3x + 2$

Solution:

Since $x = 5$ is an asymptote, f is a rational function such that the numerator is not zero and the denominator is zero at $x = 5$. Since $y = 3x + 5$ is a slant asymptote,

$$\lim_{x \to \infty} f(x) = 3x + 5.$$

Therefore,

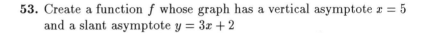

$$f(x) = 3x + 5 + \frac{1}{x - 5} = \frac{3x^2 - 10x - 24}{x - 5}$$

is one function that satisfies the requirements.

57. Determine conditions on the coefficients a, b, and c such that the graph of $f(x) = ax^3 + bx^2 + cx + d$ will resemble the accompanying graph.

Solution:

Since $\lim_{x \to \infty} f(x) = -\infty$, $a < 0$. Also, $f(x)$ is a decreasing function, and therefore $f'(x) = 3ax^2 + 2bx + c < 0$ for all x. Hence, the discriminant must be negative and you have

$$(2b)^2 - 4(3a)(c) < 0 \implies 4(b^2 - 3ac) < 0 \implies b^2 < 3ac.$$

3.7   Optimization Problems

5. Find two positive numbers such that the second number is a reciprocal of the first and their sum in minimum.

Solution:

Let $x =$ first number, $y =$ second number, and $S =$ the sum to be minimized. To minimize S, use the *primary* equation

$$S = x + y.$$

Since the second number is the reciprocal of the first, the *secondary* equation is

$$y = \frac{1}{x},$$

and therefore,

$$S = x + \frac{1}{x}.$$

Differentiation yields

$$\frac{dS}{dx} = 1 - \frac{1}{x^2}$$

$$\frac{dS}{dx} = 0 \quad \text{when} \quad x = \pm 1$$

$$\frac{d^2S}{dx^2} = \frac{2}{x^3}$$

Finally, since the second derivative is positive when $x = 1$, it follows that S is minimum when $x = 1$ and $y = 1$.

11. Find the coordinates of the point on the curve $y = \sqrt{x}$ closest to the point (4, 0).

Solution:

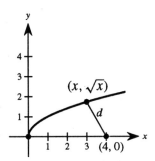

Let (x, y) to be a point on the graph of $y = \sqrt{x}$. The distance between (x, y) and the point (4, 0) is given by the *primary* equation

$$d(x) = \sqrt{(x - 4)^2 + (y - 0)^2} = [(x - 4)^2 + y^2]^{1/2}.$$

Since $y = \sqrt{x}$, you have $d(x) = [(x - 4)^2 + x]^{1/2}$.

To minimize $d(x)$, solve $d'(x) = 0$ as follows:

$$d'(x) = \frac{1}{2}[(x - 4)^2 + x]^{-1/2}[2(x - 4) + 1]$$

$$= \frac{2(x - 4) + 1}{2\sqrt{(x - 4)^2 + x}} = \frac{2x - 7}{2\sqrt{(x - 4)^2 + x}}$$

We observe that $d'(x) = 0$ when

$$2x - 7 = 0$$

$$x = \frac{7}{2} \quad \text{and} \quad y = \sqrt{\frac{7}{2}}$$

Therefore, the required point is $\left(\dfrac{7}{2}, \sqrt{\dfrac{7}{2}}\right)$.

19. An open box is to be made from a square piece of material, 24 inches on a side, by cutting equal squares from each corner and turning up the sides.

(a) Use paper and pencil to complete six rows of the following table. Use the result to guess the maximum volume.

Height	Length & Width	Volume
1	$24 - 2(1)$	$1[24 - 2(1)]^2 = 484$
2	$24 - 2(2)$	$2[24 - 2(2)]^2 = 800$

(b) Write the volume, V, as a function of x

(c) Use calculus to find the critical number of the function of part (b) and find the maximum volume.

(d) Use a graphing utility to obtain the graph of the function of part (b) and estimate the maximum volume from the graph.

Solution:

(a)

Height, x	Length & Width	Volume, V
1	$24 - 2(1)$	$1[24 - 2(1)]^2 = 484$
2	$24 - 2(2)$	$2[24 - 2(2)]^2 = 800$
3	$24 - 2(3)$	$3[24 - 2(3)]^2 = 972$
4	$24 - 2(4)$	$4[24 - 2(4)]^2 = 1024$
5	$24 - 2(5)$	$5[24 - 2(5)]^2 = 980$
6	$24 - 2(6)$	$6[24 - 2(6)]^2 = 864$

The estimate of the maximum volume is 1024 in³.

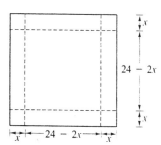

(b) Using the form of the expressions computed in part (a) and the variables assigned to the dimensions of the box (see the accompany figure), the volume is given by

$$V = x(24 - 2x)^2 = x[2(12 - x)]^2 = 4x(12 - x)^2.$$

(c) Thus, to find the maximum volume, solve $dV/dx = 0$ as follows:

$$\frac{dV}{dx} = 4[x(2)(12 - x)(-1) + (12 - x)^2(1)]$$
$$= 4(12 - x)[-2x + (12 - x)]$$
$$= 4(12 - x)(-3x + 12) = 12(12 - x)(4 - x)$$

Therefore, $dV/dx = 0$ when $x = 12$ or $x = 4$. Since 12 is not in the domain of V ($V = 0$ if $x = 12$), test $x = 4$ to determine if the volume is a maximum for this critical number.

$$\frac{d^2V}{dx^2} = 12[(12 - x)(-1) + (4 - x)(-1)] = -24(8 - x)$$

When $x = 4$, $\dfrac{d^2V}{dx^2} < 0$ and the maximum value of V is

$$V(4) = 4(4)(12 - 4)^2 = 1024 \text{ cubic inches.}$$

(d) The accompanying figure shows the graph of the function yielding the volume of the box. From the graph it also appears that the maximum occurs at the point (4, 1024).

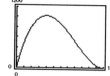

27. A right triangle is formed in the first quadrant by the x- and y-axes and a line through the point $(2, 3)$. Find the vertices of the triangle so that its area is minimum.

Solution:

The slope of the line (see the accompanying figure) is given by

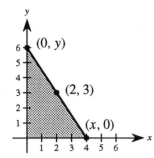

$$m = \frac{y - 3}{0 - 2} = \frac{0 - 3}{x - 2}$$

Therefore,

$$y - 3 = \frac{6}{x - 2}$$

$$y = \frac{6}{x - 2} + 3 = \frac{6 + 3x - 6}{x - 2} = \frac{3x}{x - 2}$$

The area of the triangle is

$$A = \frac{1}{2}xy = \frac{1}{2}x\left(\frac{3x}{x - 2}\right) \qquad 2 < x$$

$$= \frac{3x^2}{2(x - 2)}$$

To minimize A, solve $dA/dx = 0$ as follows:

$$\frac{dA}{dx} = \frac{2(x - 2)(6x) - (3x^2)(2)}{4(x - 2)^2}$$

$$= \frac{6x(2x - 4 - x)}{4(x - 2)^2} = \frac{3x(x - 4)}{2(x - 2)^2} = 0$$

Disregarding $x = 0$, you have

$$x = 4 \qquad \text{and} \quad y = \frac{3(4)}{4 - 2} = 6.$$

Thus the vertices are $(0, 0)$, $(4, 0)$, and $(0, 6)$.

34. A right circular cylinder is to be designed to hold V_0 cubic inches and use the minimal amount of material in its construction. Find the required dimensions.

Solution:

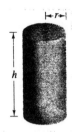

The volume of the right circular cylinder (see figure) is $V = \pi r^2 h$ and its surface area is

$$S = 2(\text{area of base}) + (\text{lateral surface})$$
$$= 2\pi r^2 + 2\pi r h = 2\pi r(r + h).$$

From the formula for the volume, you have $h = V_0/(\pi r^2)$. Thus,

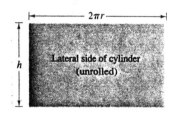

$$S = 2\pi r\left(r + \frac{V_0}{\pi r^2}\right) = 2\pi\left(r^2 + \frac{V_0}{\pi r}\right)$$
$$\frac{dS}{dr} = 2\pi\left(2r - \frac{V_0}{\pi r^2}\right) = 0$$
$$2r = \frac{V_0}{\pi r^2}$$
$$r^3 = \frac{V_0}{2\pi} \quad \text{or} \quad r = \sqrt[3]{\frac{V_0}{2\pi}}$$

Since

$$\frac{d^2S}{dr^2} = 2\pi\left(2 + \frac{2V_0}{\pi r^3}\right)$$

is positive at the critical number $r = \sqrt[3]{V_0/(2\pi)}$, this radius yields the minimum surface area of the cylinder. The corresponding height is given by

$$h = \frac{V_0}{\pi r^2} = \frac{V_0 r}{\pi r^3} = \frac{V_0 r}{\pi\left(\dfrac{V_0}{\pi r^2}\right)} = 2r.$$

Note that for a given volume the surface area is minimum when the height of the cylinder equals its diameter.

45. A wooden beam has a rectangular cross section of height h and width w, as shown in the figure. The strength S of the beam is directly proportional to the width and the square of the height. What are the dimensions of the strongest beam that can be cut from a round log of diameter 24 inches. (Hint: $S = kh^2w$, where k is the proportionality constant.)

Solution:

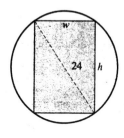

Letting S be the strength and k the constant of proportionality, you have $S = kwh^2$. Since $w^2 + h^2 = 24^2$, you have $h^2 = 24^2 - w^2$.

$$S = kw(24^2 - w^2) = k(576w - w^3)$$

Differentiating and solving $dS/dw = 0$, yields

$$\frac{dS}{dw} = k(576 - 3w^2) = 0$$
$$3w^2 = 576$$
$$w^2 = 192 \quad \text{or} \quad w = \pm 8\sqrt{3}$$

Since w is positive, if follows that $w = 8\sqrt{3}$ inches and $h = \sqrt{24^2 - (8\sqrt{3})^2} = 8\sqrt{6}$ inches will produce the strongest beam.

49. A man is in a boat 2 miles from the nearest point on the coast. He is to go to a point Q, 3 miles down the coast and 1 mile inland. If he can row at 2 mi/h and walk at 4 mi/h, toward what point on the coast should he row in order to reach point Q in the least time?

Solution:

$S = \sqrt{x^2 + 4}$

$L = \sqrt{1 + (3 - x^2)}$

From the accompanying figure, we have

$$S = \text{distance on water} = \sqrt{x^2 + 4}$$
$$L = \text{distance on land} = \sqrt{1 + (3 - x)^2} = \sqrt{x^2 - 6x + 10}$$

Total time = (time on water) + (time on land)

$$T = \frac{S}{(\text{rate on water})} + \frac{L}{(\text{rate on land})}$$
$$= \frac{\sqrt{x^2 + 4}}{2} + \frac{\sqrt{x^2 - 6x + 10}}{4}$$

To minimize T, find dT/dx as follows:

$$\frac{dT}{dx} = \left(\frac{1}{2}\right)\left(\frac{1}{2}\right)(x^2 + 4)^{-1/2}(2x)$$

$$+ \left(\frac{1}{4}\right)\left(\frac{1}{2}\right)(x^2 - 6x + 10)^{-1/2}(2x - 6)$$

$$= \frac{x}{2\sqrt{x^2 + 4}} + \frac{x - 3}{4\sqrt{x^2 - 6x + 10}}$$

We next solve the equation $dT/dx = 0$ to determine the critical numbers.

$$\frac{x}{2\sqrt{x^2 + 4}} = \frac{3 - x}{4\sqrt{x^2 - 6x + 10}}$$

$$\frac{x}{\sqrt{x^2 + 4}} = \frac{3 - x}{2\sqrt{x^2 - 6x + 10}}$$

$$\frac{x^2}{x^2 + 4} = \frac{9 - 6x + x^2}{4(x^2 - 6x + 10)}$$

$$x^2(4)(x^2 - 6x + 10) = (x^2 + 4)(9 - 6x + x^2)$$

$$4x^4 - 24x^3 + 40x^2 = 9x^2 + 36 - 6x^3 - 24x + x^4 + 4x^2$$

Collecting like terms yields the fourth degree polynomial equation

$$x^4 - 6x^3 + 9x^2 + 8x - 12 = 0.$$

Since the equation has integer coefficients, the possible integer roots are factors of the constant 12. Testing these factors yields the positive real root, $x = 1$. Thus the man should row to a point 1 mile from the nearest point on the coast.

53. A component is designed to slide a block of steel of weight W across a table and into a chute as shown in the accompanying figure. The motion of the block is resisted by a frictional force proportional to its net weight. (let k by the constant of proportionality.) Find the minimum force F needed to slide the block and find the corresponding value of θ.

Solution:

The force in the direction of motion is $F \cos \theta$ and the force tending to lift the block is $F \sin \theta$. Therefore, the net weight of the block is $W - F \sin \theta$, and

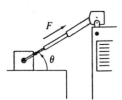

$$F \cos \theta = k(W - F \sin \theta)$$
$$F \cos \theta + kF \sin \theta = kW$$
$$F = \frac{kW}{\cos \theta + k \sin \theta}$$
$$= kW(\cos \theta + k \sin \theta)^{-1}$$

Differentiating and solving $dF/d\theta = 0$, you obtain

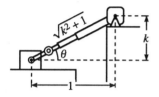

$$\frac{dF}{d\theta} = -kW(\cos \theta + k \sin \theta)^{-2}(-\sin \theta + k \cos \theta)$$
$$= \frac{-kW(k \cos \theta - \sin \theta)}{(\cos \theta + k \sin \theta)^2} = 0$$
$$\sin \theta = k \cos \theta$$
$$\tan \theta = k$$

Therefore, F is minimum when $\theta = \arctan k$ and the minimum force is

$$F = \frac{kW}{\cos \theta + k \sin \theta}$$
$$= \frac{kW}{(1/\sqrt{k^2 + 1}) + k(k/\sqrt{k^2 + 1})} = \frac{kW}{\sqrt{k^2 + 1}}$$

55. A sector with central angle θ is cut from a circle of radius 12 inches and the resulting edges are brought together to form a cone (see figure). Find the magnitude of θ so that the volume of the cone is maximum.

Solution:

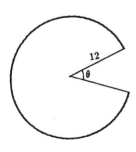

We begin by determining the radius of the cone. The circumference of the given circle is $c = 2\pi r = 24\pi$ and the length of the arc of the sector (see the accompanying figure) removed from the circle is $s = r\theta = 12\theta$. Hence, the circumference of the cone is $C = 24\pi - 12\theta$ and the radius R of the cone is

$$R = \frac{C}{2\pi} = \frac{24\pi - 12\theta}{2\pi} = \frac{6}{\pi}(2\pi - \theta)$$

We next find the height h of the cone. From the accompanying figure we observe that the slant height of the cone is the radius

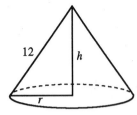

of the given circle. Since the radius of the cone has been determined, you find h by the Pythagorean Theorem.

$$h = \sqrt{12^2 - \left[\frac{6}{\pi}(2\pi - \theta)\right]^2} = \frac{6}{\pi}\sqrt{4\pi^2 - (2\pi - \theta)^2}$$

Therefore, you can determine the volume V of the cone and the value of θ for which V is maximum by solving the equation $dV/d\theta = 0$.

$$V = \frac{1}{3}\pi r^2 h$$

$$= \frac{1}{3}\pi\left(\frac{6}{\pi}\right)^3 (2\pi - \theta)^2 \sqrt{4\pi^2 - (2\pi - \theta)^2}$$

$$\frac{dV}{d\theta} = \frac{1}{3}\pi\left(\frac{6}{\pi}\right)^3 \left[(2\pi - \theta)^2 \frac{2\pi - \theta}{\sqrt{4\pi^2 - (2\pi - \theta)^2}}\right.$$

$$\left. + 2(2\pi - \theta)(-1)\sqrt{4\pi^2 - (2\pi - \theta)^2}\right]$$

$$= \frac{1}{3}\pi\left(\frac{6}{\pi}\right)^3 (2\pi - \theta) \cdot \frac{(2\pi - \theta)^2 - 2[4\pi^2 - (2\pi - \theta)^2]}{\sqrt{4\pi^2 - (2\pi - \theta)^2}}$$

$$= \frac{1}{3}\pi\left(\frac{6}{\pi}\right)^3 \frac{2\pi - \theta}{\sqrt{4\pi^2 - (2\pi - \theta)^2}}[3(2\pi - \theta)^2 - 8\pi^2]$$

Therefore, $dV/d\theta = 0$ when

$$3(2\pi - \theta)^2 = 8\pi^2$$

$$2\pi - \theta = \pm\frac{2\sqrt{2}\pi}{\sqrt{3}}$$

$$\theta = 2\pi \pm \frac{2\sqrt{2}\pi}{\sqrt{3}} = \frac{2\pi}{3}(3 \pm \sqrt{6}).$$

The maximum volume of the cone will occur when

$$\theta = \frac{2\pi}{3}(3 - \sqrt{6}) \approx 66°.$$

3.8 Newton's Method

 5. Approximate the zero of $f(x) = x^3 + x - 1$ in the interval $[0, 1]$. Use Newton's Method and continue the process until you are correct to three decimal places.

Solution:

$$f(x) = x^3 + x - 1 \qquad f'(x) = 3x^2 + 1$$

n	x_n	$f(x_n)$	$f'(x_n)$	$f(x_n)/f'(x_n)$	$x_n - [f(x_n)/f'(x_n)]$
1	0.5000	−0.3750	1.7500	−0.2143	0.7143
2	0.7143	0.0787	2.5306	0.0311	0.6832
3	0.6832	0.0021	2.4002	0.0009	0.6823
4	0.6823	0.0000	2.3967	0.0000	0.6823

Therefore, the approximate root is $x = 0.682$.

 7. Approximate the zero of $f(x) = 3\sqrt{x-1} - x$ in the interval $[1, 2]$. Use Newton's Method and continue the process until you are correct to three decimal places.

Solution:

$$f(x) = 3\sqrt{x-1} - x \qquad f'(x) = \frac{3}{2\sqrt{x-1}} - 1$$

n	x_n	$f(x_n)$	$f'(x_n)$	$f(x_n)/f'(x_n)$	$x_n - [f(x_n)/f'(x_n)]$
1	1.2000	0.1416	2.3541	0.0602	1.1398
2	1.1398	−0.0180	3.0113	−0.0060	1.1458
3	1.1458	−0.0003	2.9283	−0.0001	1.1459

Therefore, the approximate zero is $x = 1.146$.

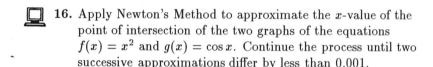

16. Apply Newton's Method to approximate the x-value of the point of intersection of the two graphs of the equations $f(x) = x^2$ and $g(x) = \cos x$. Continue the process until two successive approximations differ by less than 0.001.

Solution:

To approximate the x-value of the point of intersection, let $x^2 = \cos x$. Since this implies that $x^2 - \cos x = 0$, you must find the zeros of the function $h(x) = x^2 - \cos x$. Thus, the iterative formula for Newton's Method takes the form

$$x_{n+1} = x_n - \frac{x_n{}^2 - \cos x_n}{2x_n + \sin x_n}.$$

The calculations are shown in the table, beginning with an initial guess of $x_n = 1$.

n	x_n	$f(x_n)$	$f'(x_n)$	$f(x_n)/f'(x_n)$	$x_n - [f(x_n)/f'(x_n)]$
1	1	0.4597	2.8415	0.1618	0.8382
2	0.8382	0.0338	2.4199	0.0140	0.8242
3	0.8242	0.0003	2.3825	0.0001	0.8241

Therefore, the approximate zero of h is $x = 0.824$ Since the functions f and g are symmetric to the y axis, a second x-coordinate of the point of intersection of their graphs is $x = -0.824$.

27. Use Newton's Method to obtain a general formula for approximating $\sqrt{a}$. [Hint: Apply Newton's Method to the function $f(x) = x^2 - a$.]

Solution:

Let $f(x) = x^2 - a$. Then $f'(x) = 2x$. Since $\sqrt{a}$ is a zero of $f(x) = 0$, you can use Newton's Method to approximate $\sqrt{a}$ as follows:

$$x_{i+1} = x_i - \frac{x_i{}^2 - a}{2x_i} = \frac{x_i{}^2 + a}{2x_i}$$

For example, if $a = 2$, and $x_1 = 1$, then you approximate $\sqrt{2}$ as follows:

$$x_1 = 1$$

$$x_2 = \frac{1^2 + 2}{2(1)} = \frac{1 + 2}{2} = 1.50000$$

$$x_3 = \frac{(1.5)^2 + 2}{2(1.5)} = \frac{4.25}{3} = 1.41667$$

$$x_4 = \frac{(1.41667)^2 + 2}{2(1.41667)} = \frac{4.00697}{2.8333} = 1.41421$$

(Note: To five decimal places, $\sqrt{2} = 1.41421$.)

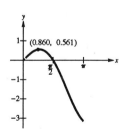   **37.** Approximate the critical number of $f(x) = x \cos x$ on the interval $[0, \pi]$ and sketch the graph of the function.

Solution:

$$f(x) = x \cos x$$

$$f'(x) = x(-\sin x) + \cos x = 0$$

$$x - \cot x = 0$$

Thus the iterative formula for Newton's Method takes the form

$$x_{n+1} = x_n - \frac{x_n - \cot x_n}{1 + \csc^2 x_n}$$

Since the critical number occurs at the x-value where $\cot x = x$ in the interval $[0, \pi]$, select $x_1 = 1$ as the initial guess and obtain the critical number $x \approx 0.860$.

To find possible points of inflection solve the equation

$$f''(x) = -(2 \sin x + x \cos x) = 0.$$

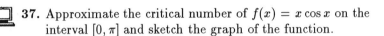

Newton's Method yields the iterative formula

$$x_{n+1} = x_n - \frac{-(2 \sin x + x \cos x)}{x \sin x - 3 \cos x}$$

Beginning with an initial guess $3\pi/4$, you obtain a possible point of inflection at $(2.289, -1.506)$.

Interval	$f(x)$	$f'(x)$	$f''(x)$	Shape of graph
$0 < x < 0.860$		$+$	$-$	increasing, concave down
$x = 0.860$	0.561	0	$-$	relative maximum
$0.860 < x < 2.289$		$-$	$-$	decreasing, concave down
$x = 2.289$	-1.506	$-$	0	point of inflection
$2.289 < x < \pi$		$-$	$+$	decreasing, concave up

3.9   Differentials

3. Find an equation of the tangent line T to the function
$f(x) = x^5$ at the point $(2, 32)$. Use this linear approximation to
completing the following table.

x	1.9	1.99	2	2.01	2.1
$f(x)$					
$T(x)$					

Solution:

Given the function $f(x) = x^5$, $f'(x) = 5x^4$ and $f'(2) = 80$.
Therefore, the equation of the tangent line is

$$y - 32 = 80(x - 2)$$
$$y = 80x - 160 + 32$$
$$T(x) = y = 80x - 128.$$

x	1.9	1.99	2	2.01	2.1
$f(x)$	24.761	31.208	32	32.808	40.841
$T(x)$	24.000	31.200	32	32.800	40.000

Notice that the closer x is to 2, the better the linear
approximation.

15. Find the differential dy of $y = x\sqrt{1 - x^2}$.

Solution:

$$y = x\sqrt{1 - x^2}$$

$$dy = [x(\tfrac{1}{2})(1 - x^2)^{-1/2}(-2x) + (1 - x^2)^{1/2}(1)]dx$$

$$= \left[\frac{-x^2}{\sqrt{1 - x^2}} + \frac{1 - x^2}{\sqrt{1 - x^2}}\right] dx = \frac{1 - 2x^2}{\sqrt{1 - x^2}}dx$$

27. The radius of a sphere is claimed to be 6 inches, with a possible error of 0.02 inches. Using differentials, approximate the maximum possible error in calculating (a) the volume of the sphere, and (b) the surface area of the sphere. (c) What is the relative error in parts (a) and (b)?

Solution:

The radius of the sphere is $r = 6 \pm 0.02$.

(a) The formula for the volume of the sphere is

$$V = \frac{4}{3}\pi r^3.$$

To approximate ΔV by dV, let $r = 6$ and $dr = \pm 0.02$.

$$dV = \frac{4}{3}\pi(3r^2)dr$$

$$dV = 4\pi(36)(\pm 0.02) = \pm 2.88\pi \, \text{in}^3$$

(b) The formula for the surface area of the sphere is $S = 4\pi r^2$. To approximate ΔS by dS, let $r = 6$ and $dr = \pm 0.02$.

$$dS = 8\pi r \, dr$$

$$dS = 8\pi(6)(\pm 0.02) = \pm 0.96\pi \, \text{in}^2$$

(c) For part (a) the relative error is approximately

$$\frac{dV}{V} = \frac{2.88\pi}{(\frac{4}{3})\pi(6^3)} = \frac{2.88}{288} = 0.01 = 1\%.$$

For (b) the relative error is approximately

$$\frac{dS}{S} = \frac{0.96\pi}{4\pi(6^2)} = \frac{0.96}{144} = 0.0067 = \frac{2}{3}\%.$$

31. The period of a pendulum is given by $T = 2\pi\sqrt{L/g}$, where L is the length of the pendulum in feet, g is the acceleration due to gravity, and T is time in seconds. Suppose that the pendulum has been subjected to an increase in temperature so that the length increases by $\frac{1}{2}\%$.

(a) What is the approximate percentage change in the period?

(b) Using the result of part (a), find the approximate error in this pendulum clock in one day.

Solution:

(a) $T = 2\pi\sqrt{\dfrac{L}{g}}$

$$dT = 2\pi\left(\frac{1}{2}\right)\left(\frac{L}{g}\right)^{-1/2}\left(\frac{1}{g}\right)dL = \frac{\pi}{g\sqrt{L/g}}dL$$

$\dfrac{dT}{T}(100) =$ percentage error

$$= \frac{(\pi/g\sqrt{L/g})dL}{2\pi\sqrt{L/g}}(100) = \frac{1}{2}\left(\frac{dL}{L}100\right)$$

$$= \frac{1}{2}(\text{percentage change in } L) = \frac{1}{2}\left(\frac{1}{2}\right) = \frac{1}{4}\%$$

(b) approximate error $= \left(\dfrac{1}{4}\%\right)(\text{number of seconds per day})$

$$= (0.0025)(60)(60)(24)$$

$$= 216 \quad \text{seconds} = 3.6 \quad \text{minutes}$$

3.10   Business and Economics Applications

13. Find the price per unit p that produces the maximum profit P if the demand function is $p = 90 - x$ and the cost function is $C = 100 + 30x$ where x is the number of units sold.

Solution:

The profit is given by

$$P = (\text{price per unit})(\text{number of units}) - (\text{cost})$$
$$= px - C$$
$$= (90 - x)(x) - (100 + 30x) = -x^2 + 60x - 100$$

To maximize P, solve $dP/dx = 0$ as follows:

$$\frac{dP}{dx} = -2x + 60 = 0$$

$$2x = 60 \quad \text{and} \quad x = 30$$

Therefore, the profit is maximum when the price is

$$p = 90 - 30 = 60.$$

23. Find the speed v in miles per hour that will minimize delivery costs on a 110-mile trip if the cost (in dollars) per hour for fuel for the van is $C = v^2/600$ and the driver is paid \$5 per hour. (Assume there are no costs other than wages and fuel.)

Solution:

Since the speed for the 110-mile is v miles per hour, the total time is $t = 110/v$ hours. Therefore, the total cost is

$$(\text{Total Cost}) = (\text{Fuel Cost}) + (\text{Wages})$$

$$C = \frac{v^2}{600}\left(\frac{110}{v}\right) + 5\left(\frac{110}{v}\right) = \frac{11}{60}v + 5(110)v^{-1}.$$

To minimize C, solve $dC/dv = 0$ as follows:

$$\frac{dC}{dv} = \frac{11}{60} - 5(110)v^{-2} = 0$$

$$\frac{11}{60} = \frac{5(110)}{v^2}$$

$$v^2 = \frac{5(110)(60)}{11} \quad \Rightarrow \quad v = 10\sqrt{30} \approx 54.8 \text{ mi/hr}$$

Thus, a speed 54.8 mi/hr will yield the minimum cost.

31. Assume that the amount of money deposited in a bank is proportional to the square of the interest rate the bank pays on this money. Furthermore, the bank can reinvest this money at 12%. Find the interest rate the bank should pay to maximize profit. (Use the simple interest formula.)

Solution:

Let
$$d = \text{amount in the bank}$$
$$i = \text{interest rate paid by the bank}$$
$$p = \text{profit}$$

The bank can take the deposited money d and reinvest to obtain 12% or $(0.12)d$. Since the bank pays out interest to its depositors, its profit is

$$P = (0.12)d - id$$

Finally, since d is proportional to the square of i, you have

$$d = ki^2$$

Thus,
$$P = (0.12)(ki^2) - i(ki^2) = k[(0.12)i^2 - i^3]$$

To maximize P, solve $dP/di = 0$ as follows:

$$\frac{dP}{di} = k(0.24i - 3i^2) = 0$$
$$ki(0.24 - 3i) = 0$$

(We disregard the critical number $i = 0$.)

$$i = \frac{0.24}{3} = 0.08$$

Thus the bank can maximize its profit by setting $i = 8\%$.

33. The ordering and transportation cost C of the components used in manufacturing a certain product is given by

$$C = 100\left(\frac{200}{x^2} + \frac{x}{x + 30}\right), \quad 1 \le x$$

where C is measured in thousands of dollars and x is the order size in hundreds. Find the order size that minimizes cost.

Solution:

$$C = 100\left(\frac{200}{x^2} + \frac{x}{x + 30}\right)$$

$$\frac{dC}{dx} = 100\left(-\frac{400}{x^3} + \frac{30}{(x + 30)^2}\right) = 1000\left[\frac{-40(x + 30)^2 + 3x^3}{x^3(x + 30)^2}\right]$$

To find the critical number of C use Newton's Method to find the zero of the function $f(x) = -40(x + 30)^2 + 3x^3$ using $x_1 = 30$ as our first estimate.

$$f(x) = 3x^3 - 40x^2 - 2400x - 36000 \quad \text{and} \quad f'(x) = 9x^2 - 80x - 2400$$

n	x_n	$f(x_n)$	$f'(x_n)$	$f(x_n)/f'(x_n)$	$x_n - [f(x_n)/f'(x_n)]$
1	30	-63000	3300	-19.091	42.091
2	42.091	$104,700$	$15,362$	6.816	40.556
3	40.556	991.398	9158.562	0.108	40.448
4	40.448	3.773	9088.305	0.0004	40.447

Therefore, if follows that the critical number is $x \approx 40.4$ and the minimum cost occurs when 40 units are ordered.

Review Exercises for Chapter 3

19. Make use of domain, range, symmetry, asymptotes, intercepts, relative extrema, or points of inflection to obtain an accurate graph of $f(x) = x\sqrt{16 - x^2}$.

Solution:

The domain of f is all real numbers in the interval $[-4, 4]$ and the graph of f is symmetric to the origin since

$$f(-x) = (-x)\sqrt{16 - (-x)^2} = -x\sqrt{16 - x^2} = -f(x)$$

$$f'(x) = x\left(\frac{1}{2}\right)(16 - x^2)^{-1/2}(-2x) + (16 - x^2)^{1/2}$$

$$= \frac{16 - 2x^2}{\sqrt{16 - x^2}}$$

$$f''(x) = \frac{\sqrt{16 - x^2}(-4x) - (16 - 2x^2)(\frac{1}{2})(16 - x^2)^{-1/2}(-2x)}{16 - x^2}$$

$$= \frac{2x(x^2 - 24)}{(16 - x^2)^{3/2}}$$

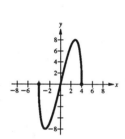

Thus, $f'(x) = 0$ when $x = \pm 2\sqrt{2}$ and undefined when $x = \pm 4$. Since $f''(2\sqrt{2}) < 0$, the graph is concave downward and $(2\sqrt{2}, 8)$ is a maximum. By symmetry, $(-2\sqrt{2}, -8)$ is a minimum. There is a point of inflection at $(0, 0)$.

23. Make use of domain, range, symmetry, asymptotes, intercepts, relative extrema, or points of inflection to obtain an accurate graph of $f(x) = x^{1/3}(x + 3)^{2/3}$.

Solution:

$$f(x) = x^{1/3}(x + 3)^{2/3} \qquad \text{Intercepts: } (0, 0), (-3, 0)$$

$$f'(x) = x^{1/3}(\tfrac{2}{3})(x + 3)^{-1/3} + (x + 3)^{2/3}(\tfrac{1}{3})x^{-2/3}$$

$$= \frac{2x^{1/3}}{3(x + 3)^{1/3}} + \frac{(x + 3)^{2/3}}{3x^{2/3}}$$

$$= \frac{2x + x + 3}{3x^{2/3}(x + 3)^{1/3}} = \frac{x + 1}{(x^3 + 3x^2)^{1/3}}$$

$$\text{Critical numbers: } x = -1, \ x = 0, \ x = -3$$

$$f''(x) = \frac{(x^3 + 3x^2)^{1/3}(1) - (x+1)(\frac{1}{3})(x^3 + 3x^2)^{-2/3}(3x^2 + 6x)}{(x^3 + 3x^2)^{2/3}}$$

$$= \frac{(x^3 + 3x^2) - (x+1)(x^2 + 2x)}{(x^3 + 3x^2)^{4/3}}$$

$$= \frac{x^3 + 3x^2 - x^3 - 2x^2 - x^2 - 2x}{(x + 3x^4)^{2/3}}$$

$$= \frac{-2x}{(x^3 + 3x^2)^{4/3}} = \frac{-2}{x^{5/3}(x+3)^{4/3}}$$

Possible point of inflection: $(0,0)$

x	$f(x)$	$f'(x)$	$f''(x)$	*Shape of graph*
$-\infty < x < -3$		$+$	$+$	increasing, concave up
$x = -3$	0	undefined	undefined	relative maximum
$-3 < x < -1$		$-$	$+$	decreasing, concave up
$x = -1$	$-\sqrt[3]{4}$	0	$+$	relative minimum
$-1 < x < 0$		$+$	$+$	increasing, concave up
$x = 0$	0	undefined	undefined	point of inflection
$0 < x < \infty$		$+$	$-$	increasing, concave down

49. For the function $f(x) = Ax^2 + Bx + C$, determine the value of c guaranteed by the Mean Value Theorem on the interval $[x_1, x_2]$.

Solution:

Since f is continuous and differentiable for all real x, the Mean Value Theorem can be applied over the specified interval. It is necessary to find all c in $[x_1, x_2]$ such that

$$f'(c) = \frac{f(x_2) - f(x_1)}{x_2 - x_1}.$$

$$f'(x) = 2Ax + B$$

$$f'(c) = 2Ac + B = \frac{f(x_2) - f(x_1)}{x_2 - x_1}$$

$$= \frac{Ax_2{}^2 + Bx_2 + C - Ax_1{}^2 - Bx_1 - C}{x_2 - x_1}$$

$$= \frac{A(x_2{}^2 - x_1{}^2) + B(x_2 - x_1)}{x_2 - x_1}$$

$$= A(x_2 + x_1) + B$$

$$2c = x_2 + x_1 \quad \Longrightarrow \quad c = \frac{x_1 + x_2}{2}$$

For a quadratic function, the required value of c is the average of x_1 and x_2.

57. Find the length of the longest pipe that can be carried level around a right-angle corner if the two intersecting corridors are of widths 4 ft and 6 ft.

Solution:

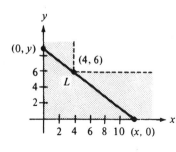

The longest pipe that will go around the corner will have a length equal to the minimum length of the hypotenuse [through the point $(4, 6)$] of the triangle whose vertices are $(0, 0), (x, 0)$, and $(0, y)$. Begin by relating x and y as follows:

$$m = \frac{y - 6}{0 - 4} = \frac{6 - 0}{4 - x}$$

$$y - 6 = \frac{-24}{4 - x}$$

$$y = \frac{24}{x - 4} + 6 = \frac{6x}{x - 4}$$

[Note that $dy/dx = -24/(x - 4)^2$.] Now the length of the hypotenuse is given by

$$L = \sqrt{x^2 + y^2}$$

To minimize L, solve $dL/dx = 0$ as follows.

$$\frac{dL}{dx} = \frac{(\frac{1}{2})[2x + (2y)(dy/dx)]}{\sqrt{x^2 + y^2}} = 0$$

$$x = -y\frac{dy}{dx} = -\left(\frac{6x}{x - 4}\right)\left[\frac{-24}{(x - 4)^2}\right]$$

$$x(x - 4)^3 = 144x$$

$$x - 4 = \sqrt[3]{144}$$

$$x = \sqrt[3]{144} + 4$$

Therefore, the minimum length of L and the maximum length of pipe are given by

$$L = \sqrt{x^2 + y^2} = \sqrt{x^2 + \frac{36x^2}{(x-4)^2}} = \frac{x}{x-4}\sqrt{(x-4)^2 + 36}$$

$$= \frac{\sqrt[3]{144} + 4}{\sqrt[3]{144}}\sqrt{144^{2/3} + 36} \approx 14.05 \text{ ft}$$

59. A hallway of width 6 ft meets a hallway of width 9 ft at right angles. Find the length of the longest pipe that can be carried level around this corner. (Hint: If L is the length of the pipe, show that

$$L = 6\csc\theta + 9\csc\left(\frac{\pi}{2} - \theta\right)$$

where θ is the angle between the pipe and the wall of the narrower hallway.)

Solution:

From the accompanying figure observe that

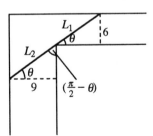

$$\csc\theta = \frac{L_1}{6} \quad \text{or} \quad L_1 = 6\csc\theta$$

$$\csc\left(\frac{\pi}{2} - \theta\right) = \frac{L_2}{9} \quad \text{or} \quad L_2 = 9\csc\left(\frac{\pi}{2} - \theta\right)$$

Therefore, the length of the pipe is given by

$$L = L_1 + L_2 = 6\csc\theta + 9\csc\left(\frac{\pi}{2} - \theta\right)$$

$$= 6\csc\theta + 9\sec\theta$$

Note that $\csc[(\pi/2) - \theta] = \sec\theta$. To maximize L, solve $dL/d\theta = 0$ as follows.

$$\frac{dL}{d\theta} = -6\csc\theta\cot\theta + 9\sec\theta\tan\theta = 0$$

$$9\sec\theta\tan\theta = 6\csc\theta\cot\theta$$

$$\frac{\sec\theta\tan\theta}{\csc\theta\cot\theta} = \frac{6}{9}$$

$$\tan^3\theta = \frac{2}{3}$$

$$\tan\theta = \frac{2^{1/3}}{3^{1/3}}$$

From the accompanying figure observe that

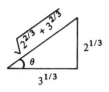

$$\csc\theta = \frac{\sqrt{2^{2/3}+3^{2/3}}}{2^{1/3}} \qquad \text{and} \qquad \sec\theta = \frac{\sqrt{2^{2/3}+3^{2/3}}}{3^{1/3}}$$

$$L = (6)\left(\frac{\sqrt{2^{2/3}+3^{2/3}}}{2^{1/3}}\right) + (9)\left(\frac{\sqrt{2^{2/3}+3^{2/3}}}{3^{1/3}}\right)$$

$$= 3\sqrt{2^{2/3}+3^{2/3}}(2^{2/3}+3^{2/3}) = 3(2^{2/3}+3^{2/3})^{3/2}$$

62. The general equation giving the height of an oscillating object attached to a spring is

$$y = A\sin\sqrt{\frac{k}{m}}t + B\cos\sqrt{\frac{k}{m}}t$$

where k is the spring constant and m is the mass of the object. Show that the maximum displacement of the object is $\sqrt{A^2+B^2}$. Show that the frequency (number of oscillations per second) is $(1/2\pi)\sqrt{k/m}$. How is the frequency changed if the stiffness k of the spring is increased? How is the frequency changed if the mass m of the object is increased?

Solution:

$$y = A\sin\sqrt{\frac{k}{m}}t + B\cos\sqrt{\frac{k}{m}}t$$

$$y' = A\sqrt{\frac{k}{m}}\cos\sqrt{\frac{k}{m}}t - B\sqrt{\frac{k}{m}}\sin\sqrt{\frac{k}{m}}t$$

Therefore, $y' = 0$ if

$$A\sqrt{\frac{k}{m}}\cos\sqrt{\frac{k}{m}}t = B\sqrt{\frac{k}{m}}\sin\sqrt{\frac{k}{m}}t$$

$$\frac{A}{B} = \frac{\sin(\sqrt{k/m})t}{\cos(\sqrt{k/m})t}$$

$$\tan\sqrt{\frac{k}{m}}t = \frac{A}{B}$$

From the accompanying figure observe that

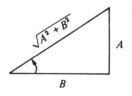

$$\sin\sqrt{\frac{k}{m}}t = \frac{A}{\sqrt{A^2+B^2}} \qquad \text{and} \qquad \cos\sqrt{\frac{k}{m}}t = \frac{B}{\sqrt{A^2+B^2}}$$

Thus, when $y' = 0$,

$$y = A\left[\frac{A}{\sqrt{A^2 + B^2}}\right] + B\left[\frac{B}{\sqrt{A^2 + B^2}}\right] = \sqrt{A^2 + B^2}$$

Since the frequency is the reciprocal of the period,

$$\text{period} = \frac{2\pi}{\sqrt{k/m}}, \qquad \text{frequency} = \frac{\sqrt{k/m}}{2\pi} = \frac{1}{2\pi}\sqrt{\frac{k}{m}}$$

Therefore, the frequency is proportional to the square root of k and inversely proportional to the square root of the mass.

71. Find the maximum profit if the demand equation is $p = 36 - 4x$ and the total cost is $C = 2x^2 + 6$.

Solution:

The profit is given by

$$P = (\text{price per unit})(\text{number of units}) - (\text{cost}) = px - C$$
$$= (36 - 4x)(x) - (2x^2 + 6) = 36x - 4x^2 - 2x^2 - 6$$
$$= -6x^2 + 36x - 6$$

To maximize P, solve $dP/dx = 0$ as follows:

$$\frac{dP}{dx} = -12x + 36 = 0$$
$$x = 3 \text{ units}$$

Thus the maximum profit is

$$P = -6(3)^2 + 36(3) - 6 = -54 + 108 - 6 = \$48.$$

4 INTEGRATION

4.1 Antiderivatives and Indefinite Integration

7. Complete the following table for the indefinite integral

Given	Rewrite	Integrate	Simplify

$$\int \frac{1}{x\sqrt{x}}\, dx$$

Solution:

Given	*Rewrite*	*Integrate*	*Simplify*
$\int \dfrac{1}{x\sqrt{x}}\, dx$	$\int x^{-3/2}\, dx$	$\dfrac{x^{-1/2}}{-\frac{1}{2}} + C$	$\dfrac{-2}{\sqrt{x}} + C$

13. Evaluate the indefinite integral $\int (x^{3/2} + 2x + 1)\, dx$ and check the result by differentiation.
Solution:

$$\int (x^{3/2} + 2x + 1)\, dx = \frac{x^{5/2}}{\frac{5}{2}} + 2\left(\frac{x^2}{2}\right) + x + C$$

$$= \frac{2x^{5/2}}{5} + x^2 + x + C$$

Check

If $y = \dfrac{2x^{5/2}}{5} + x^2 + x + C$, then

$$\frac{dy}{dx} = \left(\frac{2}{5}\right)\left(\frac{5}{2}\right)x^{3/2} + 2x + 1 + 0 = x^{3/2} + 2x + 1$$

19. Evaluate the indefinite integral

$$\int \frac{x^2 + x + 1}{x}\, dx$$

and check the result by differentiation.
Solution:

$$\int \frac{x^2 + x + 1}{x}\, dx = \int \left(\frac{x^2}{x} + \frac{x}{x} + \frac{1}{x} \right) dx$$

$$= \int \left(x + 1 + \frac{1}{x} \right) dx$$

$$= \frac{1}{2}x^2 + x + \ln x + C$$

Check

If $y = \dfrac{1}{2}x^2 + x + \ln x + C$, then

$$\frac{dy}{dx} = x + 1 + \frac{1}{x} + 0$$

$$= \frac{x^2 + x + 1}{x}$$

33. Evaluate the indefinite integral $\int (\tan^2 y + 1)\, dy$ and check the result by differentiation.
Solution:

$$\int (\tan^2 y + 1)\, dy = \int \sec^2 y\, dy = \tan y + C$$

Check

If $f(y) = \tan y + C$, then $f'(y) = \sec^2 y = \tan^2 y + 1$.

43. Find the equation for y given the derivative $dy/dx = 2x - 1$ and the point $(1, 1)$ on the curve.
Solution:

$$\frac{dy}{dx} = 2x - 1$$

$$y = \int (2x - 1)\, dx = x^2 - x + C$$

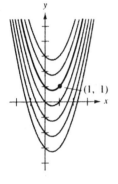

Substituting the coordinates of the solution point into the antiderivative yields

$$y = x^2 - x + C$$
$$1 = (1)^2 - (1) + C \quad \text{or} \quad C = 1.$$

Therefore, the required equation is $y = x^2 - x + 1$.

48. Find $y = f(x)$ if $f''(x) = x^2$, $f'(0) = 6$, and $f(0) = 3$.

Solution:

$$f''(x) = x^2$$

$$f'(x) = \int x^2\, dx = \frac{1}{3}x^3 + C_1$$

$$f'(0) = \frac{1}{3}(0^3) + C_1 = 6 \quad \Longrightarrow \quad C_1 = 6$$

$$f'(x) = \frac{1}{3}x^3 + 6$$

$$f(x) = \int \left(\frac{1}{3}x^3 + 6\right)\, dx = \frac{1}{12}x^4 + 6x + C_2$$

$$f(0) = \frac{1}{12}(0^4) + 6(0) + C_2 = 3 \quad \Longrightarrow \quad C_2 = 3$$

Therefore, $f(x) = \dfrac{1}{12}x^4 + 6x + 3$.

49. Find $y = f(x)$ if $f''(x) = e^x$, $f'(0) = 2$, and $f(0) = 5$.

Solution:

$$f''(x) = e^x$$

$$f'(x) = \int e^x\, dx = e^x + C_1$$

$$f'(0) = e^0 + C_1 = 2 \quad \Longrightarrow \quad C_1 = 1$$

$$f'(x) = e^x + 1$$

$$f(x) = \int (e^x + 1)\, dx = e^x + x + C_2$$

$$f(0) = e^0 + (0) + C_2 = 5 \quad \Longrightarrow \quad C_2 = 4$$

Therefore, $f(x) = e^x + x + 4$.

55. With what initial velocity must an object be thrown upward from ground level to reach a maximum height of 550 feet (approximate height of the Washington Monument)?

Solution:

If $s = f(t)$ is the position of the object at any time t, then $f'(t)$ is its velocity and $f''(t)$ its acceleration. Therefore, $f''(t) = -32$ since -32 ft/sec^2 is the acceleration due to gravity.

$$f'(t) = \int -32\, dt = -32t + C_1 = -32t + v_0$$

where v_0 is the initial velocity. Furthermore,

$$f(t) = \int (-32t + v_0)\, dt$$
$$= -16t^2 + v_0 t + C_2 = -16t^2 + v_0 t + s_0$$

where $s_0 = 0$ is the initial height. Thus,

$$s = f(t) = -16t^2 + v_0 t$$

Now since s is a maximum when $f'(t) = 0$, you have

$$-32t + v_0 = 0 \quad \text{or} \quad t = \frac{v_0}{32}$$

Finally, in order for s to attain a height of 550 ft, you must have

$$s = -16\left(\frac{v_0}{32}\right)^2 + v_0\left(\frac{v_0}{32}\right) = 550$$
$$\frac{v_0{}^2}{64} = 550$$
$$v_0{}^2 = 35,200$$
$$v_0 = \sqrt{35,200}$$
$$= 40\sqrt{22} \approx 187.617 \text{ ft/sec}$$

59. At the instant the traffic light turns green, an automobile that has been waiting at an intersection starts moving forward with a constant acceleration of 6 ft/sec^2. At the same instant a truck traveling with a constant velocity of 30 ft/sec overtakes and passes the car.

(a) How far beyond its starting point will the automobile overtake the truck?

(b) How fast will it be traveling?

Solution:

Let T(t) and A(t) represent the position functions of the truck and auto. It follows that

$$T'(t) = 30, \quad T(0) = 0$$

$$A''(t) = 6, \quad A'(0) = 0, \text{ and } A(0) = 0.$$

For the truck,

$$T(t) = \int 30\, dt = 30t + C_1$$
$$T(0) = 30(0) + C_1 = 0 \quad \Longrightarrow \quad C_1 = 0$$

For the auto,

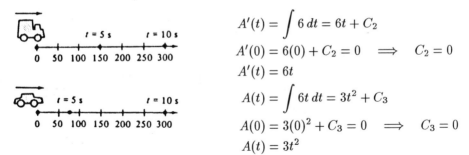

$t = 5\,s$ $t = 10\,s$

0 50 100 150 200 250 300

$t = 5\,s$ $t = 10\,s$

0 50 100 150 200 250 300

$$A'(t) = \int 6\,dt = 6t + C_2$$

$$A'(0) = 6(0) + C_2 = 0 \quad \Longrightarrow \quad C_2 = 0$$

$$A'(t) = 6t$$

$$A(t) = \int 6t\,dt = 3t^2 + C_3$$

$$A(0) = 3(0)^2 + C_3 = 0 \quad \Longrightarrow \quad C_3 = 0$$

$$A(t) = 3t^2$$

Therefore, when the auto catches up with the truck, you have

$$A(t) = T(t)$$
$$3t^2 = 30t$$
$$3t^2 - 30t = 0$$
$$3t(t - 10) = 0$$
$$t = 10 \text{ sec} \qquad \text{We disregard} \, t = 0.$$

(a)   When $t = 10$ sec, the auto will have traveled

$$A(10) = 3(10^2) = 300\text{ft}.$$

(b)   It will be traveling

$$A'(10) = 60 \text{ ft/sec} = \frac{60(3600)}{5280} \approx 41 \text{ mi/hr}.$$

4.2   Area

3. Find the sum $\displaystyle\sum_{k=0}^{4} \frac{1}{k^2 + 1}$.

Solution:

$$\sum_{k=0}^{4} \frac{1}{k^2 + 1} = \frac{1}{0^2 + 1} + \frac{1}{1^2 + 1} + \frac{1}{2^2 + 1} + \frac{1}{3^2 + 1} + \frac{1}{4^2 + 1}$$

$$= \frac{1}{1} + \frac{1}{2} + \frac{1}{5} + \frac{1}{10} + \frac{1}{17}$$

$$= \frac{170 + 85 + 34 + 17 + 10}{170}$$

$$= \frac{316}{170} = \frac{158}{85}$$

11. Write the sum

$$\left[\left(\frac{2}{n}\right)^3 - \frac{2}{n}\right]\left(\frac{2}{n}\right) + \left[\left(\frac{4}{n}\right)^3 - \frac{4}{n}\right]\left(\frac{2}{n}\right) + \cdots + \left[\left(\frac{2n}{n}\right)^3 - \frac{2n}{n}\right]\left(\frac{2}{n}\right)$$

in sigma notation.

Solution:

We begin by noting that the n terms in this sum are each of
the form

$$f(i) = \left[\left(\frac{2i}{n}\right)^3 - \frac{2i}{n}\right]\left(\frac{2}{n}\right)$$

Furthermore, observe that in the first term $i = 1$, in the second
term $i = 2$ and so on until you reach the nth term. Thus our
index i runs from 1 to n, and the sigma notation for the given
sum is

$$\sum_{i=1}^{n} f(i) = \sum_{i=1}^{n}\left[\left(\frac{2i}{n}\right)^3 - \frac{2i}{n}\right]\left(\frac{2}{n}\right)$$

$$= \frac{2}{n}\sum_{i=1}^{n}\left[\left(\frac{2i}{n}\right)^3 - \frac{2i}{n}\right].$$

19. Use the properties of sigma notation and Theorem 4.2 to
evaluate the sum

$$\sum_{i=1}^{15} \frac{1}{n^3}(i-1)^2.$$

Solution:

$$\sum_{i=1}^{15} \frac{1}{n^3}(i-1)^2 = \frac{1}{n^3}\sum_{i=1}^{15}(i^2 - 2i + 1)$$

$$= \frac{1}{n^3}\left[\sum_{i=1}^{15} i^2 - 2\sum_{i=1}^{15} i + \sum_{i=1}^{15} 1\right]$$

$$= \frac{1}{n^3}\left[\frac{15(16)(31)}{6} - 2\frac{15(16)}{2} + 15\right]$$

$$= \frac{1}{n^3}(1240 - 240 + 15) = \frac{1015}{n^3}$$

23. Find the limit of the sequence $s(n)$ as $n \to \infty$ where

$$s(n) = \frac{81}{n^4}\left[\frac{n^2(n+1)^2}{4}\right].$$

Solution:

$$\lim_{n \to \infty} s(n) = \lim_{n \to \infty} \frac{81}{n^4}\left[\frac{n^2(n+1)^2}{4}\right]$$

$$= \lim_{n \to \infty} \frac{81}{4}\left(\frac{n^4 + 2n^3 + n^2}{n^4}\right)$$

$$= \lim_{n \to \infty} \frac{81}{4}\left(1 + \frac{2}{n} + \frac{1}{n^2}\right) = \frac{81}{4}$$

29. Find a formula of the sum of n terms and then find the limit where

$$\lim_{n \to \infty} \sum_{i=1}^{n} \frac{1}{n^3}(i-1)^2.$$

Solution:

$$\sum_{i=1}^{n} \frac{1}{n^3}(i-1)^2 = \frac{1}{n^3}\sum_{i=1}^{n}(i^2 - 2i + 1)$$

$$= \frac{1}{n^3}\left[\sum_{i=1}^{n} i^2 - 2\sum_{i=1}^{n} i + \sum_{i=1}^{n} 1\right]$$

$$= \frac{1}{n^3}\left[\frac{n(n+1)(2n+1)}{6} - 2\frac{n(n+1)}{2} + n\right]$$

$$= \frac{1}{n^3}\left(\frac{n^3}{3} - \frac{n^2}{2} + \frac{n}{6}\right)$$

$$= \frac{1}{3} - \frac{1}{2n} + \frac{1}{6n^2}$$

Therefore,

$$\lim_{n \to \infty} \sum_{i=1}^{n} \frac{1}{n^3}(i-1)^2 = \lim_{n \to \infty}\left(\frac{1}{3} - \frac{1}{2n} + \frac{1}{6n^2}\right) = \frac{1}{3}.$$

37. Use the upper and lower sums to approximate the area of the region between the graph of $y = \sqrt{1 - x^2}$ and the x-axis over the interval $[0, 1]$. Use five subdivisions of equal length.

Solution:

Dividing the interval into five parts, yields

$$x_0 = 0, \quad x_1 = 0.2, \quad x_2 = 0.4, \quad x_3 = 0.6,$$
$$x_4 = 0.8, \quad x_5 = 1$$

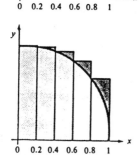

Since y is decreasing from 0 to 1, the lower sum is obtained by using the *right* endpoints of the five subintervals. Thus

$$s = 0.2\sqrt{1 - (0.2)^2} + 0.2\sqrt{1 - (0.4)^2} + 0.2\sqrt{1 - (0.6)^2}$$
$$+ 0.2\sqrt{1 - (0.8)^2} + 0.2\sqrt{1 - (1)^2}$$
$$= 0.2(\sqrt{0.96} + \sqrt{0.84} + \sqrt{0.64} + \sqrt{0.36})$$
$$\approx 0.2(0.9798 + 0.9165 + 0.8 + 0.6) \approx 0.659$$

Similarly, the upper sum is obtained by using the *left* endpoints of the five subdivisions. Thus

$$S = 0.2\sqrt{1 - 0^2} + 0.2\sqrt{1 - (0.2)^2} + 0.2\sqrt{1 - (0.4)^2}$$
$$+ 0.2\sqrt{1 - (0.6)^2} + 0.2\sqrt{1 - (0.8)^2}$$
$$= 0.2(1 + \sqrt{0.96} + \sqrt{0.84} + \sqrt{0.64} + \sqrt{0.36})$$
$$\approx 0.2(1 + 0.9798 + 0.9165 + 0.8 + 0.6) \approx 0.859$$

47. Use the limit process to find the area of the region between the graph of $y = x^2 - x^3$ and the x-axis over the interval $[-1, 1]$. Sketch the region.

Solution:

Let $\Delta x = [1 - (-1)]/n = 2/n$. Choosing right endpoints, you have

$$c_i = -1 + i\left(\frac{2}{n}\right) = -1 + \frac{2i}{n}$$

Therefore,

$$\text{area} = \lim_{n \to \infty} \sum_{i=1}^{n} f\left(-1 + \frac{2i}{1}\right)\left(\frac{2}{n}\right)$$

$$= \lim_{n \to \infty} \frac{2}{n} \sum_{i=1}^{n} \left[\left(-1 + \frac{2i}{n}\right)^2 - \left(-1 + \frac{2i}{n}\right)^3\right]$$

$$= \lim_{n \to \infty} \frac{2}{n} \sum_{i=1}^{n} \left[2 - \frac{10i}{n} + \frac{16i^2}{n^2} - \frac{8i^3}{n^3}\right]$$

$$= \lim_{n \to \infty} \left[\frac{2}{n} \sum_{i=1}^{n} 2 - \frac{20}{n^2} \sum_{i=1}^{n} i + \frac{32}{n^3} \sum_{i=1}^{n} i^2 - \frac{16}{n^4} \sum_{i=1}^{n} i^3\right]$$

$$= \lim_{n \to \infty} \left[\frac{2}{n}(2n) - \left(\frac{20}{n^2}\right)\frac{n(n+1)}{2}\right.$$

$$\left. + \left(\frac{32}{n^3}\right)\frac{n(n+1)(2n+1)}{6} - \left(\frac{16}{n^4}\right)\frac{n^2(n+1)^2}{4}\right]$$

$$= \lim_{n \to \infty} \left(4 - 10 - \frac{10}{n} + \frac{32}{3} + \frac{16}{n} + \frac{16}{3n^2} - 4 - \frac{8}{n} - \frac{4}{n^2}\right)$$

$$= 4 - 10 + \frac{32}{3} - 4 = \frac{2}{3}.$$

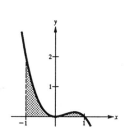

53. Use the **Midpoint Rule**

$$\text{area} \approx \sum_{i=1}^{n} f\left(\frac{x_i + x_{i-1}}{2}\right)\Delta x$$

with $n = 4$ to approximate the area of the region bounded by the graph of $f(x) = \tan x$ and the x-axis over the interval $[0, \pi/4]$.

Solution:

Let

$$\Delta x = \frac{\pi/4 - 0}{4} = \frac{\pi}{16}.$$

Dividing the interval into four parts of equal lengths yields endpoints

$$x_0 = 0, \ x_1 = \frac{\pi}{16}, \ x_2 = \frac{2\pi}{16}, \ x_3 = \frac{3\pi}{16}, \ \text{and} \ x_4 = \frac{4\pi}{16}.$$

The midpoints of these subintervals are

$$\frac{x_0 + x_1}{2} = \frac{\pi}{32}, \quad \frac{x_1 + x_2}{2} = \frac{3\pi}{32},$$

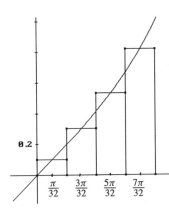

$$\frac{x_2 + x_3}{2} = \frac{5\pi}{32}, \text{ and } \frac{x_3 + x_4}{2} = \frac{7\pi}{32}.$$

$$\text{area} \approx \sum_{i=1}^{4} f\left(\frac{x_i + x_{i-1}}{2}\right) \Delta x$$

$$= \frac{\pi}{16}\left(\tan\frac{\pi}{32} + \tan\frac{3\pi}{32} + \tan\frac{5\pi}{32} + \tan\frac{7\pi}{32}\right) \approx 0.345$$

4.3   Riemann Sums and the Definite Integral

15. Sketch the region whose area is indicated by $\int_0^2 (2x + 5)\,dx$. Then use a geometric formula to evaluate the integral.

Solution:

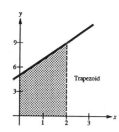

The region whose area is given by $\int_0^2 (2x + 5)\,dx$ is shown by the accompanying figure to be a trapezoid. Since the height of the trapezoid is $h = 2$ and the lengths of the two bases are $b_1 = 5$ and $b_2 = 9$, the area of the trapezoid is

$$A = h\left[\frac{b_1 + b_2}{2}\right] = 2\left[\frac{5 + 9}{2}\right] = 14.$$

23. If $\int_2^6 f(x)\,dx = 10$ and $\int_2^6 g(x)\,dx = -2$, find

(a) $\displaystyle\int_2^6 [f(x) + g(x)]\,dx$ (b) $\displaystyle\int_2^6 [g(x) - f(x)]\,dx$

(c) $\displaystyle\int_2^6 2g(x)\,dx$ (d) $\displaystyle\int_2^6 3f(x)\,dx$

Solution:

(a) $\displaystyle\int_2^6 [f(x) + g(x)]\,dx = \int_2^6 f(x)\,dx + \int_2^6 g(x)\,dx$

$$= 10 + (-2) = 8$$

(b) $\displaystyle\int_2^6 [g(x) - f(x)]\,dx = \int_2^6 g(x)\,dx - \int_2^6 f(x)\,dx$

$$= -2 - 10 = -12$$

(c) $\displaystyle\int_2^6 2g(x)\,dx = 2\int_2^6 g(x)\,dx = 2(-2) = -4$

(d) $\displaystyle\int_2^6 3f(x)\,dx = 3\int_2^6 f(x)\,dx = 3(10) = 30$

27. Evaluate the definite integral $\int_{-1}^1 x^3\,dx$.

Solution:

Let $\Delta x = [1-(-1)]/n = 2/n$. Using right-hand endpoints, you have $c_i = -1 + (2i/n)$, and the definite integral is given by the limit

$$\int_{-1}^1 x^3\,dx = \lim_{n\to\infty}\sum_{i=1}^n\left(-1+\frac{2i}{n}\right)^3\left(\frac{2}{n}\right)$$

$$= \lim_{n\to\infty}\sum_{i=1}^n\left(\frac{2}{n}\right)\left(-1+\frac{6i}{n}-\frac{12i^2}{n^2}+\frac{8i^3}{n^3}\right)$$

$$= \lim_{n\to\infty}\left(\frac{2}{n}\right)\left[-\sum_{i=1}^n 1+\frac{6}{n}\sum_{i=1}^n i-\frac{12}{n^2}\sum_{i=1}^n i^2+\frac{8}{n^3}\sum_{i=1}^n i^3\right]$$

$$= \lim_{n\to\infty}\left(\frac{2}{n}\right)\left[-n+\frac{6n(n+1)}{2n}-\frac{12n(n+1)(2n+1)}{6n^2}\right.$$
$$\left.+\frac{8n^2(n+1)^2}{4n^3}\right]$$

$$= \lim_{n\to\infty}\left[\frac{-2n}{n}+\frac{6n(n+1)}{n^2}-\frac{4n(n+1)(2n+1)}{n^3}\right.$$
$$\left.+\frac{4n^2(n+1)^2}{n^4}\right]$$

$$= -2+6-8+4 = 0$$

4.4   The Fundamental Theorem of Calculus

7. Evaluate the definite integral $\int_0^1 (2t-1)^2\,dt$.

Solution:

$$\int_0^1 (2t-1)^2\,dt = \int_0^1 (4t^2-4t+1)\,dt$$

$$= \left[\frac{4t^3}{3}-\frac{4t^2}{2}+t\right]_0^1$$

$$= \left(\frac{4}{3}-\frac{4}{2}+1\right)-(0-0+0) = \frac{4}{3}-\frac{6}{3}+\frac{3}{3} = \frac{1}{3}$$

11. Evaluate the definite integral $\displaystyle\int_1^4 \frac{u-2}{\sqrt{u}}\,du$.

Solution:

$$\int_1^4 \frac{u-2}{\sqrt{u}}\,du = \int_1^4 \left(u^{1/2} - 2u^{-1/2}\right) du$$

$$= \left[\frac{2}{3}u^{3/2} - 4u^{1/2}\right]_1^4$$

$$= \left[\frac{2}{3}(\sqrt{4})^3 - 4\sqrt{4}\right] - \left(\frac{2}{3} - 4\right) = \frac{2}{3}$$

20. Evaluate the definite integral $\displaystyle\int_0^4 |x^2 - 4x + 3|\,dx$.

Solution:

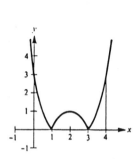

Since $x^2 - 4x + 3 = (x-1)(x-3)$, you can show that $x^2 - 4x + 3$ is negative in the interval $(1, 3)$. Therefore,

$$|x^2 - 4x + 3| = \begin{cases} x^2 - 4x + 3, & -\infty < x \le 1 \\ -(x^2 - 4x + 3), & 1 < x < 3 \\ x^2 - 4x + 3, & 3 \le x < \infty. \end{cases}$$

$$\int_0^4 |x^2 - 4x + 3|\,dx$$

$$= \int_0^1 (x^2 - 4x + 3)\,dx + \int_1^3 [-(x^2 - 4x + 3)]\,dx + \int_3^4 (x^2 - 4x + 3)\,dx$$

$$= \left[\frac{x^3}{3} - \frac{4x^2}{2} + 3x\right]_0^1 + \left[-\frac{x^3}{3} + \frac{4x^2}{2} - 3x\right]_1^3 + \left[\frac{x^3}{3} - \frac{4x^2}{2} + 3x\right]_3^4$$

$$= \left(\frac{1}{3} - 2 + 3\right) - (0) + (-9 + 18 - 9)$$

$$- \left(-\frac{1}{3} + 2 - 3\right) + \left(\frac{64}{3} - 32 + 12\right) - (9 - 18 + 9)$$

$$= \frac{4}{3} - 0 + 0 - \left(-\frac{4}{3}\right) + \frac{4}{3} - 0 = \frac{12}{3} = 4$$

25. Evaluate the definite integral $\displaystyle\int_{-\pi/3}^{\pi/3} 4\sec\phi\tan\phi\,d\phi$.

Solution:

$$\int_{-\pi/3}^{\pi/3} 4\sec\phi\tan\phi\,d\phi = \Big[4\sec\phi\Big]_{-\pi/3}^{\pi/3}$$

$$= 4[\sec\pi/3 - \sec(-\pi/3)]$$

$$= 4[2 - 2] = 0$$

31. Evaluate the definite integral $\int_{-1}^{1} (e^{\theta} + \sin \theta) \, d\theta$.

Solution:

$$\int_{-1}^{1} (e^{\theta} + \sin \theta) \, d\theta = \left[e^{\theta} - \cos \theta \right]_{-1}^{1}$$
$$= \left[e^{1} - \cos 1 \right] - \left[e^{-1} - \cos(-1) \right]$$
$$= e^{1} - e^{-1} \approx 2.350$$

43. Find the area of the region bounded by the graphs of $y = x^3 + x$, $x = 2$, and $y = 0$.

Solution:

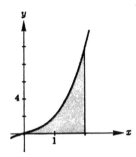

Using the accompanying figure you can see that the area of the region is

$$\text{area} = \int_{0}^{2} (x^3 + x) \, dx$$
$$= \left[\frac{1}{4}x^4 + \frac{1}{2}x^2 \right]_{0}^{2}$$
$$= (4 + 2) - (0) = 6.$$

49. Sketch the graph of $f(x) = 4 - x^2$ over the interval $[-2, 2]$. Find the average value of $f(x)$ over this interval and find all values of x in the interval for which $f(x)$ equals its average.

Solution:

The average value is given by

$$\frac{1}{b-a} \int_{a}^{b} f(x) \, dx = \frac{1}{2 - (-2)} \int_{-2}^{2} (4 - x^2) \, dx$$
$$= \frac{1}{4} \int_{-2}^{2} (4 - x^2) \, dx$$
$$= 2 \left(\frac{1}{4} \right) \int_{0}^{2} (4 - x^2) \, dx \qquad \text{(By Symmetry)}$$
$$= \frac{1}{2} \left[4x - \frac{1}{3}x^3 \right]_{0}^{2} = \frac{1}{2} \left(8 - \frac{8}{3} \right) = \frac{8}{3}$$

To find the values of x for which $f(x) = 8/3$ in the interval $[-2, 2]$, solve the equation

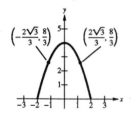

$$f(x) = 4 - x^2 = \frac{8}{3}$$

$$x^2 = \frac{4}{3}$$

$$x = \pm \frac{2}{\sqrt{3}} = \pm \frac{2\sqrt{3}}{3} \approx \pm 1.155$$

67. (a) Integrate

$$F(x) = \int_{\pi/4}^{x} \sec^2 t \, dt$$

to find F as a function of x and (b) demonstrate the second Fundamental Theorem of Calculus by finding $F'(x)$

Solution:

(a) $F(x) = \displaystyle\int_{\pi/4}^{x} \sec^2 t \, dt = \big[\tan t\big]_{\pi/4}^{x} = \tan x - 1$

(b) $F'(x) = \dfrac{d}{dx}[\tan x - 1] = \sec^2 x$

4.5   Integration by Substitution

11. Evaluate the indefinite integral $\displaystyle\int x^2(x^3 - 1)^4 \, dx$ and check the result by differentiation.

Solution:

To evaluate $\int x^2(x^3 - 1)^4 \, dx$, use the method of pattern recognition by letting $g(x) = x^3 - 1$, and $g'(x) = 3x^2$. Thus, by Theorem 4.12, you have

$$\int x^2(x^3 - 1)^4 \, dx = \int (x^3 - 1)^4 \left(\frac{1}{3}\right)(3x^2) \, dx$$

$$= \frac{1}{3} \int \underbrace{(x^3 - 1)^4}_{[g(x)]^4} \underbrace{(3x^2)}_{g'(x)} \, dx$$

$$= \frac{1}{3} \frac{[g(x)]^5}{5} + C$$

$$= \left(\frac{1}{3}\right)\left[\frac{(x^3 - 1)^5}{5}\right] + C = \frac{1}{15}(x^3 - 1)^5 + C$$

Check

If $y = \dfrac{1}{15}(x^3 - 1)^5 + C$, then

$$\frac{dy}{dx} = \tfrac{1}{15}(5)(x^3 - 1)^4(3x^2) + 0 = x^2(x^3 - 1)^4$$

13. Evaluate the indefinite integral $\displaystyle\int 5x \sqrt[3]{1 - x^2}\, dx$ and check the result by differentiation.

Solution:

To evaluate $\int 5x \sqrt[3]{1 - x^2}\, dx$, use the method of pattern recognition by letting $g(x) = 1 - x^2$, and $g'(x) = -2x$.

$$\int 5x \sqrt[3]{1 - x^2}\, dx = -\frac{5}{2} \int \underbrace{(1 - x^2)^{1/3}}_{[g(x)]^{1/3}} \underbrace{(-2x)}_{g'(x)}\, dx$$

$$= -\frac{5}{2} \frac{[g(x)]^{4/3}}{4/3} + C$$

$$= -\left(\frac{5}{2}\right)\left[\frac{(1 - x^2)^{4/3}}{4/3}\right] + C = -\frac{15}{8}(1 - x^2)^{4/3} + C$$

Check

If $y = \dfrac{-15}{8}(1 - x^2)^{4/3} + C$, then

$$\frac{dy}{dx} = -\left(\frac{15}{8}\right)\left(\frac{4}{3}\right)(1 - x^2)^{1/3}(-2x) + 0 = 5x(1 - x^2)^{1/3}$$

21. Evaluate the indefinite integral

$$\int \left(1 + \frac{1}{t}\right)^3 \left(\frac{1}{t^2}\right)\, dt$$

and check the result by differentiation.

Solution:

We evaluate this integral by changing the variable by letting $u = 1 + (1/t)$. Then,

$$du = -\frac{1}{t^2}\, dt.$$

Therefore,

$$\int \left(1+\frac{1}{t}\right)^3 \left(\frac{1}{t^2}\right) dt = -\int \left(1+\frac{1}{t}\right)^3 \left(\frac{-1}{t^2}\right) dt$$

$$= -\int u^3 \, du$$

$$= -\frac{u^4}{4} + C = -\frac{1}{4}\left(1+\frac{1}{t}\right)^4 + C$$

Check

If $y = -\frac{1}{4}\left(1+\frac{1}{t}\right)^4 + C$, then

$$\frac{dy}{dt} = \left(-\frac{1}{4}\right)(4)\left(1+\frac{1}{t}\right)^3 \left(-\frac{1}{t^2}\right) + 0 = \left(1+\frac{1}{t}\right)^3 \left(\frac{1}{t^2}\right)$$

23. Evaluate the indefinite integral $\int (1/\sqrt{2x})\, dx$ and check the result by differentiation.

Solution:

$$\int \frac{1}{\sqrt{2x}}\, dx = \int \frac{1}{\sqrt{2}\sqrt{x}}\, dx$$

$$= \frac{1}{\sqrt{2}} \int x^{-1/2}\, dx = \frac{1}{\sqrt{2}} \left(\frac{x^{1/2}}{\frac{1}{2}}\right) + C = \sqrt{2x} + C$$

Check

If $y = \sqrt{2x} + C = (2x)^{1/2} + C$, then

$$\frac{dy}{dx} = \frac{1}{2}(2x)^{-1/2}(2) + 0 = \frac{1}{\sqrt{2x}}$$

37. Evaluate $\int \frac{1}{\theta^2} \cos\frac{1}{\theta}\, d\theta$.

Solution:

If you let $u = 1/\theta$, then $du = -\frac{1}{\theta^2}\, d\theta$ and $-du = \frac{1}{\theta^2}\, d\theta$.
Therefore,

$$\int \frac{1}{\theta^2} \cos\frac{1}{\theta}\, d\theta = \int \cos\frac{1}{\theta}\left(\frac{1}{\theta^2}\, d\theta\right)$$

$$= \int \cos u(-du)$$

$$= -\int \cos u\, du = -\sin u + C = -\sin\frac{1}{\theta} + C.$$

39. Evaluate $\displaystyle\int \sin 2x \cos 2x \, dx$.

Solution:

Using the double angle identity you obtain
$2 \sin 2x \cos 2x = \sin 4x$. If you let $u = 4x$, then $du = 4 \, dx$ and
$dx = \frac{1}{4} \, du$. Therefore,

$$\int \sin 2x \cos 2x \, dx = \frac{1}{2} \int \sin 4x \, dx$$

$$= \frac{1}{2} \int \sin u \left(\frac{1}{4}\right) du$$

$$= \frac{1}{8} \int \sin u \, du = -\frac{1}{8} \cos u + C_1 = -\frac{1}{8} \cos 4x + C_1$$

We could also evaluate this integral by letting $u = \sin 2x$. Then
$du = 2 \cos 2x \, dx$ and $\cos 2x \, dx = \frac{1}{2} \, du$. Therefore,

$$\int \sin 2x \cos 2x \, dx = \int u \left(\frac{1}{2}\right) du$$

$$= \frac{1}{2} \int u \, du$$

$$= \frac{1}{4} u^2 + C_2 = \frac{1}{4} \sin^2 2x + C_2$$

Through the use of trigonometric identities it can be proved
that the results of the two methods of integration are
equivalent. This exercise shows that the method of evaluating
an integral may not be unique and the results, even though
equivalent, may appear unrelated.

45. Evaluate $\int \cot^2 x \, dx$.

Solution:

$$\int \cot^2 x \, dx = \int (\csc^2 x - 1) \, dx = -\cot x - x + C$$

51. Evaluate $\int e^x (e^x + 1)^2 \, dx$

Solution:

If you let $u = e^x + 1$, then $du = e^x \, dx$. Therefore,

$$\int e^x (e^x + 1)^2 \, dx = \int (e^x + 1)^2 (e^x) \, dx$$

$$= \int u^2 \, du$$

$$= \frac{1}{3} u^3 + C = \frac{1}{3} (e^x + 1)^3 + C.$$

55. Evaluate $\int x^2\sqrt{1-x}\,dx$ by the method shown in Example 6.

Solution:

Let $u = \sqrt{1-x}$. Then $u^2 = 1-x$, $x = 1-u^2$, and $dx = -2u\,du$. Thus,

$$\int x^2\sqrt{1-x}\,dx = \int (1-u^2)^2 u(-2u)\,du$$

$$= -\int (2u^2 - 4u^4 + 2u^6)\,du$$

$$= -\left(\frac{2u^3}{3} - \frac{4u^5}{5} + \frac{2u^7}{7}\right) + C$$

$$= \frac{-2u^3}{105}(35 - 42u^2 + 15u^4) + C$$

$$= \frac{-2}{105}(1-x)^{3/2}[35 - 42(1-x) + 15(1-x)^2] + C$$

$$= \frac{-2}{105}(1-x)^{3/2}(15x^2 + 12x + 8) + C.$$

57. Evaluate $\int \dfrac{x^2 - 1}{\sqrt{2x-1}}\,dx$ by the method shown in Example 6.

Solution:

Let $u = \sqrt{2x-1}$. Then $u^2 = 2x-1$, $x = \dfrac{u^2+1}{2}$, and $dx = u\,du$. Thus,

$$\int \frac{x^2-1}{\sqrt{2x-1}}\,dx$$

$$= \int \frac{[(u^2+1)/2]^2 - 1}{u}(u\,du)$$

$$= \frac{1}{4}\int (u^4 + 2u^2 - 3)\,du$$

$$= \frac{1}{4}\left(\frac{u^5}{5} + \frac{2u^3}{3} - 3u\right) + C$$

$$= \frac{u}{60}(3u^4 + 10u^2 - 45) + C$$

$$= \frac{1}{60}\sqrt{2x-1}[3(2x-1)^2 + 10(2x-1) - 45] + C$$

$$= \frac{1}{60}\sqrt{2x-1}(12x^2 + 8x - 52) + C$$

$$= \frac{1}{15}\sqrt{2x-1}(3x^2 + 2x - 13) + C.$$

61. Evaluate the definite integral $\int_{-1}^{1} x(x^2+1)^3 \, dx$.

Solution:

Let $u = x^2 + 1$. Then $du = 2x \, dx$, and $dx = du/2$. Furthermore, when $x = -1$, $u = 2$ and when $x = 1$, $u = 2$. Since the upper and lower limits are equal, you have

$$\int_{-1}^{1} x(x^2+1)^3 \, dx = \int_{2}^{2} u^3 \left(\frac{1}{2}\right) du = 0.$$

65. Evaluate the definite integral $\int_{1}^{9} \frac{1}{\sqrt{x}(1+\sqrt{x})^2} \, dx$.

Solution:

Let $u = 1 + \sqrt{x}$. Then $du = \frac{1}{2\sqrt{x}} \, dx$. Furthermore, if $x = 1$, then $u = 2$, and if $x = 9$, then $u = 4$. Hence,

$$\int_{1}^{9} \frac{1}{\sqrt{x}(1+\sqrt{x})^2} \, dx = 2 \int_{1}^{9} \frac{1}{(1+\sqrt{x})^2} \left(\frac{1}{2\sqrt{x}}\right) dx$$

$$= 2 \int_{2}^{4} \frac{1}{u^2} \, du$$

$$= 2 \int_{2}^{4} u^{-2} \, du$$

$$= \left[\frac{-2}{u}\right]_{2}^{4} = -\frac{1}{2} - (-1) = \frac{1}{2}$$

69. Evaluate the definite integral $\int_{0}^{7} x \sqrt[3]{x+1} \, dx$.

Solution:

Let $u = \sqrt[3]{x+1}$. Then $u^3 = x + 1$, $x = u^3 - 1$, and $dx = 3u^2 \, du$. Furthermore, if $x = 0$, then $u = 1$, and if $x = 7$, then $u = 2$. Thus,

$$\int_{0}^{7} x \sqrt[3]{x+1} \, dx = \int_{1}^{2} (u^3 - 1))(u)(3u^2 \, du)$$

$$= 3 \int_{1}^{2} (u^6 - u^3) \, du$$

$$= 3 \left[\frac{u^7}{7} - \frac{u^4}{4}\right]_{1}^{2}$$

$$= 3 \left(\frac{128}{7} - \frac{16}{4} - \frac{1}{7} + \frac{1}{4}\right)$$

$$= 3 \left(\frac{127}{7} - \frac{15}{4}\right) = 3 \left(\frac{508 - 105}{28}\right) = \frac{1209}{28},$$

73. Evaluate the definite integral $\int_{\pi/2}^{2\pi/3} \sec^2(x/2)\, dx$.

Solution:

Let $u = \dfrac{x}{2}$. Then $du = \dfrac{1}{2}\, dx$ and $dx = 2\, du$. Also, if $x = \dfrac{\pi}{2}$,

then $u = \dfrac{\pi}{4}$, and if $x = \dfrac{2\pi}{3}$, then $u = \dfrac{\pi}{3}$. Thus,

$$\int_{\pi/2}^{2\pi/3} \sec^2\left(\frac{x}{2}\right) dx = \int_{\pi/4}^{\pi/3} \sec^2 u(2)\, du$$

$$= 2\big[\tan u\big]_{\pi/4}^{\pi/3}$$

$$= 2\left[\tan\frac{\pi}{3} - \tan\frac{\pi}{4}\right] = 2(\sqrt{3} - 1)$$

75. Evaluate the definite integral $\int_{3}^{4} \dfrac{1}{x-2}\, dx$.

Solution:

Let $u = x - 2$. Then $du = dx$. Furthermore, $u = 1$ when $x = 3$ and $u = 2$ when $x = 4$. Thus,

$$\int_{3}^{4} \frac{1}{x-2}\, dx = \int_{1}^{2} \frac{1}{u}\, du$$

$$= \big[\ln u\big]_{1}^{2}$$

$$= \ln 2 - \ln 1 = \ln 2 \approx 0.693.$$

87. Determine the area of the region having the given boundaries in the accompanying figure.

$y = 2\sin x + \sin 2x$

Solution:

$$\text{area} = \int_{0}^{\pi} (2\sin x + \sin 2x)\, dx$$

$$= 2\int_{0}^{\pi} \sin x\, dx + \frac{1}{2}\int_{0}^{\pi} \sin 2x(2)\, dx$$

$$= \left[-2\cos x - \frac{1}{2}\cos 2x\right]_{0}^{\pi} = 4$$

95. The rate of depreciation, dV/dt, of a machine is inversely proportional to the square of $t + 1$ where V is the value of the machine t years after it was purchased. If the initial value of the machine was \$500,000, and its value dropped \$100,000 in the first year, estimate its value after 4 years.

Solution:

Since the rate of depreciation, dV/dt, is inversely proportional to the square of $t + 1$, it follows that

$$\frac{dV}{dt} = \frac{k}{(t+1)^2}$$

$$V = \int \frac{k}{(t+1)^2} \, dt$$

$$= k \int (t+1)^{-2} \, dt$$

$$= k \int u^{-2} \, du \qquad (u = t + 1, \ du = dt)$$

$$= k \frac{u^{-1}}{-1} + C = \frac{-k}{t+1} + C.$$

Since the initial value of the machine was \$500,000, you have

$$V(0) = \frac{-k}{0+1} + C = -k + C = 500,000.$$

During the first year the value of the machine decreased \$100,000. Therefore,

$$V(1) = \frac{-k}{1+1} + C = -\frac{1}{2}k + C = 400,000.$$

Solving the two equations simultaneously yields the solution

$$V(t) = \frac{200,000}{t+1} + 300,000.$$

The approximate value of the machine after 4 years is

$$V(4) = \frac{200,000}{4+1} + 300,000 = \$340,000.$$

Note that according to this model the value of the machine will always be greater than \$300,000 since

$$\lim_{t \to \infty} V(t) = 300,000.$$

4.6 Numerical Integration

 5. Use the Trapezoidal Rule and Simpson's Rule with n = 8 to approximate the value of $\int_0^2 x^3\,dx$. Compare these results with the exact value of the definite integral. Round your answers to four decimal places.

Solution:

(a) Trapezoidal Rule ($n = 8$)

$$\int_0^2 x^3\,dx \approx \frac{2}{2(8)}\left[0 + 2\left(\frac{1}{4}\right)^3 + 2\left(\frac{2}{4}\right)^3 + 2\left(\frac{3}{4}\right)^3 + 2\left(\frac{4}{4}\right)^3\right.$$
$$\left. + 2\left(\frac{5}{4}\right)^3 + 2\left(\frac{6}{4}\right)^3 + 2\left(\frac{7}{4}\right)^3 + 2^3\right]$$
$$= \frac{1}{8}\left[\frac{2(1^3 + 2^3 + 3^3 + 4^3 + 5^3 + 6^3 + 7^3)}{4^3} + 8\right]$$
$$= \frac{1}{8}\left[\frac{2(784)}{164} + 8\right] = \frac{65}{16} = 4.0625$$

(b) Simpson's Rule ($n = 8$)

$$\int_0^2 x^3\,dx \approx \frac{2}{3(8)}\left[0 + 4\left(\frac{1}{4}\right)^3 + 2\left(\frac{2}{4}\right)^3 + 4\left(\frac{3}{4}\right)^3 + 2\left(\frac{4}{4}\right)^3\right.$$
$$\left. + 4\left(\frac{5}{4}\right)^3 + 2\left(\frac{6}{4}\right)^3 + 4\left(\frac{7}{4}\right)^3 + 2^3\right]$$
$$= \frac{1}{12}\left[\frac{4(1^3 + 3^3 + 5^3 + 7^3) + 2(2^3 + 4^3 + 6^3)}{4^3} + 8\right]$$
$$= \frac{1}{12}\left[\frac{4(496) + 2(288)}{4^3} + 8\right]$$
$$= \frac{1}{12}\left(\frac{2560}{64} + 8\right) = \frac{1}{12}(48) = 4$$

(c) In this particular case, Simpson's Rule is exact since

$$\int_0^2 x^3\,dx = \left[\frac{x^4}{4}\right]_0^2 = \frac{16}{4} = 4.$$

13. Approximate $\int_0^1 \sqrt{x}\sqrt{1-x}\,dx$ using (a) the Trapezoidal Rule and (b) Simpson's Rule with $n = 4$.

Solution:

(a) Trapezoidal Rule $(n = 4)$

$$\int_0^1 \sqrt{x}\sqrt{1-x}\,dx$$

$$\approx \frac{1}{2(4)}\left[0 + 2\sqrt{\frac{1}{4}}\sqrt{\frac{3}{4}} + 2\sqrt{\frac{2}{4}}\sqrt{\frac{2}{4}} + 2\sqrt{\frac{3}{4}}\sqrt{\frac{1}{4}} + 0\right]$$

$$= \frac{1}{8}\left[\frac{2\sqrt{3}}{4} + \frac{2(2)}{4} + \frac{2\sqrt{3}}{4}\right] = \frac{1}{8}(1 + \sqrt{3}) \approx 0.342$$

(b) Simpson's Rule $(n = 4)$

$$\int_0^1 \sqrt{x}\sqrt{1-x}\,dx$$

$$\approx \frac{1}{3(4)}\left[0 + 4\sqrt{\frac{1}{4}}\sqrt{\frac{3}{4}} + 2\sqrt{\frac{2}{4}}\sqrt{\frac{2}{4}} + 4\sqrt{\frac{3}{4}}\sqrt{\frac{1}{4}} + 0\right]$$

$$= \frac{1}{12}\left[\frac{4\sqrt{3}}{4} + \frac{2(2)}{4} + \frac{4\sqrt{3}}{4}\right] = \frac{1}{12}(2\sqrt{3} + 1) \approx 0.372$$

15. Approximate $\int_0^{\sqrt{\pi/2}} \cos x^2\,dx$ using (a) the Trapezoidal Rule and (b) Simpson's Rule with $n = 4$.

Solution:

(a) Trapezoidal Rule with $n = 4$

$$\int_0^{\sqrt{\pi/2}} \cos x^2\,dx$$

$$\approx \frac{\sqrt{\pi/2}}{2(4)}\left[\cos(0) + 2\cos\left(\frac{\sqrt{\pi/2}}{4}\right)^2 + 2\cos\left(\frac{2\sqrt{\pi/2}}{4}\right)^2\right.$$

$$\left. + 2\cos\left(\frac{3\sqrt{\pi/2}}{4}\right)^2 + \cos\left(\frac{4\sqrt{\pi/2}}{4}\right)^2\right]$$

$$\approx 0.957$$

(b) Simpson's Rule with $n = 4$

$$\int_0^{\sqrt{\pi/2}} \cos x^2\,dx$$

$$\approx \frac{\sqrt{\pi/2}}{3(4)}\left[\cos(0) + 4\cos\left(\frac{\sqrt{\pi/2}}{4}\right)^2 + 2\cos\left(\frac{2\sqrt{\pi/2}}{4}\right)^2\right.$$

$$\left. + 4\cos\left(\frac{3\sqrt{\pi/2}}{4}\right)^2 + \cos\left(\frac{4\sqrt{\pi/2}}{4}\right)^2\right]$$

$$\approx 0.978$$

🖥 **29.** Use a symbolic differentiation utility and the error formulas to find n so that the error in the approximation of the integral $\int_0^1 \tan x^2 \, dx$ is less than 0.00001 using (a) the Trapezoidal Rule and (b) Simpson's Rule.

Solution:

(a) Begin by letting $f(x) = \tan x^2$ and finding the second derivative of f using the symbolic differentiation utility. (*Note:* The simplification of the derivatives will differ depending on which differentiation utility is used.)

$$f'(x) = 2x \sec^2 x^2$$

$$f''(x) = \frac{2(\cos x^2 + 4x^2 \sin x^2)}{\cos^3 x^2}$$

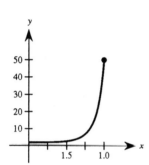

The graph of the second derivative (produced by the differentiation utility) is shown in the accompanying figure. The maximum value of $|f''(x)|$ on the interval $[0, 1]$ is $|f''(1)| \approx 50$. Thus, by Theorem 4.19, you can write

$$E \le \frac{(b-a)^3}{12n^2}|f''(1)| \le \frac{50}{12n^2}.$$

To obtain an error E that is less than 0.00001, you must choose n so that $50/(12n^2) \le 0.00001$. Thus,

$$50(100,000) \le 12n^2 \quad \Longrightarrow \quad 645.6 \approx \sqrt{\frac{50(100,000)}{12}} \le n.$$

Therefore, choose $n = 646$.

(b) Begin by letting $f(x) = \tan x^2$ and finding the fourth derivative of f using the symbolic differentiation utility. The first and second derivative are given in part (a).

$$f'''(x) = -\frac{8x(4x^2 \cos^2 x^2 - 3\sin x^2 \cos x^2 - 6x^2)}{\cos^4 x^2}$$

$$f^{(4)}(x) = -\frac{8}{\cos^5 x^2}[24x^2 \cos^3 x^2 + (16x^4 - 3)\sin x^2 \cos^2 x^2$$
$$- 36x^2 \cos x^2 - 48x^4 \sin x^2]$$

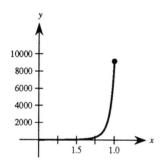

The graph of the fourth derivative (produced by the differentiation utility) is shown in the accompanying figure. The maximum value of $|f^{(4)}(x)|$ on the interval $[0, 1]$ is $|f^{(4)}(1)| \approx 9185$. Thus, by Theorem 4.19, you can write

$$E \le \frac{(b-a)^5}{180n^4}|f^{(4)}(1)| \le \frac{9185}{180n^4}.$$

To obtain an error E that is less than 0.00001, you must choose n so that $9185/(180n^4) \leq 0.00001$. Thus,

$$9185(100,000) \leq 180n^4 \implies 47.5 \approx \sqrt[4]{\frac{9185(100,000)}{180}} \leq n.$$

Therefore, choose $n = 48$.

4.7   An Alternative Development of the Natural Logarithmic Function

7. Find dy/dx given $y = \ln\left(x\sqrt{x^2 - 1}\right)$.
Solution:

$$y = \ln\left(x\sqrt{x^2 - 1}\right) = \ln x + \ln \sqrt{x^2 - 1}$$
$$= \ln x + \frac{1}{2}\ln\left(x^2 - 1\right)$$
$$\frac{dy}{dx} = \frac{1}{x} + \frac{1}{2}\left(\frac{1}{x^2 - 1}\right)(2x) = \frac{1}{x} + \frac{x}{x^2 - 1} = \frac{2x^2 - 1}{x\left(x^2 - 1\right)}$$

11. Find dy/dx given $y = \ln\sqrt{\dfrac{x+1}{x-1}}$.

Solution:

$$y = \ln\sqrt{\frac{x+1}{x-1}} = \frac{1}{2}[\ln(x+1) - \ln(x-1)]$$
$$\frac{dy}{dx} = \frac{1}{2}\left[\frac{1}{x+1} - \frac{1}{x-1}\right] = \frac{-1}{x^2 - 1}$$

17. Find dy/dx given $y = \ln\left|\dfrac{\cos x}{\cos x - 1}\right|$.
Solution:

$$y = \ln\left|\frac{\cos x}{\cos x - 1}\right| = \ln|\cos x| - \ln|\cos x - 1|$$
$$\frac{dy}{dx} = \frac{1}{\cos x}(-\sin x) - \frac{1}{\cos x - 1}(-\sin x)$$
$$= -\tan x + \frac{\sin x}{\cos x - 1}$$

23. Use Newton's Method to approximate, to three decimal places, the x-coordinate of the point of intersection of the graphs of the equations $y = \ln x$ and $y = -x$.

Solution:

Approximating the x-coordinate of the point of intersection of the graphs of $y = \ln x$ and $y = -x$ is equivalent to approximating the zero of the function $f(x) = \ln x + x$. The iterative formula for Newton's Method is

$$x_{n+1} = x_n - \frac{f(x_n)}{f'(x_n)} = x_n - \frac{\ln x_n + x_n}{(1/x_n) + 1}.$$

Using the accompanying figure, it appears appropriate to choose $x = 0.5$ as the initial estimate.

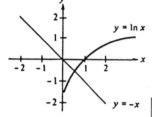

n	x_n	$f(x_n)$	$f'(x_n)$	$f(x_n)/f'(x_n)$	$x_n - \dfrac{f(x_n)}{f'(x_n)}$
1	0.5000	−0.1931	3	−0.0644	0.5644
2	0.5644	−0.0076	2.7718	−0.0027	0.5671
3	0.5671	0.0000	2.7632	0.0000	0.5671

Therefore, the approximate root is $x = 0.567$.

27. Find dy/dx by using logarithmic differentiation given

$$y = \frac{x^2\sqrt{3x - 2}}{(x - 1)^2}$$

Solution:

Begin by taking the natural logarithm of each member of the equation.

$$y = \frac{x^2\sqrt{3x - 2}}{(x - 1)^2}$$

$$\ln y = \ln \frac{x^2\sqrt{3x - 2}}{(x - 1)^2} = 2\ln x + \frac{1}{2}\ln(3x - 2) - 2\ln(x - 1)$$

$$\frac{1}{y}\frac{dy}{dx} = \frac{2}{x} + \left(\frac{1}{2}\right)\frac{3}{3x - 2} - 2\frac{1}{x - 1}$$

$$\frac{dy}{dx} = y\left[\frac{3x^2 - 15x + 8}{2x(3x - 2)(x - 1)}\right]$$

The derivative can be written in terms of x by replacing y with its equivalent in the original equation.

$$y' = \frac{3x^3 - 15x^2 + 8x}{2(x - 1)^3\sqrt{3x - 2}}$$

Review Exercises for Chapter 4

7. Find the indefinite integral $\int \dfrac{x^3 + 1}{x}\, dx$.

Solution:

$$\int \frac{x^3 + 1}{x}\, dx = \int \left(\frac{x^3}{x} + \frac{1}{x} \right) dx$$
$$= \int \left(x^2 + \frac{1}{x} \right) dx = \frac{1}{3}x^3 + \ln x + C$$

9. Find the indefinite integral $\int \dfrac{x^2}{\sqrt{x^3 + 3}}\, dx$.

Solution:

To find $\int x^2/\sqrt{x^3 + 3}\, dx$, let $u = x^3 + 3$. Then $du = 3x^2\, dx$.

$$\int \frac{x^2}{\sqrt{x^3 + 3}}\, dx = \frac{1}{3} \int (x^3 + 3)^{-1/2}(3x^2)\, dx$$
$$= \frac{1}{3} \int u^{-1/2}\, du$$
$$= \frac{2}{3} u^{1/2} + C$$
$$= \frac{2}{3}(x^3 + 3)^{1/2} + C = \frac{2}{3}\sqrt{x^3 + 3} + C$$

17. Find the indefinite integral $\int \tan^n x \sec^2 x\, dx$, $n \neq -1$.

Solution:

To find $\int \tan^n x \sec^2 x\, dx$, let $u = \tan x$ and $du = \sec^2 x\, dx$.

$$\int \tan^n x \sec^2 x\, dx = \int u^n\, du$$
$$= \frac{u^{n+1}}{n+1} + C = \frac{\tan^{n+1} x}{n+1} + C, \quad n \neq -1$$

23. An airplane taking off from a runway travels 3600 feet before lifting off. If it starts from rest, moves with constant acceleration, and makes the run in 30 seconds, with what velocity does it lift off?

Solution:

Let the position function of the plane be given by $s(t)$ where s is measured in feet and t is time in seconds. If $t = 0$ is the time the plane starts its take off roll, then $s(0) = 0$, $s(30) = 3600$, $s'(0) = 0$ and $s''(t) = a$ where a is constant.

$$s'(t) = \int a\, dt = at + C_1$$

$$s'(0) = 0 + C_1 = 0 \quad \Longrightarrow \quad C_1 = 0$$

$$s(t) = \int at\, dt = \frac{a}{2}t^2 + C_2$$

$$s(0) = 0 + C_2 = 0 \quad \Longrightarrow \quad C_2 = 0 \quad \Longrightarrow \quad s(t) = \frac{a}{2}t^2$$

$$s(30) - \frac{a}{2}(30)^2 - 3600 \quad \text{or} \quad a = \frac{3600(2)}{30^2} = 8 \text{ ft/sec}^2$$

Therefore

$$s(t) = 4t^2, \ s'(t) = v(t) = 8t \quad \text{and} \quad v(30) = 8(30) = 240 \text{ ft/sec}.$$

37. Use the Fundamental Theorem of Calculus to evaluate the definite integral

$$\int_0^3 \frac{1}{\sqrt{1+x}}\, dx$$

Solution:

If you let $u = 1 + x$, then $du = dx$. Also, when $x = 0$, $u = 1$, and when $x = 3$, $u = 4$. Therefore,

$$\int_0^3 \frac{1}{\sqrt{1+x}}\, dx = \int_1^4 u^{-1/2}\, du$$

$$= \left[2u^{1/2}\right]_1^4 = 2(2-1) = 2$$

41. Use the Fundamental Theorem of Calculus to evaluate the definite integral

$$\int_1^e \frac{\ln x}{x}\, dx.$$

Solution:

If you let $u = \ln x$, then $du = (1/x)\, dx$. Furthermore, when $x = 1$, $u = 0$ and when $x = e$, $u = 1$.

$$\int_1^e \frac{\ln x}{x}\, dx = \int_1^e \ln x \left(\frac{1}{x}\right) dx$$

$$= \int_0^1 u\, du = \left[\frac{1}{2}u^2\right]_0^1 = \frac{1}{2}$$

57. Find the average value of $f(x) = 1/\sqrt{x-1}$ over the interval $[5, 10]$. Find the values of x where the function assumes its average value and sketch the graph of the function.

Solution:

The average value is given by

$$\frac{1}{10-5}\int_5^{10} \frac{1}{\sqrt{x-1}}\, dx = \frac{1}{5}\int_5^{10} (x-1)^{-1/2}(1)\, dx$$

$$= \left[\frac{2}{5}(x-1)^{1/2}\right]_5^{10} = \frac{2}{5}$$

To find the value of x where the function assumes its mean value on $[5, 10]$, solve

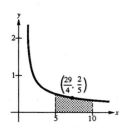

$$\frac{1}{\sqrt{x-1}} = \frac{2}{5}$$

$$\sqrt{x-1} = \frac{5}{2}$$

$$x - 1 = \frac{25}{4} \quad \Longrightarrow \quad x = \frac{29}{4}$$

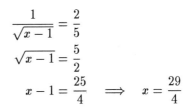

61. Use Simpson's Rule with $n = 4$ to approximate the definite integral

$$\int_1^2 \frac{1}{1 + x^3} \, dx.$$

Solution:

$$\int_1^2 \frac{1}{1 + x^3} \, dx$$

$$\approx \frac{2 - 1}{3(4)} \left(\frac{1}{1 + 1^3} + \frac{4}{1 + (1.25)^3} + \frac{2}{1 + (1.5)^3} \right.$$

$$\left. + \frac{4}{1 + (1.75)^3} + \frac{1}{1 + 2^3} \right) \approx 0.254$$

5 TRANSCENDENTAL FUNCTIONS AND DIFFERENTIAL EQUATIONS

5.1 Integrating with the Natural Logarithmic Function

5. Evaluate $\displaystyle\int \frac{x}{x^2 + 1}\, dx$.

Solution:

Letting $u = x^2 + 1$, you have $du = 2x\, dx$. By multiplying and dividing the integral by 2 yields

$$\int \frac{x}{x^2 + 1}\, dx = \frac{1}{2} \int \frac{2x}{x^2 + 1}\, dx$$
$$= \frac{1}{2} \int \frac{1}{u}\, du$$
$$= \frac{1}{2} \ln u + C$$
$$= \frac{1}{2} \ln(x^2 + 1) + C = \ln \sqrt{x^2 + 1} + C$$

[Note that absolute value signs around $(x^2 + 1)$ are unnecessary since $x^2 + 1 > 0$ for all x.]

7. Evaluate $\displaystyle\int \frac{x^2-4}{x}\,dx$.

Solution:

$$\int \frac{x^2-4}{x}\,dx = \int \left(\frac{x^2}{x} - \frac{4}{x}\right) dx$$

$$= \int \left(x - \frac{4}{x}\right) dx = \frac{x^2}{2} - 4\ln|x| + C$$

13. Evaluate $\displaystyle\int \frac{1}{\sqrt{x+1}}\,dx$.

Solution:

$$\int \frac{1}{\sqrt{x+1}}\,dx = \int (x+1)^{-1/2}(1)\,dx$$

$$= \int u^{-1/2}\,du$$

$$= \frac{u^{1/2}}{\frac{1}{2}} = 2(x+1)^{1/2} + C = 2\sqrt{x+1} + C$$

15. Evaluate $\displaystyle\int \frac{\sqrt{x}}{\sqrt{x}-3}\,dx$.

Solution:

If $u = \sqrt{x} - 3$, then $x = (u+3)^2$ and $dx = 2(u+3)\,du$. Therefore,

$$\int \frac{\sqrt{x}}{\sqrt{x}-3}\,dx = \int \frac{u+3}{u}[2(u+3)]\,du$$

$$= 2\int \frac{(u+3)^2}{u}\,du$$

$$= 2\int \frac{u^2 + 6u + 9}{u}\,du$$

$$= 2\int \left(u + 6 + \frac{9}{u}\right) du$$

$$= 2\left(\frac{u^2}{2} + 6u + 9\ln|u|\right) + C$$

$$= (\sqrt{x}-3)^2 + 12(\sqrt{x}-3) + 18\ln|\sqrt{x}-3| + C$$

17. Evaluate $\displaystyle\int \frac{2x}{(x-1)^2}\, dx$.

Solution:

$$\int \frac{2x}{(x-1)^2}\, dx = \int \frac{2x-2+2}{(x-1)^2}\, dx$$

$$= \int \frac{2(x-1)}{(x-1)^2}\, dx + 2\int \frac{1}{(x-1)^2}\, dx$$

$$= 2\int \frac{1}{x-1}\, dx + 2\int (x-1)^{-2}\, dx$$

$$= 2\int \frac{1}{u}\, du + 2\int u^{-2}\, du$$

$$= 2\ln|u| + 2\frac{u^{-1}}{-1} + C = 2\ln|x-1| - \frac{2}{x-1} + C$$

25. Evaluate $\displaystyle\int \frac{\sec x \tan x}{\sec x - 1}\, dx$.

Solution:

Letting $u = \sec x - 1$, you have $du = \sec x \tan x\, dx$. Therefore,

$$\int \frac{\sec x \tan x}{\sec x - 1}\, dx = \int \frac{1}{u}\, du = \ln|u| + C = \ln|\sec x - 1| + C$$

29. Evaluate

$$\int \frac{e^x + e^{-x}}{e^x - e^{-x}}\, dx.$$

Solution:

If $u = e^x - e^{-x}$, then $du = (e^x + e^{-x})\, dx$.

$$\int \frac{e^x + e^{-x}}{e^x - e^{-x}}\, dx = \int \frac{1}{u}\, du$$

$$= \ln|u| + C = \ln|e^x - e^{-x}| + C$$

41. Evaluate $\displaystyle\int_1^e \frac{(1+\ln x)^2}{x}\,dx$.

Solution:

Letting $u = 1 + \ln x$, you have $du = (1/x)\,dx$. When $x = 1$, $u = 1$ and when $x = e$, $u = 2$. Therefore,

$$\int_1^e \frac{(1+\ln x)^2}{x}\,dx = \int_1^e (1+\ln x)^2 \left(\frac{1}{x}\right)dx$$

$$= \int_1^2 u^2\,du = \left[\frac{u^3}{3}\right]_1^2 = \frac{7}{3}$$

49. Evaluate $\displaystyle\int_1^3 \frac{e^{3/x}}{x^2}\,dx$.

Solution:

If $u = 3/x$, then $du = (-3/x^2)\,dx$. When $x = 1$, $u = 3$ and when $x = 3$, $u = 1$. Therefore,

$$\int_1^3 \frac{e^{3/x}}{x^2}\,dx = -\frac{1}{3}\int_1^3 e^{3/x}\left(-\frac{3}{x^2}\right)dx$$

$$= -\frac{1}{3}\int_3^1 e^u\,du$$

$$= -\frac{1}{3}\left[e^u\right]_3^1 = -\frac{1}{3}(e - e^3) = \frac{e}{3}(e^2 - 1).$$

67. Find the area of the region bounded by the graphs of $y = (x^2 + 4)/x$, $y = 0$, $x = 1$, and $x = 4$.

Solution:

$$\text{area} = \int_1^4 \frac{x^2 + 4}{x}\,dx = \int_1^4 \left(x + \frac{4}{x}\right)dx$$

$$= \left[\frac{1}{2}x^2 + 4\ln x\right]_1^4$$

$$= 8 + 4\ln 4 - \frac{1}{2}$$

$$= \frac{1}{2}(15 + 16\ln 2) \approx 13.045 \text{ square units}$$

5.2   Bases other than e and Applications

6. Write the exponential equations as logarithmic equations:

(a) $27^{2/3} = 9$ (b) $16^{3/4} = 8$

Solution:

By definition,

$$a^b = x \qquad \text{if and only if} \qquad \log_a x = b$$

(a)   Letting $a = 27$, $b = \dfrac{2}{3}$, and $x = 9$ yields

$$27^{2/3} = 9 \quad \text{if and only if} \quad \log_{27} 9 = \frac{2}{3}.$$

(b)   Letting $a = 16$, $b = \dfrac{3}{4}$, and $x = 8$ yields

$$16^{3/4} = 8 \quad \text{if and only if} \quad \log_{16} 8 = \frac{3}{4}$$

13. Find x if (a) $x^2 - x = \log_5 25$ and (b) $3x + 5 = \log_2 64$.

Solution:

(a)
$$x^2 - x = \log_5 25$$
$$x^2 - x = \log_5(5^2)$$
$$x^2 - x = 2$$
$$x^2 - x - 2 = 0$$
$$(x - 2)(x + 1) = 0 \quad \implies \quad x = -1, \ 2$$

(b)
$$3x + 5 = \log_2 64$$
$$3x + 5 = \log_2(2^6)$$
$$3x + 5 = 6$$
$$3x = 1 \quad \implies \quad x = \frac{1}{3}$$

29. Find $g'(t)$ given $g(t) = t^2 2^t$.

Solution:

Using the Product Rule and Theorem 5.1 the derivative of the function $g(t) = t^2 2^t$ is

$$g'(t) = t^2 (\ln 2) 2^t + 2t(2^t) = t(2^t)(t \ln 2 + 2).$$

35. Find dy/dx given $y = \log_5 \sqrt{x^2 - 1}$.

Solution:

Begin by rewriting the given logarithmic function.

$$y = \log_5 \sqrt{x^2 - 1} = \log_5 (x^2 - 1)^{1/2} = \frac{1}{2} \log_5 (x^2 - 1)$$

Using Theorem 5.1 the derivative of the function is

$$\frac{dy}{dx} = \left(\frac{1}{2} \right) \frac{1}{(\ln 5)(x^2 - 1)} (2x) = \frac{x}{(x^2 - 1) \ln 5}.$$

37. Use logarithmic differentiation to find dy/dx given $y = x^{2/x}$.

Solution:

Taking the natural logarithm of both members of the equation yields

$$\ln y = \ln \left(x^{2/x} \right) = \frac{2}{x} \ln x$$

Differentiating implicitly yields the following.

$$\frac{1}{y} \frac{dy}{dx} = \left[\frac{2}{x} \right] \left[\frac{1}{x} \right] + (\ln x) \left[\frac{-2}{x^2} \right]$$

$$\frac{dy}{dx} = y \left[\frac{2}{x^2} - \frac{2}{x^2} \ln x \right]$$

$$= \frac{2y}{x^2} (1 - \ln x) = 2(1 - \ln x) x^{(2/x) - 2}$$

53. Assume you can earn 9% on an investment, compounded daily. Which of the following options would yield the greatest balance in 8 years? (a) $20,000 now, (b) $30,000 in 8 years, (c) $8,000 now and $20,000 in 4 years, (d) $9,000 now, $9,000 in 4 years, and $9,000 in 8 years.

Solution:

The balance in the account after 8 years is computed for each option.

(a)   A deposit of $20,000 is made now and earns interest 8 years.

$$A = 20,000 \left(1 + \frac{0.09}{365}\right)^{(365 \cdot 8)} = \$41,085.02$$

(b)   In 8 years the amount is $A = \$30,000$.

(c)   The first deposit of $8,000 earns interest 8 years. The second deposit of $20,000 in 4 years earns interest 4 years.

$$A = 8,000 \left(1 + \frac{0.09}{365}\right)^{(365 \cdot 8)} + 20,000 \left(1 + \frac{0.09}{365}\right)^{(365 \cdot 4)}$$

$$= \$45,099.32$$

(d)   Three deposits of $9,000 each are made. The first earns interest 8 years, the second 4 years, and the final deposit earns no interest.

$$A = 9,000 \left(1 + \frac{0.09}{365}\right)^{(365 \cdot 8)}$$

$$+ 9,000 \left(1 + \frac{0.09}{365}\right)^{(365 \cdot 4)} + 9,000$$

$$= \$40,387.65$$

Therefore, option (c) yields the greatest balance in 8 years.

57. A lake is stocked with 500 fish, and their population increases according to the logistics curve

$$p(t) = \frac{10,000}{1 + 19e^{-t/5}}$$

where t is measured in months. At what rate is the fish population changing at the end of 1 month and at the end of 10 months? After how many months is the population increasing most rapidly?

Solution:

Begin by finding the first and second derivative of the function.

$$p(t) = \frac{10,000}{1 + 19e^{-t/5}} = 10,000(1 + 19e^{-t/5})^{-1}$$

$$p'(t) = -10,000(1 + 19e^{-t/5})^{-2}\left(-\frac{19}{5}\right)e^{-t/5} = \frac{38,000e^{-t/5}}{(1 + 19e^{-t/5})^2}$$

$$p''(t) = 38,000\left[\frac{(1 + 19e^{-t/5})^2(-1/5)e^{-t/5}}{(1 + 19e^{-t/5})^4}\right.$$

$$\left. - \frac{e^{-t/5}(2)(1 + 19e^{-t/5})(-19/5)e^{-t/5}}{(1 + 19e^{-t/5})^4}\right]$$

$$= \frac{-38,000e^{-t/5}(1 - 19e^{-t/5})}{(1 + 19e^{-t/5})^3}$$

The rates of growth at the end of one month and ten months are

$$p'(1) \approx 113.5 \text{ fish/month, and } p'(10) \approx 403.2 \text{ fish/month.}$$

The rate of growth of the population is maximum when p' is maximum and $p'' = 0$. This occurs at the point of inflection of the graph of the function. The second derivative is zero when

$$1 - 19e^{-t/5} = 0$$

$$e^{-t/5} = \frac{1}{19}$$

$$-\frac{t}{5} = \ln\frac{1}{19} = \ln 1 - \ln 19 = -\ln 19$$

$$t = 5\ln 19 \approx 14.7 \text{ months}$$

$$p(5 \ln 19) = \frac{10,000}{1 + 19e^{-5\ln 19/5}}$$

$$= \frac{10,000}{1 + 19e^{-\ln 19}}$$

$$= \frac{10,000}{1 + 19e^{\ln(1/19)}}$$

$$= \frac{10,000}{1 + 19(1/19)} = \frac{10,000}{2} = 5000$$

Note that the rate of growth of the population is greatest when its size is one-half its the carrying capacity of the lake.

5.3   Growth and Decay

3. Solve the differential equation $y' = \sqrt{x}y$.

Solution:

$$y' = \sqrt{x}y$$

$$\frac{1}{y}y'\,dx = \sqrt{x}\,dx$$

$$\int \frac{1}{y}y'\,dx = \int \sqrt{x}\,dx$$

$$\int \frac{1}{y}\,dy = \int x^{1/2}\,dx$$

$$\ln y = \frac{2}{3}x^{3/2} + C_1$$

$$y = e^{2x^{3/2}/3 + C_1}$$

$$y = e^{C_1}e^{2x^{3/2}/3} = Ce^{2x^{3/2}/3}$$

23. If the half life of the radioactive isotope Pu^{239} is 24,360 years, find the initial amount of the isotope if after 1000 years there are 2.1 grams. How many grams will there be after 10,000 years?

Solution:

Let y represent the mass (in grams) of the isotope after t years. Since the rate of decay is proportional to y, apply the Law of Experimental Decay to conclude that $y = y_0 e^{kt}$ where y_0 is the initial amount. Since the half life of Pu^{239} is 24,360 years, you have

$$y_0 e^{24,360k} = \frac{1}{2}y_0$$

$$e^{24,360k} = \frac{1}{2}$$

$$24,360k = \ln\frac{1}{2}$$

$$k = \frac{-\ln 2}{24,360}$$

Therefore, $y = y_0 e^{\left(\frac{-\ln 2}{24,360}\right)t}$. Since $y = 2.1$ when $t = 1000$, you have

$$y_0 e^{\left(\frac{-\ln 2}{24,360}\right)1000} = 2.1$$

$$y_0 = 2.1e^{\frac{1000\ln 2}{24,360}} \approx 2.16 \text{ grams.}$$

After 10,000 years the amount of the isotope remaining will be

$$y = 2.16e^{\left(\frac{-\ln 2}{24,360}\right)10,000} \approx 1.6 \text{ grams.}$$

29. Given that an initial investment of \$750 in a savings account in which interest is compounded continuously doubles in $7\frac{3}{4}$ years, find the amount in the account after 10 years and the amount after 25 years.

Solution:

Since the interest is compounded continuously, use the formula

$$A = Pe^{rt}$$

To determine the rate, use the fact that the investment doubles in $7\frac{3}{4}$ years.

$$2(750) = 750e^{r(7.75)}$$

$$2 = e^{7.75r}$$

$$\ln 2 = \ln e^{7.75r} = 7.75r$$

$$r = \frac{\ln 2}{7.75} \approx 0.0894 = 8.94\%$$

The amount after 10 years is given by

$$A = 750e^{\left(\frac{\ln 2}{7.75}\right)10} = \$1834.37$$

The amount after 25 years is given by

$$A = 750e^{\left(\frac{\ln 2}{7.75}\right)25} = \$7016.58$$

43. The management at a certain factory has found that a worker can produce at most 30 units in a day. The learning curve for the number of units N produced per day after a new employee has worked t days is given by

$$N(t) = 30(1 - e^{kt})$$

After 20 days on the job, a particular worker produced 19 units.

(a) Find the learning curve for this worker.

(b) How many days should pass before this worker is producing 25 units per day?

Solution:

(a) Since $N(20) = 19$, it follows that

$$19 = 30(1 - e^{20k})$$
$$30e^{20k} = 11$$
$$e^{20k} = \frac{11}{30}$$
$$k = \frac{\ln(11/30)}{20} \approx -0.0502.$$

Therefore,
$$N(t) = 30(1 - e^{-0.0502t}).$$

(b) To determine the time when the worker will be producing 25 units per day solve the equation

$$25 = 30(1 - e^{-0.0502t})$$
$$e^{-0.0502t} = \frac{1}{6}$$
$$t = \frac{-\ln 6}{-0.0502} \approx 36 \, \text{days}.$$

53. A thermometer is taken from a room at $72°$ F to the outdoors where the temperature is $20°$ F. Use Newton's Law of Cooling to determine the reading on the thermometer after 5 minutes, if the reading dropped to $48°$ F after 1 minute.

Solution:

Let y be the temperature reading of the thermometer at time t. From Newton's Law of Cooling, you know that the rate of change of y is proportional to the difference between y and $20°$. This can be written as

$$\frac{dy}{dt} = k(y - 20)$$

$$\left(\frac{1}{y - 20}\right)\frac{dy}{dt} = k$$

$$\int \frac{1}{y - 20}\, dy = \int k\, dt$$

$$\ln|y - 20| = kt + C$$

Using $y = 72$ when $t = 0$ yields $C = \ln 52$, which implies that

$$kt = \ln|y - 20| - \ln 52.$$

Since $y = 48$ when $t = 1$, you know that

$$k = \ln 28 - \ln 52 = \ln \tfrac{28}{52} = \ln \frac{7}{13}.$$

Therefore,

$$\ln|y - 20| = \ln \frac{7}{13}t + \ln 52$$

$$y - 20 = e^{[\ln(7/13)]t + \ln 52}$$

$$y = e^{\ln 52}e^{[\ln(7/13)]t} + 20 = 52e^{[\ln(7/13)]t} + 20.$$

After 5 minutes the reading on the thermometer is

$$y = 52e^{5\ln(7/13)} + 20 \approx 22.35°.$$

5.4 Differential Equations: Separation of Variables

9. Determine if $y = e^{-2x}$ is a solution to the differential equation $y^{(4)} - 16y = 0$.

Solution:

Since $y = e^{-2x}$, you have

$$y' = -2e^{-2x}$$
$$y'' = 4e^{-2x}$$
$$y''' = -8e^{-2x}$$
$$y^{(4)} = 16e^{-2x}.$$

Therefore,

$$y^{(4)} - 16y = 16e^{-2x} - 16(e^{-2x}) = 0$$

and the function is a solution to the differential equation.

19. It is known that $y = Ce^{kx}$ is a solution to the differential equation

$$\frac{dy}{dx} = 0.07y.$$

Is it possible to determine C or k from the information given? If so, find its value.

Solution:

Since $y = Ce^{kx}$ is a solution to the differential equation, begin by substituting the derivative of the solution into the differential equation.

$$y = Ce^{kx}$$
$$\frac{dy}{dx} = Ce^{kx}(k) = k\left(Ce^{ky}\right) = ky = 0.07y$$

Therefore, $k = 0.07$.

27. Verify that $y = C_1 \sin 3x + C_2 \cos 3x$ is a solution to the differential equation $y'' + 9y = 0$. Then find the particular solution satisfying the initial conditions $y = 2$ and $y' = 1$ when $x = \pi/6$.

Solution:

Since $y = C_1 \sin 3x + C_2 \cos 3x$, you have

$$y' = 3C_1 \cos 3x - 3C_2 \sin 3x$$
$$y'' = -9C_1 \sin 3x - 9C_2 \cos 3x.$$

Therefore,

$$y'' + 9y = -9C_1 \sin 3x - 9C_2 \cos 3x + 9(C_1 \sin 3x + C_2 \cos 3x) = 0$$

and the function is a solution to the given differential equation. Furthermore, since $y = 2$ and $y' = 1$ when $x = \pi/6$, you have

$$2 = C_1 \sin 3\left(\frac{\pi}{6}\right) + C_2 \cos 3\left(\frac{\pi}{6}\right) \quad \Longrightarrow \quad 2 = C_1(1) + C_2(0)$$

$$1 = 3C_1 \cos 3\left(\frac{\pi}{6}\right) - 3C_2 \sin 3\left(\frac{\pi}{6}\right) \quad \Longrightarrow \quad 1 = 3C_1(0) - 3C_2(1)$$

Therefore, $C_1 = 2$, $C_2 = -1/3$, and the particular solution is

$$y = 2 \sin 3x - \frac{1}{3} \cos 3x.$$

43. Solve the differential equation $(2 + x)y' = 3y$ by separation of variables.

Solution:

Begin by separating the variables as follows.

$$(2 + x)y' = 3y$$
$$(2 + x)\frac{dy}{dx} = 3y$$
$$\frac{1}{y} dy = \frac{3}{2 + x} dx.$$

By integration you obtain

$$\int \frac{1}{y} dy = 3 \int \frac{1}{2 + x} dx$$
$$\ln |y| = 3 \ln |2 + x| + \ln C$$
$$\ln y = \ln |C(2 + x)^3|$$
$$y = C(2 + x)^3.$$

51. Solve the differential equation $y(x+1) + y' = 0$ by separation of variables and find the particular solution satisfying $y(-2) = 1$.

Solution:

Begin by separating the variables to obtain

$$y(x+1) + y' = 0$$
$$\frac{dy}{dx} = -y(x+1)$$
$$\frac{dy}{y} = -(x+1)\,dx.$$

Integration yields

$$\int \frac{dy}{y} = -\int (x+1)\,dx$$
$$\ln|y| = -\frac{(x+1)^2}{2} + C_1$$
$$y = e^{C_1 - (x+1)^2/2} = Ce^{-(x+1)^2/2}.$$

Since $y = 1$ when $x = -2$, it follows that

$$1 = Ce^{-(-2+1)^2/2} = Ce^{-1/2} \quad \text{or} \quad C = e^{1/2}$$

Therefore, the particular solution is

$$y = e^{1/2}e^{-(x+1)^2/2} = e^{[-(x+1)^2/2 + 1/2]} = e^{-(x^2+2x)/2}.$$

73. Solve the homogeneous differential equation $y' = xy/(x^2 - y^2)$.

Solution:

Letting $y = vx$, yields

$$y' = \frac{xy}{x^2 - y^2}$$
$$v + x\frac{dv}{dx} = \frac{x(vx)}{x^2 - v^2x^2} = \frac{v}{1 - v^2}$$
$$x\frac{dv}{dx} = \frac{v}{1 - v^2} - v = \frac{v^3}{1 - v^2}$$
$$x\,dv = \frac{v^3}{1 - v^2}\,dx.$$

Separating variables, you obtain

$$\frac{1 - v^2}{v^3}\, dv = \frac{dx}{x}$$

$$\int \left(v^{-3} - \frac{1}{v} \right) dv = \int \frac{dx}{x}$$

$$\frac{v^{-2}}{-2} - \ln |v| = \ln |x| + \ln |C_1|$$

$$\frac{1}{-2v^2} = \ln |v| + \ln |x| + \ln |C_1| = \ln |C_1 vx|$$

$$\frac{x^2}{-2y^2} = \ln |C_1 y|$$

$$e^{-x^2/2y^2} = C_1 y \quad \Longrightarrow \quad y = Ce^{-x^2/2y^2}.$$

77. Solve the homogeneous differential equation

$$\left(x \sec \frac{y}{x} + y \right) dx - x\, dy = 0,$$

and find the particular solution satisfying the $y(1) = 0$.

Solution:

Letting $y = vx$, yields

$$\left(x \sec \frac{vx}{x} + vx \right) dx - x(v\, dx + x\, dv) = 0$$

$$x \sec v\, dx + vx\, dx - vx\, dx - x^2 dx = 0$$

$$x \sec v\, dx = x^2\, dv$$

$$\frac{x}{x^2}\, dx = \frac{dv}{\sec v}.$$

Integration yields

$$\int \frac{dx}{x} = \int \cos v\, dv$$

$$\ln |x| = \sin v + C_1 = \sin \frac{y}{x} + C_1.$$

Since $y = 0$ when $x = 1$, you have

$$0 = \sin (0) + C_1 = C_1$$

and it follows that

$$\ln |x| = \sin \frac{y}{x} \quad \Longrightarrow \quad x = e^{\sin (y/x)}.$$

85. The rate of decomposition of radioactive radium is proportional to the amount present at a given instant. Find the percentage of a present amount that remains after 25 years if the half-life of radioactive radium is 1600 years.

Solution:

Letting y represent the percentage amount of radioactive radium present at time t, you have the differential equation

$$y' = ky$$

and the general solution has the form

$$y = Ce^{kt} = C(e^k)^t.$$

Since $y = 100\% = 1$ when $t = 0$, it follows that

$$1 = Ce^0 = C.$$

Furthermore, $y = \frac{1}{2}$ when $t = 1600$. Hence

$$\frac{1}{2} = e^{k(1600)} = (e^k)^{1600} \quad \text{or} \quad e^k = \left(\frac{1}{2}\right)^{1/1600}$$

and you have

$$y = \left(\frac{1}{2}\right)^{t/1600}$$

Finally, when $t = 25$, you have

$$y = \left(\frac{1}{2}\right)^{25/1600} = \left(\frac{1}{2}\right)^{1/64}$$

$$\approx 0.989 = 98.9\% \text{ of the original amount.}$$

95. Find the orthogonal trajectories of the family of curves $y^2 = Cx^3$, and sketch several members of each family.

Solution:

First, solve for C in the given equation and obtain

$$C = \frac{y^2}{x^3}.$$

Then, differentiating $y^2 = Cx^3$ implicitly with respect to x and substituting the expression for C given above, you have

$$2yy' = 3Cx^2 = 3\left(\frac{y^2}{x^3}\right)x^2 = \frac{3y^2}{x}.$$

Therefore,

$$\frac{dy}{dx} = \frac{3y}{2x} \qquad \text{Slope of given family}$$

Since dy/dx represents the slope of the given family of curves at (x, y), it follows that the orthogonal family has the negative reciprocal slope, and you write

$$\frac{dy}{dx} = -\frac{2x}{3y} \qquad \text{Slope of orthogonal family}$$

Now, find the orthogonal family by separating variables and integrating to obtain

$$3 \int y\, dy = -2 \int x\, dx$$
$$\frac{3}{2}y^2 = -x^2 + K$$
$$2x^2 + 3y^2 = K.$$

5.5   Inverse Trigonometric Functions and Differentiation

9. Show that the slopes of the graphs of $f(x) = \sqrt{x-4}$ and $f^{-1}(x) = x^2 + 4$ are reciprocals at the points $(5,\,1)$ and $(1,\,5)$, respectively.

Solution:

For the function $f(x) = \sqrt{x-4}$, you have

$$f'(x) = \frac{1}{2\sqrt{x-4}} \quad \text{and} \quad f'(5) = \frac{1}{2\sqrt{5-4}} = \frac{1}{2}$$

For the function $f^{-1}(x) = x^2 + 4$, you have

$$(f^{-1})'(x) = 2x \quad \text{and} \quad (f^{-1})'(1) = 2$$

Therefore, $(f^{-1})'(1) = \dfrac{1}{f'(5)}$.

11. Find the derivative of $f(x) = 2\arcsin(x-1)$.

Solution:

$$f(x) = 2\arcsin(x-1)$$
$$f'(x) = 2\frac{1}{\sqrt{1-(x-1)^2}}(1) = \frac{2}{\sqrt{2x-x^2}}$$

21. Find the derivative of $h(t) = \sin(\arccos t)$.

Solution:

From the accompanying figure you know that

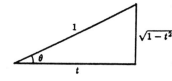

$$h(t) = \sin(\arccos t) = \sqrt{1-t^2}$$

Therefore,

$$h'(t) = \frac{1}{2}(1-t^2)^{-1/2}(-2t) = \frac{-t}{\sqrt{1-t^2}}.$$

25. Find the derivative of $f(x) = \dfrac{1}{2}\left(\dfrac{1}{2}\ln\dfrac{x+1}{x-1} + \arctan x\right)$.

Solution:

$$f(x) = \frac{1}{2}\left(\frac{1}{2}\ln\frac{x+1}{x-1} + \arctan x\right)$$

$$= \frac{1}{2}\left[\frac{1}{2}\ln(x+1) - \frac{1}{2}\ln(x-1) + \arctan x\right]$$

$$f'(x) = \frac{1}{2}\left[\frac{1}{2(x+1)} - \frac{1}{2(x-1)} + \frac{1}{1+x^2}\right]$$

$$= \frac{1}{2}\left[\frac{(x-1)-(x+1)}{2(1-x^2)} + \frac{1}{1+x^2}\right]$$

$$= \frac{1}{2}\left[\frac{-1}{x^2-1} + \frac{1}{1+x^2}\right]$$

$$= \frac{1}{2}\left[\frac{1}{1-x^2} + \frac{1}{1+x^2}\right] = \frac{1}{2}\left[\frac{1+x^2+1-x^2}{1-x^4}\right] = \frac{1}{1-x^4}$$

27. Find the derivative of $f(x) = x\arcsin x + \sqrt{1-x^2}$.

Solution:

$$f(x) = x\arcsin x + \sqrt{1-x^2}$$

$$f'(x) = x\left(\frac{1}{\sqrt{1-x^2}}\right) + \arcsin x + \frac{1}{2}(1-x^2)^{-1/2}(-2x)$$

$$= \frac{x}{\sqrt{1-x^2}} + \arcsin x + \frac{-x}{\sqrt{1-x^2}} = \arcsin x$$

31. Use a symbolic differentiation utility to find the linear approximation

$$P_1(x) = f(a) + f'(a)(x-a)$$

and quadratic approximation

$$P_2(x) = f(a) + f'(a)(x-a) + \tfrac{1}{2}f''(a)(x-a)^2$$

to the function $f(x) = \arcsin x$ when $x = 1/2$. Sketch a graph of the function and its linear and quadratic approximations.

Solution:

Begin by evaluating the function f and its first and second derivatives at $x = 1/2$.

$$f(x) \quad = \arcsin x \qquad\qquad f\left(\frac{1}{2}\right) \quad = \frac{\pi}{6}$$

$$f'(x) \quad = \frac{1}{\sqrt{1 - x^2}} \qquad f'\left(\frac{1}{2}\right) = \frac{2\sqrt{3}}{3}$$

$$f''(x) = \frac{x}{(1 - x^2)^{3/2}} \qquad f''\left(\frac{1}{2}\right) = \frac{4\sqrt{3}}{9}$$

Therefore,

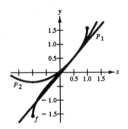

$$P_1(x) = \frac{\pi}{6} + \frac{2\sqrt{3}}{3}\left(x - \frac{1}{2}\right)$$

$$P_2(x) = \frac{\pi}{6} + \frac{2\sqrt{3}}{3}\left(x - \frac{1}{2}\right) + \frac{4\sqrt{3}}{9}\left(x - \frac{1}{2}\right)^2$$

The graphs of f, P_1, and P_2, produced by a graphing utility, are shown in the accompanying figure.

5.6   Inverse Trigonometric Functions: Integration and Completing the Square

1. Evaluate $\displaystyle\int_0^{1/6} \frac{1}{\sqrt{1 - 9x^2}}\, dx$.

Solution:

If you let $a = 1$ and $u = 3x$, then $du = 3\, dx$. Thus,

$$\int_0^{1/6} \frac{1}{\sqrt{1 - 9x^2}}\, dx = \frac{1}{3}\int_0^{1/6}\left(\frac{1}{\sqrt{1 - (3x)^2}}\right)(3)\, dx$$

$$= \frac{1}{3}\Big[\arcsin(3x)\Big]_0^{1/6}$$

$$= \frac{1}{3}\left[\arcsin\left(\frac{1}{2}\right) - \arcsin 0\right] = \frac{1}{3}\left(\frac{\pi}{6}\right) = \frac{\pi}{18}.$$

7. Evaluate $\displaystyle\int \frac{x^3}{x^2+1}\,dx$.

Solution:

Since the degree of the numerator is greater than the degree of the denominator, you must divide to obtain

$$\frac{x^3}{x^2+1} = x - \frac{x}{x^2+1}$$

Therefore,

$$
\begin{aligned}
\int \frac{x^3}{x^2+1}\,dx &= \int \left(x - \frac{x}{x^2+1} \right) dx \\
&= \int x\,dx - \frac{1}{2}\int \frac{2x}{x^2+1}\,dx \\
&= \frac{x^2}{2} - \frac{1}{2}\ln\left(x^2+1\right) + C \\
&= \frac{1}{2}x^2 - \frac{1}{2}\ln\left(x^2+1\right) + C.
\end{aligned}
$$

15. Evaluate $\displaystyle\int_0^{1/\sqrt{2}} \frac{\arcsin x}{\sqrt{1-x^2}}\,dx$.

Solution:

If you let $u = \arcsin x$, then $du = \dfrac{1}{\sqrt{1-x^2}}\,dx$. Thus,

$$
\begin{aligned}
\int_0^{1/\sqrt{2}} \frac{\arcsin x}{\sqrt{1-x^2}}\,dx &= \int_0^{1/\sqrt{2}} (\arcsin x)^1 \frac{1}{\sqrt{1-x^2}}\,dx \\
&= \left[\frac{(\arcsin x)^2}{2} \right]_0^{1/\sqrt{2}} \\
&= \frac{1}{2}\left\{ \left[\arcsin\left(\frac{1}{\sqrt{2}}\right) \right]^2 - [\arcsin(0)]^2 \right\} \\
&= \frac{1}{2}\left[\left(\frac{\pi}{4}\right)^2 - (0)^2 \right] = \frac{\pi^2}{32} \approx 0.308.
\end{aligned}
$$

23. Evaluate $\displaystyle\int \frac{e^{2x}}{4 + e^{4x}}\, dx$.

Solution:

If you let $a = 2$ and $u = e^{2x}$, then $du = 2e^{2x}\, dx$. Thus,

$$\int \frac{e^{2x}}{4 + e^{4x}}\, dx = \frac{1}{2} \int \frac{1}{2^2 + (e^{2x})^2}(2e^{2x})\, dx$$

$$= \frac{1}{2} \int \frac{du}{a^2 + u^2}$$

$$= \frac{1}{2a} \arctan \frac{u}{a} + C = \frac{1}{4} \arctan \frac{e^{2x}}{2} + C$$

29. Evaluate $\displaystyle\int \frac{2x}{x^2 + 6x + 13}\, dx$.

Solution:

$$\int \frac{2x}{x^2 + 6x + 13}\, dx$$

$$= \int \frac{(2x + 6) - 6}{x^2 + 6x + 13}\, dx$$

$$= \int \frac{2x + 6}{x^2 + 6x + 13}\, dx - \int \frac{6}{x^2 + 6x + 13}\, dx$$

$$= \int \frac{2x + 6}{x^2 + 6x + 13}\, dx - \int \frac{6}{(x^2 + 6x + 9) + 4}\, dx$$

$$= \int \frac{2x + 6}{x^2 + 6x + 13}\, dx - 6 \int \frac{1}{(x + 3)^2 + 2^2}\, dx$$

$$= \ln (x^2 + 6x + 13) - 3 \arctan \left(\frac{x + 3}{2} \right) + C$$

31. Evaluate $\displaystyle\int \frac{1}{\sqrt{-x^2 - 4x}}\, dx$.

Solution:

$$\int \frac{1}{\sqrt{-x^2 - 4x}}\, dx = \int \frac{1}{\sqrt{-(x^2 + 4x)}}\, dx$$

$$= \int \frac{1}{\sqrt{4 - (x^2 + 4x + 4)}}\, dx$$

$$= \int \frac{1}{\sqrt{2^2 - (x + 2)^2}}\, dx = \arcsin \left(\frac{x + 2}{2} \right) + C$$

35. Evaluate $\displaystyle\int_2^3 \frac{2x-3}{\sqrt{4x-x^2}}\,dx$.

Solution:

$$\int_2^3 \frac{2x-3}{\sqrt{4x-x^2}}\,dx = \int_2^3 \frac{(2x-4)+1}{\sqrt{4x-x^2}}\,dx$$

$$= \int_2^3 \frac{2x-4}{\sqrt{4x-x^2}}\,dx + \int_2^3 \frac{1}{\sqrt{4x-x^2}}\,dx$$

$$= -\int_2^3 (4x-x^2)^{-1/2}(4-2x)\,dx$$

$$+ \int_2^3 \frac{1}{\sqrt{4-(x^2-4x+4)}}\,dx$$

$$= -\int_2^3 (4x-x^2)^{-1/2}(4-2x)\,dx$$

$$+ \int_2^3 \frac{1}{\sqrt{2^2-(x-2)^2}}\,dx$$

$$= \left[-\frac{(4x-x^2)^{1/2}}{\frac{1}{2}} + \arcsin\left(\frac{x-2}{2}\right)\right]_2^3$$

$$= -2\sqrt{3}+\frac{\pi}{6}-(-4+0) = 4-2\sqrt{3}+\frac{\pi}{6} \approx 1.059$$

43. Evaluate $\displaystyle\int \sqrt{e^t-3}\,dt$.

Solution:

If $u = \sqrt{e^t-3}$, then $e^t = u^2+3$, $t = \ln(u^2+3)$, and $dt = 2u/(u^2+3)\,du$. Therefore,

$$\int \sqrt{e^t-3}\,dt = \int \frac{2u^2}{u^2+3}\,du$$

Now since the numerator and denominator are of equal degree, divide to obtain

$$2\int \frac{u^2}{u^2+3}\,du = 2\int \left[1-\frac{3}{u^2+3}\right]du$$

$$= 2\left[\int du - 3\int \frac{1}{(\sqrt{3})^2+u^2}\,du\right]$$

$$= 2\left[u - 3\left(\frac{1}{\sqrt{3}}\right)\arctan\left(\frac{u}{\sqrt{3}}\right)\right]+C$$

$$= 2\sqrt{e^t-3} - 2\sqrt{3}\arctan\left(\frac{\sqrt{e^t-3}}{\sqrt{3}}\right)+C$$

51. Find the area of the region bounded by the graphs of

$$y = \frac{1}{x^2 - 2x + 5}, \ y = 0, \ x = 1, \text{ and } x = 3.$$

Solution:

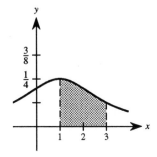

From the accompanying figure, you can see that the area is given by

$$A = \int_1^3 \frac{1}{x^2 - 2x + 5}\, dx$$

$$= \int_1^3 \frac{1}{(x^2 - 2x + 1) + 5 - 1}\, dx$$

$$= \int_1^3 \frac{1}{(x - 1)^2 + 2^2}\, dx$$

$$= \frac{1}{2} \arctan \frac{x - 1}{2}\Big]_1^3 = \frac{1}{2}\left[\arctan 1 - \arctan 0\right] = \frac{\pi}{8}$$

5.7   Hyperbolic Functions

3. Evaluate (a) csch (ln 2) and (b) coth (ln 5). If the functional value is not a rational number, give the answer to three decimal place accuracy.

Solution:

(a) $\text{csch}\,(\ln 2) = \dfrac{1}{\sinh\,(\ln 2)} = \dfrac{2}{e^{\ln 2} - e^{-\ln 2}} = \dfrac{2}{2 - \left(\frac{1}{2}\right)} = \dfrac{4}{3}$

(b) $\coth\,(\ln 5) = \dfrac{\cosh\,(\ln 5)}{\sinh\,(\ln 5)} = \dfrac{e^{\ln 5} + e^{-\ln 5}}{e^{\ln 5} - e^{-\ln 5}} = \dfrac{5 + \left(\frac{1}{5}\right)}{5 - \left(\frac{1}{5}\right)} = \dfrac{13}{12}$

19. If $y = \ln\left(\tanh\dfrac{x}{2}\right)$, find y' and simplify.

Solution:

$$y = \ln\left(\tanh\frac{x}{2}\right)$$

$$y' = \frac{1}{\tanh(x/2)}\left[\frac{1}{2}\operatorname{sech}^2\left(\frac{x}{2}\right)\right]$$

$$= \frac{1}{2}\left[\frac{\cosh(x/2)}{\sinh(x/2)}\right]\left[\frac{1}{\cosh^2(x/2)}\right]$$

$$= \frac{1}{2\sinh(x/2)\cosh(x/2)} = \frac{1}{\sinh x} = \operatorname{csch} x$$

25. If $y = x^{\cosh x}$, find y' and simplify.

Solution:

Using logarithmic differentiation yields

$$y = x^{\cosh x}$$

$$\ln y = \ln\left(x^{\cosh x}\right) = (\cosh x)(\ln x)$$

$$\frac{y'}{y} = (\cosh x)\left(\frac{1}{x}\right) + (\ln x)(\sinh x)$$

$$y' = y\left[\frac{\cosh x}{x} + (\ln x)(\sinh x)\right]$$

$$= \frac{y}{x}[\cosh x + x(\sinh x)\ln x].$$

43. Evaluate $\displaystyle\int \frac{\operatorname{csch}(1/x)\coth(1/x)}{x^2}\,dx$

Solution:

If you let $u = \dfrac{1}{x}$, then $du = -\dfrac{1}{x^2}\,dx$ and

$$\int \frac{\operatorname{csch}(1/x)\coth(1/x)}{x^2}\,dx = -\int \operatorname{csch}\frac{1}{x}\coth\frac{1}{x}\left(-\frac{1}{x^2}\right)dx$$

$$= -\int \operatorname{csch} u\,\coth u\,du$$

$$= \operatorname{csch} u + C = \operatorname{csch}\frac{1}{x} + C$$

57. If $y = \sinh^{-1}(2x) - \sqrt{1 + 4x^2}$, find y' and simplify.

Solution:

$$y = 2x \sinh^{-1}(2x) - \sqrt{1 + 4x^2}$$

$$y' = 2x \left[\frac{1}{\sqrt{1 + (2x)^2}} \right](2) + 2\sinh^{-1}(2x) - \frac{1}{2}(1 + 4x^2)^{-1/2}(8x)$$

$$= \frac{4x}{\sqrt{1 + 4x}} + 2\sinh^{-1}(2x) - \frac{4x}{\sqrt{1 + 4x^2}} = 2\sinh^{-1}(2x)$$

59. Find dy/dx for the tractrix $y = a\,\text{sech}^{-1}(x/a) - \sqrt{a^2 - x^2}$.

Solution:

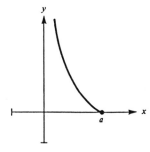

Note that the domain of this function restricts x to the interval $(0, a)$ where $0 < a$.

$$y = a\,\text{sech}^{-1}\left(\frac{x}{a}\right) - \sqrt{a^2 - x^2}$$

$$\frac{dy}{dx} = a\left[\frac{-1}{|x/a|\sqrt{(1 - (x/a)^2}} \right]\left(\frac{1}{a}\right) - \frac{1}{2}(a^2 - x^2)^{-1/2}(-2x)$$

$$= \frac{-1}{|x/a|\sqrt{(a^2 - x^2)/a^2}} + \frac{x}{\sqrt{a^2 - x^2}}$$

$$= \frac{-a^2}{x\sqrt{a^2 - x^2}} + \frac{x}{\sqrt{a^2 - x^2}}$$

(Note that the absolute value signs can be deleted since $0 < x$.)

$$\frac{dy}{dx} = \frac{-a^2 + x^2}{x\sqrt{a^2 - x^2}} = \frac{-(a^2 - x^2)}{x\sqrt{a^2 - x^2}} = \frac{-\sqrt{a^2 - x^2}}{x}$$

61. Evaluate $\displaystyle\int \frac{1}{\sqrt{1 + e^{2x}}}\,dx$.

Solution:

If you let $u = e^x$, then $du = e^x\,dx$ and you have

$$\int \frac{1}{\sqrt{1 + e^{2x}}}\,dx = \int \frac{e^x}{e^x\sqrt{1 + (e^x)^2}}\,dx = \int \frac{du}{u\sqrt{1 + u^2}}$$

$$= -\ln\left(\frac{1 + \sqrt{1 + e^{2x}}}{e^x}\right) + C = -\text{csch}^{-1}(e^x) + C.$$

63. Evaluate $\displaystyle\int \frac{1}{\sqrt{x}\sqrt{1+x}}\,dx$.

Solution:

If you let $u = \sqrt{x}$, then $du = \dfrac{1}{2\sqrt{x}}\,dx$ and you have

$$\int \frac{1}{\sqrt{x}\sqrt{1+x}}\,dx = 2\int \frac{1}{\sqrt{1+(\sqrt{x})^2}}\left(\frac{1}{2\sqrt{x}}\right)dx$$
$$= 2\int \frac{du}{\sqrt{1+u^2}}$$
$$= 2\sinh^{-1}\sqrt{x} + C$$
$$= 2\ln(\sqrt{x}+\sqrt{1+x}) + C.$$

Review Exercises for Chapter 5

9. Evaluate $\displaystyle\int_0^{\pi/3} \sec\theta\,d\theta$.

Solution:

$$\int_0^{\pi/3} \sec\theta\,d\theta = \Big[\ln|\sec\theta + \tan\theta|\Big]_0^{\pi/3}$$
$$= \ln(2+\sqrt{3}) - \ln(1+0) = \ln(2+\sqrt{3})$$

21. Evaluate $\log_5(\frac{1}{5})$ without using a calculator.

Solution:

$$\log_5 \frac{1}{5} = \log_5 5^{-1} = (-1)\log_5 5 = -1$$

33. Find the first derivative of the function $y = x^{2x+1}$.

Solution:

Taking the natural logarithm of both members of the equation yields
$$\ln y = \ln x^{2x+1} = (2x+1)\ln x.$$

Differentiating implicitly yields

$$\frac{1}{y}\frac{dy}{dx} = (2x+1)\left(\frac{1}{x}\right) + (\ln x)2$$

$$\frac{dy}{dx} = y\left(\frac{2x+1}{x} + 2\ln x\right) = x^{2x+1}\left(\frac{2x+1}{x} + 2\ln x\right)$$

41. Evaluate the integral $\displaystyle\int x(3^{x^2})\,dx$.

Solution:

If you let $u = x^2$, then $du = 2x\,dx$. Thus,

$$\int x(3^{x^2})\,dx = \frac{1}{2}\int (3^{x^2})(2x)\,dx$$

$$= \frac{1}{2}\int 3^u\,du$$

$$= \frac{1}{2}\left(\frac{1}{\ln 3}\right)3^u + C = \frac{1}{2\ln 3}3^{x^2} + C$$

45. How large a deposit, at 7% interest compounded continuously, must be made to obtain a balance of $10,000$ in 15 years?

Solution:

Using the formula for continuous interest yields

$$A = Pe^{rt}$$

$$10,000 = Pe^{(0.07)(15)}$$

$$P = \frac{10,000}{e^{1.05}} \approx \$3499.38$$

55. Solve the differential equation $\dfrac{dy}{dx} = \dfrac{x^2 + 3}{x}$.

Solution:

$$\frac{dy}{dx} = \frac{x^2 + 3}{x}$$

$$dy = \frac{x^2 + 3}{x}\,dx$$

$$\int dy = \int \frac{x^2 + 3}{x}\,dx$$

$$y = \int \left(x + \frac{3}{x}\right)\,dx$$

$$y = \frac{1}{2}x^2 + 3\ln|x| + C$$

63. Solve the differential equation $\dfrac{dy}{dx} = \dfrac{x^2 + y^2}{2xy}$.

Solution:

Letting $y = vx$, yields

$$\frac{dy}{dx} = \frac{x^2 + y^2}{2xy}$$

$$v + x\frac{dv}{dx} = \frac{x^2 + (vx)^2}{2x(vx)} = \frac{1 + v^2}{2v}$$

$$x\frac{dv}{dx} = \frac{1 + v^2}{2v} - v = \frac{1 - v^2}{2v}.$$

Separating variables, you obtain

$$\frac{2v}{1 - v^2}\,dv = \frac{1}{x}\,dx$$

$$-\int \frac{1}{1 - v^2}(-2v)\,dv = \frac{1}{x}\,dx$$

$$-\ln|(1 - v^2)| = \ln|x| + \ln|C|$$

$$0 = \ln|Cx(1 - v^2)|$$

$$1 = Cx\left[1 - \left(\frac{y}{x}\right)^2\right]$$

$$x = C(x^2 - y^2)$$

75. Find dy/dx given $y = \tan(\arcsin x)$.

Solution:

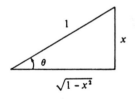

Begin by sketching a triangle to represent θ, "the angle whose sine is x." This yields

$$\theta = \arcsin x$$

$$y = \tan(\arcsin x) = \tan \theta = \frac{x}{\sqrt{1 - x^2}}.$$

Therefore,

$$\frac{dy}{dx} = \frac{\sqrt{1 - x^2}(1) - x(\frac{1}{2})(1/\sqrt{1 - x^2})(-2x)}{1 - x^2}$$

$$= \frac{(1 - x^2) + x^2}{(1 - x^2)\sqrt{1 - x^2}} = (1 - x^2)^{-3/2}$$

81. Find dy/dx given $y = 2x - \cosh \sqrt{x}$.

Solution:

$$y = 2x - \cosh \sqrt{x}$$

$$\frac{dy}{dx} = 2 - \sinh \sqrt{x} \left(\frac{1}{2\sqrt{x}}\right) = 2 - \frac{\sinh \sqrt{x}}{2\sqrt{x}}$$

85. Evaluate $\displaystyle\int \frac{1}{e^{2x} + e^{-2x}}\, dx$.

Solution:

If you let $u = e^{2x}$, then $du = 2e^{2x}\, dx$. Thus,

$$\int \frac{1}{e^{2x} + e^{-2x}}\, dx = \int \left(\frac{1}{e^{2x} + e^{-2x}}\right)\left(\frac{e^{2x}}{e^{2x}}\right) dx$$

$$= \int \frac{e^{2x}}{e^{4x} + 1}\, dx$$

$$= \frac{1}{2} \int \left[\frac{1}{1 + (e^{2x})^2}\right](2e^{2x})\, dx$$

$$= \frac{1}{2} \int \frac{du}{1 + u^2} = \frac{1}{2} \arctan(e^{2x}) + C$$

6 APPLICATIONS OF INTEGRATION

6.1 Area of a Region between Two Curves

15. Sketch the region bounded by the graphs of $f(x) = x^2 + 2x + 1$ and $g(x) = 3x + 3$ and find the area of this region by means of a definite integral.

Solution:

The points of intersection f and g are found by solving

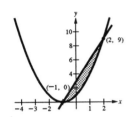

$$f(x) = g(x)$$
$$x^2 + 2x + 1 = 3x + 3$$
$$x^2 - x - 2 = 0$$
$$(x - 2)(x + 1) = 0 \implies x = -1, 2$$

Since $x^2 + 2x + 1 \leq 3x + 3$ for $-1 \leq x \leq 2$, you have

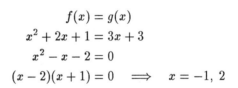

$$\text{area} = \int_{-1}^{2} [(3x + 3) - (x^2 + 2x + 1)]dx$$
$$= \int_{-1}^{2} (-x^2 + x + 2)dx$$
$$= \left[\frac{-x^3}{3} + \frac{x^2}{2} + 2x \right]_{-1}^{2}$$
$$= \left(\frac{-8}{3} + 2 + 4 \right) - \left(\frac{1}{3} + \frac{1}{2} - 2 \right)$$
$$= \frac{-16 + 12 + 24 - 2 - 3 + 12}{6} = \frac{27}{6} = \frac{9}{2}$$

29. Sketch the region bounded by the graphs of
$f(y) = y^2 + 1$, $g(y) = 0$, $y = -1$, and $y = 2$ and find the area of
this region by means of a definite integral.

Solution:

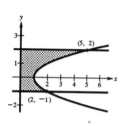

Using horizontal representative rectangles, you have

$$\text{area} = \int_{-1}^{2} (y^2 + 1)dy = \left[\frac{y^3}{3} + y \right]_{-1}^{2}$$

$$= \left(\frac{8}{3} + 2 \right) - \left(\frac{-1}{3} - 1 \right)$$

$$= \frac{8 + 6 + 1 + 3}{3} = 6.$$

35. Sketch the region bounded by the graphs of $f(x) = 2\sin x$ and
$g(x) = \tan x$ on the interval $-\pi/3 \leq x \leq \pi/3$ and find the area
of the region.

Solution:

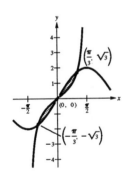

Since f and g are symmetric to the origin, the area of the
region bounded by their graphs for $-\pi/3 \leq x \leq \pi/3$ is twice the
area of the region bounded by their graphs for $0 \leq x \leq \pi/3$.
Since $\tan x \leq 2\sin x$ for $0 \leq x \leq \pi/3$, you have

$$\text{area} = 2\int_{0}^{\pi/3} (2\sin x - \tan x)dx$$

$$= 2\left[-2\cos x + \ln|\cos x| \right]_{0}^{\pi/3}$$

$$= 2\left[-2\left(\frac{1}{2} \right) + \ln\left(\frac{1}{2} \right) - (-2) \right] = 2(1 - \ln 2) \approx 0.614$$

43. Use integration to find the area of the triangle with vertices
$(0,0)$, $(a,0)$, and (b,c).

Solution:

From the accompanying figure, observe that the triangular
region is bounded by $y = (c/b)x$, $y = [c/(b - a)](x - a)$, and

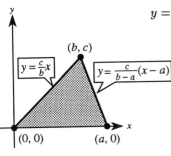

$y = 0$. Therefore, the area is given by

$$\text{area} = \int_0^b \frac{c}{b}x \, dx + \int_b^a \left(\frac{c}{b-a}\right)(x-a)dx$$

$$= \frac{c}{b}\left[\frac{x^2}{2}\right]_0^b + \left(\frac{c}{b-a}\right)\left[\frac{x^2}{2} - ax\right]_b^a$$

$$= \frac{1}{2}bc + \left(\frac{c}{b-a}\right)\left[\left(\frac{a^2}{2} - a^2\right) - \left(\frac{b^2}{2} - ab\right)\right]$$

$$= \frac{1}{2}bc + \frac{1}{2}(a-b)c = \frac{1}{2}ac$$

49. Find the area of the region bounded by the graph of $f(x) = x^3$ and the tangent line to the graph at the point $(1,1)$.

Solution:

Since $f(x) = x^3$, $f'(x) = 3x^2$, and $f'(1) = 3$ (slope of the tangent line), the equation of the tangent line to the graph of f is given by

$$y - 1 = 3(x - 1)$$
$$y = 3x - 2.$$

The x-coordinates of the points of intersection of the tangent line and the function of are the solutions to the equation

$$x^3 = 3x - 2$$
$$x^3 - 3x + 2 = 0$$
$$(x - 1)(x^2 + x - 2) = 0$$
$$(x - 1)^2(x + 2) = 0 \quad \Longrightarrow \quad x = -2, 1$$

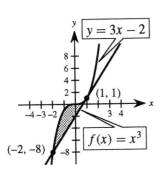

Therefore the points of intersection are given by $(1,1)$ and $(-2,-8)$. (See the accompanying figure.)

$$\text{area} = \int_{-2}^1 [x^3 - (3x - 2)]dx$$

$$= \int_{-2}^1 (x^3 - 3x + 2)dx$$

$$= \left[\frac{1}{4}x^4 - \frac{3}{2}x^2 + 2x\right]_{-2}^1$$

$$= \left(\frac{1}{4} - \frac{3}{2} + 2\right) - (4 - 6 - 4) = \frac{27}{4}$$

63. Find the consumer surplus and producer surplus of the demand function $p_1 = 50 - 0.5x$ and supply function $p_2 = 0.125x$. The consumer surplus and producer surplus are represented by the areas shown in the accompanying figure.

Solution:

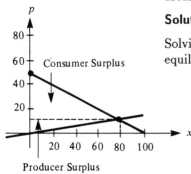

Solving the equations simultaneously yields the point of equilibrium to be $(80, 10)$. Therefore,

$$\text{Consumer Surplus} = \int_0^{80} [(50 - 0.5x) - 10]\,dx$$

$$= \left[40x - 0.25x^2\right]_0^{80} = 1600$$

$$\text{Producer Surplus} = \int_0^{80} (10 - 0.125x)\,dx$$

$$= \left[10x - 0.0625x^2\right]_0^{80} = 400$$

6.2   Volume: The Disc Method

7. Find the volume of the solid generated by revolving the region bounded by $y = x^2$ and $y = x^3$ about the x-axis.

Solution:

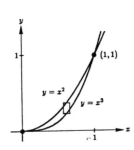

From the accompanying figure you have

$$R(x) = x^2 \qquad \text{Outer radius}$$
$$r(x) = x^3 \qquad \text{Inner radius.}$$

Now, integrating between 0 and 1, you have

$$V = \pi \int_a^b ([R(x)]^2 - [r(x)]^2)\,dx$$

$$= \pi \int_0^1 (x^4 - x^6)\,dx$$

$$= \pi \left[\frac{x^5}{5} - \frac{x^7}{7}\right]_0^1 = \pi \left(\frac{1}{5} - \frac{1}{7}\right) = \frac{2\pi}{35}$$

13. Find the volume of the solid generated by revolving the region
bounded by $y = \sqrt{x}$, $y = 0$, and $x = 4$ about:
(a)   the x-axis   (b)   the y-axis
(c)   the line $x = 4$ (d)   the line $x = 6$

Solution:

(a)   From the accompanying figure, you have
$$R(x) = y = \sqrt{x}.$$

$$V = \pi \int_a^b \left([R(x)]^2 - [r(x)]^2\right) dx$$

$$= \pi \int_0^4 \left(\sqrt{x}\right)^2 dx$$

$$= \pi \int_0^4 x\, dx = \pi \left[\frac{x^2}{2}\right]_0^4 = \frac{16\pi}{2} = 8\pi$$

(b)   From the accompanying figure, you have
$$R(y) = 4 \qquad \text{Outer radius}$$
$$r(y) = y^2 \qquad \text{Inner radius.}$$

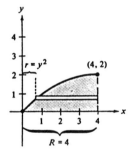

$$V = \pi \int_a^b \left([R(y)]^2 - [r(y)]^2\right) dy$$

$$= \pi \int_0^2 [4^2 - (y^2)^2]\, dy$$

$$= \pi \int_0^2 (16 - y^4)\, dy$$

$$= \left[16y - \frac{y^5}{5}\right]_0^2 = \pi\left[32 - \frac{32}{5}\right] = \frac{128\pi}{5}$$

(c)   From the accompanying figure, you have
$$R(y) = 4 - y^2.$$

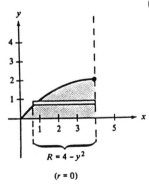

$$V = \pi \int_a^b \left([R(y)]^2 - [r(y)]^2\right) dy$$

$$= \pi \int_0^2 (4 - y^2)^2\, dy$$

$$= \pi \int_0^2 (16 - 8y^2 + y^4)\, dy$$

$$= \pi\left[16y - \frac{8y^3}{3} + \frac{y^5}{5}\right]_0^2 = \pi\left[32 - \frac{64}{3} + \frac{32}{5}\right] = \frac{256\pi}{15}$$

(d)   From the accompanying figure, you have
$$R(y) = 6 - y^2 \qquad \text{Outer radius}$$
$$r(y) = 2 \qquad \text{Inner radius.}$$

$$V = \pi \int_a^b \left([R(y)]^2 - [r(y)]^2\right) dy$$

$$= \pi \int_0^2 \left[(6 - y^2)^2 - 2^2\right] dy$$

$$= \pi \int_0^2 (32 - 12y^2 + y^4)\, dy$$

$$= \pi \left[32y - 4y^3 + \frac{y^5}{5}\right]_0^2 = \pi \left[64 - 32 + \frac{32}{5}\right] = \frac{192\pi}{5}$$

15. Find the volume of the solid formed by revolving the region
bounded by $y = x^2$ and $y = 4x - x^2$ about:
(a)   the x-axis   (b)   the line $y = 6$

Solution:

The points of intersection of the graphs of the two function are
$(0, 0)$ and $(2, 4)$.

(a)   From the accompanying figure, you have
$$R(x) = 4x - x^2 \qquad \text{Outer radius}$$
$$r(x) = x^2 \qquad \text{Inner radius.}$$

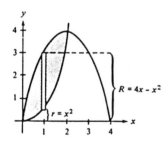

$$V = \pi \int_a^b \left([R(x)]^2 - [r(x)]^2\right) dx$$

$$= \pi \int_0^2 \left[(4x - x^2)^2 - (x^2)^2\right] dx$$

$$= \pi \int_0^2 (16x^2 - 8x^3)\, dx$$

$$= \pi \left[\frac{16}{3}x^3 - 2x^4\right]_0^2 = 8\pi \left[\frac{16}{3} - 4\right] = \frac{32\pi}{3}$$

(b)   From the accompanying figure, you have
$$R(x) = 6 - (4x - x^2) \qquad \text{Outer radius}$$
$$r(x) = 6 - x^2 \qquad \text{Inner radius.}$$

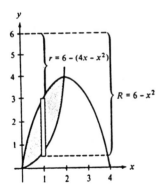

$$V = \pi \int_a^b ([R(x)]^2 - [r(x)]^2)\,dx$$

$$= \pi \int_0^2 [(6 - x^2)^2 - (6 - 4x + x^2)^2]\,dx$$

$$= 8\pi \int_0^2 (x^3 - 5x^2 + 6x)\,dx$$

$$= 8\pi \left[\frac{x^4}{4} - \frac{5}{3}x^3 + 3x^2 \right]_0^2$$

$$= 32\pi \left(1 - \frac{10}{3} + 3 \right) = \frac{64\pi}{3}$$

29. Find the volume of the solid generated by revolving the region bounded by the graphs of $y = e^{-x}$, $y = 0$, $x = 0$, and $x = 1$ about the x-axis.

Solution:

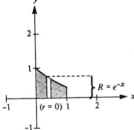

$$V = \pi \int_0^1 (e^{-x})^2\,dx$$

$$= \pi \int_0^1 e^{-2x}\,dx = \left[-\frac{\pi}{2}e^{-2x} \right]_0^1 = \frac{\pi}{2}\left(1 - \frac{1}{e^2} \right) \approx 1.358$$

39. Use the Disc Method to verify that the volume of a sphere of radius r is $\frac{4}{3}\pi r^3$.

Solution:

Let $y = \sqrt{r^2 - x^2}$ and let the region bounded by $y = \sqrt{r^2 - x^2}$ and $y = 0$ be revolved about the x-axis. The resulting solid of revolution is a sphere with volume

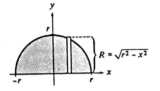

$$V = \pi \int_{-r}^{r} (\sqrt{r^2 - x^2})^2\,dx$$

$$= \pi \int_{-r}^{r} (r^2 - x^2)\,dx$$

$$= \pi \left[r^2 x - \frac{x^3}{3} \right]_{-r}^{r}$$

$$= \pi \left[\left(r^3 - \frac{r^3}{3} \right) - \left(-r^3 + \frac{r^3}{3} \right) \right] = \frac{4\pi r^3}{3}.$$

45. The tank on a water tower is a sphere of radius 50 ft. Determine the depth of the water when the tank is filled to one-fourth and three-fourth its total capacity.

Solution:

The total volume of the sphere is

$$V = \frac{4}{3}\pi r^3 = \frac{4\pi(50)^3}{3} = \frac{500,000\pi}{3} \text{ ft}^3$$

The volume of the portion filled with water is

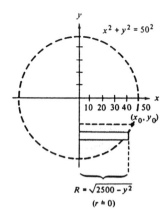

$$\frac{1}{4}V = \frac{125,000\pi}{3}$$

$$= \pi \int_{-50}^{y_0} (\sqrt{2500 - y^2})^2 \, dy$$

$$= \pi \int_{-50}^{y_0} (2500 - y^2) \, dy$$

$$= \pi \left[2500y - \frac{y^3}{3}\right]_{-50}^{y_0}$$

$$= \pi \left[\left(2500y_0 - \frac{y_0^3}{3}\right) - \left(-125,000 + \frac{125,000}{3}\right)\right]$$

$$= \pi \left[2500y_0 - \frac{y_0^3}{3} + \frac{250,000}{3}\right]$$

Simplifying yields the equation

$$y_0^3 - 7500y_0 - 125,000 = 0.$$

Using Newton's Method with an initial guess of $y_0 = -20$ yields the approximate root $y_0 = -17.36$. When the tank is one-fourth full the approximate depth of the water is $[-17.36 - (-50)] = 32.64$ feet. By symmetry, the tank is three-fourth full when the depth of the water is 82.64 feet.

53. The base of a solid is bounded by $y = x^3$, $y = 0$, and $x = 1$. Find the volume of the solid for the following cross sections (taken perpendicular to the y- axis).

(a) squares

(b) semicircles

(c) equilateral triangles

(d) trapezoids for which $h = b_1 = \frac{1}{2}b_2$, where b_1 and b_2 are the lengths of the upper and lower bases, respectively

(e) semiellipses whose heights are twice the length of their bases

Solution:

The base of the solid is shown in the accompanying figure. Since the cross sections are taken perpendicular to the y-axis, the base of each cross section is given by $(1 - x) = (1 - \sqrt[3]{y})$.

(a) The cross sections are squares whose sides are given by

$$s = (1 - \sqrt[3]{y}).$$

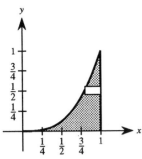

y

Base of Cross Section $= 1 - \sqrt[3]{y}$

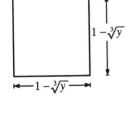

$1 - \sqrt[3]{y}$

Thus,

$$A(y) = s^2 = (1 - \sqrt[3]{y})^2$$

and

$$V = \int_0^1 (1 - \sqrt[3]{y})^2\, dy$$

$$= \int_0^1 (1 - 2y^{1/3} + y^{2/3})\, dy$$

$$= \left[y - \frac{2y^{4/3}}{\frac{4}{3}} + \frac{y^{5/3}}{\frac{5}{3}} \right]_0^1 = 1 - \frac{3}{2} + \frac{3}{5} = \frac{1}{10}.$$

(b) The cross sections are semicircles whose radii are given by

$$r = \left(\frac{1}{2}\right)(1 - \sqrt[3]{y}).$$

$1 - \sqrt[3]{y}$

Thus,

$$A(y) = \left(\frac{1}{2}\right)\pi\left[\left(\frac{1}{2}\right)(1 - \sqrt[3]{y})\right]^2 = \frac{\pi}{8}(1 - \sqrt[3]{y})^2$$

and

$$V = \frac{\pi}{8}\int_0^1 (1 - \sqrt[3]{y})^2\, dy$$

$$= \frac{\pi}{8}\left(\frac{1}{10}\right) = \frac{\pi}{80}. \qquad \text{From part (a)}$$

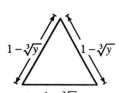

(c) The cross sections are equilateral triangles whose sides are given by

$$1 - \sqrt[3]{y}.$$

Thus,

$$A(y) = \frac{1}{2}bh = \frac{1}{2}(1 - \sqrt[3]{y})\left[\frac{\sqrt{3}}{2}(1 - \sqrt[3]{y})\right] = \frac{\sqrt{3}}{4}(1 - \sqrt[3]{y})^2$$

and

$$V = \frac{\sqrt{3}}{4}\int_0^1 (1 - \sqrt[3]{y})^2 \, dy$$

$$= \frac{\sqrt{3}}{4}\left(\frac{1}{10}\right) = \frac{\sqrt{3}}{40}. \qquad \text{From part (a)}$$

(d) The cross sections are trapezoids for which $h = b_1 = b_2/2$, where

$$b_2 = 1 - \sqrt[3]{y}.$$

Thus,

$$A(y) = \left(\frac{b_1 + b_2}{2}\right)h = \left(\frac{3}{4}\right)(b_2)\left(\frac{1}{2}\right)(b_2) = \frac{3}{8}(1 - \sqrt[3]{y})^2$$

and

$$V = \frac{3}{8}\int_0^1 (1 - \sqrt[3]{y})^2 \, dy$$

$$= \frac{3}{8}\left(\frac{1}{10}\right) = \frac{3}{80}. \qquad \text{From part (a)}$$

(e) The cross sections are semiellipses whose heights are twice the lengths of their bases. Thus $a = 2b$, where

$$b = 1 - \sqrt[3]{y}.$$

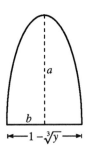

Thus,

$$A(y) = \left(\frac{1}{2}\right)\pi(a)\left(\frac{b}{2}\right) = \frac{\pi}{2}b^2 = \frac{\pi}{2}(1 - \sqrt[3]{y})^2$$

and

$$V = \frac{\pi}{2}\int_0^1 (1 - \sqrt[3]{y})^2 \, dy$$

$$= \frac{\pi}{2}\left(\frac{1}{10}\right) = \frac{\pi}{20}. \qquad \text{From part (a)}$$

55. Two planes cut to the axis of a right circular cylinder to form a wedge. One plane is perpendicular to the axis of the cylinder and the second makes an angle of θ degrees with the first.

(a) Find the volume of the wedge if $\theta = 45°$.

(b) Find the volume of the wedge for an arbitrary but fixed angle $\theta = \theta_0$. Assuming the cylinder has sufficient length, how does the volume of the wedge change as θ_0 increases in the interval $0 < \theta_0 < 90°$?

Solution:

(a) Since $\theta = 45°$, the cross sections are isosceles right triangles for which the base and height are equal. Thus,

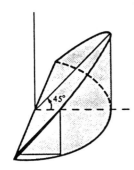

$$A(x) = \frac{1}{2}bh$$

$$= \frac{1}{2}(\sqrt{r^2 - x^2})(\sqrt{r^2 - x^2}) = \frac{1}{2}(r^2 - x^2)$$

and

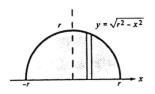
$$y = \sqrt{r^2 - x^2}$$

$$V = \frac{1}{2}\int_{-r}^{r}(r^2 - x^2)\,dx$$

$$= \frac{1}{2}\left[r^2 x - \frac{x^3}{3}\right]_{-r}^{r}$$

$$= \frac{1}{2}\left[\left(r^3 - \frac{r^3}{3}\right) - \left(-r^3 + \frac{r^3}{3}\right)\right] = \frac{2r^3}{3} \text{ in}^3.$$

(b) When $\theta = \theta_0$ ($0 < \theta_0 < 90°$), the cross sections are right triangles with base $\sqrt{r^2 - x^2}$ and height $\sqrt{r^2 - x^2}\tan\theta_0$. Thus,

$$A(x) = \frac{1}{2}bh$$

$$= \frac{1}{2}(\sqrt{r^2 - x^2})(\sqrt{r^2 - x^2})\tan\theta_0$$

$$= \frac{1}{2}(r^2 - x^2)\tan\theta_0.$$

Since the integration is with respect to x, you can use the result of part (a) and determine that the volume is

$$V = \frac{2r^3}{3}(\tan\theta_0).$$

As θ_0 increases, the volume of the wedge increases.

6.3   Volume: The Shell Method

7. Use the Shell Method to find the volume generated by revolving the region bounded by $y = x^2$ and $y = 4x - x^2$ about the y-axis.

Solution:

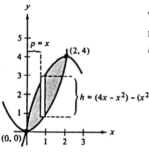

The distance from the center of the rectangle to the axis of revolution is $p(x) = x$, and the height of the rectangle is $h(x) = (4x - x^2) - (x^2)$.

$$V = 2\pi \int_a^b p(x)h(x)\,dx$$

$$= 2\pi \int_0^2 x[(4x - x^2) - (x^2)]\,dx$$

$$= 2\pi \int_0^2 x(4x - 2x^2)\,dx$$

$$= 4\pi \int_0^2 (2x^2 - x^3)\,dx$$

$$= 4\pi \left[\frac{2x^3}{3} - \frac{x^4}{4}\right]_0^2 = 4\pi \left[\frac{16}{3} - 4\right] = \frac{16\pi}{3}$$

13. Use the Shell Method to find the volume generated by revolving the region bounded by $y = x$, $y = 0$, and $x = 2$ about the x-axis.

Solution:

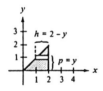

The distance from the center of the rectangle to the axis of revolution is $p(y) = y$, and the height of the rectangle is $h(y) = 2 - y$.

$$V = 2\pi \int_a^b p(y)h(y)\,dy$$

$$= 2\pi \int_0^2 y(2 - y)\,dy$$

$$= 2\pi \int_0^2 (2y - y^2)\,dy = 2\pi \left[y^2 - \frac{y^3}{3}\right]_0^2 = 2\pi \left(4 - \frac{8}{3}\right) = \frac{8\pi}{3}$$

17. Use the Shell Method to find the volume generated by revolving the region bounded by $y = x^2$ and $y = 4x - x^2$ about the line $x = 4$.

Solution:

The distance from the center of the rectangle to the axis of revolution is $p(x) = 4 - x$, and the height of the rectangle is $h(x) = (4x - x^2) - (x^2)$.

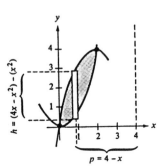

$$V = 2\pi \int_a^b p(x)h(x)\,dx$$

$$= 2\pi \int_0^2 (4 - x)[(4x - x^2) - (x^2)]\,dx$$

$$= 2\pi \int_0^2 (4 - x)(4x - 2x^2)\,dx$$

$$= 4\pi \int_0^2 (8x - 6x^2 + x^3)\,dx$$

$$= 4\pi \left[4x^2 - 2x^3 + \frac{x^4}{4} \right]_0^2 = 4\pi[16 - 16 + 4] = 16\pi$$

21. Use the Disc or Shell Method to find the volume of the solid generated by revolving the region bounded by $y = x^3$, $y = 0$, and $x = 2$ about:
(a)   the x-axis   (b)   the y-axis
(c)   the line $x = 4$ (d)   the line $y = 8$

Solution:

(a)   Disc Method

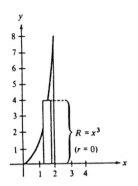

$$V = \pi \int_a^b \left([R(x)]^2 - [r(x)]^2\right)dx$$

$$= \pi \int_0^2 [(x^3)^2 - (0)^2]\,dx$$

$$= \pi \int_0^2 x^6\,dx$$

$$= \pi \left[\frac{x^7}{7} \right]_0^2 = \frac{128\pi}{7}$$

(b)   Shell Method

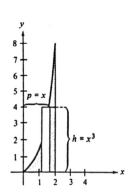

$$V = 2\pi \int_1^b p(x)h(x)\,dx$$

$$= 2\pi \int_0^2 x(x^3)\,dx$$

$$= 2\pi \int_0^2 x^4\,dx$$

$$= 2\pi \left[\frac{x^5}{5}\right]_0^2 = \frac{64\pi}{5}$$

(c)   Shell Method

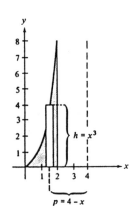

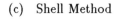

$$V = 2\pi \int_a^b p(x)h(x)\,dx$$

$$= 2\pi \int_0^2 (4-x)x^3\,dx$$

$$= 2\pi \int_0^2 (4x^3 - x^4)\,dx$$

$$= 2\pi \left[x^4 - \frac{x^5}{5}\right]_0^2 = \frac{96\pi}{5}$$

(d)   Disc Method

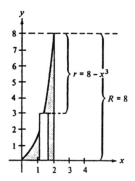

$$V = \pi \int_a^b ([R(x)]^2 - [r(x)]^2)\,dx$$

$$= \pi \int_0^2 [8^2 - (8 - x^3)^2]\,dx$$

$$= \pi \int_0^2 (16x^3 - x^6)\,dx$$

$$= \pi \left[4x^4 - \frac{x^7}{7}\right]_0^2 = \frac{320\pi}{7}$$

27. A solid is generated by revolving the region bounded by $y = \frac{1}{2}x^2$ and $y = 2$ about the y-axis. A hole, centered along the axis of revolution, is drilled through this solid so that $\frac{1}{4}$ of the volume is removed. Find the diameter of the hole.

Solution:

The total volume of the solid is given by

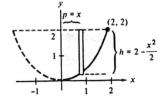

$$V = 2\pi \int_0^2 x\left(2 - \frac{x^2}{2}\right) dx$$

$$= 2\pi \int_0^2 \left(2x - \frac{x^3}{2}\right) dx$$

$$= 2\pi \left[x^2 - \frac{x^4}{8}\right]_0^2 = 2\pi(4 - 2) = 4\pi$$

If a hole drilled in the center with a radius of x_0 removes $\frac{1}{4}$ of this volume, you have

$$\left(\frac{3}{4}\right)V = 2\pi \int_{x_0}^2 x\left(2 - \frac{x^2}{2}\right) dx$$

$$3\pi = 2\pi \left[x^2 - \frac{x^4}{8}\right]_{x_0}^2$$

$$3\pi = 2\pi \left[(4 - 2) - \left(x_0{}^2 - \frac{x_0{}^4}{8}\right)\right]$$

$$3 = 4 - 2x_0{}^2 + \left(\frac{x_0{}^4}{4}\right)$$

$$x_0{}^4 - 8x_0{}^2 + 4 = 0$$

$$x_0{}^2 = \frac{8 \pm \sqrt{64 - 16}}{2}$$

$$x_0 = \sqrt{4 - 2\sqrt{3}}$$

(Since $0 < x_0 < 2$, you are not interested in $x_0 = \sqrt{4 + 2\sqrt{3}} \approx 2.7$.) Finally, since

$$\text{radius} = x_0 = \sqrt{4 - 2\sqrt{3}} = 0.732$$

you have

$$\text{diameter} = 2x_0 = 2\sqrt{4 - 2\sqrt{3}} \approx 1.464$$

6.4   Arc Length and Surfaces of Revolution

7. Find the arc length of $y = (x^4/8) + (1/4x^2)$ between $x = 1$ and $x = 2$.

Solution:

$$y = \frac{x^4}{8} + \frac{1}{4x^2} \qquad y' = \frac{x^3}{2} - \frac{1}{2x^3}$$

$$s = \int_1^2 \sqrt{1 + (y')^2}\,dx = \int_1^2 \sqrt{1 + \left(\frac{x^3}{2} - \frac{1}{2x^3}\right)^2}\,dx$$

$$= \int_1^2 \sqrt{1 + \frac{x^6}{4} - \frac{1}{2} + \frac{1}{4x^6}}\,dx = \int_1^2 \sqrt{\frac{x^6}{4} + \frac{1}{2} + \frac{1}{4x^6}}\,dx$$

$$= \int_1^2 \sqrt{\left(\frac{x^3}{2} + \frac{1}{2x^3}\right)^2}\,dx$$

(Note that $0 < \dfrac{x^3}{2} + \dfrac{1}{2x^3}$ for $1 \le x \le 2$.)

$$s = \int_1^2 \left(\frac{x^3}{2} + \frac{1}{2x^3}\right)dx = \left[\frac{x^4}{8} - \frac{1}{4x^2}\right]_1^2$$

$$= \left(\frac{16}{8} - \frac{1}{16}\right) - \left(\frac{1}{8} - \frac{1}{4}\right) = \frac{32 - 1 - 2 + 4}{16} = \frac{33}{16}$$

23. A fleeing object leaves the origin and moves up the y-axis. At the same time a pursuer leaves the point $(1,0)$ and moves always toward the fleeing object. If the pursuer's speed is twice that of the fleeing object, the equation of the path is

$$y = \frac{1}{3}(x^{3/2} - 3x^{1/2} + 2).$$

How far has the fleeing object traveled when it is caught? Show that the pursuer traveled twice as far.

Solution:

The y-intercept of $y = \frac{1}{3}(x^{3/2} - 3x^{1/2} + 2)$ is $(0, \frac{2}{3})$. Therefore, the fleeing object traveled from $(0,0)$ to $(0, \frac{2}{3})$, a distance of $\frac{2}{3}$.

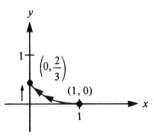

$$y = \frac{1}{3}(x^{3/2} - 3x^{1/2} + 2)$$

$$y' = \frac{1}{3}\left[\frac{3}{2}x^{1/2} - \frac{3}{2}x^{-1/2}\right]$$

$$= \frac{1}{2}\left(\sqrt{x} - \frac{1}{\sqrt{x}}\right) = \frac{x-1}{2\sqrt{x}}$$

$$1 + (y')^2 = 1 + \left(\frac{x-1}{2\sqrt{x}}\right)^2 = \frac{4x + (x^2 - 2x + 1)}{4x}$$

$$= \frac{x^2 + 2x + 1}{4x} = \frac{(x+1)^2}{4x}$$

Therefore, the distance traveled by the pursuer is given by

$$s = \int_0^1 \sqrt{1 + (y')^2}\,dx$$

$$= \frac{1}{2}\int_0^1 \frac{x+1}{\sqrt{x}}\,dx$$

$$= \frac{1}{2}\int_0^1 (x^{1/2} + x^{-1/2})\,dx$$

$$= \frac{1}{2}\left[\frac{2}{3}x^{3/2} + 2x^{1/2}\right]_0^1 = \frac{4}{3}$$

Thus the pursuer traveled a distance of $\frac{4}{3}$, twice the distance of the fleeing object.

31. Find the area of the surface formed by revolving the graph of $y = (x^3/6) + (1/2x)$, $1 \le x \le 2$, about the x-axis.
Solution:

$$y = \frac{x^3}{6} + \frac{1}{2x}$$

$$y' = \frac{1}{2}x^2 - \frac{1}{2x^2} = \frac{x^4 - 1}{2x^2}$$

$$1 + (y')^2 = 1 + \left(\frac{x^4 - 1}{2x^2}\right)^2$$

$$= \frac{4x^4 + (x^8 - 2x^4 + 1)}{4x^2}$$

$$= \frac{x^8 + 2x^4 + 1}{4x^4} = \left(\frac{x^4 + 1}{2x^2}\right)^2$$

$$S = 2\pi \int_1^2 y\sqrt{1 + (y')^2}\, dx$$

$$= 2\pi \int_1^2 \left(\frac{x^3}{6} + \frac{1}{2x}\right)\left(\frac{x^4 + 1}{2x^2}\right) dx$$

$$= 2\pi \int_1^2 \left(\frac{x^5}{12} + \frac{x}{3} + \frac{1}{4x^3}\right) dx$$

$$= 2\pi \left[\frac{x^6}{72} + \frac{x^2}{6} - \frac{1}{8x^2}\right]_1^2 = \frac{47\pi}{16}$$

33. Find the area of the surface formed by revolving the graph of $y = \sqrt[3]{x} + 2, 1 \le x \le 8$, about the y-axis.

Solution:

$$y = \sqrt[3]{x} + 2 \qquad y' = \frac{1}{3}x^{-2/3}$$

$$S = 2\pi \int_1^8 x\sqrt{1 + (y')^2}\, dx$$

$$= 2\pi \int_1^8 x\sqrt{1 + \left(\frac{1}{3x^{2/3}}\right)^2}\, dx$$

$$= 2\pi \int_1^8 x\sqrt{\frac{9x^{4/3} + 1}{9x^{4/3}}}\, dx = 2\pi \int_1^8 \frac{x}{3x^{2/3}}\sqrt{9x^{4/3} + 1}\, dx$$

$$= \frac{2\pi}{3}\left(\frac{1}{12}\right) \int_1^8 (9x^{4/3} + 1)^{1/2}(12x^{1/3})\, dx$$

$$= \frac{\pi}{18}\left[\frac{(9x^{4/3} + 1)^{3/2}}{3/2}\right]_1^8 = \frac{\pi}{27}[145\sqrt{145} - 10\sqrt{10}] \approx 199.48$$

37. A right circular cone is generated by revolving the region
bounded by $y = hx/r, y = h$, and $x = 0$ about the y-axis. Verify
that the lateral surface area of the cone is $S = \pi r \sqrt{r^2 + h^2}$.

Solution:

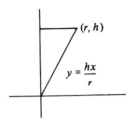

The distance between the y-axis and the graph of the line
$y = hx/r$ is

$$r(y) = g(y) = \frac{ry}{h}$$

and since $g'(y) = r/h$, the surface area is given by

$$S = 2\pi \int_0^h r(y) \sqrt{1 + [g'(y)]^2}\,dy$$

$$= 2\pi \int_0^h \left(\frac{ry}{h}\right) \sqrt{1 + \left(\frac{r}{h}\right)^2}\,dy$$

$$= \frac{2\pi r \sqrt{r^2 + h^2}}{h^2} \int_0^h y\,dy$$

$$= \frac{2\pi r \sqrt{r^2 + h^2}}{h^2} \left[\frac{1}{2}y^2\right]_0^h = \pi r \sqrt{r^2 + h^2}$$

39. Find the area of the zone of a sphere formed by revolving the
graph of $y = \sqrt{9 - x^2}$, $0 \le x \le 2$, about the y-axis.
Solution:

$$y = \sqrt{9 - x^2} \qquad y' = \frac{-x}{\sqrt{9 - x^2}}$$

$$S = 2\pi \int_0^2 x \sqrt{1 + \left(\frac{-x}{\sqrt{9 - x^2}}\right)^2}\,dx$$

$$= 2\pi \int_0^2 x \sqrt{\frac{9}{9 - x^2}}\,dx$$

$$= 2\pi \int_0^2 3x(9 - x^2)^{-1/2}\,dx$$

$$= -3\pi \int_0^2 (9 - x^2)^{-1/2}(-2x)\,dx$$

$$= -3\pi \left[\frac{(9 - x^2)^{1/2}}{1/2}\right]_0^2 = -6\pi(\sqrt{5} - 3) = 6\pi(3 - \sqrt{5}) \approx 14.40$$

6.5   Work

7. A force of 60 lb stretches a spring 1 ft. How much work is done in stretching the spring from 9 in to 15 in?

Solution:

Let $F(x)$ be the force required to stretch a spring x units. By Hooke's Law you know that $F(x) = kx$. Since 60 lb stretches the spring 1 ft(12 in), you have

$$60 = k(12) \quad \text{or} \quad k = 5$$

Therefore, the work done by stretching the spring from 9 to 15 in is

$$W = \int_9^{15} \underbrace{F(x)}_{\text{(force)}} \underbrace{dx}_{\text{(distance)}}$$

$$= \int_9^{15} 5x \, dx$$

$$= \left[\frac{5x^2}{2} \right]_9^{15} = \frac{5}{2}(225 - 81) = 360 \text{ in} \cdot \text{lb}$$

13. Neglecting air resistance, determine the work done in propelling a 10-ton satellite to a height of (a) 11,000 miles and (b) 22,000 miles above the earth.

Solution:

Because the weight of a body varies inversely as the square of its distance from the center of the earth, the force $F(x)$ exerted by gravity is

$$F(x) = \frac{C}{x^2}.$$

Because the satellite weighs 10 tons on the surface of the earth and the radius of the earth is approximately 4000 miles, you have

$$10 = \frac{C}{(4000)^2} \quad \Longrightarrow \quad C = 160,000,000.$$

(a) $$W = \int_{4000}^{15,000} \frac{160,000,000}{x^2} \, dx$$

$$= \left[-\frac{160,000,000}{x} \right]_{4000}^{15,000}$$

$$= 2.93 \times 10^4 \text{ mi} \cdot \text{tons} \approx 3.10 \times 10^{11} \text{ ft} \cdot \text{lb}$$

(b) $$W = \int_{4000}^{26,000} \frac{160,000,000}{x^2} \, dx$$

$$= \left[-\frac{160,000,000}{x} \right]_{4000}^{26,000}$$

$$= 3.38 \times 10^4 \text{ mi} \cdot \text{tons} \approx 3.57 \times 10^{11} \text{ ft} \cdot \text{lb}$$

19. A hemispherical tank of radius 6 ft is positioned so that its base is circular. How much work is required to fill the tank with water through a hole in the base if the water source is at the base?

Solution:

To fill the tank with water from a hole in the bottom, all the water is not moved to a height of 6 ft. Some of the water must be moved 6 ft, some 5 ft, some 4 ft, and so on. In general, the "disc" of water that must be moved y feet has a volume of

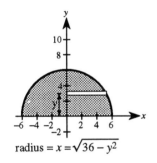

radius $= x = \sqrt{36 - y^2}$

$$\pi x^2 \Delta y = \pi (\sqrt{36 - y^2})^2 \Delta y = \pi (36 - y^2) \Delta y \text{ cubic feet}$$

The weight of the disc of water is

$$62.4(\pi)(36 - y^2)\Delta y \text{ pounds}$$

Thus, the work done in filling the tank from the bottom is

$$W = \int_0^6 \underbrace{(y)}_{\text{(distance)}} \underbrace{[62.4\pi (36 - y^2) \, dy}_{\text{(force: weight of water)}}$$

$$= 62.4\pi \int_0^6 (36y - y^3) \, dy$$

$$= 62.4\pi \left[18y^2 - \frac{y^4}{4} \right]_0^6$$

$$= 62.4\pi (648 - 324) = 20,217.6\pi \text{ ft} \cdot \text{lb}$$

21. An open tank has the shape of a right circular cone. If the tank is 8 ft across the top and has a height of 6 ft, how much work is required to empty the tank of water by pumping the water over the top edge?

Solution:

A disc of water at height y has to be moved $(6 - y)$ feet up and has a volume of $\pi x^2 \Delta y$. To find x in terms of y, solve the equation

$$\frac{6 - 0}{4 - 0} = \frac{y - 0}{x - 0}$$

$$x = \frac{2y}{3}$$

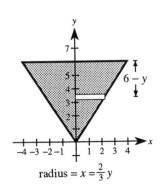

radius $= x = \frac{2}{3} y$

Thus the work done in moving the water over the top of the tank is

$$W = \int_0^6 \underbrace{(6 - y)}_{\text{(distance)}} \underbrace{\left[62.4\pi \left(\frac{2y}{3} \right)^2 dy \right]}_{\text{(force: weight of water)}}$$

$$= \frac{4}{9}(62.4)\pi \int_0^6 (6y^2 - y^3)dy$$

$$= \frac{83.2}{3}\pi \left[2y^3 - \frac{y^4}{4} \right]_0^6$$

$$= \frac{83.2}{3}\pi(432 - 324) = 2995.2\pi \text{ ft} \cdot \text{lb}$$

23. A cylindrical gasoline tank 3 feet in diameter and 4 feet long is carried on the back of a truck and is used to fuel tractors in the field. The axis of the tank is horizontal. Find the work done to pump the entire contents of the full tank into a tractor if the opening in the tractor tank is 5 ft above the top of the tank in the truck. Assume gasoline weighs 42 pounds per cubic foot.

Solution:

A layer of gasoline at height y (see the accompanying figure) is lifted $\left(\frac{13}{2} - y \right)$ feet and has a volume $V = lwh = 4(2x)\Delta y$. To find x in terms of y, solve for x in the equation of the circle representing a cross-section of the tank and obtain

$$x^2 + y^2 = \frac{9}{4}$$

$$x^2 = \frac{9}{4} - y^2$$

$$x = \sqrt{\frac{9}{4} - y^2}$$

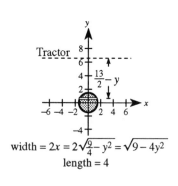

width $= 2x = 2\sqrt{\frac{9}{4} - y^2} = \sqrt{9 - 4y^2}$

length $= 4$

Thus the work done is

$$W = \int_{-1.5}^{1.5} \underbrace{\left(\frac{13}{2} - y\right)}_{\text{(distance)}} \underbrace{\left[42(4)\left(2\sqrt{\frac{9}{4} - y^2}\right) dy\right]}_{\text{(force: weight of water)}}$$

$$= 336\left[\frac{13}{2}\int_{-1.5}^{1.5}\sqrt{\frac{9}{4} - y^2}\, dy - \int_{-1.5}^{1.5} y\sqrt{\frac{9}{4} - y^2}\, dy\right]$$

$$= 336\left[\frac{13}{2}\int_{-1.5}^{1.5}\sqrt{\frac{9}{4} - y^2}\, dy - \int_{-1.5}^{1.5} y\sqrt{\frac{9}{4} - y^2}\, dy\right]$$

The first integral represents the area of a semicircle of radius 3/2 and the second integral is zero since the integrand is odd and the limits of integration are symmetric to the origin. Therefore,

$$W = 336\left(\frac{13}{2}\right)\left(\frac{1}{2}\right)(\pi)\left(\frac{3}{2}\right)^2 = 2457\pi \text{ ft} \cdot \text{lb}.$$

29. A chain 15 ft long and weighing 3 lb/ft is suspended vertically from a height of 15 ft. How much work is required to take the bottom of the chain and raise it to the 15-ft level, leaving the chain doubled but still hanging vertically?

Solution:

A small piece of chain of length Δy at height y must be moved so that it is y feet from the top. Therefore, the distance moved (as seen in the accompanying figure) is $15 - 2y$. (For example, the chain at an initial height of 7.5 is moved 0 ft.) The weight of a piece of chain of length Δy is

$$\frac{3 \text{ pounds}}{\text{foot}}\Delta y \text{ feet} = 3\Delta y \text{ pounds}.$$

Finally, since you are only moving chain that has an initial height between 0 and 7.5 ft, you have

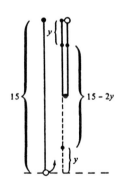

$$W = \int_0^{7.5} \underbrace{(15 - 2y)}_{\text{(distance)}}\underbrace{(3\, dy)}_{\text{(force)}}$$

$$= 3\left[15y - y^2\right]_0^{7.5} = 3(112.5 - 56.25) = 168.75 \text{ ft} \cdot \text{lb}$$

6.6 Fluid Pressure and Fluid Force

7. Find the force on a vertical side of a tank if the tank is full of water and the side has the shape of a trapezoid, as in the accompanying figure.

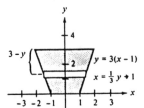

Solution:

The force against a representative rectangle of length $2x$ is

$$\Delta F = (\text{density})(\text{depth})(\text{area}) = (62.4)(3 - y)(2x\,\Delta y)$$

$$= (62.4)(3 - y)(2)\left(\frac{1}{3}y + 1\right)\Delta y$$

Since y ranges from 0 to 3, the total force is

$$F = \int_0^3 (62.4)(3 - y)(2)\left(\frac{1}{3}y + 1\right)dy$$

$$= 124.8\int_0^3 \left(3 - \frac{1}{3}y^2\right)dy$$

$$= 124.8\left[3y - \frac{1}{9}y^3\right]_0^3 = 748.8 \text{ lb}$$

13. Find the total force on the vertical plate submerged in water as shown in the accompanying figure.

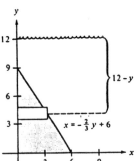

Solution:

The force against a representative rectangle of length x is

$$\Delta F = (\text{density})(\text{depth})(\text{area}) = (62.4)(12 - y)\left(-\frac{2}{3}y + 6\right)\Delta y$$

Since y ranges from 0 to 9, the total force is

$$F = \int_0^9 62.4(12 - y)\left(-\frac{2}{3}y + 6\right)dy$$

$$= 62.4\int_0^9 \left(\frac{2}{3}y^2 - 14y + 72\right)dy$$

$$= 62.4\left[\frac{2}{9}y^3 - 7y^2 + 72y\right]_0^9 = 15,163.2 \text{ lb}$$

19. A cylindrical gasoline tank is placed so that the axis of the cylinder is horizontal. If the tank is half full, find the force on a circular end of the tank, assuming that the diameter is 3 ft and gasoline has a density of 42 lb/ft^3.

Solution:

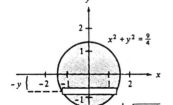

The force against a representative rectangle of length $2x$ is

$$\Delta F = (\text{density})(\text{depth})(\text{area}) = (42)(-y)(2x\,\Delta y)$$
$$= (42)(-y)(2)\left(\frac{1}{2}\right)(9-4y^2)^{1/2}\Delta y = -42y(9-4y^2)^{1/2}\Delta y$$

Since y varies from $-\frac{3}{2}$ to 0, the total force is

$$F = \int_{-3/2}^{0} (-42)y(9-4y^2)^{1/2}\,dy$$
$$= \frac{42}{8}\int_{-3/2}^{0} (9-4y^2)^{1/2}(-8y)\,dy$$
$$= \frac{21}{4}\left(\frac{2}{3}\right)\left[(9-4y^2)^{3/2}\right]_{-3/2}^{0} = 94.5 \text{ lb}$$

25. A swimming pool is 20 ft wide, 40 ft long, 4 ft deep at one end, and 8 ft deep at the other. The bottom is an inclined plane. Find the total force on each of the vertical walls of the pool.

Solution:

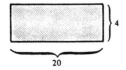

The force on the rectangular shallow end is given by

$$F = \int_{0}^{4} 62.4(4-y)(20)\,dy = 1,248\int_{0}^{4}(4-y)\,dy$$
$$= 1,248\left[4y - \frac{y^2}{2}\right]_{0}^{4} = 1248(8) = 9,948 \text{ lb}$$

and for the deep end,

$$F = \int_{0}^{8} 62.4(8-y)(20)\,dy = 1,248\int_{0}^{8}(8-y)\,dy$$
$$= 1,248\left[8y - \frac{y^2}{2}\right]_{0}^{8} = 1,248(32) = 39,936 \text{ lb.}$$

The force on each side is given by

$$F = \int_{0}^{4} 62.4(8-y)(x)\,dy + \int_{4}^{8} 62.4(8-y)40\,dy$$

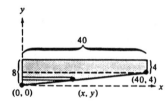

where x and y are related by

$$\frac{4-0}{40-0} = \frac{y-0}{x-0} \quad \text{or} \quad x = 10y.$$

Thus,

$$F = \int_0^4 62.4(8-y)(10y)\,dy + \int_4^8 62.4(8-y)(40)\,dy$$

$$= 624\int_0^4 (8y - y^2)\,dy + 2,496\int_4^8 (8-y)\,dy$$

$$= 624\left[4y^2 - \frac{y^3}{3}\right]_0^4 + 2,496\left[8y - \frac{y^2}{2}\right]_4^8$$

$$= 624\left(64 - \frac{64}{3}\right) + 2,496[(64-32)-(32-8)]$$

$$= 624\left(\frac{128}{3}\right) + 2,496(8) = 46,592 \text{ lb}$$

6.7   Moments, Centers of Mass, and Centroids

9. Find the center of mass for the masses located at the given points.

m_i	3	4	2	1	6
(x_i, y_i)	$(-2,-3)$	$(-1,0)$	$(7,1)$	$(0,0)$	$(-3,0)$

Solution:

$$\bar{x} = \frac{m_1x_1 + m_2x_2 + m_3x_3 + m_4x_4 + m_5x_5}{m_1 + m_2 + m_3 + m_4 + m_5}$$

$$= \frac{3(-2) + 4(-1) + 2(7) + 1(0) + 6(-3)}{3+4+2+1+6} = -\frac{7}{8}$$

$$\bar{y} = \frac{m_1y_1 + m_2y_2 + m_3y_3 + m_4y_4 + m_5y_5}{m_1 + m_2 + m_3 + m_4 + m_5}$$

$$= \frac{3(-3) + 4(0) + 2(1) + 1(0) + 6(0)}{3+4+2+1+6} = -\frac{7}{16}$$

19. Find M_x, M_y, and $(\overline{x}, \overline{y})$ for the lamina of uniform density ρ bounded by $x = 4 - y^2$ and $x = 0$.

Solution:

Since the region is symmetric with respect to the x-axis, you know that

$$M_x = 0 \text{ and } \overline{y} = \frac{M_x}{m} = 0.$$

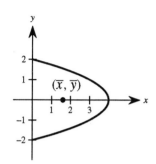

To find $\overline{x}$, observe that x is a function of y, and use the formula

$$\overline{x} = \frac{\displaystyle\int_a^b \left[\frac{f(y) + g(y)}{2}\right][f(y) - g(y)]\,dy}{m} = \frac{M_y}{m}$$

where $f(y) = 4 - y^2$, $g(y) = 0$, $a = -2$, and $b = 2$.

$$m = \rho \int_{-2}^{2} (4 - y^2)\,dy = \rho\left[4y - \frac{y^3}{3}\right]_{-2}^{2} = \frac{32\rho}{3}$$

$$M_y = \frac{\rho}{2} \int_{-2}^{2} (4 - y^2)^2\,dy$$

$$= \frac{\rho}{2} \int_{-2}^{2} (16 - 8y^2 + y^4)\,dy$$

$$= \frac{\rho}{2}\left[16y - \frac{8y^3}{3} + \frac{y^5}{5}\right]_{-2}^{2}$$

$$= \frac{\rho}{2}\left[\left(32 - \frac{64}{3} + \frac{32}{5}\right) - \left(-32 + \frac{64}{3} - \frac{32}{5}\right)\right]$$

$$= \frac{\rho}{2}\left(\frac{512}{15}\right) = \frac{256\rho}{15}$$

$$\overline{x} = \frac{M_y}{m} = \frac{256\rho/15}{32\rho/3} = \frac{8}{5}$$

Therefore, $(\overline{x},\ \overline{y}) = \left(0,\ \dfrac{8}{5}\right)$.

23. Find the centroid of the region bounded by $y = f(x) = x$ and $y = g(x) = x^2$.

Solution:

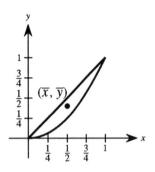

The two graphs intersect at the points $(0, 0)$ and $(1, 1)$. The area is given by

$$A = \int_0^1 [f(x) - g(x)]\, dx$$

$$= \int_0^1 (x - x^2)\, dx = \left[\frac{1}{2}x^2 - \frac{1}{3}x^3\right]_0^1 = \frac{1}{6}$$

$$\overline{x} = \frac{\displaystyle\int_0^1 x[f(x) - g(x)]\, dx}{A}$$

$$= \frac{\displaystyle\int_0^1 x(x - x^2)\, dx}{\frac{1}{6}}$$

$$= 6\int_0^1 (x^2 - x^3)\, dx = 6\left[\frac{1}{3}x^3 - \frac{1}{4}x^4\right]_0^1 = \frac{1}{2}$$

$$\overline{y} = \frac{\dfrac{1}{2}\displaystyle\int_0^1 \left[\dfrac{f(x) + g(x)}{2}\right][f(x) - g(x)]\, dx}{A}$$

$$= \frac{\dfrac{1}{2}\displaystyle\int_0^1 [(x)^2 - (x^2)^2]\, dx}{\frac{1}{6}}$$

$$= 3\int_0^1 (x^2 - x^4)\, dx = 3\left[\frac{1}{3}x^3 - \frac{1}{5}x^5\right]_0^1 = \frac{2}{5}$$

Therefore, $(\overline{x}, \overline{y}) = \left(\dfrac{1}{2}, \dfrac{2}{5}\right)$.

27. Find the centroid of the triangle with vertices $(-a, 0)$, $(a, 0)$, and (b, c). Show that it is at the point of intersection of the medians. (Assume that $-a < b < a$.)

Solution:

The equation of the line containing $(-a, 0)$ and (b, c) is

$$y = \left(\frac{c}{b + a}\right)(x + a).$$

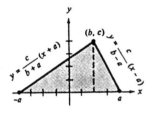

The equation of the line containing $(a, 0)$, and (b, c) is

$$y = \left(\frac{c}{b-a}\right)(x-a).$$

Since the area of the triangle is

$$A = \left(\frac{1}{2}\right)(2a)(c) = ac,$$

you have

$$\bar{x} = \frac{\displaystyle\int_{-a}^{b} x\left(\frac{c}{b+a}\right)(x+a)\,dx + \int_{b}^{a} x\left(\frac{c}{b-a}\right)(x-a)\,dx}{ac}$$

$$= \frac{1}{ac}\left[\frac{c}{b+a}\int_{-a}^{b}(x^2+ax)\,dx + \frac{c}{b-a}\int_{a}^{b}(x^2-ax)\,dx\right]$$

$$= \frac{1}{ac}\left(\frac{c}{b+a}\left[\frac{x^3}{3}+\frac{ax^2}{2}\right]_{-a}^{b} + \frac{c}{b-a}\left[\frac{x^3}{3}-\frac{ax^2}{2}\right]_{b}^{a}\right)$$

$$= \frac{1}{ac}\left[\frac{c}{b+a}\left(\frac{b^3}{3}+\frac{ab^2}{2}+\frac{a^3}{3}-\frac{a^3}{2}\right)\right.$$

$$\left. +\frac{c}{b-a}\left(\frac{a^3}{3}-\frac{a^3}{2}-\frac{b^3}{3}+\frac{ab^2}{2}\right)\right]$$

$$= \frac{2b^3+3ab^2-a^3}{6a(b+a)} + \frac{-2b^3+3ab^2-a^3}{6a(b-a)}$$

$$= \frac{(2b^2+ab-a^2)(a+b)}{6a(b+a)} + \frac{(-2b^2+ab+a^2)(b-a)}{6a(b-a)}$$

$$= \frac{2ab}{6a} = \frac{b}{3}$$

$$\bar{y} = \frac{\displaystyle\frac{1}{2}\int_{-a}^{b}\left[\frac{c}{b+a}(x+a)\right]^2 dx + \frac{1}{2}\int_{b}^{a}\left[\frac{c}{b-a}(x-a)\right]^2 dx}{ac}$$

$$= \frac{1}{2ac}\left\{\frac{c^2}{(b+a)^2}\left[\frac{(x+a)^3}{3}\right]_{-a}^{b} + \frac{c^2}{(b-a)^2}\left[\frac{(x-a)^3}{3}\right]_{b}^{a}\right\}$$

$$= \frac{1}{2ac}\left\{\frac{c^2}{(b+a)^2}\left[\frac{(b+a)^3}{3}\right] - \frac{c^2}{(b-a)^2}\left[\frac{(b-a)^3}{3}\right]\right\}$$

$$= \frac{1}{2ac}\left[\frac{c^2(b+a)}{3} - \frac{c^2(b-a)}{3}\right] = \frac{c}{3}$$

From Exercise 58, Section P.4, you know that the point $(b/3, \; c/3)$ is the intersection of the medians of the triangle.

37. The manufacturer of glass for a window in a conversion van needs to approximate its center of mass. A coordinate system is superimposed on a prototype of the glass (see figure). The measurements (in centimeters) for the right half of the symmetric piece of glass are given in the following table.

x	0	10	20	30	40
y	30	29	26	20	0

Approximate the center of mass of the glass.

Solution:

From the symmetry of the glass you have $\bar{x} = 0$. To approximate the mass of the glass, use its symmetry with respect to the y-axis and Simpson's Rule with $n = 4$ to obtain

$$m = 2\rho \int_0^{40} y\, dx$$

$$\approx 2\rho \left[\frac{40 - 0}{3(4)}\right][30 + 4(29) + 2(26) + 4(20) + 0] \approx 1853.33\rho.$$

To approximate M_x, use the symmetry of the glass and Simpson's Rule with $n = 4$ to obtain

$$M_x = 2\rho \int_0^{40} \left(\frac{y}{2}\right) y\, dx = \rho \int_0^{40} y^2\, dx$$

$$\approx \rho \left[\frac{40 - 0}{3(4)}\right][30^2 + 4(29^2) + 2(26^2) + 4(20^2) + 0^2] \approx 24,053.33\rho.$$

Therefore, $\bar{y} = \dfrac{M_x}{m} \approx \dfrac{24,053.33\rho}{1853.33\rho} \approx 12.98.$

39. A plate of uniform density is formed by a circle and a square as shown in the accompanying figure. Introduce an appropriate rectangular coordinate system and find the coordinates of the center of mass.

Solution:

Although a coordinate system may be introduced in many different ways, the one shown in the accompanying figure is a fairly natural choice. Since both the circle and square have a uniform density, their masses are proportional to their areas, π and 4, respectively. (For simplicity, assume the density to be 1 unit of mass per 1 unit of area.) Again, because of the uniform density, both the circle and the square have their

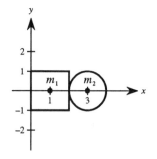

centers of mass at their geometrical centers, $(3,0)$ and $(1,0)$, respectively. Therefore, you can find the center of mass of the plate by considering a mass of π centered at $(3,0)$ and a mass of 4 centered at $(1,0)$. Thus

$$\bar{x} = \frac{\pi(3) + 4(1)}{\pi + 4} = \frac{3\pi + 4}{\pi + 4} \quad \text{and} \quad \bar{y} = \frac{\pi(0) + 4(0)}{\pi + 4} = 0$$

Review Exercises for Chapter 6

 8. Sketch and find the area of the region bounded by the graphs of $y = x^2 - 4x + 3$, $y = x^3$, and $x = 0$.

Solution:

The points of intersection are given by

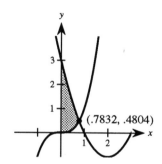
(.7832, .4804)

$$x^3 = x^2 - 4x + 3$$
$$x^3 - x^2 + 4x - 3 = 0$$

To approximate the required root of the cubic equation, apply Newton's Method to the function

$$f(x) = x^3 - x^2 + 4x - 3$$

By letting $x_1 = 1$, you have

$$x_2 = x_1 - \frac{f(x_1)}{f'(x_1)} = 0.8000$$

$$x_3 = x_2 - \frac{f(x_2)}{f'(x_2)} = 0.7833$$

$$x_4 = x_3 - \frac{f(x_3)}{f'(x_3)} = 0.7832$$

Since $x^3 \leq x^2 - 4x + 3$ for $0 \leq x \leq 0.7832$, you have

$$A \approx \int_0^{0.7832} (x^2 - 4x + 3 - x^3)\, dx$$

$$= \left[\frac{x^3}{3} - 2x^2 + 3x - \frac{x^4}{4} \right]_0^{0.7832}$$

$$\approx 0.1601 - 1.2268 + 2.3496 - 0.0941 \approx 1.189$$

15. Sketch and find the area of the region bounded by the graphs of $x = y^2 - 2y$ and $x = 0$. Set up integrals for finding the area of the region by using vertical and horizontal representative rectangles. Find the area of evaluating the easier integral.

Solution:

To find the y-intercepts of the graph, solve the equation

$$y^2 - 2y = 0$$
$$y(y-2) = 0 \quad \Longrightarrow \quad y = 0,\ 2.$$

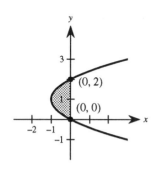

The graph of the equation is a parabola opening to the right as shown in the accompanying figure. To find the area of the region by using vertical representative rectangles, begin by solving the equation for y.

$$x = y^2 - 2y \quad \Longrightarrow \quad 0 = y^2 - 2y - x$$

Using the Quadratic Formula with $a = 1$, $b = -2$, and $c = -x$, you have

$$y = \frac{-(-2) \pm \sqrt{(-2)^2 - 4(1)(-x)}}{2(1)}$$

$$= \frac{1}{2}[2 \pm \sqrt{4 + 4x}] = 1 \pm \sqrt{1 + x}$$

Therefore, the top half and bottom halves of the parabola are given by $f(x) = 1 + \sqrt{1 + x}$ and $g(x) = 1 - \sqrt{1 + x}$, respectively. Thus, the area of the region is given by

$$A = \int_{-1}^{0} [(1 + \sqrt{1 + x}) - (1 - \sqrt{1 + x})]\, dx = \int_{-1}^{0} (2\sqrt{1 + x})\, dx.$$

Using horizontal representative rectangles and the fact that $y^2 - 2y \leq 0$ for $0 \leq y \leq 2$, you have

$$A = \int_{0}^{2} [0 - (y^2 - 2y)]\, dy$$

$$= \int_{0}^{2} (-y^2 + 2y)\, dy$$

$$= \left[\frac{-y^3}{3} + y^2 \right]_{0}^{2} = -\frac{8}{3} + 4 = \frac{4}{3}$$

21. Find the volume of the solid generated by revolving the region bounded by the graph of $x^2/16 + y^2/9 = 1$ about:

(a)   the y-axis (oblate spheroid). (b)   the x-axis (prolate spheroid).

Solution:

(a)   Revolving the first quadrant portion of the ellipse about the y-axis will generate one-half the solid. Solving the equation for y yields

$$y = \frac{3}{4}\sqrt{16 - x^2}.$$

Therefore, using the symmetry and the Shell Method you have

$$V = 2\pi \int_{-4}^{4} x f(x)\, dx$$

$$= 4\pi \int_{0}^{4} x \left(\frac{3}{4}\right) \sqrt{16 - x^2}\, dx$$

$$= 3\pi \int_{0}^{4} x\sqrt{16 - x^2}\, dx$$

$$= -\frac{3\pi}{2} \int (16 - x^2)^{1/2}(-2x)\, dx$$

$$= -\frac{3\pi}{2} \left(\frac{2}{3}\right) \left[(16 - x^2)^{3/2}\right]_{0}^{4} = -\pi[0 - 16^{3/2}] = 64\pi$$

(b)   Revolving the first quadrant portion of the ellipse about the x-axis will generate one-half the solid. Solving the equation for y^2 yields

$$y^2 = \frac{9}{16}(16 - x^2).$$

Therefore, using the symmetry and the Disc Method you have

$$V = \pi \int_{-4}^{4} [f(x)]^2\, dx$$

$$= 2\pi \int_{0}^{4} \frac{9}{16}(16 - x^2)\, dx$$

$$= \frac{9\pi}{8} \left[16x - \frac{1}{3}x^3\right]_{0}^{4}$$

$$= \frac{9\pi}{8} \left[64 - \frac{64}{3}\right] = 48\pi$$

31. Find the arc length of the graph of $f(x) = \frac{4}{5}x^{5/4}$ from $x = 0$ to $x = 4$.

Solution:

Since $f'(x) = x^{1/4}$, you have

$$s = \int_0^4 \sqrt{1 + [f'(x)]^2}\, dx = \int_0^4 \sqrt{1 + \sqrt{x}}\, dx$$

Let $u = \sqrt{1 + \sqrt{x}}$. Then $u^2 = 1 + \sqrt{x}$, $x = (u^2 - 1)^2$, and $dx = 2(u^2 - 1)(2u)\, du$. If $x = 0$, then $u = 1$, and if $x = 4$, then $u = \sqrt{3}$. Thus,

$$s = \int_0^4 \sqrt{1 + \sqrt{x}}\, dx$$

$$= \int_1^{\sqrt{3}} u(2)(u^2 - 1)(2u)\, du$$

$$= 4 \int_0^{\sqrt{3}} (u^4 - u^2)\, du$$

$$= 4\left[\frac{u^5}{5} - \frac{u^3}{3}\right]_1^{\sqrt{3}} = \frac{8}{15}(6\sqrt{3} + 1) \approx 6.076$$

37. A water well has an 8-in casing (diameter) and is 175 ft deep. If the water is 25 ft from the top of the well, determine the amount of work done in pumping it dry, assuming that no water enters the well while it is being pumped.

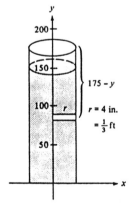

Solution:

A disk of water at height y must be lifted $(175 - y)$ feet and has a volume of $\pi(\frac{1}{3})^2 \Delta y$. Thus, the work done in lifting the water to the top of the well is

$$W = \int_0^{150} \underbrace{(175 - y)}_{\text{(distance)}} \underbrace{\left[62.4\pi\left(\frac{1}{3}\right)^2 dy\right]}_{\text{(force: weight of water)}}$$

$$= \frac{62.4\pi}{9} \int_0^{150} (175 - y)\, dy$$

$$= \frac{62.4\pi}{9} \left[175y - \frac{1}{2}y^2\right]_0^{150}$$

$$= 104{,}000\pi \text{ ft} \cdot \text{lb} \approx 163.4 \text{ ft} \cdot \text{ton}$$

42. Show that the force against any vertical region in a liquid is the product of the density ρ of the liquid, the area of the region, and the depth of the centroid of the region.

Solution:

The force is given by

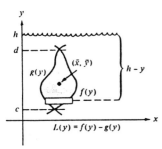

$$F = \int_c^d \rho(h-y)[f(y) - g(y)]\,dy$$

$$= \rho\left\{ h\int_c^d [f(y) - g(y)]\,dy - \int_c^d y[f(y) - g(y)]\,dy \right\}$$

$$= \rho\left\{ \int_c^d [f(y) - g(y)]\,dy \right\}\left\{ h - \frac{\int_c^d y[f(y) - g(y)]\,dy}{\int_c^d [f(y) - g(y)]\,dy} \right\}$$

$$= \rho(\text{area})(h - \bar{y}) = \rho(\text{area})(\text{depth of centroid})$$

45. Find the centroid of the region bounded by the graphs of $\sqrt{x} + \sqrt{y} = \sqrt{a}$, $x = 0$, and $y = 0$.

Solution:

Solving the equation $\sqrt{x} + \sqrt{y} = \sqrt{a}$ for y yields $y = (\sqrt{a} - \sqrt{x})^2$.

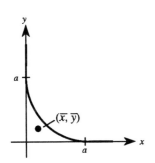

$$A = \int_0^a (\sqrt{a} - \sqrt{x})^2\,dx$$

$$= \int_0^a (a - 2\sqrt{a}\,x^{1/2} + x)\,dx$$

$$= \left[ax - \frac{4}{3}\sqrt{a}\,x^{3/2} + \frac{1}{2}x^2 \right]_0^a = \frac{a^2}{6}$$

$$\bar{x} = \frac{1}{(a^2/6)} \int_0^a x(\sqrt{a} - \sqrt{x})^2\,dx$$

$$= \frac{6}{a^2} \int_0^a (ax - 2\sqrt{a}\,x^{3/2} + x^2)\,dx$$

$$= \frac{6}{a^2} \left[\frac{ax^2}{2} - \frac{4}{5}\sqrt{a}\,x^{5/2} + \frac{1}{3}x^3 \right]_0^a = \frac{1}{5}a$$

By symmetry, you have $\bar{y} = \dfrac{a}{5}$. Therefore, $(\bar{x}, \bar{y}) = \left(\dfrac{a}{5}, \dfrac{a}{5} \right)$.

7 INTEGRATION TECHNIQUES, L'HÔPITAL'S RULE, AND IMPROPER INTEGRALS

7.1 Basic Integration Formulas

11. Select the basic integration formula you would use to evaluate

$$\int t \sin t^2 \, dt$$

and identify u and a when appropriate.

Solution:

If you let $u = t^2$, then $du = 2t \, dt$ and

$$\int t \sin t^2 \, dt = \frac{1}{2} \int \sin(t^2)(2t) \, dt = \frac{1}{2} \int \sin u \, du.$$

19. Evaluate $\displaystyle\int \frac{t^2 - 3}{-t^3 + 9t + 1} \, dt.$

Solution:

If you let $u = -t^3 + 9t + 1$, then
$du = (-3t^2 + 9) \, dt = -3(t^2 - 3) \, dt.$ Thus,

$$\int \frac{t^2 - 3}{-t^3 + 9t + 1} \, dt = -\frac{1}{3} \int \frac{1}{-t^3 + 9t + 1}[-3(t^2 - 3)] \, dt$$

$$= -\frac{1}{3} \int \frac{1}{u} \, du = -\frac{1}{3} \ln|-t^3 + 9t + 1| + C$$

37. Evaluate $\int \dfrac{2t-1}{t^2+4}\,dt$.

Solution:

$$\int \frac{2t-1}{t^2+4}\,dt = \int \frac{2t}{t^2+4}\,dt - \int \frac{1}{t^2+2^2}\,dt$$

$$= \ln(t^2+4) - \frac{1}{2}\arctan\frac{t}{2} + C$$

39. Evaluate $\int \dfrac{-1}{\sqrt{1-(2t-1)^2}}\,dt$.

Solution:

If you let $u = 2t - 1$, then $du = 2\,dt$. Thus,

$$\int \frac{-1}{\sqrt{1-(2t-1)^2}}\,dt = \frac{-1}{2}\int \frac{2}{\sqrt{1-(2t-1)^2}}\,dt$$

$$= \frac{-1}{2}\int \frac{1}{\sqrt{a^2-u^2}}\,du$$

$$= -\frac{1}{2}\arcsin(2t-1) + C$$

43. Evaluate $\int \dfrac{3}{\sqrt{6x-x^2}}\,dx$.

Solution:

By completing the square you have

$$\int \frac{3}{\sqrt{6x-x^2}}\,dx = 3\int \frac{1}{\sqrt{9-(9-6x+x^2)}}\,dx$$

$$= 3\int \frac{1}{\sqrt{9-(x-3)^2}}\,dx$$

$$= 3\arcsin\left(\frac{x-3}{3}\right) + C$$

51. Solve the differential equation $(4 + \tan^2 x)y' = \sec^2 x$.

Solution:

$$(4 + \tan^2 x)y' = \sec^2 x$$

$$y' = \frac{\sec^2 x}{4 + \tan^2 x}$$

$$\int y' \, dx = \int \frac{\sec^2 x}{4 + \tan^2 x} \, dx$$

If you let $u = \tan x$, then $du = \sec^2 x \, dx$ and you have

$$y = \int \frac{\sec^2 x}{4 + \tan^2 x} \, dx$$

$$= \int \frac{1}{a^2 + u^2} \, du$$

$$= \frac{1}{2} \arctan\left(\frac{\tan x}{2}\right) + C$$

56. Evaluate $\displaystyle\int_1^e \frac{1 - \ln x}{x} \, dx$.

Solution:

If you let $u = 1 - \ln x$, then $du = \dfrac{1}{x} \, dx$ and you have

$$\int_1^e \frac{1 - \ln x}{x} \, dx = -\int_1^e (1 - \ln x)^1 \left(\frac{-1}{x}\right) dx$$

$$= -\left[\frac{(1 - \ln x)^2}{2}\right]_1^e$$

$$= -\frac{1}{2}[(1 - \ln e)^2 - (1 - \ln 1)^2]$$

$$= -\frac{1}{2}[(0)^2 - 1] = \frac{1}{2}.$$

71. The region bounded by $y = e^{-x^2}$, $y = 0$, and $x = b$ is revolved around the y-axis.

(a) Find the volume of the solid generated if $b = 1$.

(b) Find b so that the volume of the genereated solid is $\frac{4}{3}$ cubic units.

Solution:

(a) By using the Shell Method you have

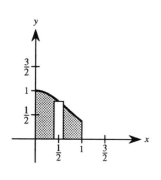

$$V = 2\pi \int_a^b p(x)h(x)\,dx$$

$$= 2\pi \int_0^b xe^{-x^2}\,dx$$

$$= -\pi \int_0^b e^{-x^2}(-2x\,dx) \qquad (\text{Let } u = -x^2,\ du = -2x\,dx)$$

$$= \left[-\pi e^{-x^2}\right]_0^b = \pi(1 - e^{-b^2}).$$

When $b = 1$, $V = \pi(1 - e^{-1}) \approx 1.986$.

(b) Setting the expression for the volume equal to 4/3 and solving for b yields

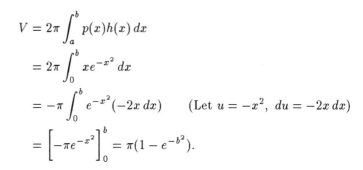

$$\pi(1 - e^{-b^2}) = \frac{4}{3}$$

$$3\pi - 3\pi e^{-b^2} = 4$$

$$e^{-b^2} = \frac{3\pi - 4}{3\pi}$$

$$-b^2 = \ln\left(\frac{3\pi - 4}{3\pi}\right),$$

$$b = \sqrt{\ln\left(\frac{3\pi - 4}{3\pi}\right)^{-1}} = \sqrt{\ln\left(\frac{3\pi}{3\pi - 4}\right)} \approx 0.743$$

7.2   Integration by Parts

9. Evaluate $\int x e^{-2x}\, dx$.

Solution:

Letting $u = x$ and $dv = e^{-2x}\, dx$, you have

$$dv = e^{-2x}\, dx \quad\Longrightarrow\quad v = \int e^{-2x}\, dx = -\frac{1}{2} e^{-2x}$$

$$u = x \quad\Longrightarrow\quad du = dx$$

Therefore, you have

$$\int u\, dv = uv - \int v\, du$$

$$\int x e^{-2x}\, dx = x\left(-\frac{1}{2} e^{-2x}\right) - \int -\frac{1}{2} e^{-2x}\, dx$$

$$= -\frac{x}{2} e^{-2x} - \frac{1}{4} e^{-2x} + C$$

$$= -\frac{1}{4e^{2x}}(2x + 1) + C$$

15. Evaluate $\int t \ln(t + 1)\, dt$.

Solution:

Letting $u = \ln(t + 1)$ and $dv = t\, dt$, you have

$$dv = t\, dt \quad\Longrightarrow\quad v = \int t\, dt = \frac{t^2}{2}$$

$$u = \ln(t + 1) \quad\Longrightarrow\quad du = \frac{1}{t+1}\, dt \quad .$$

Therefore, you have

$$\int u\, dv = uv - \int v\, du$$

$$\int t \ln(t + 1)\, dt = \frac{t^2}{2} \ln(t + 1) - \frac{1}{2} \int \frac{t^2}{t+1}\, dt$$

$$= \frac{t^2}{2} \ln(t + 1) - \frac{1}{2} \int \left(t - 1 + \frac{1}{t+1}\right) dt$$

$$= \frac{t^2}{2} \ln(t + 1) - \frac{1}{2} \left[\frac{t^2}{2} - t + \ln(t + 1)\right] + C$$

$$= \frac{1}{4}[2(t^2 - 1)\ln(t + 1) - t^2 + 2t] + C$$

23. Evaluate $\int x\sqrt{x-1}\,dx$.

Solution:

Letting $u = x$ and $dv = \sqrt{x-1}\,dx$, you have

$$dv = \sqrt{x-1}\,dx \quad \Longrightarrow \quad v = \int \sqrt{x-1}\,dx = \frac{2}{3}(x-1)^{3/2}$$
$$u = x \qquad\qquad \Longrightarrow \quad du = dx$$

Therefore,

$$\int u\,dv = uv - \int v\,du$$
$$\int x\sqrt{x-1}\,dx = \frac{2}{3}x(x-1)^{3/2} - \int \frac{2}{3}(x-1)^{3/2}\,dx$$
$$= \frac{2}{3}x(x-1)^{3/2} - \frac{4}{15}(x-1)^{5/2} + C$$
$$= \frac{2}{15}(x-1)^{3/2}[5x - 2(x-1)] + C$$
$$= \frac{2}{15}(x-1)^{3/2}(3x+2) + C$$

27. Evaluate $\int \arctan x\,dx$.

Solution:

Let

$$dv = dx \qquad\qquad \Longrightarrow \quad v = \int dx = x$$
$$u = \arctan x \quad \Longrightarrow \quad du = \frac{1}{1+x^2}\,dx$$

Therefore, you have

$$\int \arctan x\,dx = x\arctan x - \int \frac{x}{1+x^2}\,dx$$
$$= x\arctan x - \frac{1}{2}\ln(1+x^2) + C$$

29. Evaluate $\int e^{2x} \sin x \, dx$.

Solution:

Let
$$dv = \sin x \, dx \quad \Longrightarrow \quad v = \int \sin x \, dx = -\cos x$$
$$u = e^{2x} \quad \Longrightarrow \quad du = 2e^{2x} \, dx$$

$$\int e^{2x} \sin x \, dx = -e^{2x} \cos x + \int \cos x (2e^{2x}) \, dx$$

Using integration by parts again, let
$$dv = \cos x \, dx \quad \Longrightarrow \quad v = \int \cos x \, dx = \sin x$$
$$u = 2e^{2x} \quad \Longrightarrow \quad du = 4e^{2x} \, dx$$

$$\int e^{2x} \sin x \, dx = -e^{2x} \cos x + 2e^{2x} \sin x - \int 4e^{2x} \sin x \, dx$$

Adding the integral in the right hand member of the equation to both members of the equation yields

$$5 \int e^{2x} \sin x \, dx = -e^{2x} \cos x + 2e^{2x} \sin x + C_1$$

Finally, dividing both members of the equation by 5 yields

$$\int e^{2x} \sin x \, dx = \frac{e^{2x}}{5}(2 \sin x - \cos x) + C$$

38. Evaluate $\int_0^1 x \arcsin x^2 \, dx$.

Solution:

Let
$$dv = x \, dx \quad \Longrightarrow \quad v = \int x \, dx = \frac{x^2}{2}$$
$$u = \arcsin x^2 \quad \Longrightarrow \quad du = \frac{2x}{\sqrt{1 - x^4}} \, dx$$

$$\int x \arcsin x^2 \, dx = \frac{x^2 \arcsin x^2}{2} - \int \frac{x^2}{2}\left(\frac{2x}{\sqrt{1-x^4}}\right) dx$$

$$= \frac{x^2 \arcsin x^2}{2} - \int \frac{x^3}{\sqrt{1-x^4}} \, dx$$

$$= \frac{x^2 \arcsin x^2}{2} + \frac{1}{4}\int (1-x^4)^{-1/2}(-4x^3) \, dx$$

$$= \frac{x^2 \arcsin x^2}{2} + \frac{1}{4}\left[\frac{(1-x^4)^{1/2}}{\frac{1}{2}}\right] + C$$

$$= \frac{1}{2}[x^2 \arcsin x^2 + \sqrt{1-x^4}] + C$$

Finally,

$$\int_0^1 x \arcsin x^2 \, dx = \frac{1}{2}\left[x^2 \arcsin x^2 + \sqrt{1-x^4}\right]_0^1$$

$$= \frac{1}{2}\left(\frac{\pi}{2} + 0 - 0 - 1\right) = \frac{1}{2}\left(\frac{\pi}{2} - 1\right)$$

41. Evaluate $\displaystyle\int_0^{\pi/2} x \cos x \, dx$.

Solution:

Let

$$dv = \cos \, dx \implies v = \int \cos \, dx = \sin x$$

$$u = x \implies du = dx$$

$$\int x \cos x \, dx = x \sin x - \int \sin x \, dx$$

$$= x \sin x + \cos x + C$$

Finally,

$$\int_0^{\pi/2} x \cos x \, dx = \left[x \sin x + \cos x\right]_0^{\pi/2}$$

$$= \frac{\pi}{2} - 1$$

43. Use the tabular method for repeated application of Integration by Parts to evaluate

$$\int x^2 e^{2x}\, dx.$$

Solution:

Begin by letting $u = x^2$ and $dv = v'\, dx = e^{2x}\, dx$. We next create a table consisting of three columns as follows:

Alternate Signs	u and its Derivatives	v' and its Antiderivatives
$+$	x^2	e^{2x}
$-$	$2x$	$\frac{1}{2}e^{2x}$
$+$	2	$\frac{1}{4}e^{2x}$
$-$	0	$\frac{1}{8}e^{2x}$

Finally, the solution is given by multiplying the signed products of the diagonal entries of the table to obtain

$$\int x^2 e^{2x}\, dx = \frac{1}{2}x^2 e^{2x} - 2x\left(\frac{1}{4}\right)e^{2x} + 2\left(\frac{1}{8}\right)e^{2x} + C$$

$$= \frac{e^{2x}}{4}(2x^2 - 2x + 1) + C$$

61. Use integration by parts to verify the formula

$$\int x^n \ln x\, dx = \frac{x^{n+1}}{(n+1)^2}[-1 + (n+1)\ln x] + C.$$

Solution:

Let

$$dv = x^n\, dx \quad \Longrightarrow \quad v = \int x^n\, dx = \frac{x^{n+1}}{n+1}$$

$$u = \ln x \quad \Longrightarrow \quad du = \frac{1}{x}\, dx$$

$$\int x^n \ln x\, dx = \frac{x^{n+1}}{n+1}\ln x - \int \frac{x^{n+1}}{n+1}\left(\frac{1}{x}\right)dx$$

$$= \frac{x^{n+1}}{n+1}\ln x - \frac{1}{n+1}\int x^n\, dx$$

$$= \frac{x^{n+1}}{n+1}\ln x - \frac{x^{n+1}}{(n+1)^2} + C$$

$$= \frac{x^{n+1}}{(n+1)^2}[-1 + (n+1)\ln x] + C$$

73. Given the region bounded by the graphs of $y = \ln x$, $y = 0$, and $x = e$, find

(a) the area of the region.

(b) the volume of the solid generated by revolving the region about the x-axis.

(c) the volume of the solid generated by revolving the region about the y-axis.

(d) the centroid of the region.

Solution:

(a) From the accompanying figure you have

$$A = \int_1^e \ln x \, dx$$

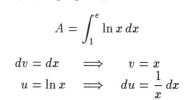

$$dv = dx \implies v = x$$
$$u = \ln x \implies du = \frac{1}{x} \, dx$$

Therefore,

$$A = \int_1^e \ln x \, dx = x \ln x - \int dx$$
$$= \left[x \ln x - x \right]_1^e = 1$$

(b) Using the Disk Method you have

$$V = \pi \int_1^e (\ln x)^2 \, dx$$

Let

$$dv = dx \implies v = x$$
$$u = (\ln x)^2 \implies du = \frac{2 \ln x}{x} \, dx$$

Therefore,

$$\int (\ln x)^2 \, dx = x(\ln x)^2 - 2 \int \ln x \, dx$$
$$= x(\ln x)^2 - 2x(\ln x - 1) \qquad \text{From part (a)}$$

Finally,

$$V = \pi \int_1^e (\ln x)^2 \, dx$$
$$= \pi \left[x(\ln x)^2 - 2x(\ln x - 1) \right]_1^e = \pi(e - 2) \approx 2.257$$

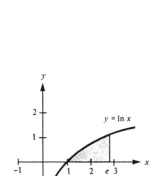

(c) Using the Shell Method you have

$$V = 2\pi \int_1^e x \ln x \, dx$$

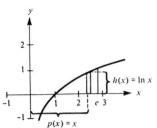

Let

$$dv = x \, dx \quad \Longrightarrow \quad v = \frac{x^2}{2}$$

$$u = \ln x \quad \Longrightarrow \quad du = \frac{1}{x} \, dx$$

Therefore,

$$\int x \ln x \, dx = \frac{x^2}{2} \ln x - \frac{1}{2} \int x \, dx = \frac{x^2}{2} \ln x - \frac{x^2}{4}$$

Finally,

$$V = 2\pi \int_1^e x \ln x \, dx$$

$$= 2\pi \left[\frac{x^2}{4} (2 \ln x - 1) \right]_1^e = \frac{\pi}{2}(e^2 + 1) \approx 13.177$$

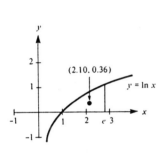

(d) $\displaystyle \bar{x} = \frac{1}{A} \int_1^e x \ln x \, dx$

$$= \left[\frac{x^2}{4} (2 \ln x - 1) \right]_1^e \qquad \text{From part (c)}$$

$$= \frac{1}{4}(e^2 + 1) \approx 2.097$$

$$\bar{y} = \frac{1}{2A} \int_1^e (\ln x)^2 \, dx$$

$$= \frac{1}{2} \left[x(\ln x)^2 - 2x(\ln x - 1) \right]_1^e \qquad \text{From part (b)}$$

$$= \frac{e - 2}{2} \approx 0.359$$

7.3   Trigonometric Integrals

5. Evaluate $\int \sin^5 x \cos^2 x \, dx$.

Solution:

$$\int \sin^5 x \cos^2 x \, dx = \int \sin x (\sin^2 x)^2 \cos^2 x \, dx$$

$$= \int \sin x (1 - \cos^2 x)^2 \cos^2 x \, dx$$

$$= -\int (\cos^2 x - 2\cos^4 x + \cos^6 x)(-\sin x) \, dx$$

$$= -\frac{1}{3}\cos^3 x + \frac{2}{5}\cos^5 x - \frac{1}{7}\cos^7 x + C$$

9. Evaluate $\int x \sin^2 x \, dx$.

Solution:

Use Integration by Part by letting $u = x$ and $dv = \sin^2 x \, dx$. Then $du = dx$ and

$$v = \int \sin^2 x \, dx$$

$$= \int \frac{1 - \cos 2x}{2} \, dx$$

$$= \frac{1}{2}\left(x - \frac{1}{2}\sin 2x\right) = \frac{1}{4}(2x - \sin 2x).$$

Therefore,

$$\int u \, dv = uv - \int v \, du$$

$$\int x \sin 2x \, dx = \frac{1}{4}x(2x - \sin 2x) - \frac{1}{4}\int (2x - \sin 2x) \, dx$$

$$= \frac{1}{4}(2x - \sin 2x) - \frac{1}{4}\left(x^2 + \frac{1}{2}\cos 2x\right) + C$$

$$= \frac{1}{8}(2x^2 - 2x \sin 2x - \cos 2x) + C$$

21. Evaluate $\displaystyle\int \tan^5 \frac{x}{4}\, dx$.

Solution:

$$\int \tan^5 \frac{x}{4}\, dx = \int \tan^2 \frac{x}{4} \tan^3 \frac{x}{4}\, dx$$

$$= \int \left(\sec^2 \frac{x}{4} - 1 \right) \tan^3 \frac{x}{4}\, dx$$

$$= \int \tan^3 \frac{x}{4} \sec^2 \frac{x}{4}\, dx - \int \tan^3 \frac{x}{4}\, dx$$

$$= \tan^4 \frac{x}{4} - \int \tan^2 \frac{x}{4} \tan \frac{x}{4}\, dx$$

$$= \tan^4 \frac{x}{4} - \int \left(\sec^2 \frac{x}{4} - 1 \right) \tan \frac{x}{4}\, dx$$

$$= \tan^4 \frac{x}{4} - \int \tan \frac{x}{4} \sec^2 \frac{x}{4}\, dx + \int \tan \frac{x}{4}\, dx$$

$$= \tan^4 \frac{x}{4} - 2 \tan^2 \frac{x}{4} - 4 \ln \left| \cos \frac{x}{4} \right| + C$$

27. Evaluate $\displaystyle\int \sec^6 4x \tan 4x\, dx$.

Solution:

$$\int \sec^6 4x \tan 4x\, dx$$

$$= \int (\sec^2 4x)(\sec^2 4x)^2 \tan 4x\, dx$$

$$= \frac{1}{4} \int \tan 4x (\tan^2 4x + 1)^2 (4 \sec^2 4x)\, dx$$

$$= \frac{1}{4} \int (\tan^5 4x + 2 \tan^3 4x + \tan 4x)(4 \sec^2 4x)\, dx$$

$$= \frac{1}{4} \left[\frac{\tan^6 4x}{6} + \frac{\tan^4 4x}{2} + \frac{\tan^2 4x}{2} \right] + C$$

$$= \frac{1}{24} (\tan^2 4x)(\tan^4 4x + 3 \tan^2 4x + 3) + C$$

or

$$\int \sec^6 4x \tan 4x\, dx = \frac{1}{4} \int \sec^5 4x (4 \sec 4x \tan 4x)\, dx$$

$$= \frac{1}{24} \sec^6 4x + C_1$$

(See Exercise 67 for a comparison of the two methods.)

31. Solve the differential equation $\dfrac{dr}{d\theta} = \sin^4 \pi\theta$.

Solution:

$$\frac{dr}{d\theta} = \sin^4 \pi\theta$$

$$\int \frac{dr}{d\theta}\, d\theta = \sin^4 \pi\theta \, d\theta$$

$$r = \int (\sin^2 \pi\theta)^2 \, d\theta$$

$$= \int \left(\frac{1 - \cos 2\pi\theta}{2}\right)^2 d\theta$$

$$= \frac{1}{4}\int (1 - 2\cos 2\pi\theta + \cos^2 2\pi\theta)\, d\theta$$

$$= \frac{1}{4}\int \left(1 - 2\cos 2\pi\theta + \frac{1 + \cos 4\pi\theta}{2}\right) d\theta$$

$$= \frac{1}{8}\int (3 - 4\cos 2\pi\theta + \cos 4\pi\theta)\, d\theta$$

$$= \frac{1}{8}\left(3\theta - \frac{2}{\pi}\sin 2\pi\theta + \frac{1}{4\pi}\sin 4\pi\theta\right) + C$$

$$= \frac{1}{32\pi}(12\pi\theta - 8\sin 2\pi\theta + \sin 4\pi\theta) + C$$

35. Evaluate $\displaystyle\int \sin 3x \cos 2x \, dx$.

Solution:

Using the trigonometric identity
$\sin u \cos v = \frac{1}{2}[\sin(u+v) + \sin(u-v)]$, you have

$$\int \sin 3x \cos 2x \, dx = \int \frac{1}{2}(\sin 5x + \sin x)\, dx$$

$$= \frac{1}{2}\left(\frac{1}{5}\right)\int \sin 5x(5)\, dx + \frac{1}{2}\int \sin x \, dx$$

$$= -\frac{1}{10}\cos 5x - \frac{1}{2}\cos x + C$$

$$= -\frac{1}{10}(\cos 5x + 5\cos x) + C$$

51. Evaluate $\displaystyle\int_0^{\pi/4} \tan^3 x \, dx$.

Solution:

$$\int_0^{\pi/4} \tan^3 x \, dx = \int_0^{\pi/4} (\sec^2 x - 1)(\tan x) \, dx$$

$$= \int_0^{\pi/4} \tan x \sec^2 x \, dx - \int_0^{\pi/4} \tan x \, dx$$

$$= \left[\frac{1}{2} \tan^2 x + \ln|\cos x| \right]_0^{\pi/4} = \frac{1}{2}(1 - \ln 2)$$

67. Evaluate $\int \sec^4 3x \tan^3 3x \, dx$ in two ways and show that the results differ only by a constant.

Solution:

Method 1: Since there is an odd power of the tangent, write

$$\int \sec^4 3x \tan^3 3x \, dx = \int \sec^3 3x \tan^2 3x (\sec 3x \tan 3x) \, dx$$

$$= \int \sec^3 3x (\sec^2 3x - 1) \sec 3x \tan 3x \, dx$$

$$= \frac{1}{3} \int (\sec^5 3x - \sec^3 3x)(3 \sec 3x \tan 3x \, dx)$$

$$= \frac{1}{3} \left(\frac{1}{6} \sec^6 3x - \frac{1}{4} \sec^4 3x \right) + C$$

Method 2: Since the power of the secant is even, write

$$\int \sec^4 3x \tan^3 3x \, dx = \int \sec^2 3x \tan^3 3x (\sec^2 3x) \, dx$$

$$= \int (1 + \tan^2 3x) \tan^3 3x (\sec^2 3x) \, dx$$

$$= \frac{1}{3} \int (\tan^3 3x + \tan^5 3x)(3 \sec^2 3x) \, dx$$

$$= \frac{1}{3} \left(\frac{\tan^4 3x}{4} + \frac{\tan^6 3x}{6} \right) + C$$

Comparing the results of the two methods, you have

$$\frac{1}{3}\left[\frac{1}{4}\tan^4 3x + \frac{1}{6}\tan^6 3x\right] + C$$

$$= \frac{1}{3}\left[\frac{1}{4}(\sec^2 3x - 1)^2 + \frac{1}{6}(\sec^2 3x - 1)^3\right] + C$$

$$= \frac{1}{3}\left[\frac{1}{4}(\sec^4 3x - 2\sec^2 3x + 1)\right.$$

$$\left. + \frac{1}{6}(\sec^6 3x - 3\sec^4 3x + 3\sec^2 3x - 1)\right] + C$$

$$= \frac{1}{3}\left[\sec^6 3x - \frac{1}{4}\sec^4 3x + \frac{1}{4} - \frac{1}{6}\right] + C$$

Therefore, the results differ only by the constant $\frac{1}{3}(\frac{1}{4} - \frac{1}{6})$.

71. Find (a) the volume of the solid generated by revolving the region bounded by the graphs of $y = \sin x$, $y = 0$, $x = 0$, and $x = \pi$ about the x-axis and (b) find the centroid of the region.

Solution:

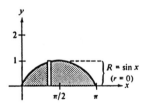

(a) $$V = \pi \int_0^\pi R^2\, dx = \pi \int_0^\pi \sin^2 x\, dx$$

$$= \pi \int_0^\pi \frac{1 - \cos 2x}{2}\, dx$$

$$= \frac{\pi}{2}\int_0^\pi (1 - \cos 2x)\, dx$$

$$= \frac{\pi}{2}\left[x - \frac{1}{2}\sin 2x\right]_0^\pi = \frac{\pi}{2}\left[\pi - \frac{1}{2}(0) - 0\right] = \frac{\pi^2}{2}$$

(b) By symmetry $\bar{x} = \dfrac{\pi}{2}$.

The area of the region is given by

$$A = \int_0^\pi \sin x\, dx = \left[-\cos x\right]_0^\pi = 2$$

Therefore,

$$\bar{y} = \frac{1}{2A} \int_0^\pi \sin^2 x \, dx$$

$$= \frac{1}{8} \int_0^\pi (1 - \cos 2x) \, dx = \left[\frac{1}{8} \left(x - \frac{1}{2} \sin 2x \right) \right]_0^\pi = \frac{\pi}{8}$$

Thus, $(\bar{x}, \bar{y}) = \left(\dfrac{\pi}{2}, \dfrac{\pi}{8} \right)$.

75. Use integration by parts to verify the formula

$$\int \cos^m x \sin^n x \, dx$$

$$= -\frac{\cos^{m+1} x \sin^{n-1} x}{m+n} + \frac{n-1}{m+n} \int \cos^m x \sin^{n-2} x \, dx$$

Solution:

Let $dv = \cos^m x \sin x \, dx$ and $u = \sin^{n-1} x$. Then
$v = (-\cos^{m+1} x)/(m+1)$ and $du = (n-1)\sin^{n-2} x(\cos x) \, dx$.
Therefore,

$$\int \cos^m x \sin^n x \, dx$$

$$= -\frac{\sin^{n-1} x \cos^{m+1} x}{m+1} + \frac{n-1}{m+1} \int \sin^{n-2} x \cos^{m+2} x \, dx$$

$$= -\frac{\sin^{n-1} x \cos^{m+1} x}{m+1} + \frac{n-1}{m+1} \int \sin^{n-2} x \cos^m x (1 - \sin^2 x) \, dx$$

$$= -\frac{\sin^{n-1} x \cos^{m+1} x}{m+1} + \frac{n-1}{m+1} \int \sin^{n-2} x \cos^m x \, dx$$

$$- \frac{n-1}{m+1} \int \sin^n x \cos^m x \, dx$$

Now observe that the last integral is a multiple of the original. Adding yields

$$\frac{m+n}{m+1} \int \cos^m x \sin^n x \, dx$$

$$= -\frac{\sin^{n-1} x \cos^{m+1} x}{m+1} + \frac{n-1}{m+1} \int \cos^m x \sin^{n-2} x \, dx$$

$$\int \cos^m x \sin^n x \, dx$$

$$= -\frac{\cos^{m+1} x \sin^{n-1} x}{m+n} + \frac{n-1}{m+n} \int \cos^m x \sin^{n-2} x \, dx$$

7.4   Trigonometric Substitution

7. Evaluate $\displaystyle\int \frac{\sqrt{25 - x^2}}{x}\, dx$.

Solution:

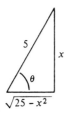

Let $x = 5\sin\theta$. Then $\sqrt{25 - x^2} = 5\cos\theta$ and $dx = 5\cos\theta\, d\theta$. Thus,

$$
\int \frac{\sqrt{25 - x^2}}{x}\, dx = \int \frac{5\cos\theta}{5\sin\theta} 5\cos\theta\, d\theta
$$

$$
= 5\int \frac{\cos^2\theta}{\sin\theta}\, d\theta
$$

$$
= 5\int \frac{1 - \sin^2\theta}{\sin\theta}\, d\theta
$$

$$
= 5\int (\csc\theta - \sin\theta)\, d\theta
$$

$$
= 5(\ln|\csc\theta - \cot\theta| + \cos\theta) + C
$$

$$
= 5\left(\ln\left|\frac{5}{x} - \frac{\sqrt{25 - x^2}}{x}\right| + \frac{\sqrt{25 - x^2}}{5}\right) + C
$$

$$
= 5\ln\left|\frac{5 - \sqrt{25 - x^2}}{x}\right| + \sqrt{25 - x^2} + C
$$

11. Evaluate $\int x^3\sqrt{x^2 - 4}\, dx$.

Solution:

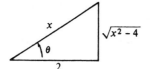

Let $x = 2\sec\theta$. Then $\sqrt{x^2 - 4} = 2\tan\theta$ and $dx = 2\sec\theta\tan\theta\, d\theta$. Thus,

$$\int x^3 \sqrt{x^2 - 4}\, dx$$

$$= \int (2\sec\theta)^3 (2\tan\theta)(2\sec\theta\tan\theta)\, d\theta$$

$$= 32 \int \sec^4\theta \tan^2\theta\, d\theta$$

$$= 32 \int \sec^2\theta \tan^2\theta \sec^2\theta\, d\theta$$

$$= 32 \int (\tan^2\theta + 1)\tan^2\theta \sec^2\theta\, d\theta$$

$$= 32 \int (\tan^4\theta + \tan^2\theta)\sec^2\theta\, d\theta$$

$$= 32 \left[\frac{1}{5}\tan^5\theta + \frac{1}{3}\tan^3\theta \right] + C$$

$$= 32 \left[\frac{1}{5}\left(\frac{\sqrt{x^2-4}}{2}\right)^5 + \frac{1}{3}\left(\frac{\sqrt{x^2-4}}{2}\right)^3 \right] + C$$

$$= \frac{1}{15}(x^2 - 4)^{3/2}(3x^2 + 8) + C$$

15. Evaluate $\displaystyle\int \frac{1}{(1+x^2)^2}\, dx.$

Solution:

Let $x = \tan\theta$. Then $1 + x^2 = \sec^2\theta$ and $dx = \sec^2\theta\, d\theta$. Thus,

$$\int \frac{1}{(1+x^2)^2}\, dx = \int \frac{1}{\sec^4\theta}(\sec^2\theta)\, d\theta$$

$$= \int \cos^2\theta\, d\theta$$

$$= \frac{1}{2}\int (1 + \cos 2\theta)\, d\theta$$

$$= \frac{1}{2}\left(\theta + \frac{1}{2}\sin 2\theta\right) + C$$

$$= \frac{1}{2}(\theta + \sin\theta\cos\theta) + C$$

$$= \frac{1}{2}\left(\arctan x + \frac{x}{\sqrt{1+x^2}} \cdot \frac{1}{\sqrt{1+x^2}}\right) + C$$

$$= \frac{1}{2}\left(\arctan x + \frac{x}{1+x^2}\right) + C$$

27. Evaluate $\displaystyle\int \frac{1}{x\sqrt{4x^2+9}}\, dx$.

Solution:

Let $2x = 3\tan\theta$. Then $x = \frac{3}{2}\tan\theta$, $\sqrt{4x^2+9} = 3\sec\theta$, and $dx = \frac{3}{2}\sec^2\theta\, d\theta$.

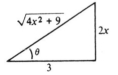

$$\int \frac{1}{x\sqrt{4x^2+9}}\, dx$$

$$= \int \frac{\frac{3}{2}\sec^2\theta\, d\theta}{\frac{3}{2}\tan\theta(3\sec\theta)} = \frac{1}{3}\int \csc\theta\, d\theta$$

$$= \frac{1}{3}\ln|\csc\theta - \cot\theta| + C$$

$$= \frac{1}{3}\ln\left|\frac{\sqrt{4x^2+9}-3}{2x}\right| + C$$

$$= -\frac{1}{3}\ln\left|\frac{2x}{\sqrt{4x^2+9}-3}\right| + C$$

$$= -\frac{1}{3}\ln\left|\left(\frac{2x}{\sqrt{4x^2+9}-3}\right)\left(\frac{\sqrt{4x^2+9}+3}{\sqrt{4x^2+9}+3}\right)\right| + C$$

$$= -\frac{1}{3}\ln\left|\frac{3+\sqrt{4x^2+9}}{2x}\right| + C$$

35. Evaluate $\displaystyle\int \frac{1}{4+4x^2+x^4}\, dx$.

Solution:

$$\int \frac{1}{4+4x^2+x^4}\, dx = \int \frac{1}{(2+x^2)^2}\, dx$$

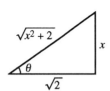

Let $x = \sqrt{2}\tan\theta$. Then $2 + x^2 = 2\sec^2\theta$ and $dx = \sqrt{2}\sec^2\theta\, d\theta$.

Thus,

$$\int \frac{1}{(2+x^2)^2}\, dx = \int \frac{\sqrt{2}\sec^2\theta\, d\theta}{(2\sec^2\theta)^2} = \frac{\sqrt{2}}{4}\int \cos^2\theta\, d\theta$$

$$= \frac{\sqrt{2}}{4}\int \frac{1+\cos 2\theta}{2}\, d\theta$$

$$= \frac{\sqrt{2}}{8}\left(\theta + \frac{1}{2}\sin 2\theta\right) + C$$

$$= \frac{\sqrt{2}}{8}(\theta + \sin\theta\cos\theta) + C$$

$$= \frac{\sqrt{2}}{8}\left[\arctan\frac{x}{\sqrt{2}} + \left(\frac{x}{\sqrt{x^2+2}}\right)\left(\frac{\sqrt{2}}{\sqrt{x^2+2}}\right)\right] + C$$

$$= \frac{1}{4}\left[\frac{x}{x^2+2} + \frac{1}{\sqrt{2}}\arctan\frac{x}{\sqrt{2}}\right] + C$$

37. Evaluate $\int \operatorname{arcsec} 2x\, dx$.

Solution:

Let

$$dv = dx \qquad \Longrightarrow \qquad v = x$$

$$u = \operatorname{arcsec} x \qquad \Longrightarrow \qquad du = \frac{1}{x\sqrt{4x^2-1}}\, dx$$

Therefore,

$$\int \operatorname{arcsec} 2x\, dx$$

$$= uv - \int v\, du$$

$$= x\operatorname{arcsec} 2x - \int x\left(\frac{1}{x\sqrt{4x^2-1}}\right) dx$$

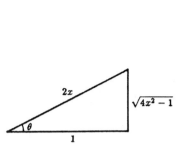

$$= x\operatorname{arcsec} 2x - \frac{1}{2}\int \frac{2}{\sqrt{(2x)^2-1^2}}\, dx \qquad (\text{Let } 2x = \sec\theta)$$

$$= x\operatorname{arcsec} 2x - \int \frac{(1/2)\sec\theta\tan\theta\, d\theta}{\tan\theta}$$

$$= x\operatorname{arcsec} 2x - \frac{1}{2}\int \sec\theta\, d\theta$$

$$= x\operatorname{arcsec} 2x - \frac{1}{2}\ln|\sec\theta + \tan\theta| + C$$

$$= x\operatorname{arcsec} 2x - \frac{1}{2}\ln|2x + \sqrt{4x^2-1}| + C$$

53. The region bounded by the circle $(x - 3)^2 + y^2 = 1$ is revolved about the y-axis. The resulting doughnut-shaped solid is called a torus. Find the volume of the solid.

Solution:

Shell Method

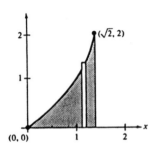

$$V = 2\pi \int_2^4 x\left[2\sqrt{1 - (x - 3)^2}\right] dx = 4\pi \int_2^4 x\sqrt{1 - (x - 3)^2}\, dx$$

Let $x - 3 = \sin\theta$. Then $\sqrt{1 - (x - 3)^2} = \cos\theta$ and $dx = \cos\theta\, d\theta$. Also, when $x = 2$, $\sin\theta = -1$ and $\theta = -\pi/2$. When $x = 4$, $\sin\theta = 1$ and $\theta = \pi/2$. Therefore,

$$V = 4\pi \int_{-\pi/2}^{\pi/2} (3 + \sin\theta)(\cos\theta)(\cos\theta)\, d\theta$$

$$= 4\pi\left[\int_{-\pi/2}^{\pi/2} 3\cos^2\theta\, d\theta + \int_{-\pi/2}^{\pi/2} \cos^2\theta \sin\theta\, d\theta\right]$$

$$= 4\pi\left[\int_{-\pi/2}^{\pi/2} \frac{3}{2}(1 + \cos 2\theta)\, d\theta + \int_{-\pi/2}^{\pi/2} \cos^2\theta \sin\theta\, d\theta\right]$$

$$= 4\pi\left[\frac{3}{2}(\theta + \frac{1}{2}\sin 2\theta) - \frac{1}{3}\cos^3\theta\right]_{-\pi/2}^{\pi/2} = 6\pi^2$$

63. Find the surface area of the solid generated by revolving the region bounded by $y = x^2$, $y = 0$, $x = 0$, and $x = \sqrt{2}$ about the x-axis.

Solution:

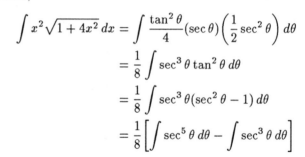

$$S = 2\pi \int_a^b y\sqrt{1 + (y')^2}\, dx = 2\pi \int_0^{\sqrt{2}} x^2\sqrt{1 + 4x^2}\, dx$$

Using trigonometric substitution, let $2x = \tan\theta$. Then $\sqrt{1 + 4x^2} = \sec\theta$, $x^2 = \frac{1}{4}\tan^2\theta$, and $dx = \frac{1}{2}\sec^2\theta\, d\theta$. Therefore,

$$\int x^2\sqrt{1 + 4x^2}\, dx = \int \frac{\tan^2\theta}{4}(\sec\theta)\left(\frac{1}{2}\sec^2\theta\right) d\theta$$

$$= \frac{1}{8}\int \sec^3\theta \tan^2\theta\, d\theta$$

$$= \frac{1}{8}\int \sec^3\theta(\sec^2\theta - 1)\, d\theta$$

$$= \frac{1}{8}\left[\int \sec^5\theta\, d\theta - \int \sec^3\theta\, d\theta\right]$$

Now using Integration by Parts to evaluate $\int \sec^5 \theta \, d\theta$, let

$$dv = \sec^2 \theta \, d\theta \quad \Longrightarrow \quad v = \tan \theta$$
$$u = \sec^3 \theta \quad \Longrightarrow \quad du = 3 \sec^3 \theta \tan \theta \, d\theta$$

Thus,

$$\int \sec^5 \theta \, d\theta = \sec^3 \theta \tan \theta - 3 \int \sec^3 \theta \tan^2 \theta \, d\theta$$

$$= \sec^3 \theta \tan \theta - 3 \int \sec^3 \theta (\sec^2 \theta - 1) \, d\theta$$

$$= \sec^3 \theta \tan \theta - 3 \int \sec^5 \theta \, d\theta + 3 \int \sec^3 \theta \, d\theta$$

$$4 \int \sec^5 \theta \, d\theta = \sec^3 \theta \tan \theta + 3 \int \sec^3 \theta \, d\theta$$

$$\int \sec^5 \theta \, d\theta = \frac{1}{4} \left(\sec^3 \theta \tan \theta + 3 \int \sec^3 \theta \, d\theta \right)$$

Hence,

$$\int x^2 \sqrt{1 + 4x^2} \, dx$$

$$= \frac{1}{8} \left[\frac{1}{4} \sec^3 \theta \tan \theta + \frac{3}{4} \int \sec^3 \theta \, d\theta - \int \sec^3 \theta \, d\theta \right]$$

$$= \frac{1}{32} \left(\sec^3 \theta \tan \theta - \int \sec^3 \theta \, d\theta \right)$$

Using the formula for $\int \sec^3 \theta \, d\theta$ from Section 7.2, you have

$$\int x^2 \sqrt{1 + 4x^2} \, dx$$

$$= \frac{1}{32} \left[\sec^3 \theta \tan \theta - \frac{1}{2} (\sec \theta \tan \theta + \ln | \sec \theta + \tan \theta |) \right] + C$$

Finally, when $x = 0$, $\theta = 0$, and when $x = \sqrt{2}$, $\theta = \arctan 2\sqrt{2}$. Therefore,

$$S = 2\pi \int_0^{\sqrt{2}} x^2 \sqrt{1 + 4x^2} \, dx$$

$$= \frac{\pi}{16} \left[\sec^3 \theta \tan \theta - \frac{1}{2} (\sec \theta \tan \theta + \ln | \sec \theta + \tan \theta |) \right]_0^{\arctan 2\sqrt{2}}$$

$$= \frac{\pi}{16} \left[51\sqrt{2} - \frac{1}{2} \ln (2\sqrt{2} + 3) \right]$$

$$= \frac{\pi}{32} [102\sqrt{2} - \ln (2\sqrt{2} + 3) \approx 13.989$$

7.5 Partial Fractions

13. Evaluate $\displaystyle\int \frac{x^2 + 12x + 12}{x^3 - 4x}\, dx$.

Solution:

$$\frac{x^2 + 12x + 12}{x^3 - 4x} = \frac{x^2 + 12x + 12}{x(x - 2)(x + 2)} = \frac{A}{x} + \frac{B}{x - 2} + \frac{C}{x + 2}$$

Multiplying by $(x)(x - 2)(x + 2)$, yields

$$x^2 + 12x + 12 = A(x - 2)(x + 2) + B(x)(x + 2) + C(x)(x - 2).$$

If $x = 0$, then $12 = A(-2)(2)$ or $A = -3$.

If $x = 2$, then $4 + 24 + 12 = B(2)(4)$ or $B = 5$

If $x = -2$, then $4 - 24 + 12 = C(-2)(-4)$ or $C = -1$

Thus,

$$\int \frac{x^2 + 12x + 12}{x^3 - 4x}\, dx = \int \left(\frac{-3}{x} + \frac{5}{x - 2} + \frac{-1}{x + 2} \right) dx$$

$$= -3 \ln |x| + 5 \ln |x - 2| - \ln |x + 2| + C$$

$$= \ln \left| \frac{(x - 2)^5}{x^3(x + 2)} \right| + C$$

25. Evaluate $\displaystyle\int \frac{x^2 + 5}{x^3 - x^2 + x + 3}\, dx$.

Solution:

Since $x^3 - x^2 + x + 3 = (x + 1)(x^2 - 2x + 3)$, you have

$$\frac{x^2 + 5}{x^3 - x^2 + x + 3} = \frac{A}{x + 1} + \frac{Bx + C}{x^2 - 2x + 3}.$$

Multiplying by $(x + 1)(x^2 - 2x + 3)$ yields

$$x^2 + 5 = A(x^2 - 2x + 3) + (Bx + C)(x + 1).$$

If $x = -1$, then $1 + 5 = A(1 + 2 + 3)$ or $A = 1$.

Therefore,

$$x^2 + 5 = x^2 - 2x + 3 + Bx^2 + Bx + Cx + C$$
$$2x + 2 = Bx^2 + (B + C)x + C$$

Equating coefficients, yields $B = 0$, $B + C = 2$, and $C = 2$.
Finally,

$$\int \frac{x^2}{x^3 - x^2 + x + 3}\, dx$$

$$= \int \left(\frac{1}{x + 1} + \frac{2}{x^2 - 2x + 3} \right) dx$$

$$= \int \frac{1}{x + 1}\, dx + 2 \int \frac{1}{(x - 1)^2 + 2}\, dx$$

$$= \ln|x + 1| + 2 \left(\frac{1}{\sqrt{2}} \right) \arctan \left(\frac{x - 1}{\sqrt{2}} \right) + C$$

$$= \ln|x + 1| + \sqrt{2} \arctan \left(\frac{x - 1}{\sqrt{2}} \right) + C$$

27. Solve the differential equation $\dfrac{dy}{dx} = \dfrac{x^4}{(x - 1)^3}$.

Solution:

$$\frac{dy}{dx} = \frac{x^4}{(x - 1)^3}$$

$$\int \frac{dy}{dx}\, dx = \int \frac{x^4}{(x - 1)^3}\, dx$$

$$y = \int \frac{x^4}{(x - 1)^3}\, dx$$

Division yields

$$\frac{x^4}{(x - 1)^3} = x + 3 + \frac{6x^2 - 8x + 3}{(x - 1)^3}$$

Furthermore, by partial fractions you have

$$\frac{6x^2 - 8x + 3}{(x - 1)^3} = \frac{A}{x - 1} + \frac{B}{(x - 1)^2} + \frac{C}{(x - 1)^3}$$

Multiplying by $(x-1)^3$, yields

$$6x^2 - 8x + 3 = A(x-1)^2 + B(x-1) + C$$
$$= A(x^2 - 2x + 1) + B(x-1) + C$$
$$= Ax^2 + (B - 2A)x + (A - B + C).$$

Now by equating coefficients of like terms, you have the three equations

$$6 = A, \quad -8 = B - 2A, \quad 3 = A - B + C.$$

Solving these equations yields

$$A = 6, \quad B = 4, \quad C = 1.$$

Thus,

$$y = \int \frac{x^4}{(x-1)^3} \, dx$$
$$= \int \left[x + 3 + \frac{6}{x-1} + \frac{4}{(x-1)^2} + \frac{1}{(x-1)^3} \right] dx$$
$$= \frac{x^2}{2} + 3x + 6 \ln|x-1| - \frac{4}{x-1} - \frac{1}{2(x-1)^2} + C$$

33. Evaluate $\displaystyle\int_1^2 \frac{x+1}{x(x^2+1)} \, dx$.

Solution:

$$\frac{x+1}{x(x^2+1)} = \frac{A}{x} + \frac{Bx+C}{x^2+1}$$

Multiplying by $(x)(x^2+1)$ yields

$$x + 1 = A(x^2+1) + (Bx+C)(x) = Ax^2 + A + Bx^2 + Cx$$
$$= (A+B)x^2 + Cx + A.$$

By equating coefficients, you have $A + B = 0$, $C = 1$, and $A = 1$. Thus, $B = -1$, and you obtain

$$\int_1^2 \frac{x+1}{x(x^2+1)}\,dx$$

$$= \int_1^2 \left(\frac{1}{x} + \frac{-x+1}{x^2+1} \right) dx$$

$$= \int_1^2 \frac{1}{x}\,dx - \frac{1}{2}\int_1^2 \frac{2x}{x^2+1}\,dx + \int_1^2 \frac{1}{x^2+1}\,dx$$

$$= \left[\ln|x| - \frac{1}{2}\ln|x^2+1| + \arctan x \right]_1^2$$

$$= \ln 2 - \frac{1}{2}\ln 5 + \arctan 2 - \ln 1 + \frac{1}{2}\ln 2 - \arctan 1$$

$$= \frac{3}{2}\ln 2 - \frac{1}{2}\ln 5 + \arctan 2 - \frac{\pi}{4}$$

$$= \frac{1}{2}\ln\frac{8}{5} + \arctan 2 - \frac{\pi}{4} \approx 0.557$$

47. Evaluate $\displaystyle\int \frac{e^x}{(e^x-1)(e^x+4)}\,dx$ by letting $u = e^x$.

Solution:

If $u = e^x$, then $du = e^x\,dx$, and

$$\int \frac{e^x}{(e^x-1)(e^x+4)}\,dx = \int \frac{du}{(u-1)(u+4)}$$

Using partial fractions yields

$$\frac{1}{(u-1)(u+4)} = \frac{A}{u-1} + \frac{B}{u+4}.$$

Multiplying by $(u-1)(u+4)$ yields

$$1 = A(u+4) + B(u-1).$$

If $u = 1$, then $1 = A(5)$ or $A = \dfrac{1}{5}$.

If $u = -4$, then $1 = B(-5)$ or $B = -\dfrac{1}{5}$.

Therefore,

$$\int \frac{du}{(u-1)(u+4)} = \frac{1}{5}\int \left(\frac{1}{u-1} - \frac{1}{u+4}\right) du$$

$$= \frac{1}{5}(\ln|u-1| - \ln|u+4|) + C$$

$$= \frac{1}{5}\ln\left|\frac{u-1}{u+4}\right| + C$$

$$= \frac{1}{5}\ln\left|\frac{e^x - 1}{e^x + 4}\right| + C$$

57. A single infected individual enters a community of n individuals susceptible to the disease. Let x be the number of newly infected individuals after time t. The common Epidemic Model assumes that the disease spreads at a rate proportional to the product of the total number infected and the number of susceptible not yet infected. Thus, $dx/dt = k(x+1)(n-x)$, and you obtain

$$\int \frac{1}{(x+1)(n-x)}\, dx = \int k\, dt.$$

Solve for x as a function of t.

Solution:

$$\frac{1}{(x+1)(n-x)} = \frac{A}{x+1} + \frac{B}{n-x}$$

Multiplying by $(x+1)(n-x)$ yields

$$1 = A(n-x) + B(x+1).$$

If $x = -1$, then $1 = A(n+1)$ or $A = \dfrac{1}{n+1}$.

If $x = n$, then $1 = B(n+1)$ or $B = \dfrac{1}{n+1}$.

Therefore,

$$\int \frac{1}{(x+1)(n-x)}\, dx = \int k\, dt$$

$$\frac{1}{n+1}\int \left(\frac{1}{x+1} + \frac{1}{n-x}\right) dx = \int k\, dt$$

$$\frac{1}{n+1}[\ln(x+1) - \ln(n-x)] = kt + C$$

$$\left(\frac{1}{n+1}\right)\ln\left(\frac{x+1}{n-x}\right) = kt + C.$$

When $t = 0$, $x = 0$. Thus,

$$\left(\frac{1}{n+1}\right)\ln\left(\frac{1}{n}\right) = C.$$

Therefore,

$$\ln\left(\frac{x+1}{n-x}\right) = k(n+1)t + \ln\left(\frac{1}{n}\right)$$

$$\frac{x+1}{n-x} = \frac{1}{n}e^{k(n+1)t}$$

$$x = \frac{n[e^{(n+1)kt} - 1]}{e^{(n+1)kt} + n}.$$

7.6   Integration by Tables and Other Techniques

9. Use an integration table to evaluate

$$\int \frac{1}{\sqrt{x}(1 - \cos\sqrt{x})}\, dx.$$

Solution:

Consider the form

$$\int \frac{1}{1 \pm \cos u}\, du = -\cot u \pm \csc u + C$$

with $u = \sqrt{x}$ and $du = (1/2\sqrt{x})\, dx$. Then,

$$\int \frac{1}{\sqrt{x}(1 - \cos\sqrt{x})}\, dx = 2\int \frac{(1/2\sqrt{x})}{1 - \cos\sqrt{x}}\, dx$$

$$= 2(-\cot\sqrt{x} - \csc\sqrt{x}) + C$$

$$= -2(\cot\sqrt{x} + \csc\sqrt{x}) + C.$$

11. Use an integration table to evaluate

$$\int \frac{1}{1 + e^{2x}} \, dx.$$

Solution:

Consider the form

$$\int \frac{1}{1 + e^u} \, du = u - \ln(1 + e^u) + C$$

where $u = 2x$ and $du = 2 \, dx$. Then,

$$\int \frac{1}{1 + e^{2x}} \, dx = \frac{1}{2} \int \frac{2}{1 + e^{2x}} \, dx$$

$$= \frac{1}{2}[2x - \ln(1 + e^{2x})] + C = x - \frac{1}{2}\ln(1 + e^{2x}) + C$$

21. Use an integration table to evaluate

$$\int \frac{1}{x^2 \sqrt{x^2 - 4}} \, dx.$$

Solution:

Consider the form

$$\int \frac{1}{u^2 \sqrt{u^2 \pm a^2}} \, du = \mp \frac{\sqrt{u^2 \pm a^2}}{a^2 u} + C$$

where $u = x$ and $a = 2$. Then,

$$\int \frac{1}{x^2 \sqrt{x^2 - 4}} \, dx = \int \frac{1}{u^2 \sqrt{u^2 - a^2}} \, du$$

$$= \frac{\sqrt{u^2 - a^2}}{a^2 u} + C = \frac{\sqrt{x^2 - 4}}{4x} + C.$$

23. Use an integration table to evaluate

$$\int \frac{2x}{(1-3x)^2}\,dx.$$

Solution:

Consider the form

$$\int \frac{u}{(a+bu)^2}\,du = \frac{1}{b^2}\left(\frac{a}{a+bu} + \ln|a+bu|\right) + C$$

where $u = x$, $a = 1$, $b = -3$, and $a + bu = 1 - 3x$. Then,

$$\int \frac{2x}{(1-3x)^2}\,dx = 2\int \frac{x}{(1-3x)^2}\,dx = 2\int \frac{u}{(a+bu)^2}\,du$$

$$= 2\left(\frac{1}{b^2}\right)\left(\frac{1}{a+bu} + \ln|a+bu|\right) + C$$

$$= \frac{2}{9}\left[\frac{1}{1-3x} + \ln|1-3x|\right] + C.$$

37. Use an integration table to evaluate

$$\int \frac{\ln x}{x(3 + 2\ln x)}\,dx.$$

Solution:

Consider the form

$$\int \frac{u}{a+bu} = \frac{1}{b^2}(bu - a\ln|a+bu|) + C$$

where $u = \ln x$, $a = 3$, $b = 2$, and $du = (1/x)\,dx$. Then,

$$\int \frac{\ln x}{x(3+2\ln x)}\,dx = \int \frac{\ln x}{3+2\ln x}\left(\frac{1}{x}\right)\,dx$$

$$= \frac{1}{4}[2\ln|x| - 3\ln(3 + 2\ln|x|)] + C.$$

43. Use an integration table to evaluate

$$\int \frac{x^3}{\sqrt{4 - x^2}}\, dx.$$

Solution:

Consider the form

$$\int \frac{u}{\sqrt{a + bu}}\, du = \frac{-2(2u - bu)}{3b^2}\sqrt{a + bu} + C$$

where $u = x^2$, $a = 4$, $b = -1$, $du = 2x\, dx$, and $\sqrt{a + bu} = \sqrt{4 - x^2}$. Then,

$$\int \frac{x^2}{\sqrt{4 - x^2}}\, dx = \frac{1}{2}\int \frac{x^2}{\sqrt{4 - x^2}}\,(2x)\, dx = \frac{1}{2}\int \frac{u}{\sqrt{a + bu}}\, du$$

$$= \frac{1}{2}\left[\frac{(-2)(8 + x^2)}{3}\right]\sqrt{4 - x^2} + C$$

$$= -\left(\frac{x^2 + 8}{3}\right)\sqrt{4 - x^2} + C.$$

47. Verify the formula

$$\int \frac{u^2}{(a + bu)^2}\, du = \frac{1}{b^3}\left(bu - \frac{a^2}{a + bu} - 2a\ln|a + bu|\right) + C$$

by the method of partial fractions.

Solution:

Since the numerator and denominator are of the same degree, begin by dividing

$$\frac{u^2}{(a + bu)^2} = \frac{1}{b^2} - \frac{(2a/b)u + (a^2/b^2)}{(a + bu)^2}.$$

Now use partial fractions to obtain

$$\frac{(2a/b)u + (a^2/b^2)}{(a + bu)^2} = \frac{A}{a + bu} + \frac{B}{(a + bu)^2}.$$

Multiplying by $(a + bu)^2$ yields

$$\left(\frac{2a}{b}\right)u + \frac{a^2}{b^2} = A(a + bu) + B = bAu + (aA + B).$$

Equating the coefficients of like terms, you have

$$bA = \frac{2a}{b} \quad \text{and} \quad A = \frac{2a}{b^2}$$

$$aA + B = \frac{a^2}{b^2} \quad \text{and} \quad B = \frac{a^2}{b^2} - a\left(\frac{2a}{b^2}\right) = -\frac{a^2}{b^2}.$$

Therefore,

$$\int \frac{u^2}{(a+bu)^2}\,du$$

$$= \frac{1}{b^2}\int du - \frac{2a}{b^2}\left(\frac{1}{b}\right)\int \frac{b}{a+bu}\,du + \frac{a^2}{b^2}\left(\frac{1}{b}\right)\int \frac{b}{(a+bu)^2}\,du$$

$$= \left(\frac{1}{b^2}\right)u - \frac{2a}{b^3}(\ln|a+bu|) - \frac{a^2}{b^3}\left(\frac{1}{a+bu}\right) + C$$

$$= \frac{1}{b^3}\left[bu - \frac{a^2}{a+bu} - 2a\ln|a+bu|\right] + C$$

55. Evaluate $\displaystyle\int_0^{\pi/2} \frac{1}{1+\sin\theta+\cos\theta}\,d\theta$.

Solution:

Let $u = \dfrac{\sin\theta}{1+\cos\theta}$. Then $\cos\theta = \dfrac{1-u^2}{1+u^2}$, $\sin\theta = \dfrac{2u}{1+u^2}$, and $d\theta = \dfrac{2\,du}{1+u^2}$. Furthermore, when $\theta = \pi/2$, $u = 1$, and when $\theta = 0$, $u = 0$.

$$\int_0^{\pi/2} \frac{1}{1+\sin\theta+\cos\theta}\,d\theta$$

$$= \int_0^1 \frac{[2/(1+u^2)]\,du}{1+[(2u)/(1+u^2)]+[(1-u^2)/(1+u^2)]}$$

$$= \int_0^1 \frac{1}{u+1}\,du = \Big[\ln|u+1|\Big]_0^1 = \ln 2$$

7.7   Indeterminate Forms and L'Hôpital's Rule

9. Evaluate $\lim\limits_{x \to 0} \dfrac{\sqrt{4 - x^2} - 2}{x}$.

Solution:

Since a direct substitution of $x = 0$ yields the indeterminate form $0/0$, apply L'Hôpital's Rule to obtain

$$\lim_{x \to 0} \frac{\sqrt{4 - x^2} - 2}{x} = \lim_{x \to 0} \frac{(\frac{1}{2})(4 - x^2)^{-1/2}(-2x)}{1}$$

$$= \lim_{x \to 0} \frac{-x}{\sqrt{4 - x^2}} = \frac{0}{2} = 0.$$

13. Evaluate $\lim\limits_{x \to 0^+} \dfrac{e^x - (1 + x)}{x^n}$ where $n = 1, 2, 3 \ldots$.

Solution:

Case 1: $n = 1$ (Apply L'Hôpital's Rule once.)

$$\lim_{x \to 0^+} \frac{e^x - (1 + x)}{x} = \lim_{x \to 0^+} \frac{e^x - 1}{1} = 0$$

Case 2: $n = 2$ (Apply L'Hôpital's Rule twice.)

$$\lim_{x \to 0^+} \frac{e^x - (1 + x)}{x^2} = \lim_{x \to 0^+} \frac{e^x - 1}{2x} = \lim_{x \to 0^+} \frac{e^x}{2} = \frac{1}{2}$$

Case 3: $n \geq 3$ (Apply L'Hôpital's Rule twice.)

$$\lim_{x \to 0^+} \frac{e^x - (1 + x)}{x^n} = \lim_{x \to 0^+} \frac{e^x - 1}{nx^{n-1}} = \lim_{x \to 0^+} \frac{e^2}{n(n - 1)x^{n-2}} = \infty$$

25. Evaluate $\lim\limits_{x \to \infty} \dfrac{\ln x}{x}$.

Solution:

Since direct substitution leads to the indeterminate form ∞/∞, use L'Hôpital's Rule.

$$\lim_{x \to \infty} \frac{\ln x}{x} = \lim_{x \to \infty} \frac{1/x}{1} = \lim_{x \to \infty} \frac{1}{x} = 0$$

29. Identify the indeterminate form (if any) and evaluate
$\lim\limits_{x \to \infty} x \sin \dfrac{1}{x}$.

Solution:

Since direct substitution yields the indeterminate form $\infty \cdot 0$, use L'Hôpital's Rule.

$$
\begin{aligned}
\lim_{x \to \infty} x \sin \frac{1}{x} &= \lim_{x \to \infty} \frac{\sin(1/x)}{1/x} \\
&= \lim_{x \to \infty} \frac{(-1/x^2)\cos(1/x)}{-1/x^2} \\
&= \lim_{x \to \infty} \cos \frac{1}{x} = 1
\end{aligned}
$$

31. Identify the indeterminate form (if any) and evaluate $\lim\limits_{x \to 0^+} x^{1/x}$.

Solution:

Since 0^∞ is not an indeterminate form, you have

$$\lim_{x \to 0^+} x^{1/x} = 0^\infty = 0.$$

33. Identify the indeterminate form (if any) and evaluate $\lim\limits_{x \to \infty} x^{1/x}$.

Solution:

Since direct substitution yields the indeterminate form ∞^0, use L'Hôpital's Rule. Begin by taking the natural logarithm of both members of the equation

$$y = \lim_{x \to \infty} x^{1/x}.$$

You obtain

$$
\begin{aligned}
\ln y &= \ln \left[\lim_{x \to \infty} x^{1/x} \right] \\
&= \lim_{x \to \infty} \left[\frac{1}{x} \ln x \right] \\
&= \lim_{x \to \infty} \frac{\ln x}{x} = \lim_{x \to \infty} \frac{1/x}{1} = 0.
\end{aligned}
$$

Finally, as $\ln y \longrightarrow 0$, you know that $y \longrightarrow 1$ and conclude that

$$\lim_{x \to \infty} x^{1/x} = 1.$$

47. Use L'Hôpital's Rule to evaluate $\lim\limits_{x\to\infty}\dfrac{(\ln x)^n}{x^m}$, where $0 < n,\ m$.

Solution:

$$\lim_{x\to\infty}\frac{(\ln x)^n}{x^m} = \lim_{x\to\infty}\left[\frac{n(\ln x)^{n-1}(1/x)}{mx^{m-1}}\right] = \lim_{x\to\infty}\left[\frac{n(\ln x)^{n-1}}{mx^m}\right]$$

If $n - 1 \le 0$, this limit is zero. If $n - 1 > 0$, repeat L'Hôpital's Rule to obtain

$$\lim_{x\to\infty}\left[\frac{n(n-1)(\ln x)^{n-2}}{m^2 x^m}\right]$$

Again, if $n - 2 \le 0$, this limit is zero. If $n - 2 > 0$, repeated applications of L'Hôpital's Rule will eventually yield a form where the numerator approaches a finite number and the denominator approaches infinity. Thus, in every case the limit is 0.

63. The velocity of an object falling in a resisting medium such as air or water (if the downward direction is positive) is given by

$$v = \frac{32}{k}\left(1 - e^{-kt} + \frac{v_0 k e^{-kt}}{32}\right)$$

where v_0 is the initial velocity, t is the time, and k is the resistance constant of the medium. Use L'Hôpital's Rule to find the formula for the velocity of a falling body in a vacuum by fixing v_0 and t and letting k approach zero.

Solution:

$$\lim_{k\to 0}\frac{32}{k}\left(1 - e^{-kt} + \frac{v_0 k e^{-kt}}{32}\right)$$

$$= \lim_{k\to 0}\frac{32(1 - e^{-kt})}{k} + \lim_{k\to 0}\frac{32}{k}\cdot\frac{v_0 k e^{-kt}}{32}$$

$$= \lim_{k\to 0}\frac{32te^{-kt}}{1} + v_0 = 32t + v_0$$

(Apply L'Hôpital's Rule to the first limit the right-hand member of the equation.)

7.8   Improper Integrals

3. Determine the divergence or convergence of

$$\int_0^2 \frac{1}{(x-1)^2} \, dx$$

and evaluate the integral if it converges.

Solution:

$$\int_0^2 \frac{1}{(x-1)^2} \, dx$$

$$= \lim_{b \to 1^-} \int_0^b (x-1)^{-2} \, dx + \lim_{c \to 1^+} \int_c^2 (x-1)^{-2} \, dx$$

$$= \lim_{b \to 1^-} \left[\frac{-1}{x-1} \right]_0^b + \lim_{c \to 1^+} \left[\frac{-1}{x-1} \right]_c^2$$

$$= \lim_{b \to 1^-} \left[\frac{-1}{b-1} - 1 \right] + \lim_{c \to 1^+} \left[-1 - \frac{-1}{c-1} \right] = \infty$$

Therefore, the improper integral diverges.

13. Determine the divergence or convergence of

$$\int_0^\infty e^{-x} \cos x \, dx$$

and evaluate the integral if it converges.

Solution:

Since

$$\int e^{-x} \cos x \, dx = \frac{1}{2} e^{-x} (- \cos x + \sin x) + C,$$

you have

$$\int_0^\infty e^{-x} \cos x \, dx = \lim_{b \to \infty} \int_0^b e^{-x} \cos x \, dx$$

$$= \lim_{b \to \infty} \frac{1}{2} \left[\frac{(- \cos x + \sin x)}{e^x} \right]_0^b$$

$$= \frac{1}{2} [0 - (-1)] = \frac{1}{2}.$$

23. Determine the divergence or convergence of

$$\int_0^8 \frac{1}{\sqrt[3]{8-x}} \, dx$$

and evaluate the integral if it converges.

Solution:

$$\int_0^8 \frac{1}{\sqrt[3]{8-x}} \, dx = \lim_{b \to 8-} \int_0^b \frac{1}{\sqrt[3]{8-x}} \, dx$$

$$= \lim_{b \to 8-} \left[-\frac{(8-x)^{2/3}}{\frac{2}{3}} \right]_0^b$$

$$= \lim_{b \to 8-} \left[\frac{3}{2}(8-b)^{2/3} + \frac{3}{2}(8-0)^{3/2} \right]$$

$$= -\frac{3}{2}(0) + \frac{3}{2}(4) = 6$$

29. Determine the divergence or convergence of

$$\int_2^4 \frac{1}{\sqrt{x^2-4}} \, dx$$

and evaluate the integral if it converges.

Solution:

$$\int_2^4 \frac{1}{\sqrt{x^2-4}} \, dx = \lim_{a \to 2+} \int_a^4 \frac{1}{\sqrt{x^2-4}} \, dx$$

$$= \lim_{a \to 2+} \left[\ln |x + \sqrt{x^2-4}| \right]_a^4$$

$$= \lim_{a \to 2+} \left[\ln(4 + \sqrt{12}) - \ln |a + \sqrt{a^2-4}| \right]$$

$$= \ln(4 + \sqrt{12}) - \ln(2 + 0)$$

$$= \ln \left(\frac{4 + 2\sqrt{3}}{2} \right) = \ln(2 + \sqrt{3})$$

35. Use mathematical induction to verify that the following integral converges for any positive integer n.

$$\int_0^\infty x^n e^{-x}\, dx$$

Solution:

When $n = 1$, the integral converges, since

$$
\begin{aligned}
\int_0^\infty x e^{-x}\, dx &= \lim_{b\to\infty} \int_0^b x e^{-x}\, dx \\
&= \lim_{b\to\infty}\left[-e^{-x}(x+1)\right]_0^b \quad \text{(Integration by parts)} \\
&= \lim_{b\to\infty}\left[-e^{-b}(b+1)+1\right] \\
&= 0 + 1 = 1 \quad \text{(L'Hôpital's Rule)}
\end{aligned}
$$

Now assume that the integral converges for $n = k$ and verify that it converges for $n = k + 1$.

$$\int_0^\infty x^{k+1} e^{-x}\, dx$$

$$
\begin{aligned}
&= \lim_{b\to\infty} \int_0^b x^{k+1} e^{-x}\, dx \\
&= \lim_{b\to\infty}\left[-x^{k+1}e^{-x} - \frac{k+1}{-1}\int_0^b x^k e^{-x}\, dx\right]_0^\infty \quad \text{(Integration by parts)} \\
&= 0 + (k+1)\int_0^\infty x^k e^{-x}\, dx \quad \text{(L'Hôpital's Rule)} \\
&= (k+1)\int_0^\infty x^k e^{-x}\, dx
\end{aligned}
$$

Therefore, you have shown the integral for $n = k + 1$ converges if the integral for $n = k$ converges. Combining this with the results for $n = 1$, if follows by mathematical induction that the integral converges for any positive integer n.

45. Use the result of Exercise 36 to determine if $\displaystyle\int_0^\infty e^{-x^2}\,dx$ converges.

Solution:

On $[1, \infty)$ you have

$$x \le x^2$$

$$e^x \le e^{x^2}$$

$$\frac{1}{e^x} \ge \frac{1}{e^{x^2}}$$

$$e^{-x} \ge e^{-x^2}$$

Therefore, by Exercise 36 you have

$$\int_0^\infty e^{-x^2}\,dx = \int_0^1 e^{-x^2}\,dx + \int_1^\infty e^{-x^2}\,dx$$

$$< \int_0^1 e^{-x^2}\,dx + \int_1^\infty e^{-x}\,dx$$

$$= \int_0^1 e^{-x^2}\,dx + \Big[-e^{-x}\Big]_1^\infty = \int_0^1 e^{-x^2}\,dx + e^{-1}$$

Therefore, the given integral converges.

57. Sketch the graph of the hypocycloid of four cusps, $x^{2/3} + y^{2/3} = 1$, and find its perimeter.

Solution:

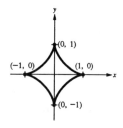

By symmetry the perimeter of this figure is four times the length of the arc in the first quadrant. Thus,

$$s = 4\int_0^1 \sqrt{1 + (y')^2}\,dx.$$

By implicit differentiation you have

$$x^{2/3} + y^{2/3} = 1$$

$$\frac{2}{3}x^{-1/3} + \frac{2}{3}y^{-1/3}y' = 0$$

$$y' = -\frac{y^{1/3}}{x^{1/3}}.$$

Therefore,

$$s = 4 \int_0^1 \sqrt{1 + \left(-\frac{y^{1/3}}{x^{1/3}}\right)^2} \, dx$$

$$= 4 \int_0^1 \sqrt{\frac{x^{2/3} + y^{2/3}}{x^{2/3}}} \, dx$$

$$= 4 \lim_{b \to 0+} \int_b^1 \sqrt{\frac{1}{x^{2/3}}} \, dx$$

$$= 4 \lim_{b \to 0+} \int_b^1 x^{-1/3} \, dx = 4 \lim_{b \to 0+} \left[\frac{x^{2/3}}{2/3}\right]_b^1 = 4\left(\frac{3}{2}\right) = 6.$$

Review Exercises for Chapter 7

5. Evaluate $\int \sec^4 \frac{x}{2} \, dx$.

Solution:

$$\int \sec^4 \frac{x}{2} \, dx = \int \sec^2 \frac{x}{2} \sec^2 \frac{x}{2} \, dx$$

$$= \int \left[\tan^2 \frac{x}{2} + 1\right] \sec^2 \frac{x}{2} \, dx$$

$$= 2 \int \tan^2 \frac{x}{2} \sec^2 \frac{x}{2} \left(\frac{1}{2}\right) dx + 2 \int \sec^2 \frac{x}{2} \left(\frac{1}{2}\right) dx$$

$$= \frac{2}{3} \tan^3 \frac{x}{2} + 2 \tan \frac{x}{2} + C$$

$$= \frac{2}{3} \left[\tan^3 \frac{x}{2} + 3 \tan \frac{x}{2}\right] + C$$

9. Evaluate $\displaystyle\int \frac{x^2 + 2x}{x^3 - x^2 + x - 1}\, dx.$

Solution:

Since $x^3 - x^2 + x - 1 = (x - 1)(x^2 + 1)$, you have

$$\frac{x^2 + 2x}{(x - 1)(x^2 + 1)} = \frac{A}{x - 1} + \frac{Bx + C}{x^2 + 1}.$$

Multiplying by $(x - 1)(x^2 + 1)$ yields

$$x^2 + 2x = A(x^2 + 1) + (Bx + C)(x - 1).$$

If $x = 1$, then $1 + 2 = A(2) + (B + C)(0)$ or $A = \dfrac{3}{2}.$

Furthermore,

$$
\begin{aligned}
x^2 + 2x &= Ax^2 + A + Bx^2 - Bx + Cx - C \\
&= (A + B)x^2 + (C - B)x + (A - C).
\end{aligned}
$$

Now by equating coefficients, you have

$$1 = A + B = \frac{3}{2} + B \quad \text{or} \quad B = -\frac{1}{2}$$

$$2 = C - B = C + \frac{1}{2} \quad \text{or} \quad C = \frac{3}{2}.$$

Therefore,

$$
\begin{aligned}
\int \frac{x^2 + 2x}{(x - 1)(x^2 + 1)}\, dx \\
= \int \left[\frac{3/2}{x - 1} + \frac{(-x/2) + (3/2)}{x^2 + 1} \right] dx \\
= \frac{3}{2} \int \frac{1}{x - 1}\, dx - \frac{1}{2} \int \frac{x}{x^2 + 1}\, dx + \frac{3}{2} \int \frac{1}{x^2 + 1}\, dx \\
= \frac{3}{2} \ln |x - 1| - \frac{1}{4} \ln (x^2 + 1) + \frac{3}{2} \arctan x + C \\
= \frac{1}{4}[6 \ln |x - 1| - \ln(x^2 + 1) + 6 \arctan x] + C.
\end{aligned}
$$

13. Evaluate $\displaystyle\int \frac{1}{1 - \sin\theta}\, d\theta$.

Solution:

$$\int \frac{1}{1 - \sin\theta}\, d\theta = \int \frac{1}{1 - \sin\theta} \left(\frac{1 + \sin\theta}{1 + \sin\theta} \right) d\theta$$

$$= \int \frac{1 + \sin\theta}{1 - \sin^2\theta}\, d\theta = \int \frac{1 + \sin\theta}{\cos^2\theta}\, d\theta$$

$$= \int \left(\frac{1}{\cos^2\theta} + \frac{\sin\theta}{\cos\theta\cos\theta} \right) d\theta$$

$$= \int \left(\sec^2\theta + \sec\theta\tan\theta \right) d\theta$$

$$= \tan\theta + \sec\theta + C$$

15. Evaluate $\displaystyle\int \frac{\ln(2x)}{x^2}\, dx$.

Solution:

Using Integration by Parts, let

$$dv = \frac{1}{x^2}\, dx \quad \Longrightarrow \quad v = -\frac{1}{x}$$

$$u = \ln(2x) \quad \Longrightarrow \quad du = \frac{1}{x}$$

$$\int \frac{\ln(2x)}{x^2}\, dx = -\frac{\ln(2x)}{x} - \int \frac{1}{x}\left(-\frac{1}{x} \right) dx$$

$$= -\frac{\ln(2x)}{x} + \int x^{-2}\, dx$$

$$= -\frac{\ln(2x)}{x} - \frac{1}{x} + C = -\frac{1}{x}(1 + \ln 2x) + C$$

35. Evaluate $\displaystyle\int \frac{x^{1/4}}{1+x^{1/2}}\,dx$.

Solution:

Let $u = x^{1/4}$. Then $u^4 = x$, $u^2 = x^{1/2}$, $dx = 4u^3\,du$, and

$$
\int \frac{x^{1/4}}{1+x^{1/2}}\,dx = \int \frac{u}{1+u^2}(4u^3\,du) = 4\int \frac{u^4}{1+u^2}\,du
$$

$$
= 4\int \left(u^2 - 1 + \frac{1}{1+u^2}\right)\,du
$$

$$
= 4\left(\frac{u^3}{3} - u + \arctan u\right) + C
$$

$$
= \frac{4}{3}(x^{3/4} - 3x^{1/4} + 3\arctan x^{1/4}) + C.
$$

43. Integrate $\displaystyle\int \frac{x^3}{\sqrt{4+x^2}}\,dx$ by the each of the specified methods.

(a) Trigonometric substitution

(b) Substitution: $u^2 = 4 + x^2$

(c) By parts: $dv = (x/\sqrt{4+x^2})\,dx$

Solution:

(a) Let $x = 2\tan\theta$. Then $\sqrt{4+x^2} = 2\sec\theta$ and $dx = 2\sec^2\theta\,d\theta$. Thus,

$$
\int \frac{x^3}{\sqrt{4+x^2}}\,dx = \int \frac{(2\tan\theta)^3}{2\sec\theta}(2\sec^2\theta\,d\theta)
$$

$$
= 8\int \tan^3\theta\sec\theta\,d\theta
$$

$$
= 8\int (\sec^2\theta - 1)\tan\theta\sec\theta\,d\theta
$$

$$
= 8\left(\frac{1}{3}\sec^3\theta - \sec\theta\right) + C
$$

$$
= \frac{8}{3}\sec\theta(\sec^2\theta - 3) + C
$$

$$
= \frac{\sqrt{4+x^2}}{3}(x^2 - 8) + C.
$$

(b) Let $u^2 = 4 + x^2$. Thus, $2u\,du = 2x\,dx$ and you have

$$\int \frac{x^3}{\sqrt{4+x^2}}\,dx = \int \frac{x^2(x\,dx)}{\sqrt{4+x^2}}$$

$$= \int \frac{(u^2-4)(u\,du)}{u}$$

$$= \int (u^2 - 4)\,du$$

$$= \frac{1}{3}u^3 - 4u + C$$

$$= \frac{u}{3}(u^2 - 12) + C = \frac{\sqrt{4+x^2}}{3}(x^2 - 8) + C.$$

(c)   Let

$$dv = \frac{x}{\sqrt{4+x^2}}\,dx \quad \Longrightarrow \quad v = \int \frac{x}{\sqrt{4+x^2}}\,dx = \sqrt{4+x^2}$$

$$u = x^2 \qquad\qquad \Longrightarrow \quad du = 2x\,dx$$

$$\int u\,dv = uv - \int v\,du$$

$$\int \frac{x^3}{\sqrt{4+x^2}}\,dx = x^2\sqrt{4+x^2} - \int 2x\sqrt{4+x^2}\,dx$$

$$= x^2\sqrt{4+x^2} - \frac{2}{3}(4+x^2)^{3/2} + C$$

$$= \frac{\sqrt{4+x^2}}{3}(x^2 - 8) + C$$

63. Use L'Hôpital's Rule to evaluate $\lim\limits_{x \to \infty} (\ln x)^{2/x}$.

Solution:

Begin by taking the natural logarithm of both members of the equation

$$y = \lim_{x \to \infty} (\ln x)^{2/x}.$$

Then you obtain

$$\ln y = \ln \left[\lim_{x \to \infty} (\ln x)^{2/x}\right]$$

$$= \lim_{x \to \infty} \left[\frac{2}{x}\ln(\ln x)\right]$$

$$= 2\lim_{x \to \infty} \frac{(1/\ln x)(1/x)}{1} = 0.$$

Finally, as $\ln y \longrightarrow 0$, you know that $y \longrightarrow 1$ and conclude that

$$\lim_{x \to \infty} (\ln x)^{2/x} = 1.$$

8 INFINITE SERIES

8.1 Sequences

19. Simplify $\dfrac{(2n-1)!}{(2n+1)!}$.

Solution:

$$\frac{(2n-1)!}{(2n+1)!} = \frac{(2n-1)(2n-2)\cdots 3\cdot 2\cdot 1}{(2n+1)(2n)(2n-1)(2n-2)\cdots 3\cdot 2\cdot 1} = \frac{1}{(2n+1)(2n)}$$

23. Write an expression for the nth term of the sequence
$-1,\ 2,\ 7,\ 14,\ 23,\ldots$.

Solution:

Compare the terms of the sequence $-1,\ 2,\ 7,\ 14,\ 23,\ldots$ with the sequence of squares

$$1^2,\ 2^2,\ 3^2,\ 4^2,\ 5^2,\ldots = 1,\ 4,\ 9,\ 16,\ 25,\ldots.$$

Observe that each term of the given sequence is two less than the sequence of squares. Thus, you write the nth term of the given sequence as $a_n = n^2 - 2$.

27. Write an expression for the nth term of the sequence

$$2, \; -1, \; \frac{1}{2}, \; \frac{-1}{2}, \; \frac{1}{8}, \ldots.$$

Solution:

First observe that the denominators are powers of 2, where for $n = 4$ the denominator is 2^2. This implies that the nth term has a denominator 2^{n-2}. Note further that the signs alternate starting with a positive sign, and thus the nth term is

$$a_n = (-1)^{n-1}\left(\frac{1}{2^{n-2}}\right) = \frac{(-1)^{n-1}}{2^{n-2}}.$$

33. Write an expression for the nth term of the sequence

$$1, \; -\frac{1}{1\cdot 3}, \; \frac{1}{1\cdot 3\cdot 5}, \; -\frac{1}{1\cdot 3\cdot 5\cdot 7}, \ldots.$$

Solution:

The denominator of the nth term is the product of the first n positive odd integers. Note further that the signs alternate starting with a positive sign, and thus the nth term is

$$a_n = \frac{(-1)^{n-1}}{1\cdot 3\cdot 5\cdots(2n-1)}.$$

A second form of the nth term is obtained when you multiply the numerator and denominator by the n missing even integers. Then,

$$\begin{aligned}
a_n &= \frac{(-1)^{n-1}}{1\cdot 3\cdot 5\cdots(2n-1)} \\
&= \frac{(-1)^{n-1}2\cdot 4\cdot 6\cdot 8\cdots(2n)}{1\cdot 2\cdot 3\cdot 4\cdot 5\cdots(2n-1)(2n)} \\
&= \frac{(-1)^{n-1}2^n(1\cdot 2\cdot 3\cdot 4\cdots n)}{1\cdot 2\cdot 3\cdot 4\cdot 5\cdots(2n-1)(2n)} = \frac{(-1)^{n-1}2^n n!}{(2n)!}.
\end{aligned}$$

37. Given $a_n = (-1)^n[n/(n+1)]$, determine if the sequence $\{a_n\}$ converges or diverges. If it converges, find its limit.

Solution:

Since

$$\lim_{n \to \infty} a_n = \lim_{n \to \infty} (-1)^n \left(\frac{n}{n+1} \right) = \pm 1,$$

the limit does not exist and the sequence $\{a_n\}$ diverges.

47. Given $a_n = (n+1)!/n!$, determine if the sequence $\{a_n\}$ converges or diverges. If it converges, find its limit.

Solution:

Since

$$\lim_{n \to \infty} a_n = \lim_{n \to \infty} \frac{(n+1)!}{n!}$$
$$= \lim_{n \to \infty} \frac{(n+1)(n)(n-1) \cdots 3 \cdot 2 \cdot 1}{n(n-1) \cdots 3 \cdot 2 \cdot 1}$$
$$= \lim_{n \to \infty} (n+1) = \infty,$$

the sequence $\{a_n\}$ diverges.

53. Given $a_n = [1 + (k/n)]^n$, determine if the sequence $\{a_n\}$ converges or diverges. If it converges, find its limit.

Solution:

$$\lim_{n \to \infty} a_n = \lim_{n \to \infty} \left(1 + \frac{k}{n} \right)^n = \lim_{n \to \infty} \left[\left(1 + \frac{k}{n} \right)^{n/k} \right]^k$$

Now let $u = k/n$. Then as n approaches infinity, u approaches zero, and

$$\lim_{n \to \infty} a_n = \lim_{u \to 0} \left[(1+u)^{1/u} \right]^k = e^k$$

Therefore the sequence converges to e^k.

59. Determine if the sequence $a_n = (-1)^n \left(\dfrac{1}{n}\right)$ is monotonic. Discuss the boundedness of the sequence.

Solution:

Writing out the first few terms of the sequence you have

$$a_1 = -1, \ a_2 = \frac{1}{2}, \ a_3 = -\frac{1}{3}, \ a_4 = \frac{1}{4}, \cdots.$$

Because of the alternating signs, observe that the terms are neither nondecreasing nor nonincreasing. Therefore, the sequence is *not* monotonic. The sequence is bounded since $-1 \le a_n \le 1$ for all n.

73. A government program that currently costs taxpayers $2.5 billion per year is to be cut back by 20% per year.

(a) Write an expression for the amount budgeted for this program after n years.

(b) Compute the budgets for the first 4 years.

(c) Determine the convergence or divergence of the sequence of reduced budgets. If the sequence converges, find its limit.

Solution:

(a) If you let A_n be the amount budgeted after n years, you have

$$A_1 = 2.5 - 0.2(2.5) = 0.8(2.5) \text{ billion}$$
$$A_2 = A_1 - 0.2A_1 = 0.8A_1 = 0.8^2(2.5) \text{ billion}$$
$$A_3 = A_2 - 0.2A_2 = 0.8A_2 = 0.8^2(2.5) \text{ billion}$$
$$\vdots$$
$$A_n = 2.5(0.8)^n \text{ billion}$$

(b) $\quad A_1 = 2.5(0.8) = \$2 \text{ billion}$

$A_2 = 2.5(0.8)^2 = \$1.6 \text{ billion}$

$A_3 = 2.5(0.8)^3 = \$1.28 \text{ billion}$

$A_4 = 2.5(0.8)^4 = \$1.024 \text{ billion}$

(c) $\lim\limits_{n\to\infty} 2.5(0.8)^n = 0$ and therefore the sequence converges.

8.2   Series and Convergence

3. Find the first five terms of the sequence of partial sums for the series

$$3 - \frac{9}{2} + \frac{27}{4} - \frac{81}{8} + \frac{243}{16} - \cdots.$$

Solution:

$$S_1 = 3$$

$$S_2 = 3 - \frac{9}{2} = -\frac{3}{2} = -1.5$$

$$S_3 = 3 - \frac{9}{2} + \frac{27}{4} = \frac{21}{4} = 5.25$$

$$S_4 = 3 - \frac{9}{2} + \frac{27}{4} - \frac{81}{8} = -\frac{39}{8} = -4.875$$

$$S_5 = 3 - \frac{9}{2} + \frac{27}{4} - \frac{81}{8} + \frac{243}{16} = \frac{165}{16} = 10.3125$$

11. Verify that the series $\displaystyle\sum_{n=0}^{\infty} 3\left(\frac{3}{2}\right)^n$ diverges.

Solution:

Writing out a few terms of the series you obtain

$$\sum_{n=0}^{\infty} 3\left(\frac{3}{2}\right)^n = 3 + \frac{9}{2} + \frac{27}{4} + \frac{81}{8} + \cdots.$$

The series is geometric with common ration $r = \frac{3}{2}$. Since $|r| \geq 1$, the series diverges.

15. Verify that the infinite series $\displaystyle\sum_{n=1}^{\infty} \frac{2^n + 1}{2^{n+1}}$ diverges.

Solution:

$$\lim_{n\to\infty} a_n = \lim_{n\to\infty} \frac{2^n - 1}{2^{n+1}} = \lim_{n\to\infty} \frac{1 - (1/2^n)}{2} = \frac{1}{2} \neq 0$$

Therefore, by Theorem 8.9 the series diverges.

17. Verify that the infinite series

$$\sum_{n=0}^{\infty} 2\left(\frac{3}{4}\right)^n = 2 + \frac{3}{2} + \frac{9}{8} + \frac{27}{32} + \frac{81}{128} + \cdots$$

converges.

Solution:

The series

$$\sum_{n=2}^{\infty} 2\left(\frac{3}{4}\right)^n$$

is geometric with $a = 2$ and $r = \frac{3}{4}$. Since $|r| < 1$, the series converges by Theorem 8.6.

21. Verify that the infinite series $\displaystyle\sum_{n=1}^{\infty} \frac{1}{n(n+1)}$ converges.

Solution:

Using partial fractions you can write

$$a_n = \frac{1}{n(n+1)} = \frac{1}{n} - \frac{1}{n+1}.$$

Therefore,

$$\sum_{n=1}^{\infty} \frac{1}{n(n+1)} = \sum_{n=1}^{\infty} \left(\frac{1}{n} - \frac{1}{n+1}\right)$$

$$= \left(1 - \frac{1}{2}\right) + \left(\frac{1}{2} - \frac{1}{3}\right) + \left(\frac{1}{3} - \frac{1}{4}\right) + \cdots$$

where each term after the first term of 1 cancels. Thus, the series converges to 1.

27. Find the sum of the series $1 + 0.1 + 0.01 + 0.001 + \cdots$.

Solution:

The series

$$1 + 0.1 + 0.01 + 0.001 + \cdots = \sum_{n=0}^{\infty} (0.1)^n$$

is geometric with $a = 1$ and $r = 0.1$. Therefore, the sum is

$$S = \frac{a}{1 - r} = \frac{1}{1 - 0.1} = \frac{10}{9}$$

33. Find the sum of the series $\displaystyle\sum_{n=1}^{\infty}\frac{4}{n(n+2)}$.

Solution:

Using partial fractions you have

$$\frac{4}{n(n+2)}=\frac{A}{n}+\frac{B}{n+2}.$$

Multiplying by $n(n+2)$, yields

$$4=A(n+2)+Bn.$$

If $n=0$, then $4=2A$ or $A=2$. If $n=-2$, then $4=-2B$ or $B=-2$. Thus,

$$\sum_{n=1}^{\infty}\frac{4}{n(n+2)}=\sum_{n=1}^{\infty}\left[\frac{2}{n}-\frac{2}{n+2}\right]$$

$$=\left[2-\frac{2}{3}\right]+\left[\frac{2}{2}-\frac{2}{4}\right]+\left[\frac{2}{3}-\frac{2}{5}\right]+\left[\frac{2}{4}-\frac{2}{6}\right]$$

$$+\left[\frac{2}{5}-\frac{2}{7}\right]+\cdots=2+1=3.$$

45. Determine the convergence or divergence of the series

$$\sum_{n=1}^{\infty}\frac{3n-1}{2n+1}.$$

Solution:

Since

$$\lim_{n\to\infty}a_n=\lim_{n\to\infty}\frac{3n-1}{2n+1}=\frac{3}{2}\neq 0$$

it follows by Theorem 8.9 that the given series diverges.

75. A deposit of $50 is made at the end of each month in an account at an annual interest rate 7%. Use the results of Exercise 74 to find the balance A after twenty years if the interest is compounded (a) monthly and (b) continuously.

Solution:

(a) The balance A in the account at the end of the 20 years if the interest is compounded monthly is

$$A = 50 + 50\left(1 + \frac{0.07}{12}\right) + 50\left(1 + \frac{0.07}{12}\right)^2$$

$$+ \cdots + 50\left(1 + \frac{0.07}{12}\right)^{239}$$

$$= 50\left(\frac{12}{0.07}\right)\left[\left(1 + \frac{0.07}{12}\right)^{(12)(20)} - 1\right] = \$26,046.33.$$

(b) The balance A in the account at the end of the 20 years if the interest is compounded continuously is

$$A = 50 + 50e^{1(0.07)/12} + 50e^{2(0.07)/12} + \cdots + 50e^{239(0.07)/12}$$

$$= \frac{50[e^{(0.07)(20)} - 1]}{e^{0.07/12} - 1} = \$26,111.12.$$

8.3   The Integral Test and p-Series

7. Determine the convergence of divergence of the p-series

$$\frac{\ln 2}{2} + \frac{\ln 3}{3} + \frac{\ln 4}{4} + \frac{\ln 5}{5} + \cdots$$

by the Integral Test.

Solution:

First observe that the nth term of the series is

$$a_n = \frac{\ln n}{n}$$

Since

$$f(x) = \frac{\ln x}{x}$$

satisfies the conditions for the Integral Test, you have

$$\int_1^\infty \frac{\ln x}{x}\, dx = \int_1^\infty (\ln x) \frac{1}{x}\, dx = \left[\frac{(\ln x)^2}{2} \right]_1^\infty = \infty.$$

Therefore, the given series diverges.

17. Determine the convergence or divergence of the p-series

$$1 + \frac{1}{2\sqrt{2}} + \frac{1}{3\sqrt{3}} + \frac{1}{4\sqrt{4}} + \frac{1}{5\sqrt{5}} + \cdots.$$

Solution:

Since

$$1 + \frac{1}{2\sqrt{2}} + \frac{1}{3\sqrt{3}} + \frac{1}{4\sqrt{4}} + \frac{1}{5\sqrt{5}} + \cdots = \sum_{n=1}^\infty \frac{1}{n\sqrt{n}}$$

$$= \sum_{n=1}^\infty \frac{1}{n^{3/2}},$$

you have a p-series with $p = \frac{3}{2} > 1$. Therefore, by Theorem 8.11 the series converges.

29. Determine the convergence or divergence of the series

$$\sum_{n=1}^\infty \left(1 + \frac{1}{n} \right)^n.$$

Solution:

From the definition of the irrational number e, you have

$$\lim_{n \to \infty} a_n = \lim_{n \to \infty} \left(1 + \frac{1}{n} \right)^n = e \neq 0.$$

Thus, by Theorem 8.9, the series diverges.

33. Find the values of p for which the series

$$\sum_{n=2}^{\infty} \frac{1}{n(\ln n)^p}$$

converges.

Solution:

Since the conditions of the Theorem 8.10 are met, use the Integral Test to find the values of p for which the series converges. If $u = \ln x$ then $du = (1/x)\, dx$. Thus,

$$\int_{2}^{\infty} \frac{1}{x(\ln x)^p}\, dx = \lim_{b \to \infty} \int_{2}^{b} (\ln x)^{-p} \frac{1}{x}\, dx$$

$$= \lim_{b \to \infty} \left[\frac{(\ln x)^{-p+1}}{-p+1} \right]_{2}^{b}$$

$$= \lim_{b \to \infty} \left[\left(\frac{1}{1-p} \right) \frac{1}{(\ln b)^{p-1}} - \left(\frac{1}{1-p} \right) \frac{1}{(\ln 2)^{p-1}} \right]$$

$$= 0 + \frac{1}{(p-1)(\ln 2)^{p-1}} \quad \text{for} \quad p - 1 > 0.$$

Therefore, the given series converges if $p > 1$.

43. Approximate the sum of the series $\displaystyle\sum_{n=1}^{\infty} n e^{-n^2}$ using the first four terms. Include an estimate of the maximum error for your approximation.

Solution:

Since $f(x) = x e^{-x^2}$ satisfies the conditions for the Integral Test, you have

$$\int_{1}^{\infty} x e^{-x^2}\, dx = -\frac{1}{2} \int_{1}^{\infty} e^{-x^2}(-2x)\, dx$$

$$= -\frac{1}{2} \lim_{b \to \infty} \left[e^{-x^2} \right]_{1}^{b}$$

$$= -\frac{1}{2} \lim_{b \to \infty} [e^{-b^2} - e^{-1}] = \frac{1}{2e}.$$

Thus, the series converges and has a sum S. Furthermore, since

$$R_4 \leq \int_{4}^{\infty} x e^{-x^2}\, dx = \frac{1}{2e^{16}} \approx 5.6 \times 10^{-8}$$

and $S_4 = e^{-1} + 2e^{-2} + 3e^{-3} + 4e^{-4} \approx 0.40488$, it follows that $S = 0.4049$.

8.4   Comparisons of Series

---

3. Use the Direct Comparison Test to determine the convergence or divergence of the series

$$\sum_{n=2}^{\infty} \frac{1}{n-1}.$$

Solution:

This series resembles $\displaystyle\sum_{n=2}^{\infty} \frac{1}{n}$, a divergent p-series.

Term-by-term comparison yields

$$a_n = \frac{1}{n} < \frac{1}{n-1} = b_n.$$

Thus, it follows by the Direct Comparison Test that the series diverges.

9. Use the Direct Comparison Test to determine the convergence or divergence of the series

$$\sum_{n=0}^{\infty} \frac{1}{n!}.$$

Solution:

Compare the given series to the convergent p-series $\displaystyle\sum_{n=1}^{\infty} \frac{1}{n^2}$.

If $n > 3$, then $n^2 < n!$ and $\dfrac{1}{n!} < \dfrac{1}{n^2}$. Hence, by the Direct Comparison Test, the given series converges.

13. Use the Limit Comparison Test to determine the convergence or divergence of the series

$$\sum_{n=1}^{\infty} \frac{n}{n^2+1}.$$

Solution:

This series can be compared with the series

$$\sum_{n=1}^{\infty} \frac{1}{n}$$

since the degree of the denominator is only one greater than the degree of the numerator. Furthermore, since

$$\lim_{n\to\infty} \frac{n/(n^2+1)}{1/n} = \lim_{n\to\infty} \frac{n^2}{n^2+1} = 1$$

and since the series

$$\sum_{n=1}^{\infty} \frac{1}{n}$$

diverges, it follows that the given series also diverges by the Limit Comparison Test.

21. Use the Limit Comparison Test to determine the convergence or divergence of the series

$$\sum_{n=1}^{\infty} \frac{1}{n\sqrt{n^2+1}}.$$

Solution:

For "large" values of n

$$\frac{1}{n\sqrt{n^2+1}} \approx \frac{1}{n\sqrt{n^2}} = \frac{1}{n^2}.$$

By comparing the given series to the convergent p-series

$$\sum_{n=1}^{\infty} \frac{1}{n^2}$$

we have

$$\lim_{n\to\infty} \frac{a_n}{b_n} = \lim_{n\to\infty} \left(\frac{1}{n\sqrt{n^2+1}}\right)\left(\frac{n^2}{1}\right) = \lim_{n\to\infty} \frac{1}{\sqrt{1+(1/n)}} = 1.$$

You can conclude by the Limit Comparison Test that the given series converges.

31. Use one of the tests studied thus far to determine the convergence or divergence of the series

$$\sum_{n=1}^{\infty} \frac{n}{2n+3}$$

Solution:

Since

$$\lim_{n \to \infty} a_n = \lim_{n \to \infty} \frac{n}{2n+3} = \frac{1}{2} \neq 0,$$

it follows that the series diverges by the nth-Term Divergence Test.

37. Use the Polynomial Test of Exercise 36 to determine the convergence or divergence of the series

$$\frac{1}{2} + \frac{2}{5} + \frac{3}{10} + \frac{4}{17} + \frac{5}{26} + \cdots .$$

Solution:

The nth term of the series is

$$a_n = \frac{n}{n^2+1}$$

and the series can be written

$$\sum_{n=1}^{\infty} \frac{n}{n^2+1} = \sum_{n=1}^{\infty} \frac{P(n)}{Q(n)}.$$

In this case $P(n)$ has degree $k = 1$ and $Q(n)$ has degree $k = 2$. Hence, it follows from the Polynomial Test that the given series diverges.

8.5   Alternating Series

3. Use the Alternating Series Test to determine the convergence or divergence of

$$\sum_{n=1}^{\infty} \frac{(-1)^{n+1}}{2n-1}.$$

Solution:

Observe that

$$\lim_{n \to \infty} \frac{1}{2n - 1} = 0$$

and

$$a_{n+1} = \frac{1}{2(n+1) - 1} = \frac{1}{2n + 1} < \frac{1}{2n - 1} = a_n$$

for all n. Therefore, by the Alternating Series Test the given series converges.

---

9. Use the Alternating Series Test to determine the convergence or divergence of

$$\sum_{n=1}^{\infty} \frac{(-1)^{n+1}(n+1)}{\ln(n+1)}.$$

Solution:

Since

$$\lim_{n \to \infty} a_n = \lim_{n \to \infty} \frac{n+1}{\ln(n+1)} \qquad \text{(L'Hôpital's Rule)}$$
$$= \infty \neq 0,$$

the series diverges.

---

19. Use the Alternating Series Test to determine the convergence or divergence of

$$\sum_{n=1}^{\infty} \frac{2(-1)^{n+1}}{e^n + e^{-n}}$$

Solution:

Begin by using differentiation to establish that $a_{n+1} \leq a_n$.

$$f(x) = \frac{2}{e^x + e^{-x}}$$
$$f'(x) = \frac{-2(e^x - e^{-x})}{(e^x + e^{-x})^2}$$

Observe that $f'(x)$ is negative for $x > 0$. Hence, f is a decreasing function and it follows that $a_{n+1} \leq a_n$ for $n \geq 1$. Since

$$\lim_{n \to \infty} \frac{2}{e^n + e^{-n}} = 0,$$

the series converges by the Alternating Series Test.

23. Examine

$$\sum_{n=1}^{\infty} \frac{(-1)^{n+1}}{\sqrt{n}}$$

for conditional convergence or absolute convergence.

Solution:

In this case, the Alternating Series Test indicates that the given series converges. However, the series

$$\sum_{n=1}^{\infty} \left| \frac{(-1)^{n+1}}{\sqrt{n}} \right| = \sum_{n=1}^{\infty} \frac{1}{\sqrt{n}}$$

is a divergent p-series with $p = \frac{1}{2}$. Therefore, it follows that the given series is conditionally convergent.

29. Examine

$$\sum_{n=2}^{\infty} \frac{(-1)^n n}{n^3 - 1}$$

for conditional convergence or absolute convergence.

Solution:

In this case, the Alternating Series Test indicates that the given series converges. Also, the series

$$\sum_{n=2}^{\infty} \left| \frac{(-1)^{n+1} n}{n^3 - 1} \right| = \sum_{n=2}^{\infty} \frac{n}{n^3 - 1}$$

converges by the Limit Comparison Test with the series $\sum_{n=2}^{\infty} \frac{1}{n^2}$. Therefore, it follows that the given series is absolutely convergent.

37. Approximate the sum of the series

$$\sum_{n=0}^{\infty} \frac{(-1)^n}{n!}$$

with an error less that 0.001. (The actual sum is $1/e$.)

Solution:

The series is alternating, and since

$$\lim_{n \to \infty} \frac{1}{n!} = 0 \quad \text{and} \quad \frac{1}{(n+1)!} < \frac{1}{n!}$$

the given series converges. By Theorem 8.15 the error, R_N, after N terms is such that $|R_N| \leq a_{N+1}$. Therefore, to ensure an error less than 0.001, choose N sufficiently large so that

$$\frac{1}{(N+1)!} \leq 0.001$$

Since $1/6! = 1/720 = 0.0013888$ and $1/7! = 1/5040 < 0.001$, choose $N = 6$ and obtain the approximation

$$\sum_{n=0}^{6} \frac{(-1)^n}{n!} = 1 - 1 + \frac{1}{2} - \frac{1}{6} + \frac{1}{24} - \frac{1}{120} + \frac{1}{720} \approx 0.368.$$

This approximation is within 0.001 of the actual sum.

43. Find the number of terms necessary to approximate the sum of the series

$$\sum_{n=0}^{\infty} \frac{(-1)^{n+1}}{2n^3 - 1}$$

with an error less than 0.001.

Solution:

By Theorem 8.15 the error, R_N, after N terms is such that $|R_N| \leq a_{N+1}$. Therefore, to insure an error less than 0.001, choose N sufficiently large so that

$$\frac{1}{2(N+1)^3 - 1} \leq 0.001$$

$$1000 \leq 2(N+1)^3 - 1$$

$$\frac{1001}{2} \leq (N+1)^3$$

$$7.9 \approx \sqrt[3]{\frac{1001}{2}} \leq N + 1$$

$$6.9 \leq N.$$

Therefore, choose $N = 7$.

8.6   The Ratio and Root Tests

7. Use the Ratio Test to test for convergence or divergence of

$$\sum_{n=0}^{\infty} \frac{3^n}{n!}.$$

Solution:

Since

$$
\begin{aligned}
\lim_{n \to \infty} \left| \frac{a_{n+1}}{a_n} \right| &= \lim_{n \to \infty} \left| \frac{3^{n+1}/(n+1)!}{3^n/n!} \right| \\
&= \lim_{n \to \infty} \left| \frac{3^{n+1}}{1 \cdot 2 \cdot 3 \cdot 4 \cdots n \cdot (n+1)} \cdot \frac{1 \cdot 2 \cdot 3 \cdot 4 \cdots n}{3^n} \right| \\
&= \lim_{n \to \infty} \left(\frac{3}{n+1} \right) = 0 < 1,
\end{aligned}
$$

you conclude by the Ratio Test that the series

$$\sum_{n=0}^{\infty} \frac{3^n}{n!}$$

converges.

21. Use the Ratio Test to test for convergence or divergence of

$$\sum_{n=0}^{\infty} \frac{4n}{3^n + 1}.$$

Solution:

Since

$$
\begin{aligned}
\lim_{n \to \infty} \left| \frac{a_{n+1}}{a_n} \right| &= \lim_{n \to \infty} \left[\frac{4^{n+1}}{3^{n+1} + 1} \cdot \frac{3^n + 1}{4^n} \right] \\
&= \lim_{n \to \infty} \left[\frac{4(3^n + 1)}{3^{n+1} + 1} \right] \\
&= 4 \lim_{n \to \infty} \left[\frac{1 + (1/3^n)}{3 + (1/3^n)} \right] = \frac{4}{3} > 1,
\end{aligned}
$$

the series diverges by the Ratio Test.

23. Use the Ratio Test to test for convergence or divergence of

$$\sum_{n=0}^{\infty} \frac{(-1)^{n+1} n!}{1 \cdot 3 \cdot 5 \cdots (2n+1)}.$$

Solution:

Since

$$\lim_{n \to \infty} \left| \frac{a_{n+1}}{a_n} \right|$$

$$= \lim_{n \to \infty} \left[\frac{(n+1)!}{1 \cdot 3 \cdot 5 \cdots (2n+1)(2n+3)} \cdot \frac{1 \cdot 3 \cdot 5 \cdots (2n+1)}{n!} \right]$$

$$= \lim_{n \to \infty} \frac{(n+1)}{(2n+3)} = \frac{1}{2},$$

the series converges by the Ratio Test.

31. Use the Root Test to test for convergence or divergence of

$$\sum_{n=1}^{\infty} (2 \sqrt[n]{n} + 1)^n.$$

Solution:

Using the Root Test, you have

$$\lim_{n \to \infty} \sqrt[n]{a_n} = \lim_{n \to \infty} \sqrt[n]{(2 \sqrt[n]{n} + 1)^n} = \lim_{n \to \infty} (2 \sqrt[n]{n} + 1)$$

Now let $y = \sqrt[n]{n} = n^{1/n}$. Then $\ln y = (1/n) \ln n$ and

$$\lim_{n \to \infty} \left(\frac{\ln n}{n} \right) = \lim_{n \to \infty} \left(\frac{1/n}{1} \right) \qquad \text{(L'Hôpital's Rule)}$$

$$= 0.$$

Therefore, as n approaches infinity, $\ln y$ approaches zero and $y = \sqrt[n]{n}$ approaches one. Thus,

$$\lim_{n \to \infty} (2 \sqrt[n]{n} + 1) = 2 + 1 = 3$$

and the series diverges.

43. Test for convergence or divergence using any appropriate test from this chapter for the series

$$\sum_{n=1}^{\infty} \frac{10n + 3}{n2^n}.$$

Identify the test used.

Solution:

Compare the given series to the convergent geometric series

$$\sum_{n=0}^{\infty} \frac{1}{2^n}.$$

Since

$$\lim_{n \to \infty} \frac{(10n + 3)/n2^n}{1/2^n} = \lim_{n \to \infty} \frac{10n + 3}{n} = 10,$$

the Limit Comparison Test implies that the given series also converges.

45. Test for convergence or divergence using any appropriate test from this chapter for the series

$$\sum_{n=1}^{\infty} \frac{\cos(n)}{2^n}.$$

Identify the test used.

Solution:

Even though some of the terms of this series are negative, it is not an alternating series. Use the Comparison Test and observe that

$$\left| \frac{\cos(n)}{2^n} \right| \le \frac{1}{2^n}.$$

Since the series

$$\sum_{n=0}^{\infty} \frac{1}{2^n}$$

is a convergent geometric series, it follows from the Comparison Test that

$$\sum_{n=1}^{\infty} \left| \frac{\cos(n)}{2^n} \right|$$

also converges. Finally, by Theorem 8.16, since

$$\sum_{n=1}^{\infty} \left| \frac{\cos(n)}{2^n} \right|$$

converges, the series

$$\sum_{n=1}^{\infty} \frac{\cos(n)}{2^n}$$

also converges.

51. Test for convergence or divergence using any appropriate test from this chapter for the series

$$\sum_{n=1}^{\infty} \frac{(-3)^n}{3 \cdot 5 \cdot 7 \cdots (2n+1)}.$$

Identify the test used.

Solution:

Use the Ratio Test to test for convergence or divergence and obtain

$$\lim_{n \to \infty} \left| \frac{a_{n+1}}{a_n} \right|$$

$$= \lim_{n \to \infty} \left| \frac{(-3)^{n+1}}{3 \cdot 5 \cdot 7 \cdots (2n+1)[2(n+1)+1]} \cdot \frac{3 \cdot 5 \cdot 7 \cdots (2n+1)}{(-3)^n} \right|$$

$$= \lim_{n \to \infty} \left| \frac{-3}{2(n+1)+1} \right| = 0 < 1.$$

Therefore, by the Ratio Test the given series converges.

8.7   Taylor Polynomials and Approximations

9. Find the Maclaurin polynomial of degree 5 for $f(x) = \sin x$.

Solution:

$$
\begin{array}{ll}
f(x) = \sin x & f(0) \;\; = 0 \\
f'(x) = \cos x & f'(0) \;\; = 1 \\
f''(x) = -\sin x & f''(0) \;\; = 0 \\
f'''(x) = -\cos x & f'''(0) = -1 \\
f^{(4)}(x) = \sin x & f^{(4)}(0) \;\; = 0 \\
f^{(5)}(x) = \cos x & f^{(5)}(0) \;\; = 1
\end{array}
$$

Therefore, the expansion yields

$$
\begin{aligned}
P_5(x) &= f(0) + f'(0)x + \frac{f''(0)}{2!}x^2 + \frac{f'''(0)}{3!} + \frac{f^{(4)}(0)}{4!}x^4 + \frac{f^{(5)}(0)}{5!}x^5 \\
&= x - \frac{x^3}{6} + \frac{x^5}{120}.
\end{aligned}
$$

13. Find the Maclaurin polynomial of degree 4 for $f(x) = \dfrac{1}{x+1}$.

Solution:

$$
\begin{array}{ll}
f(x) = \dfrac{1}{x+1} & f(0) = 1 \\[2mm]
f'(x) = \dfrac{-1}{(x+1)^2} & f'(0) = -1 \\[2mm]
f''(x) = \dfrac{2}{(x+1)^3} & f''(0) = 2 \\[2mm]
f'''(x) = \dfrac{-6}{(x+1)^4} & f'''(0) = -6 \\[2mm]
f^{(4)}(x) = \dfrac{24}{(x+1)^5} & f^{(4)}(0) = 24
\end{array}
$$

Therefore, the expansion yields

$$
\begin{aligned}
P_4(x) &= f(0) + f'(0)x + \frac{f''(0)}{2!}x^2 + \frac{f'''(0)}{3!}x^3 + \frac{f^{(4)}(0)}{4!}x^4 \\
&= 1 - x + x^2 - x^3 + x^4.
\end{aligned}
$$

19. Find the Taylor polynomial of degree 4, centered at $c = 1$ for $f(x) = \ln x$.

Solution:

$$
\begin{aligned}
f(x) &= \ln x & f(1) &= 0 \\
f'(x) &= \frac{1}{x} & f'(1) &= 1 \\
f''(x) &= -\frac{1}{x^2} & f''(1) &= -1 \\
f'''(x) &= \frac{2}{x^3} & f'''(1) &= 2 \\
f^{(4)}(x) &= -\frac{6}{x^4} & f^{(4)}(1) &= -6
\end{aligned}
$$

Therefore, the expansion yields

$$
\begin{aligned}
P_4(x) &= f(1) + f'(1)(x-1) + \frac{f''(1)}{2!}(x-1)^2 \\
&\quad + \frac{f'''(1)}{3!}(x-1)^3 + \frac{f^{(4)}(1)}{4!}(x-1)^4 \\
&= (x-1) - \frac{1}{2}(x-1)^2 + \frac{1}{3}(x-1)^3 - \frac{1}{4}(x-1)^4.
\end{aligned}
$$

25. (a) Find the Taylor polynomial $P(x)$ of degree 3 for
$$f(x) = \arcsin x.$$

(b) Complete the accompanying table for $f(x)$ and $P_3(x)$.

x	-1	-0.75	-0.50	-0.25	0	0.25	0.50	0.75	1
$f(x)$									
$P_3(x)$									

(c) Sketch the graphs of $f(x)$ and $P_3(x)$.

Solution:

(a)

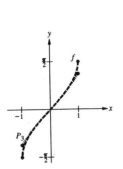

$$
\begin{aligned}
f(x) &= \arcsin x & f(0) &= 0 \\
f'(x) &= \frac{1}{\sqrt{1-x^2}} & f'(0) &= 1 \\
f''(x) &= \frac{x}{(1-x^2)^{3/2}} & f''(0) &= 0 \\
f'''(x) &= \frac{2x^2 + 1}{(1-x^2)^{5/2}} & f'''(0) &= 1
\end{aligned}
$$

$$P_3(x) = f(0) + f'(0)x + \frac{f''(0)}{2!}x^2 + \frac{f'''(0)}{3!}x^3 = x + \frac{x^3}{6}$$

(b)

x	-1.0	-0.75	-0.50	-0.25	0	0.25	0.50	0.75	1.0
$f(x)$	-1.571	-0.848	-0.524	-0.253	0	0.253	0.524	0.848	1.571
$P_3(x)$	-1.167	-0.820	-0.521	-0.253	0	0.253	0.521	0.820	1.167

29. Approximate $f(1.2)$ for $f(x) = \ln x$ using the polynomial found in Exercise 19.

Solution:

$$f(x) \approx P_4(x) = (x-1) - \frac{1}{2}(x-1)^2 + \frac{1}{3}(x-1)^3 - \frac{1}{4}(x-1)^4$$

$$f(1.2) \approx P_4(1.2) = (1.2-1) - \frac{1}{2}(1.2-1)^2$$

$$+ \frac{1}{3}(1.2-1)^3 - \frac{1}{4}(1.2-1)^4 \approx 0.1823$$

(The actual functional value accurate to four decimal places is also 0.1823.)

 33. Use Taylor's Theorem to determine the accuracy of the approximation

$$\arcsin(0.4) \approx 0.4 + \frac{(0.4)^3}{2 \cdot 3}.$$

Solution:

Using Taylor's Theorem, you have

$$\arcsin(0.4) \approx 0.4 + \frac{(0.4)^3}{2 \cdot 3} + R_3(x) = 0.4 + \frac{(0.4)^3}{2 \cdot 3} + \frac{f^{(4)}(z)}{4!}x^4$$

where $0 < z < 0.4$. Using a symbolic differentiation utility to find the $f^{(4)}(x)$ and sketch its graph, yields

$$f^{(4)}(x) = \frac{3x(2x^2 + 3)}{(1 - x^2)^{3/2}}.$$

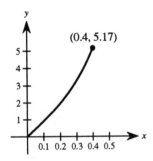

(0.4, 5.17)

From the accompanying graph of the fourth derivative, it follows that $f^{(4)}(x) < 5.2$ in the interval $[0, 0.4]$. Therefore,

$$0 < R_3(0.4) = \frac{f^{(4)}(z)}{4!}(0.4)^4 < \frac{5.2}{4!}(0.4)^4 \approx 0.00555.$$

Since

$$\arcsin(0.4) \approx 0.4 + \frac{(0.4)^3}{2 \cdot 3} = 0.41067,$$

you have

$$0.41067 < \arcsin(0.4) = 0.41067 + R_3(0.4) < 0.41067 + 0.00555$$
$$0.41067 < \arcsin(0.4) < 0.41622$$

39. For what values of $x < 0$ can e^x be replaced by the polynomial $1 + x + (x^2/2!) + (x^3/3!)$ if the error cannot exceed 0.001?

Solution:

By Theorem 8.19

$$e^x = 1 + x + \frac{x^2}{2!} + \frac{x^3}{3!} + R_3$$

where

$$R_3 = \frac{f^{(4)}(z)}{4!}x^4 = \frac{e^z}{4!}x^4$$

For $z < 0$ you have

$$R_3 = \frac{e^z x^4}{4!} < \frac{x^4}{4!}$$

and you wish to find $x < 0$ such that

$$\frac{x^4}{4!} < 0.001$$
$$x^4 < 24(0.001) = 0.024$$
$$|x| < (0.024)^{1/4} \approx 0.3936$$

Therefore, for values of x such that $-0.39 < x < 0$, you have

$$e^x \approx 1 + x + \frac{x^2}{2!} + \frac{x^3}{3!}.$$

8.8   Power Series

9. Find the interval of convergence for the power series

$$\sum_{n=1}^{\infty} \frac{(-1)^n x^n}{n}.$$

Include a check for convergence at the endpoints of the interval.

Solution:

Letting $u_n = (-1)^n x^n / n$ produces

$$\lim_{n \to \infty} \left| \frac{u_{n+1}}{u_n} \right| = \lim_{n \to \infty} \left| \frac{(-1)^{n+1} x^{n+1} / (n+1)}{(-1)^2 x^n / n} \right|$$

$$= \lim_{n \to \infty} \left[\frac{x^{n+1}}{n+1} \cdot \frac{n}{x} \right]$$

$$= \lim_{n \to \infty} \left| \frac{nx}{n+1} \right| = |x|.$$

By the Ratio Test, the series converges if $|x| < 1$. Hence, the radius of convergence is 1 and that the interval of convergence includes $-1 < x < 1$. When $x = -1$, you have the harmonic series

$$\sum_{n=1}^{\infty} \frac{1}{n}$$

which diverges. When $x = 1$, you have the alternating series

$$\sum_{n=1}^{\infty} \frac{(-1)^n}{n}$$

which converges. Thus the interval of convergence is $-1 < x \leq 1$.

13. Find the interval of convergence for the power series

$$\sum_{n=0}^{\infty} (2n)! \left(\frac{x}{2}\right)^n.$$

Include a check for convergence at the endpoints of the interval.

Solution:

Letting $u_n = (2n)! \left(\frac{x}{2}\right)^n$ produces

$$\lim_{n \to \infty} \left| \frac{u_{n+1}}{u_n} \right| = \lim_{n \to \infty} \left| \frac{(2n+2)!/(x/2)^{n+1}}{(2n)!/(x/2)^n} \right|$$

$$= \lim_{n \to \infty} \left| \frac{(2n+2)(2n+1)x}{2} \right| = \infty$$

for any real $x \neq 0$. Therefore, by the Ratio Test it follows that the series converges only when $x = 0$.

17. Find the interval of convergence for the power series

$$\sum_{n=1}^{\infty} \frac{(-1)^{n+1}(x-5)^n}{n5^n}.$$

Include a check for convergence at the endpoints of the interval.

Solution:

Letting $u_n = (-1)^{n+1}(x-5)^n/(n5^n)$ produces

$$\lim_{n \to \infty} \left| \frac{u_{n+1}}{u_n} \right| = \lim_{n \to \infty} \left| \frac{n5^n(x-5)}{(n+1)5^{n+1}} \right|$$

$$= \lim_{n \to \infty} \left| \frac{x-5}{5} \left(\frac{n}{n+1} \right) \right| = \left| \frac{x-5}{5} \right|.$$

By the Ratio Test, the series converges if $|(x-5)/5| < 1$, or $|x - 5| < 5$. Hence, the radius of convergence is $R = 5$, and since the series is centered at $x = 5$, the series will converge in the interval $(0, 10)$. Furthermore, when $x = 0$ you have the series

$$\sum_{n=1}^{\infty} \frac{(-1)^{n+1}(-1)^n}{n} = \sum_{n=1}^{\infty} \frac{(-1)^{2n+1}}{n} = -\sum_{n=1}^{\infty} \frac{1}{n}$$

which diverges $(p = 1)$. When $x = 10$, you have the series

$$\sum_{n=1}^{\infty} \frac{(-1)^{n+1}}{n}$$

which converges by the Alternating Series Test. Hence, the interval of convergence of the given series is $0 < x \leq 10$.

27. Find the interval of convergence for the power series

$$\sum_{n=1}^{\infty} \frac{k(k+1)(k+2)\cdots(k+n-1)x^n}{n!} \quad (k \geq 1).$$

Include a check for convergence at the endpoints of the interval.

Solution:

Since

$$\lim_{n\to\infty} \left|\frac{u_{n+1}}{u_n}\right|$$

$$= \lim_{n\to\infty} \left|\frac{k(k+1)\cdots(k+n-1)(k+n)x^{n+1}}{(n+1)!} \cdot \frac{n!}{k(k+1)\cdots(k+n-1)x^n}\right|$$

$$= \lim_{n\to\infty} \left|\frac{(k+n)x}{n+1}\right| = |x|,$$

the radius of convergence is $R = 1$. Since the series is centered at $x = 0$, it will converge on the interval $(-1, 1)$. To test for convergence at the endpoints, note that for $k \geq 1$, you have

$$\lim_{n\to\infty} a_n = \lim_{n\to\infty} \left[\left(\frac{k}{1}\right)\left(\frac{k+1}{2}\right)\left(\frac{k+2}{3}\right)\cdots\left(\frac{k+n-1}{n}\right)\right] \neq 0$$

Thus, for $x = \pm 1$ the series diverges, and it follows that the interval of convergence is $-1 < x < 1$.

31. Find the interval of convergence (check for convergence at the endpoints) of (a) $f(x)$, (b) $f'(x)$, and (d) $\int f(x)\,dx$ if

$$f(x) = \sum_{n=0}^{\infty} \left(\frac{x}{2}\right)^n.$$

Solution:

(a) The given series is geometric with $r = x/2$ and converges if

$$\left|\frac{x}{2}\right| < 1 \quad \text{or} \quad -2 < x < 2$$

(b) $f'(x) = \sum_{n=1}^{\infty} n\left(\frac{x}{2}\right)^{n-1}\left(\frac{1}{2}\right) = \sum_{n=1}^{\infty} \left(\frac{n}{2}\right)\left(\frac{x}{2}\right)^{n-1}$

Therefore the series for $f'(x)$ diverges for $x = \pm 2$, and its interval of convergence is $-2 < x < 2$.

(c)

$$f''(x) = \sum_{n=2}^{\infty} \left(\frac{n}{2}\right)(n-1)\left(\frac{x}{2}\right)^{n-2}\left(\frac{1}{2}\right)$$

$$= \sum_{n=2}^{\infty} \frac{n(n-1)}{4}\left(\frac{x}{2}\right)^{n-2}$$

Therefore the series for $f''(x)$ diverges for $x = \pm 2$, and its interval of convergence is $-2 < x < 2$.

(d) The series for $\displaystyle\int f(x)\,dx$ is

$$\sum_{n=0}^{\infty} \frac{2}{n+1}\left(\frac{x}{2}\right)^{n+1}.$$

and it converges (Alternating Series Test) for $x = -2$ and diverges [Limit Comparison Test with $\sum_{n=1}^{\infty}(1/n)$] for $x = 2$. Therefore, its interval of convergence is $-2 \le x < 2$.

8.9   Representation of Functions by Power Series

5. Find a power series centered at $c = 5$ for $f(x) = 1/(2-x)$, and find the interval of convergence.

Solution:

Writing $f(x)$ in the form $a/(1-r)$, you have

$$\frac{1}{2-x} = \frac{1}{-3-x+5} = \frac{-1/3}{1-[(x-5)/(-3)]} = \frac{a}{1-r}$$

which implies that $a = -1/3$ and $r = (x-5)/(-3)$. Therefore,

$$\frac{1}{2-x} = \sum_{n=0}^{\infty} ar^n$$

$$= -\frac{1}{3}\sum_{n=0}^{\infty}\left(\frac{x-5}{-3}\right)^n = \sum_{n=0}^{\infty}\frac{(x-5)^n}{(-3)^{n+1}}.$$

Since

$$\lim_{n\to\infty}\left|\frac{u_{n+1}}{u_n}\right| = \lim_{n\to\infty}\left|\frac{(x-5)^{n+1}}{(-3)^{n+2}}\cdot\frac{(-3)^{n+1}}{(x-5)^n}\right| = \left|\frac{x-5}{3}\right|,$$

the radius of convergence is $R = 3$. Since the series is centered at $c = 5$, it converges in the interval $(2, 8)$. Finally, since the series diverges at both endpoints, the interval of convergence is $2 < x < 8$.

13. Find a power series centered at $c = 0$ for $f(x) = 3x/(x^2+x-2)$, and find the interval of convergence.

Solution:

Using the method for finding partial fractions (Section 7.5), you have

$$\frac{3x}{x^2+x-2} = \frac{2}{x+2} + \frac{1}{x-1} = \frac{1}{1-[-x/2]} - \frac{1}{1-x}.$$

Therefore, you have the difference of the sum of two geometric series. Using the fact that

$$\sum_{n=0}^{\infty} ar^n = \frac{a}{1-r},$$

you can write

$$\frac{1}{1-[-x/2]} - \frac{1}{1-x} = \sum_{n=0}^{\infty}\left(\frac{x}{-2}\right)^n - \sum_{n=0}^{\infty} x^n$$

$$= \sum_{n=0}^{\infty}\left[\frac{x^n}{(-2)^n} - x^n\right] = \sum_{n=0}^{\infty}\left[\frac{1}{(-2)^n} - 1\right]x^n.$$

By the Ratio Test you have

$$\lim_{n\to\infty}\left|\frac{u_{n+1}}{u_n}\right| = \lim_{n\to\infty}\left|\frac{[1/(-2)^{n+1}]-1}{[1/(-2)^n]-1}\cdot\frac{x^{n+1}}{x^n}\right| = |x|$$

and the series converges if $|x| < 1$. Finally, when $x = \pm 1$, you have

$$\lim_{n\to\infty} a_n = \lim_{n\to\infty}\left[\frac{1}{(-2)^n} - 1\right] \neq 0.$$

Thus, the series diverges when $x = \pm 1$ and the interval of convergence is $-1 < x < 1$.

15. Find a power series centered at $c = 0$ for $f(x) = 2/(1 - x^2)$, and find the interval of convergence.

Solution:

Letting $u = x^2$, you have

$$\frac{2}{1 - x^2} = \frac{2}{1 - u}$$

$$= 2 \sum_{n=0}^{\infty} u^n$$

$$= 2 \sum_{n=0}^{\infty} (x^2)^n = 2 \sum_{n=0}^{\infty} x^{2n}.$$

The series will converge if $x^2 < 1$ or $-1 < x < 1$.

21. Use the power series

$$\frac{1}{x + 1} = \sum_{n=0}^{\infty} (-1)^n x^n$$

to determine a power series representation about $c = 0$ for $f(x) = \ln(x + 1)$. Specify the interval of convergence.

Solution:

Since

$$\int \frac{1}{x + 1}\, dx = \ln(x + 1) + C$$

you can integrate the power series for $1/(x + 1)$ to obtain the series for $\ln(x + 1)$.

$$\frac{1}{x + 1} = \sum_{n=0}^{\infty} (-1)^n x^n$$

$$\ln(x + 1) = \sum_{n=0}^{\infty} \frac{(-1)^n x^{n+1}}{n + 1} + C$$

By substituting $x = 0$ on both sides of this equation, yields $C = 0$. Furthermore, by the Ratio Test

$$\lim_{n \to \infty} \left| \frac{u_{n+1}}{u_n} \right| = \lim_{n \to \infty} \left| \frac{x^{n+2}}{n + 2} \cdot \frac{n + 1}{x^{n+1}} \right| = |x|,$$

and the series converges if $|x| < 1$. At $x = -1$ you have the divergent series

$$\sum_{n=0}^{\infty} \frac{(-1)^{2n+1}}{n+1} = -\sum_{n=0}^{\infty} \frac{1}{n+1}.$$

At $x = 1$ you have the convergent alternating series

$$\sum_{n=0}^{\infty} \frac{(-1)^n}{n+1}.$$

Therefore,

$$\ln(x+1) = \sum_{n=0}^{\infty} \frac{(-1)^n x^{n+1}}{n+1}, \quad \text{for} \quad -1 < x \le 1.$$

31. Use the series for the function $f(x) = \arctan x$ to approximate (with $R_N \le 0.001$)

$$\int_0^{1/2} \frac{\arctan x^2}{x} \, dx$$

Solution:

From Example 5, you have

$$\arctan x = \sum_{n=0}^{\infty} \frac{(-1)^n x^{2n+1}}{2n+1}$$

Therefore,

$$\arctan x^2 = \sum_{n=0}^{\infty} \frac{(-1)^n (x^2)^{(2n+1)}}{2n+1}$$

$$= \sum_{n=0}^{\infty} \frac{(-1)^n (x^{(4n+2)})}{2n+1}$$

and

$$\frac{\arctan x^2}{x} = \sum_{n=0}^{\infty} \frac{(-1)^n x^{4n+1}}{2n+1}.$$

Thus,

$$\int_0^{1/2} \frac{\arctan x^2}{x}\, dx = \left[\sum_{n=0}^{\infty} \frac{(-1)^n x^{4n+2}}{(2n+1)(4n+2)}\right]_0^{1/2}$$

$$= \sum_{n=0}^{\infty} \frac{(-1)^n (1/2)^{4n+2}}{(2n+1)(4n+2)}$$

$$\approx \left(\frac{1}{2}\right)\left(\frac{1}{2}\right)^2 = \frac{1}{8}.$$

Note that the second term of the series is $-(\frac{1}{18})(\frac{1}{2})^6 < 0.001$. Therefore, by Theorem 8.15 one term is sufficient to approximate the integral.

43. Find the sum of the series

$$\sum_{n=1}^{\infty} (-1)^{n+1} \frac{2^n}{5^n n}.$$

Explain how you found the sum.

Solution:

In Example 4 it was shown that

$$\ln x = \sum_{n=0}^{\infty} (-1)^n \frac{(x-1)^{n+1}}{n+1} = \sum_{n=1}^{\infty} (-1)^{n-1} \frac{(x-1)^n}{n}.$$

Letting $x = 7/5$ in the series yields

$$\sum_{n=1}^{\infty} (-1)^{n-1} \frac{([7/5]-1)^n}{n} = \sum_{n=1}^{\infty} (-1)^{n-1} \frac{(2/5)^n}{n} = \sum_{n=1}^{\infty} (-1)^{n-1} \frac{2^n}{5^n n}.$$

Therefore, the sum of the series is

$$\ln \frac{7}{5} \approx 0.3365.$$

8.10   Taylor and Maclaurin Series

7. Find the Taylor Series centered at $c = 0$ for the function $f(x) = \sin 2x$.

Solution:

Since

$$
\begin{aligned}
f(x) &= \sin 2x & f(0) &= 0 \\
f'(x) &= 2\cos 2x & f'(0) &= 2 \\
f''(x) &= -4\sin 2x & f''(0) &= 0 \\
f'''(x) &= -8\cos 2x & f'''(0) &= -8 = -2^3 \\
f^{(4)}(x) &= 16\sin 2x & f^{(4)}(0) &= 0 \\
f^{(5)}(x) &= 32\cos 2x & f^{(5)}(0) &= 32 = 2^5 \quad,
\end{aligned}
$$

you can see that the signs alternate and that $|f^{(n)}(0)| = 2^n$ if n is odd. Therefore the Taylor Series is

$$
\begin{aligned}
\sin 2x \\
&= f(0) + f'(0)x + \frac{f''(0)x^2}{2!} + \frac{f'''(0)x^3}{3!} + \frac{f^{(4)}(0)x^4}{4!} + \cdots \\
&= \frac{2x}{1!} - \frac{2^3 x^3}{3!} + \frac{2^5 x^5}{5!} - \cdots + \frac{(-1)^n (2x)^{2n+1}}{(2n+1)!} + \cdots \\
&= \sum_{n=0}^{\infty} \frac{(-1)^n (2x)^{2n+1}}{(2n+1)!}
\end{aligned}
$$

Note that you could have arrived at the same result by substituting $2x$ into the series for $\sin x$ as follows:

$$
\begin{aligned}
\sin x &= x - \frac{x^3}{3!} + \frac{x^5}{5!} - \frac{x^7}{7!} + \cdots \\
\sin (2x) &= (2x) - \frac{(2x)^3}{3!} + \frac{(2x)^5}{5!} - \frac{(2x)^7}{7!} + \cdots \\
&= \sum_{n=0}^{\infty} \frac{(-1)^n (2x)^{2n+1}}{(2n+1)!}
\end{aligned}
$$

13. Use the binomial series to find the power series centered at $c = 0$ for the function $f(x) = 1/\sqrt{4 + x^2}$.

Solution:

Consider f in the form

$$f(x) = \frac{1}{\sqrt{4 + x^2}} = \frac{1}{2\sqrt{1 + (x/2)^2}} = \frac{1}{2}\left[1 + \left(\frac{x}{2}\right)^2\right]^{-1/2}$$

which is similar to the binomial form $(1 + x)^{-k}$. Since

$$(1 + x)^{-k} = 1 - kx + \frac{k(k + 1)x^2}{2!} - \frac{k(k + 1)(k + 2)x^3}{3!} + \cdots$$

you have, for $k = \dfrac{1}{2}$,

$$
\begin{aligned}
(1 + x)^{-1/2} &= 1 - \frac{1}{2}x + \frac{(\frac{1}{2})(\frac{3}{2})x^2}{2!} - \frac{(\frac{1}{2})(\frac{3}{2})(\frac{5}{2})x^3}{3!} + \cdots \\
&= 1 - \frac{x}{2} + \frac{1 \cdot 3x^2}{2^2 2!} - \frac{1 \cdot 3 \cdot 5x^3}{2^3 3!} + \cdots \\
&= 1 + \sum_{n=1}^{\infty} \frac{(-1)^{n+1} 1 \cdot 3 \cdot 5 \cdots (2n - 1)x^n}{2^n n!}.
\end{aligned}
$$

Now substituting $(x/2)^2$ for x, yields

$$
\begin{aligned}
f(x) &= \frac{1}{\sqrt{4 + x^2}} = \frac{1}{2}\left[1 + \left(\frac{x}{2}\right)^2\right]^{-1/2} \\
&= \frac{1}{2}\left[1 + \sum_{n=1}^{\infty} \frac{(-1)^{n+1} 1 \cdot 3 \cdot 5 \cdots (2n - 1)(x/2)^{2n}}{2^n n!}\right] \\
&= \frac{1}{2} + \sum_{n=1}^{\infty} \frac{(-1)^{n+1} 1 \cdot 3 \cdot 5 \cdots (2n - 1)x^{2n}}{2^{3n+1} n!} \\
&= \frac{1}{2}\left[1 + \sum_{n=1}^{\infty} \frac{(-1)^{n+1} 1 \cdot 3 \cdot 5 \cdots (2n - 1)x^{2n}}{2^{3n} n!}\right].
\end{aligned}
$$

17. Find the power series for $f(x) = e^{x^2/2}$ by using the series for e^x.

Solution:

Since $\quad e^x = 1 + x + \dfrac{x^2}{2!} + \dfrac{x^3}{3!} + \dfrac{x^4}{4!} + \dfrac{x^5}{5!} + \cdots,$

you can substitute $x^2/2$ for x and obtain the series

$$e^{x^2/2} = 1 + \frac{x^2}{2} + \frac{(x^2/2)^2}{2!} + \frac{(x^2/2)^3}{3!} + \frac{(x^2/2)^4}{4!} + \cdots$$

$$= 1 + \frac{x^2}{2} + \frac{x^4}{2^2 2!} + \frac{x^6}{2^3 3!} + \frac{x^8}{2^4 4!} + \cdots$$

$$= \sum_{n=0}^{\infty} \frac{x^{2n}}{2^n n!}$$

23. Find the power series for $f(x) = (\sin x)/x$ by using the series for $\sin x$.

Solution:

Since $\quad \sin x = x - \dfrac{x^3}{3!} + \dfrac{x^5}{5!} - \dfrac{x^7}{7!} + \dfrac{x^9}{9!} - \cdots$

you can divide by x to obtain

$$\frac{\sin x}{x} = 1 - \frac{x^2}{3!} + \frac{x^4}{5!} - \frac{x^6}{7!} + \frac{x^8}{9!} - \cdots = \sum_{n=0}^{\infty} \frac{(-1)^n x^{2n}}{(2n+1)!}$$

29. Use the power series for e^x to show that

$$g(x) = \frac{1}{2i}(e^{ix} - e^{-ix}) = \sin x.$$

Solution:

Begin by observing the following powers of i:

$$i = \sqrt{-1} \qquad\qquad i^5 = i^4 \cdot i = 1$$
$$i^2 = -1 \qquad\qquad i^6 = i^4 \cdot i^2 = -1$$
$$i^3 = i^2 \cdot i = -i \qquad\qquad i^7 = i^4 \cdot i^3 = -i$$
$$i^4 = i^2 \cdot i^2 = 1 \qquad\qquad i^8 = 1^4 \cdot i^4 = 1$$

Since $e^x = 1 + x + \dfrac{x^2}{2!} + \dfrac{x^3}{3!} + \dfrac{x^4}{4!} + \dfrac{x^5}{5!} + \cdots$

you can substitute (ix) and $(-ix)$ for x and obtain the series for e^{ix} and e^{-ix}.

$$e^{ix} = 1 + (ix) + \frac{(ix)^2}{2!} + \frac{(ix)^3}{3!} + \frac{(ix)^4}{4!} + \frac{(ix)^5}{5!} + \cdots$$

$$= 1 + ix - \frac{x^2}{2!} - \frac{ix^3}{3!} + \frac{x^4}{4!} + \frac{ix^5}{5!} - \cdots$$

$$e^{-ix} = 1 + (-ix) + \frac{(-ix)^2}{2!} + \frac{(-ix)^3}{3!} + \frac{(-ix)^4}{4!} + \frac{(-ix)^5}{5!} + \cdots$$

$$= 1 - ix - \frac{x^2}{2!} + \frac{ix^3}{3!} + \frac{x^4}{4!} - \frac{ix^5}{5!} - \cdots$$

Therefore, subtracting the series for e^{ix} and e^{-ix} yields

$$e^{ix} - e^{-ix} = 2ix - \frac{2ix^3}{3!} + \frac{2ix^5}{5!} - .$$

and

$$\frac{e^{ix} - e^{-ix}}{2i} = x - \frac{x^3}{3!} + \frac{x^5}{5!} - \cdots$$

$$= \sum_{n=0}^{\infty} \frac{(-1)^n x^{2n+1}}{(2n+1)!} = \sin x$$

45. Use power series to approximate

$$\int_0^1 \frac{\sin x}{x}\, dx$$

with an error less than 0.0001. (Assume the integrand is defined to be 1 when $x = 0$.)

Solution:

From Exercise 23 you have

$$\frac{\sin x}{x} = \sum_{n=0}^{\infty} \frac{(-1)^n x^{2n}}{(2n+1)!}.$$

Therefore

$$\int_0^1 \frac{\sin x}{x}\, dx = \left[\sum_{n=0}^{\infty} \frac{(-1)^n x^{2n+1}}{(2n+1)(2n+1)!} \right]_0^1$$

$$= \sum_{n=0}^{\infty} \frac{(-1)^n}{(2n+1)(2n+1)!}$$

$$\approx 1 - \frac{1}{3 \cdot 3!} + \frac{1}{5 \cdot 5!} \approx 0.9461$$

By the Alternating Series Remainder, you have

$$|R_3| \le a_4 = \frac{1}{7 \cdot 7!} \approx 0.00003 < 0.0001.$$

Review Exercises for Chapter 8

7. Determine the convergence or divergence of $\{a_n\}$ when $a_n = \sqrt{n+1} - \sqrt{n}$.

Solution:

$$\lim_{n \to \infty} a_n = \lim_{n \to \infty} [\sqrt{n+1} - \sqrt{n}]$$

$$= \lim_{n \to \infty} \left[(\sqrt{n+1} - \sqrt{n}) \frac{\sqrt{n+1} + \sqrt{n}}{\sqrt{n+1} + \sqrt{n}} \right]$$

$$= \lim_{n \to \infty} \frac{1}{\sqrt{n+1} + \sqrt{n}} = 0$$

Therefore the sequence converges to 0.

15. Find the first five terms of the sequence of partial sums for the series

$$\sum_{n=1}^{\infty} \frac{(-1)^{n+1}}{(2n)!}.$$

Solution:

Since

$$\sum_{n=1}^{\infty} \frac{(-1)^{n+1}}{(2n)!} = \frac{1}{2!} - \frac{1}{4!} + \frac{1}{6!} - \frac{1}{8!} + \frac{1}{10!} - \cdots$$

$$= \frac{1}{2} - \frac{1}{24} + \frac{1}{720} - \frac{1}{40,320} + \frac{1}{3,628,800} - \cdots$$

you have

$$S_1 = \frac{1}{2} = 0.5$$

$$S_2 = \frac{1}{2} - \frac{1}{24} \approx 0.45833$$

$$S_3 = \frac{1}{2} - \frac{1}{24} + \frac{1}{720} \approx 0.45972$$

$$S_4 = \frac{1}{2} - \frac{1}{24} + \frac{1}{720} - \frac{1}{40,320} \approx 0.45970$$

$$S_5 = \frac{1}{2} - \frac{1}{24} + \frac{1}{720} - \frac{1}{40,320} + \frac{1}{3,628,800} \approx 0.45970$$

19. Find the sum of the infinite series

$$\sum_{n=0}^{\infty} \left(\frac{1}{2^n} - \frac{1}{3^n} \right).$$

Solution:

Begin by writing the given series as the difference of two geometric series

$$\sum_{n=0}^{\infty} \left(\frac{1}{2^n} - \frac{1}{3^n} \right) = \sum_{n=0}^{\infty} \left(\frac{1}{2} \right)^n - \sum_{n=0}^{\infty} \left(\frac{1}{3} \right)^n.$$

Since these two geometric series have the values $a = 1$ and $r = \frac{1}{2}$ and $\frac{1}{3}$, respectively, it follows that the sum of the given series is

$$\sum_{n=0}^{\infty} \left(\frac{1}{2^n} - \frac{1}{3^n} \right) = \frac{1}{1 - (\frac{1}{2})} - \frac{1}{1 - (\frac{1}{3})} = 2 - \frac{3}{2} = \frac{1}{2}.$$

29. Determine the convergence or divergence of the series

$$\sum_{n=1}^{\infty} \frac{1}{(n^3 + 2n)^{1/3}}$$

Solution:

Since the denominator is the cube root of a third-degree polynomial, compare the given series to the divergent harmonic series

$$\sum_{n=1}^{\infty} \frac{1}{n}.$$

Thus,

$$\lim_{n \to \infty} \frac{[1/(n^3 + 2n)^{1/3}]}{1/n} = \lim_{n \to \infty} \frac{n}{(n^3 + 2n)^{1/3}}$$

$$= \lim_{n \to \infty} \frac{1}{[1 + (2/n^2)]^{1/3}} = 1,$$

and it follows by the Limit Comparison Test that the given series diverges.

45. Find the interval of convergence of the power series

$$\sum_{n=0}^{\infty} n!(x - 2)^n.$$

Solution:

Since

$$\lim_{n \to \infty} \left| \frac{u_{n+1}}{u_n} \right| = \lim_{n \to \infty} \left| \frac{(n + 1)!(x - 2)^{n+1}}{n!(x - 2)^2} \right|$$

$$= \lim_{n \to \infty} |(n + 1)(x - 2)| = \infty$$

for all $x \neq 2$, it follows that the radius of convergence is $R = 0$. Therefore, the given series will converge only at $x = 2$.

49. Find the power series for $f(x) = 3^x$ centered at $c = 0$.

Solution:

Since $3^x = e^{\ln(3^x)} = e^{x(\ln 3)}$, you can substitute $x(\ln 3)$ for x in the series

$$e^x = 1 + x + \frac{x^2}{2!} + \frac{x^3}{3!} + \frac{x^4}{4!} + \cdots$$

to obtain

$$3^x = e^{x(\ln 3)}$$
$$= 1 + (\ln 3)x + \frac{(\ln 3)^2 x^2}{2!} + \frac{(\ln 3)^3 x^3}{3!} + \frac{(\ln 3)^4 x^4}{4!} + \cdots$$
$$= \sum_{n=0}^{\infty} \frac{(x \ln 3)^n}{n!}.$$

9 CONIC SECTIONS

9.1 Parabolas

---

9. Find the vertex, focus, and directrix of the parabola given by $y^2 = -6x$, and sketch its graph.

Solution:

Begin by writing the equation of this parabola in standard form.

$$(y - k)^2 = 4p(x - h)$$
$$y^2 = -6x$$
$$(y - 0)^2 = 4\left(-\frac{3}{2}\right)(x - 0)$$

Thus, $k = 0$, $h = 0$, and $p = -\frac{3}{2}$. It follows that

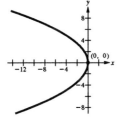

$$\text{vertex, } (h, k): \quad (0, 0)$$
$$\text{focus, } (h + p, k): \quad \left(-\frac{3}{2}, 0\right)$$
$$\text{directrix, } (x = h - p): \quad x = \frac{3}{2}$$

17. Find the vertex, focus, and directrix of the parabola given by $y = \frac{1}{4}(x^2 - 2x + 5)$, and sketch its graph.

Solution:

Begin by writing the equation of this parabola in standard form.

$$(x - h)^2 = 4p(y - k)$$
$$\frac{1}{4}(x^2 - 2x + 5) = y$$
$$x^2 - 2x + 5 = 4y$$
$$x^2 - 2x + 1 = 4y - 4$$
$$(x - 1)^2 = 4(1)(y - 1)$$

Thus, $h = 1$, $k = 1$, and $p = 1$. You can conclude that

$$
\begin{array}{lc}
\text{vertex, } (h, k): & (1, 1) \\
\text{focus, } (h, k + p): & (1, 2) \\
\text{directrix, } (y = k - p): & y = 0
\end{array}
$$

19. Find the vertex, focus, and directrix of the parabola given by $y^2 - 4y - 4x = 0$, and sketch its graph.

Solution:

Begin by writing the equation of this parabola in standard form.

$$(y - k)^2 = 4p(x - h)$$
$$y^2 - 4y = 4x$$
$$y^2 - 4y + 4 = 4x + 4$$
$$(y - 2)^2 = 4(1)(x + 1)$$

Thus, $h = -1$, $k = 2$, and $p = 1$. You can conclude that

$$
\begin{array}{lc}
\text{vertex, } (h, k): & (-1, 2) \\
\text{focus, } (h + p, k): & (0, 2) \\
\text{directrix, } (x = h - p): & x = -2
\end{array}
$$

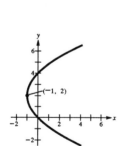

29. Find the equation of a parabola with vertex at $(3, 2)$ and focus at $(1, 2)$.

Solution:

Since the vertex and focus lie on a horizontal line, the axis of this parabola must be horizontal and its standard form is

$$(y - k)^2 = 4p(x - h)$$

Since the vertex lies at $(3, 2)$, you have $h = 3$ and $k = 2$. Furthermore, the *directed distance* from the focus to the vertex is

$$p = 1 - 3 = -2.$$

Thus, the standard equation is

$$(y - 2)^2 = 4(-2)(x - 3)$$
$$y^2 - 4y + 4 = -8x + 24$$
$$y^2 - 4y + 8x - 20 = 0.$$

37. Find an equation for the parabola whose axis is parallel to the y-axis and passes through the points $(0, 3)$, $(3, 4)$, $(4, 11)$.

Solution:

Since the axis of the parabola is vertical, the standard form is

$$(x - h)^2 = 4p(y - k).$$

For this exercise, a more convenient form for this equation is

$$y = ax^2 + bx + c.$$

Substituting the values of the given coordinates into this equation, you obtain three equations:

(i) $3 = a(0)^2 + b(0) + c$

(ii) $4 = a(3)^2 + b(3) + c$

(iii) $11 = a(4)^2 + b(4) + c.$

Simplification yields

(i) $3 = c$

(ii) $4 = 9a + 3b + c \rightarrow 1 = 9a + 3b$

(iii) $11 = 16a + 4b + c \rightarrow 8 = 16a + 4b.$

Solving (ii) and (iii) for a and b yields

(ii) $\qquad\qquad 9a + 3b = 1 \;\rightarrow\; 9a + 3b = 1$

(iii) $\qquad\qquad 4a + b = 2 \;\rightarrow\; 12a + 3b = 6$

$$-3a = -5$$
$$a = \frac{5}{3}$$
$$b = -\frac{14}{3}$$

Finally,

$$y = ax^2 + bx + c$$
$$y = \frac{5}{3}x^2 - \frac{14}{3}x + 3$$
$$3y = 5x^2 - 14x + 9$$
$$5x^2 - 14x - 3y + 9 = 0.$$

53. In mountainous areas reception of radio and television is sometimes poor. Consider an idealized case where a hill is given by the graph of the parabola $y = x - x^2$, a transmitter is located at the point $(-1,\ 1)$, and a receiver is located on the other side of the hill at the point $(x_0,\ 0)$. What is the closest the receiver can be to the hill so that the reception is unobstructed?

Solution:

The position of the receiver is the x-intercept of the tangent line to the curve $y = x - x^2$ that passes through the point $(-1,1)$ (see figure). The point of tangency has coordinates $(x,\ y) = (x,\ x - x^2)$. You now can determine the slope of the tangent line in two ways. First, use the formula for the slope of a line through two points and obtain

$$m = \frac{(x - x^2) - 1}{x - (-1)} = \frac{x - x^2 - 1}{x + 1}.$$

The slope of the line is also given by $y' = 1 - 2x$. Equating these two expressions for the slope of the tangent line and solving the resulting equation gives the x-coordinate of the point of tangency.

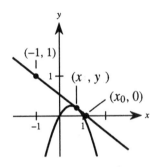

$$\frac{x - x^2 - 1}{x + 1} = 1 - 2x$$
$$x - x^2 - 1 = (1 - 2x)(1 + x)$$
$$x^2 + 2x - 2 = 0 \quad \Longrightarrow \quad x = \sqrt{3} - 1$$

The value of the derivative at this value of x is

$$y' = 1 - 2(\sqrt{3} - 1) = 3 - 2\sqrt{3},$$

and the equation of the tangent line is

$$y - 1 = (3 - 2\sqrt{3})[x - (\sqrt{3} - 1)]$$
$$y = (3 - 2\sqrt{3})x + 2(2 - \sqrt{3}).$$

Since the receiver is located at the x-intercept of the line, set $y = 0$ and solve for x to obtain

$$x_0 = x = \frac{2\sqrt{3}}{3}.$$

The distance between the base of the hill and the receiver is

$$\frac{2\sqrt{3}}{3} - 1 \approx 0.155.$$

61. A cable of a parabolic suspension bridge is suspended between two towers that are 400 feet apart and 50 feet above the roadway as shown in the accompanying figure. The cable touches the roadway midway between the towers. Find the length of the cable.

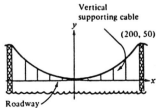

Vertical supporting cable

(200, 50)

Roadway

Solution:

Since the axis of the parabola is vertical, the standard form is

$$(x - h)^2 = 4p(y - k).$$

Since the vertex is $(0, 0)$, $h = 0$ and $k = 0$. Therefore

$$x^2 = 4py$$

Since the parabola passes through the point $(200, 50)$, you have

$$200^2 = 4p(50) \quad \text{or} \quad 200 = p.$$

Therefore,

$$x^2 = 4(200)y \quad \text{or} \quad y = \frac{1}{800}x^2.$$

$$y' = \frac{1}{400}x$$

$$s = \int_{-200}^{200} \sqrt{1 + (y')^2}\, dx$$

$$= 2\int_{0}^{200} \sqrt{1 + \left(\frac{x}{400}\right)^2}\, dx$$

$$= \frac{1}{200}\int_{0}^{200} \sqrt{400^2 + x^2}\, dx$$

$$= \frac{1}{400}\left[x\sqrt{400^2 + x^2} + 400^2 \ln\left(x + \sqrt{400^2 + x^2}\right)\right]_{0}^{200}$$

$$= 100\left[\sqrt{5} + 4\ln\left(\frac{1 + \sqrt{5}}{2}\right)\right] \approx 416.1 \text{ ft}$$

9.2   Ellipses

13. Find the center, foci, vertices, eccentricity, and sketch the graph of the ellipse given by $x^2 + 4y^2 = 4$.

Solution:

In standard form you have

$$x^2 + 4y^2 = 4$$

$$\frac{x^2}{4} + y^2 = 1$$

$$\frac{(x-0)^2}{2^2} + \frac{(y-0)^2}{1^2} = 1$$

$$\frac{(x-h)^2}{a^2} + \frac{(y-k)^2}{b^2} = 1$$

Thus, $h = 0$, $k = 0$, $a = 2$, $b = 1$, and $c = \sqrt{2^2 - 1^2} = \sqrt{3}$. If follows that

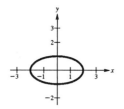

$$
\begin{aligned}
\text{center, } (h, k): &\quad (0, 0) \\
\text{foci, } (h \pm c, k): &\quad (\pm\sqrt{3}, 0) \\
\text{vertices, } (h \pm a, k): &\quad (\pm 2, 0)
\end{aligned}
$$

$$e = \frac{c}{a} = \frac{\sqrt{3}}{2}$$

19. Find the center, foci, vertices, eccentricity, and sketch the graph of the ellipse given by $9x^2 + 4y^2 + 36x - 24y + 36 = 0$.

Solution:

In standard form you have

$$9x^2 + 4y^2 + 36x - 24y + 36 = 0$$
$$9x^2 + 36x + 4y^2 - 24y = -36$$
$$9(x^2 + 4x + 4) + 4(y^2 - 6y + 9) = -36 + 36 + 36$$
$$9(x+2)^2 + 4(y-3)^2 = 36$$
$$\frac{(x+2)^2}{4} + \frac{(y-3)^2}{9} = 1$$
$$\frac{(x+2)^2}{2^2} + \frac{(y-3)^2}{3^2} = 1$$
$$\frac{(x-h)^2}{b^2} + \frac{(y-k)^2}{a^2} = 1$$

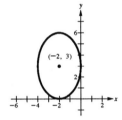

Thus, $h = -2$, $k = 3$, $a = 3$, $b = 2$, and $c = \sqrt{3^2 - 2^2} = \sqrt{5}$. You can conclude that

$$\begin{aligned}
\text{center, } (h,k): &\quad (-2,3) \\
\text{foci, } (h, k \pm c): &\quad (-2, 3 \pm \sqrt{5}) \\
\text{vertices, } (h, k \pm a): &\quad (-2, 3 \pm 3)
\end{aligned}$$

$$e = \frac{c}{a} = \frac{\sqrt{5}}{3}$$

25. Find the center, foci, vertices, and eccentricity of the ellipse given by $12x^2 + 20y^2 - 12x + 40y - 37 = 0$. Use a graphing utility to graph the ellipse.

Solution:

In standard form you have

$$12x^2 + 20y^2 - 12x + 40y - 37 = 0$$
$$12(x^2 - x + \tfrac{1}{4}) + 20(y^2 + 2y + 1) = 37 + 3 + 20$$
$$12(x - \tfrac{1}{2})^2 + 20(y+1)^2 = 60$$
$$\frac{(x - \tfrac{1}{2})^2}{5} + \frac{(y+1)^2}{3} = 1$$

Thus $h = 1/2$, $k = -1$, $a = \sqrt{5}$, $b = \sqrt{3}$, and $c = \sqrt{5-3} = \sqrt{2}$. We conclude that

$$\begin{aligned}
\text{center, } (h, k) &: \quad (\tfrac{1}{2}, -1) \\
\text{foci, } (h \pm c, k) &: \quad (\tfrac{1}{2} \pm \sqrt{2}, -1) \\
\text{vertices, } (h \pm a, k) &: \quad (\tfrac{1}{2} \pm \sqrt{5}, -1)
\end{aligned}$$

$$e = \frac{c}{a} = \frac{\sqrt{2}}{\sqrt{5}} = \frac{\sqrt{10}}{5}$$

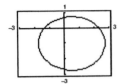

(*Note:* It may be necessary to solve for y in the equation of the ellipse and obtain two explicit functions to graph on your graphing utility.)

31. Find an equation for the ellipse with vertices $(5, 0)$ and $(-5, 0)$ and eccentricity $\frac{3}{5}$.

Solution:

Since the vertices lie on the x-axis, the standard form for the ellipse is

$$\frac{(x-h)^2}{a^2} + \frac{(y-k)^2}{b^2} = 1$$

Since the center is the midpoint of the line segment connecting the vertices, it follows that $(h, k) = (0, 0)$. Furthermore, since a is the distance from the center to the vertices, $a = 5$. Also,

$$e = \frac{c}{a} = \frac{c}{5} = \frac{3}{5}$$

Therefore $c = 3$ and $b^2 = a^2 - c^2 = 25 - 9 = 16$. Finally, the equation is

$$\frac{x^2}{25} + \frac{y^2}{16} = 1.$$

33. Find an equation of the ellipse with vertices $(3, 1)$ and $(3, 9)$, and minor axis of length 6.

Solution:

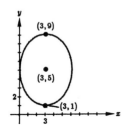

Since the vertices lie on a vertical line (see the accompanying figure), the standard form of the equation of the ellipse is

$$\frac{(x - h)^2}{b^2} + \frac{(y - k)^2}{a^2} = 1$$

Since the center of the ellipse is the midpoint of the line segment connecting the vertices, you have

$$(h, k) = \left(\frac{3 + 3}{2}, \frac{1 + 9}{2}\right) = (3, 5)$$

The length of the major axis is $2a = 8$ and therefore, $a = 4$. Furthermore, since the length of the minor axis is 6, $2b = 6$ or $b = 3$. Finally, the equation is

$$\frac{(x - 3)^2}{9} + \frac{(y - 5)^2}{16} = 1.$$

59. A particle is traveling clockwise on the elliptical orbit given by

$$\frac{x^2}{10^2} + \frac{y^2}{5^2} = 1$$

The particle leaves the orbit at the point $(-8, 3)$ and travels in a straight line tangent to the ellipse. At which point will the particle cross the y-axis?

Solution:

To find the slope of the tangent line at $(-8, 3)$, differentiate implicitly as follows:

$$\frac{x^2}{100} + \frac{y^2}{25} = 1$$

$$\frac{x}{50} + \frac{2yy'}{25} = 0$$

$$x + 4yy' = 0$$

$$y' = \frac{-x}{4y}$$

Thus, at $(-8, 3)$ the slope is

$$m = \frac{-(-8)}{4(3)} = \frac{8}{12} = \frac{2}{3}.$$

The equation of the tangent line is

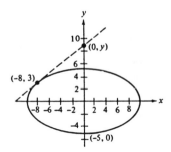

$$(y - 3) = \frac{2}{3}[x - (-8)]$$

$$3y - 9 = 2x + 16$$

$$3y = 2x + 25$$

Finally, when $x = 0$,

$$3y = 25 \quad \text{or} \quad y = \frac{25}{3}$$

and the required point is $(0, 25/3)$.

61. Find the dimensions of the rectangle of maximum area that can be inscribed in the ellipse

$$\frac{x^2}{a^2} + \frac{y^2}{b^2} = 1.$$

Solution:

Let (x, y) be a vertex of the rectangle located on the ellipse in Quadrant I (see the accompanying figure). Solving the equation of the ellipse for y, yields

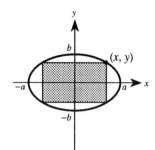

$$\frac{y^2}{b^2} = 1 - \frac{x^2}{a^2} = \frac{a^2 - x^2}{a^2}$$

$$y^2 = \frac{b^2}{a^2}(a^2 - x^2)$$

$$y = \pm\frac{b}{a}\sqrt{a^2 - x^2}.$$

Therefore, the dimensions of the rectangle are:

$$\text{length} = 2x$$

$$\text{width} = 2y = \frac{2b}{a}\sqrt{a^2 - x^2}$$

The area of the rectangle is given by

$$A = lw = 2x\left[\frac{2b}{a}\sqrt{a^2 - x^2}\right] = \frac{4b}{a}(x\sqrt{a^2 - x^2}),$$

and

$$\frac{dA}{dx} = \frac{4b}{a}\left[\frac{-x^2}{\sqrt{a^2-x^2}} + \sqrt{a^2-x^2}\right]$$

$$= \frac{4b}{a}\left[\frac{a^2-2x^2}{\sqrt{a^2-x^2}}\right]$$

Thus, $\dfrac{dA}{dx} = 0$ when $x = \dfrac{a}{\sqrt{2}}$ and the dimensions of the rectangle of maximum area are

$$\text{length} = 2x = \sqrt{2}a$$

$$\text{width} = \frac{2b}{a}\sqrt{a^2-x^2} = \frac{2b}{a}\sqrt{a^2 - \frac{a^2}{2}} = \sqrt{2}b$$

65. Consider the region bounded by the graph of the ellipse

$$\frac{x^2}{4} + \frac{y^2}{1} = 1.$$

Find (a) its area, (b) the volume and surface area of the solid generated by revolving the region about its major axis (prolate spheroid), and (c) the volume and surface area of the solid generated by revolving the region about its minor axis (oblate spheroid).

Solution:

(a) Using the symmetry of the region you have

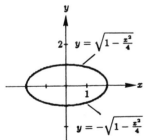

$$\text{area} = 4\int_0^2 \sqrt{1 - \frac{x^2}{4}}\, dx$$

$$= 2\int_0^2 \sqrt{4 - x^2}\, dx$$

$$= \left[\sqrt{4 - x^2} + 4\arcsin\frac{x}{2}\right]_0^2 \qquad \text{(Trigonometric substitution)}$$

$$= 4\arcsin 1 = 2\pi$$

(b) Using the disc method and the symmetry of the region, yields

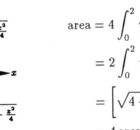

$$V = 2\pi\int_0^2 \left(\sqrt{1 - \frac{x^2}{4}}\right)^2 dx$$

$$= \frac{\pi}{2}\int_0^2 (4 - x^2)\, dx = \frac{\pi}{2}\left[4x - \frac{1}{3}x^3\right]_0^2 = \frac{8\pi}{3}$$

Since $y = \sqrt{1 - \dfrac{x^2}{4}} = \dfrac{1}{2}\sqrt{4 - x^2}$, differentiation yields

$$y' = \dfrac{x}{2\sqrt{4 - x^2}}.$$

Therefore,

$$\sqrt{1 + \left(\dfrac{dy}{dx}\right)^2} = \sqrt{1 + \dfrac{x^2}{4(4 - x^2)}}$$

$$= \dfrac{16 - 3x^2}{2\sqrt{4 - x^2}} = \dfrac{\sqrt{16 - 3x^2}}{4y}$$

and

$$S = 2(2\pi) \int_0^2 y\sqrt{1 + \left(\dfrac{dy}{dx}\right)^2}\, dx$$

$$= 4\pi \int_0^2 y\dfrac{\sqrt{16 - 3x^2}}{4y}\, dx$$

$$= \pi \int_0^2 \sqrt{4^2 - (\sqrt{3}x)^2}\, dx \quad \text{(Trigonometric substitution)}$$

$$= \dfrac{\pi}{2\sqrt{3}}\left[\sqrt{3}x\sqrt{16 - 3x^2} + 16 \arcsin\left(\dfrac{\sqrt{3}x}{4}\right)\right]_0^2$$

$$= \dfrac{2\pi}{9}(9 + 4\sqrt{3}\pi) \approx 21.48$$

(c) Using the Shell Method and the symmetry of the region you have

$$V = 2(2\pi) \int_0^2 x\left(\dfrac{1}{2}\sqrt{4 - x^2}\right) dx$$

$$= 2\pi\left(-\dfrac{1}{2}\right) \int_0^2 (4 - x^2)^{1/2}(-2x)\, dx$$

$$= -\dfrac{2\pi}{3}\left[(4 - x^2)^{3/2}\right]_0^2 = \dfrac{16\pi}{3}$$

Since $x = 2\sqrt{1 - y^2}$, it follows that $\dfrac{dx}{dy} = \dfrac{-2y}{\sqrt{1 - y^2}}$.

Therefore,

$$\sqrt{1 + \left(\dfrac{dx}{dy}\right)^2} = \sqrt{1 + \dfrac{4y^2}{1 - y^2}}$$

$$= \dfrac{\sqrt{1 + 3y^2}}{\sqrt{1 - y^2}} = \dfrac{2\sqrt{1 + 3y^2}}{x}.$$

and

$$S = 2(2\pi) \int_0^1 x \sqrt{1 + \left(\frac{dx}{dy}\right)^2} \, dy$$

$$= 4\pi \int_0^1 x \, \frac{2\sqrt{1+3y^2}}{x} \, dy$$

$$= 8\pi \int_0^1 \sqrt{1+3y^2} \, dy \quad \text{(Trigonometric substitution)}$$

$$= \frac{8\pi}{2\sqrt{3}} \left[\sqrt{3}y\sqrt{1+3y^2} + \ln\left|\sqrt{3}y + \sqrt{1+3y^2}\right| \right]_0^1$$

$$= \frac{4\pi}{3}[6 + \sqrt{3}\ln(2+\sqrt{3})] \approx 34.69$$

9.3   Hyperbolas

9. Find the center, vertices, and foci of the hyperbola given by $y^2 - (x^2/4) = 1$ and sketch its graph, using asymptotes as an aid.

Solution:

Writing the equation in standard form, yields

$$y^2 - \frac{x^2}{4} = 1$$

$$\frac{(y-0)^2}{1^2} - \frac{(x-0)^2}{2^2} = 1$$

$$\frac{(y-k)^2}{a^2} - \frac{(x-h)^2}{b^2} = 1$$

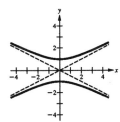

Thus, $h = 0$, $k = 0$, $a = 1$, $b = 2$, and $c = \sqrt{1^2 + 2^2} = \sqrt{5}$. We conclude that

$$\begin{aligned} \text{center, } (h,k): \quad & (0,0) \\ \text{vertices, } (h, k \pm a): \quad & (0,\pm 1) \\ \text{foci, } (h, k \pm c): \quad & (0, \pm\sqrt{5}) \end{aligned}$$

Finally, the asymptotes are given by

$$y = k \pm \frac{a}{b}(x - h) = \pm\frac{x}{2}.$$

15. Find the center, vertices, and foci of the hyperbola given by $[(x-1)^2/4] - [(y+2)^2/1] = 1$ and sketch its graph, using asymptotes as an aid.

Solution:

$$\frac{(x-1)^2}{4} - \frac{(y+2)^2}{1} = 1$$

$$\frac{(x-h)^2}{a^2} - \frac{(y+k)^2}{b^2} = 1$$

Thus, $h = 1$, $k = -2$, $a = 2$, $b = 1$, and $c = \sqrt{4+1} = \sqrt{5}$. We conclude that

$$\begin{aligned} \text{center, } (h,k): &\quad (1,-2) \\ \text{vertices, } (h \pm a, k): &\quad (-1,-2), (3,-2) \\ \text{foci, } (h \pm c, k): &\quad (1 \pm \sqrt{5}, -2) \end{aligned}$$

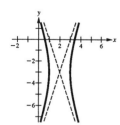

Finally, the asymptotes are given by

$$y = k \pm \frac{b}{a}(x-h) = -2 \pm \frac{1}{2}(x-1)$$

$$y = \frac{1}{2}x - \frac{5}{2} \quad \text{and} \quad y = -\frac{1}{2}x - \frac{3}{2}.$$

19. Find the center, vertices, and foci of the hyperbola given by $9x^2 - y^2 - 36x - 6y + 18 = 0$ and sketch its graph, using asymptotes as an aid.

Solution:

Writing the equation in standard form, yields

$$\begin{aligned} 9x^2 - y^2 - 36x - 6y + 18 &= 0 \\ 9x^2 - 36x - (y^2 + 6y) &= -18 \\ 9(x^2 - 4x + 4) - (y^2 + 6y + 9) &= -18 + 36 - 9 \\ 9(x-2)^2 - (y+3)^2 &= 9 \\ \frac{(x-2)^2}{1^2} - \frac{(y+3)^2}{3^2} &= 1 \end{aligned}$$

Thus, $h = 2$, $k = -3$, $a = 1$, $b = 3$, and $c = \sqrt{1^2 + 3^2} = \sqrt{10}$. We conclude that

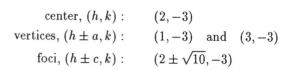

$$\begin{aligned} \text{center, } (h,k): &\quad (2,-3) \\ \text{vertices, } (h \pm a, k): &\quad (1,-3) \quad \text{and} \quad (3,-3) \\ \text{foci, } (h \pm c, k): &\quad (2 \pm \sqrt{10}, -3) \end{aligned}$$

Finally, the asymptotes are given by

$$y = k \pm \frac{b}{a}(x - h) = -3 \pm \frac{3}{1}(x - 2)$$
$$y = 3x - 9 \quad \text{and} \quad y = -3x + 3.$$

31. Find an equation for the hyperbola with vertices $(-1, 0)$ and $(1, 0)$ and whose asymptotes are given by $y = \pm 3x$.

Solution:

Since the vertices lie on a horizontal line, the standard form is

$$\frac{(x - h)^2}{a^2} - \frac{(y - k)^2}{b^2} = 1.$$

The center of the hyperbola lies at the midpoint of the line segment connecting the vertices. Thus,

$$(h, k) = \left(\frac{1 + (-1)}{2}, \frac{0 + 0}{2} \right) = (0, 0)$$

and you have $h = 0$ and $k = 0$. Since the asymptotes are of the form

$$y = k \pm \frac{b}{a}(x - h) = \pm 3x$$

you have

$$\pm \frac{b}{a} = \pm 3 \quad \text{or} \quad b = 3a.$$

Finally, since $a = 1$, you have $b = 3$ and the equation is

$$\frac{x^2}{1^2} - \frac{y^2}{3^2} = 1 \quad \text{or} \quad x^2 - \frac{y^2}{9} = 1.$$

37. Find an equation of the hyperbola such that for any point on the hyperbola, the difference of its distances from the points (2, 2) and (10, 2) is 6.

Solution:

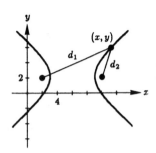

From the accompanying figure and the distance formula you have

$$d_1 - d_2 = 6$$

$$\sqrt{(x-2)^2 + (y_2-2)^2} - \sqrt{(x-10)^2 + (y-2)^2} = 6.$$

We now isolate the radicals one at a time, square each member of the resulting equation, and simplify.

$$\sqrt{(x-2)^2 + (y-2)^2} = 6 + \sqrt{(x-10)^2 + (y-2)^2}$$

$$(x-2)^2 + (y-2)^2 = 36 + 12\sqrt{(x-10)^2 + (y-2)^2}$$
$$+ (x-10)^2 + (y-2)^2$$

$$(x-2)^2 - (x-10)^2 - 36 = 12\sqrt{(x-10)^2 + (y-2)^2}$$

$$4x - 33 = 3\sqrt{(x-10)^2 + (y-2)^2}$$

$$16x^2 - 264x + 1089 = 9(x^2 - 20x + 100 + y^2 - 4y + 4)$$

$$7x^2 - 9y^2 - 84x + 36y + 153 = 0$$

$$7(x^2 - 12x) - 9(y^2 - 4y) = -153$$

$$7(x-6)^2 - 9(y-2)^2 = 63$$

$$\frac{(x-6)^2}{9} - \frac{(y-2)^2}{7} = 1$$

47. Find the volume and the surface area of the solid generated by revolving the region bounded by the graphs of $x^2 - y^2 = 1, y = 0$, and $x = 2$ about the x−axis.

Solution:

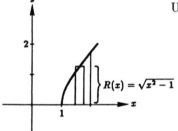

Using the Disc Method to find the volume, yields

$$V = \pi \int_1^2 \left(\sqrt{x^2 - 1}\right)^2 dx$$

$$= \pi \int_1^2 (x^2 - 1)\, dx$$

$$= \pi \left(\frac{x^3}{3} - x\right)\Big]_1^2 = \pi \left[\left(\frac{8}{3} - 2\right) - \left(\frac{1}{3} - 1\right)\right] = \frac{4\pi}{3}$$

To find the surface area, begin by differentiating the equation for the hyperbola implicitly to obtain y'.

$$x^2 - y^2 = 1$$

$$2x - 2y\, y' = 0 \implies y' = \frac{x}{y}$$

Therefore,

$$
\begin{aligned}
\sqrt{1 + (y')^2} &= \sqrt{1 + \left(\frac{x}{y}\right)^2} \\
&= \sqrt{\frac{y^2 + x^2}{y^2}} \\
&= \frac{\sqrt{(x^2 - 1) + x^2}}{y} \qquad \text{(equation of the hyperbola)} \\
&= \frac{\sqrt{2x^2 - 1}}{y}.
\end{aligned}
$$

Thus, the surface area is

$$
\begin{aligned}
S &= 2\pi \int_1^2 y\sqrt{1 + (y')^2}\, dx \\
&= 2\pi \int_1^2 y\left(\frac{\sqrt{2x^2 - 1}}{y}\right) dx \\
&= \sqrt{2}\pi \int_1^2 \sqrt{(\sqrt{2}x)^2 - 1}\sqrt{2}\, dx \\
&= \frac{\sqrt{2}\pi}{2}\left[\sqrt{2}x\sqrt{2x^2 - 1} - \ln\sqrt{2}x + \sqrt{2x^2 - 1})\right]_1^2 \\
&= \pi(2\sqrt{7} - 1) + \frac{\sqrt{2}\pi}{2}\ln\left(\frac{\sqrt{2} + 1}{2\sqrt{2} + \sqrt{7}}\right) \approx 11.66.
\end{aligned}
$$

49. Show that an equation of the tangent line to

$$\frac{x^2}{a^2} - \frac{y^2}{b^2} = 1$$

at the point (x_0, y_0) is

$$\frac{x_0}{a^2}x - \frac{y_0}{b^2}y = 1.$$

Solution:

To find the slope of the tangent line at the point (x_0, y_0), differentiate the equation for the hyperbola implicitly.

$$\frac{x^2}{a^2} - \frac{y^2}{b^2} = 1$$

$$\frac{2x}{a^2} - \frac{2y\,y'}{b^2} = 0 \implies y' = \frac{b^2 x}{a^2 y}$$

Therefore, an equation of the tangent line is

$$y - y_0 = \frac{b^2 x_0}{a^2 y_0}(x - x_0)$$

$$a^2 y_0 y - a^2 y_0^2 = b^2 x_0 x - b^2 x_0^2$$

$$b^2 x_0^2 - a^2 y_0^2 = b^2 x_0 x - a^2 y_0 y$$

$$a^2 b^2 = b^2 x_0 x - a^2 y_0 y \qquad \text{(equation of the hyperbola)}$$

$$1 = \frac{x_0 x}{a^2} - \frac{y_0 y}{b^2}.$$

9.4   Rotation and the General Second-Degree Equation

3. Rotate the axes to eliminate the xy term in the equation $x^2 - 10xy + y^2 + 1 = 0$. Sketch the graph of the resulting equation, showing both sets of axes.

Solution:

From the equations

$$x^2 - 10xy + y^2 + 1 = 0$$

$$Ax^2 + Bxy + Cy^2 + Dx + Ey + F = 0$$

you have $A = 1$, $B = -10$, $C = 1$, $D = 0$, $E = 0$, and $F = 1$.
Thus,

$$\cot 2\theta = \frac{A - C}{B} = 0 \quad \text{or} \quad 2\theta = \frac{\pi}{2} \quad \text{and} \quad \theta = \frac{\pi}{4}.$$

Therefore, $\sin \theta = \cos \theta = \sqrt{2}/2$ and

$$x = x' \cos \theta - y' \sin \theta = \frac{\sqrt{2}}{2}x' - \frac{\sqrt{2}}{2}y'$$

$$y = x' \sin \theta + y' \cos \theta = \frac{\sqrt{2}}{2}x' + \frac{\sqrt{2}}{2}y'$$

Substitution into $x^2 - 10xy + y^2 + 1 = 0$ yields

$$\left(\frac{\sqrt{2}}{2}x' - \frac{\sqrt{2}}{2}y'\right)^2 - 10\left(\frac{\sqrt{2}}{2}x' - \frac{\sqrt{2}}{2}y'\right)\left(\frac{\sqrt{2}}{2}x' + \frac{\sqrt{2}}{2}y'\right)$$
$$+ \left(\frac{\sqrt{2}}{2}x' + \frac{\sqrt{2}}{2}y'\right)^2 + 1 = 0$$

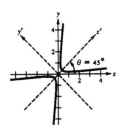

After expanding and combining like terms you have

$$-4(x')^2 + 6(y')^2 + 1 = 0 \quad \text{or} \quad \frac{(x')^2}{1/4} - \frac{(y')^2}{1/6} = 1.$$

7. Rotate the axes to eliminate the xy term in the equation
$5x^2 - 2xy + 5y^2 - 12 = 0$. Sketch the graph of the resulting
equation, showing both sets of axes.

Solution:

From the equations

$$5x^2 - 2xy + 5y^2 - 12 = 0$$
$$Ax^2 + Bxy + Cy^2 + Dx + Ey + F = 0$$

you have $A = 5$, $B = -2$, $C = 5$, $D = 0$, $E = 0$, and $F = -12$.
Thus

$$\cot 2\theta = \frac{A - C}{B} = 0 \quad \text{or} \quad 2\theta = \frac{\pi}{2} \quad \text{and} \quad \theta = \frac{\pi}{4}.$$

Therefore, $\sin\theta = \cos\theta = \dfrac{\sqrt{2}}{2}$ and

$$x = x'\cos\theta - y'\sin\theta = \frac{\sqrt{2}}{2}x' - \frac{\sqrt{2}}{2}y'$$

$$y = x'\sin\theta + y'\cos\theta = \frac{\sqrt{2}}{2}x' + \frac{\sqrt{2}}{2}y'.$$

Substituting into $5x^2 - 2xy + 5y^2 - 12 = 0$ yields

$$5\left(\frac{\sqrt{2}}{2}x' - \frac{\sqrt{2}}{2}y'\right)^2 - 2\left(\frac{\sqrt{2}}{2}x' - \frac{\sqrt{2}}{2}y'\right)\left(\frac{\sqrt{2}}{2}x' + \frac{\sqrt{2}}{2}y'\right)$$

$$+ 5\left(\frac{\sqrt{2}}{2}x' + \frac{\sqrt{2}}{2}y'\right)^2 - 12 = 0.$$

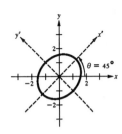

After expanding and combining like terms you have

$$4(x')^2 + 6(y')^2 - 12 = 0$$

$$\frac{(x')^2}{3} + \frac{(y')^2}{2} = 1$$

11. Rotate the axes to eliminate the xy term in the equation $9x^2 + 24xy + 16y^2 + 90x - 130y = 0$. Sketch the graph of the resulting equation, showing both sets of axes.

Solution:

From the equations

$$9x^2 + 24xy + 16y^2 + 90x - 130y = 0$$

$$Ax^2 + Bxy + Cy^2 + Dx + Ey + F = 0$$

you have $A = 9$, $B = 24$, $C = 16$, $D = 90$, $E = -130$, and $F = 0$. Thus.

$$\cot 2\theta = \frac{A - C}{B} = \frac{9 - 16}{24} = \frac{-7}{24}.$$

From the identity

$$\cot 2\theta = \frac{\cot^2\theta - 1}{2\cot\theta}$$

you have

$$\frac{\cot^2\theta - 1}{2\cot\theta} = \frac{-7}{24}$$

$$24\cot^2\theta - 24 = -14\cot\theta$$

$$12\cot^2\theta + 7\cot\theta - 12 = 0$$

$$(4\cot\theta - 3)(3\cot\theta + 4) = 0$$

$$\cot\theta = \frac{3}{4} \text{ or } -\frac{4}{3}$$

Since $0 < \theta < 90°$, choose $\cot\theta = \frac{3}{4}$ and therefore, $\theta \approx 53.13°$. Since $\cot\theta = \frac{3}{4}$, $\sin\theta = \frac{4}{5}$ and $\cos\theta = \frac{3}{5}$. Therefore, using the equations

$$x = x'\cos\theta - y'\sin\theta = \frac{3}{5}x' - \frac{4}{5}y'$$

$$y = x'\sin\theta + y'\cos\theta = \frac{4}{5}x' + \frac{3}{5}y'$$

and

$$9x^2 + 24xy + 16y^2 + 90x - 130y = 0$$

you have

$$9\left(\frac{3}{5}x' - \frac{4}{5}y'\right)^2 + 24\left(\frac{3}{5}x' - \frac{4}{5}y'\right)\left(\frac{4}{5}x' + \frac{3}{5}y'\right)$$

$$+ 16\left(\frac{4}{5}x' + \frac{3}{5}y'\right)^2 + 90\left(\frac{3}{5}x' - \frac{4}{5}y'\right) - 130\left(\frac{4}{5}x' + \frac{3}{5}y'\right) = 0.$$

After expanding and combining like terms you have

$$25(x')^2 - 50x' - 150y' = 0$$

$$(x')^2 - 2x' - 6y' = 0$$

$$(x' - 1)^2 = 4\left(\frac{3}{2}\right)\left(y' + \frac{1}{6}\right).$$

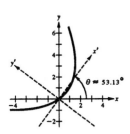

Review Exercises for Chapter 9

15. Analyze the equation $3x^2 + 2y^2 - 12x + 12y + 29 = 0$ and sketch its graph.

Solution:

In standard form the equation is
$$3x^2 + 2y^2 - 12x + 12y + 29 = 0$$
$$3(x^2 - 4x) + 2(y^2 + 6y) = -29$$
$$3(x^2 - 4x + 4) + 2(y^2 + 6y + 9) = -29 + 12 + 18$$
$$3(x - 2)^2 + 2(y + 3)^2 = 1$$
$$\frac{(x - 2)^2}{(1/\sqrt{3})^2} + \frac{(y + 3)^2}{(1/\sqrt{2})^2} = 1$$

Therefore the graph is an ellipse where
$h = 2$, $k = -3$, $a = 1/\sqrt{2}$, $b = 1/\sqrt{3}$, and
$c = \sqrt{(1/2) - (1/3)} = 1/\sqrt{6}$.

We conclude that

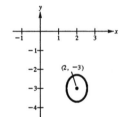

$$\text{center, } (h, k): \qquad (2, -3)$$

$$\text{vertices, } (h, k \pm a): \qquad \left(2, -3 \pm \frac{\sqrt{2}}{2}\right)$$

$$\text{foci, } (h, k \pm c): \qquad \left(2, -3 \pm \frac{\sqrt{6}}{6}\right)$$

25. Find an equation of the parabola with vertex at $(0, 2)$ and directrix $x = -3$.

Solution:

Since the directrix is vertical, the axis of the parabola is horizontal and its form is
$$(y - k)^2 = 4p(x - h).$$

The parabola opens to the right since the vertex is to the right of the directrix and thus $p > 0$. Therefore, $p = 3$ (distance between the vertex and directrix), $h = 0$, and $k = 2$.
$$(y - 2)^2 = 4(3)(x - 0)$$
$$y^2 - 4y + 4 = 12x$$
$$y^2 - 4y - 12x + 4 = 0$$

38. The ellipse $(x^2/a^2) + (y^2/b^2) = 1$ is revolved about its minor axis to form an oblate spheriod. Show that the volume of the spheroid is $\frac{4}{3}\pi a^2 b$ and its surface area is $2\pi a^2 + \pi(b^2/e)\ln[(1+e)/(1-e)]$.

Solution:

Solving for x as a function of y yields

$$x = \frac{a}{b}\sqrt{b^2 - y^2} \quad \text{and} \quad \frac{dx}{dy} = \left(\frac{a}{b}\right)\frac{-y}{\sqrt{b^2 - y^2}}.$$

Revolving about the $y-$axis and using the disc method, you have

$$V = \pi \int_{-b}^{b} \left(\frac{a}{b}\sqrt{b^2 - y^2}\right)^2 dy = \frac{2\pi a^2}{b^2}\int_0^b (b^2 - y^2)\, dy$$

$$= \frac{2\pi a^2}{b^2}\left[b^2 y - \frac{y^3}{3}\right]_0^b = \frac{2\pi a^2}{b^2}\left(b^3 - \frac{b^3}{3}\right) = \frac{4\pi a^2 b}{3}$$

$$S = 4\pi \int_0^b x\sqrt{1 + \left(\frac{dx}{dy}\right)^2}\, dy$$

$$= 4\pi \int_0^b \frac{a}{b}\sqrt{b^2 - y^2}\sqrt{1 + \frac{a^2 y^2}{b^2(b^2 - y^2)}}\, dy$$

$$= \frac{4\pi a}{b^2}\int_0^b \sqrt{b^4 + (a^2 - b^2)y^2}\, dy = \frac{4\pi a}{b^2}\int_0^b \sqrt{b^4 + c^2 y^2}\, dy$$

$$= \frac{4\pi a}{b^2}\left[\frac{1}{2c}\left(cy\sqrt{b^4 + c^2 y^2} + b^4 \ln|cy + \sqrt{b^4 + c^2 y^2}|\right)\right]_0^b$$

$$= \frac{2\pi a}{b^2 c}[bc\sqrt{b^4 + c^2 b^2} + b^4 \ln|bc + \sqrt{b^4 + b^2 c^2}| - b^4 \ln b^2]$$

$$= \frac{2\pi a}{b^2 c}\left[bc\sqrt{b^4 + (a^2 - b^2)b^2} + b^4 \ln\left(\frac{bc + \sqrt{b^4 + (a^2 - b^2)b^2}}{b^2}\right)\right]$$

$$= \frac{2\pi a}{b^2 c}\left[ab^2 c + b^4 \ln\left(\frac{cb + ab}{b^2}\right)\right] = 2\pi a^2 + \frac{2\pi ab^2}{c}\ln\left(\frac{a+c}{b}\right)$$

$$= 2\pi a^2 + \frac{\pi b^2}{c/a}\ln\frac{(a+c)^2}{b^2}.$$

Note that

$$\frac{(a+c)^2}{a^2 - c^2} = \frac{(a+c)(a+c)}{(a+c)(a-c)} = \frac{a+c}{a-c} = \frac{1 - (c/a)}{1 - (c/a)} = \frac{1+e}{1-e}.$$

Thus,

$$S = 2\pi a^2 + \pi\frac{b^2}{e}\ln\left(\frac{1+e}{1-e}\right).$$

43. Consider a fire truck with a water tank 16 feet long whose vertical cross section are ellipses as described by the equation

$$\frac{x^2}{16} + \frac{y^2}{9} = 1.$$

Find the depth of water in the tank if it is $\frac{3}{4}$ full (by volume) and the truck is on level ground.

Solution:

The truck will be carrying $\frac{3}{4}$ of its total capacity when the water covers $\frac{3}{4}$ of the area of a cross section of the tank. One-half of this area will be below the major axis and $\frac{1}{4}$ above the major axis of the ellipse

$$\frac{x^2}{16} + \frac{y^2}{9} = 1.$$

The total area is given by

$$A = \pi ab = \pi(4)(3) = 12\pi$$

$$\frac{1}{4}A = 3\pi$$

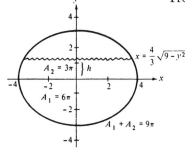

From the accompanying figure you have

$$\int_0^h 2\left(\frac{4}{3}\sqrt{9 - y^2}\right) dy = 3\pi$$

$$\int_0^y \sqrt{3^2 - y^2}\, dy = \frac{9\pi}{8}$$

$$\frac{1}{2}\left[y\sqrt{9 - y^2} + 9 \arcsin \frac{y}{3}\right]_0^h = \frac{9\pi}{8}$$

$$h\sqrt{9 - h^2} + 9 \arcsin \frac{h}{3} = \frac{9\pi}{4}$$

Find h such that

$$f(h) = \sqrt{9 - h^2} + \arcsin\left(\frac{h}{3}\right) - \frac{9\pi}{4} = 0$$

By Newton's Method, with an initial estimate of $h = 1$, you have

$$h_{n+1} = h_n - \frac{f(h_n)}{f'(h_n)}$$

$$= h_n - \frac{h_n\sqrt{9 - h_n^2} + 9\arcsin(h_n/3) - (9\pi/4)}{2\sqrt{9 - h_n^2}}$$

n	1	2	3	4
h_n	1.0000	1.2089	1.2119	1.2119

We conclude that $h \approx 1.212$ and therefore the total height of the water in the tank is $3 + 1.212 = 4.212$ feet.

49. Find an equation of the hyperbola with foci $(\pm 4, 0)$ where the absolute value of the difference of the distances from a point on the hyperbola to the foci is 4.

Solution:

We first observe that the center of the hyperbola is at the origin and therefore $(h, k) = (0, 0)$. Since the foci are $(\pm 4, 0)$, you know that the transverse axis is horizontal and $c = 4$. Also, from the definition of a hyperbola you know that the absolute value of the difference of the distances from a point on the hyperbola to the foci is $2a$. Thus $2a = 4$ or $a = 2$. Since $b^2 = c^2 - a^2$, it follows that $b^2 = 16 - 4 = 12$. Therefore, an equation of the hyperbola is

$$\frac{x^2}{4} - \frac{y^2}{12} = 1.$$

10 PLANE CURVES, PARAMETRIC EQUATIONS, AND POLAR COORDINATES

10.1 Plane Curves and Parametric Equations

7. Sketch the curve represented by the parametric equations $x = t^3$, $y = t^2/2$ and write the corresponding rectangular equation by eliminating the parameter.

Solution:

For values of t on the interval $[-\frac{3}{2}, \frac{3}{2}]$, the parametric equations yield the points (x, y) shown in the following table.

t	$-\dfrac{3}{2}$	-1	$-\dfrac{1}{2}$	0	$\dfrac{1}{2}$	1	$\dfrac{3}{2}$
x	$-\dfrac{27}{8}$	-1	$-\dfrac{1}{8}$	0	$\dfrac{1}{8}$	1	$\dfrac{27}{8}$
y	$\dfrac{9}{8}$	$\dfrac{1}{2}$	$\dfrac{1}{8}$	0	$\dfrac{1}{8}$	$\dfrac{1}{2}$	$\dfrac{9}{8}$

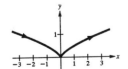

By plotting these points in the order of increasing t and using the continuity of the parametric equations, you obtain the curve shown in the accompanying figure.

Since $x = t^3$ and $y = t^2/2$, you have

$$t = x^{1/3} \quad \text{or} \quad y = \frac{1}{2}(x^{1/3})^2 = \frac{1}{2}x^{2/3}.$$

351

23. Sketch the curve represented by the parametric equations $x = 4\sec\theta$, $y = 3\tan\theta$ and write the corresponding rectangular equation by eliminating the parameter.

Solution:

For values of θ in the interval $[-\frac{\pi}{3}, \frac{4\pi}{3}]$, the parametric equations yield the points (x, y) shown in the following table.

θ	$-\dfrac{\pi}{3}$	$-\dfrac{\pi}{6}$	0	$\dfrac{\pi}{6}$	$\dfrac{\pi}{3}$	$\dfrac{\pi}{2}$	$\dfrac{2\pi}{3}$	$\dfrac{5\pi}{6}$	π	$\dfrac{7\pi}{6}$	$\dfrac{4\pi}{3}$
x	8	$\dfrac{8}{\sqrt{3}}$	4	$\dfrac{8}{\sqrt{3}}$	8	undef.	-8	$-\dfrac{8}{\sqrt{3}}$	-4	$-\dfrac{8}{\sqrt{3}}$	-8
y	$-3\sqrt{3}$	$-\sqrt{3}$	0	$\sqrt{3}$	$3\sqrt{3}$	undef.	$-3\sqrt{3}$	$-\sqrt{3}$	0	$\sqrt{3}$	$3\sqrt{3}$

By plotting these points in the order of increasing θ you obtain the curve shown in the accompanying figure.

Since the parametric equations involve secants and tangents, use the identity $\sec^2\theta - \tan^2\theta = 1$. Therefore,

$$\frac{x}{4} = \sec\theta \quad \text{and} \quad \frac{y}{3} = \tan\theta$$

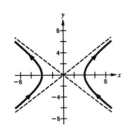

or

$$\frac{x^2}{16} - \frac{y^2}{9} = \sec^2\theta - \tan^2\theta = 1$$

The graph of this equation is a hyperbola centered at the origin with vertices $(\pm 4, 0)$.

29. Describe any differences in the curves represented by the given sets of parametric equations. Are the orientations the same? Are the curves smooth?

(a) $x = t$
$\quad\;\; y = 2t + 1$

(b) $x = \cos\theta$
$\quad\;\; y = 2\cos\theta + 1$

(c) $x = e^{-t}$
$\quad\;\; y = 2e^{-t} + 1$

(d) $x = e^t$
$\quad\;\; y = 2e^t + 1$

Solution:

By eliminating the parameter in each part of this exercise, you get $y = 2x + 1$. The graphs differ in their orientation and domains. In each case the derivatives of $x = f(t)$ and $y = g(t)$ are continuous and not simultaneously zero. Therefore, the graphs are smooth.

(a) $x = t$
$-\infty < x < \infty$

$y = 2t + 1$
$-\infty < y < \infty$

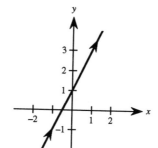

(b) $x = \cos\theta$
$-1 \le x \le 1$

$y = 2\cos\theta + 1$
$-1 \le y \le 3$

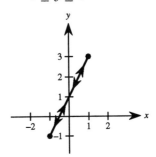

(c) $x = e^{-t}$
$x > 0$

$y = 2e^{-t} + 1$
$y > 1$

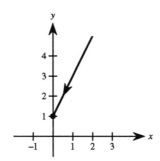

(d) $x = e^{t}$
$x > 0$

$y = 2e^{t} + 1$
$y > 1$

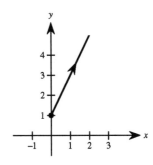

35. Eliminate the parameter from the parametric equations
$x = h + r\cos\theta$ and $y = k + r\sin\theta$, and obtain the standard
form of a circle.

Solution:

Since the parametric equations involve sines and cosines, use
the identity $\sin^2\theta + \cos^2\theta = 1$.

$$x = h + r\cos\theta \qquad\qquad y = k + r\sin\theta$$

$$\frac{x - h}{r} = \cos\theta \qquad\qquad \frac{y - k}{r} = \sin\theta$$

$$\frac{(x - h)^2}{r^2} = \cos^2\theta \qquad\qquad \frac{(y - k)^2}{r^2} = \sin^2\theta$$

$$\frac{(x-h)^2}{r^2} + \frac{(y-k)^2}{r^2} = \cos^2\theta + \sin^2\theta = 1$$

$$\frac{(x-h)^2}{r^2} + \frac{(y-k)^2}{r^2} = 1$$

$$(x-h)^2 + (y-k)^2 = r^2$$

45. Find a set of parametric equation for the hyperbola with vertices at $(\pm 4, 0)$, and foci at $(\pm 5, 0)$.

Solution:

The midpoint of the line segment joining the vertices is the center of the hyperbola. Therefore, $(h, k) = (0, 0)$. The transverse axis of the hyperbola is horizontal with $a = 4$ and $c = 5$. Since $b^2 = c^2 - a^2$, you have $b = 3$. Using the result of Exercise 37 a set of parametric equations are

$$x = h + a\sec\theta \quad y = k + b\tan\theta$$

$$x = 4\sec\theta \qquad , y = 3\tan\theta.$$

 53. Use a graphing utility to graph the curve (a prolate cycloid) represented by $x = \theta - \frac{3}{2}\sin\theta$, and $y = 1 - \frac{3}{2}\cos\theta$.

Solution:

Select the parametric equations mode of your graphing utility and enter the set of parametric equations. This parametric mode may require that you use the parameter t rather than θ. It is also necessary to set the interval for the parameter t and some graphing utilities also require the corresponding intervals for x and y. For example, with the Texas Instruments TI-81 you can begin with the following settings and then use the *square* feature.

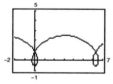

Tmin $= -2$
Tmax $= 14$
Tstep 0.2
Xmin $= -1$
Xmax $= 14$
Xscl $= 1$
Ymin $= -2$
Ymax $= 3$
Yscl $= 1$

10.2   Parametric Equations and Calculus

1. For the parametric equations $x = 2t$, $y = 3t - 1$, find dy/dx and d^2y/dx^2 and evaluate there two derivatives when $t = 3$.

Solution:

Since $x = 2t$ and $y = 3t - 1$, you have

$$\frac{dy}{dx} = \frac{dy/dt}{dx/dt} = \frac{3}{2} \quad \text{for all values of } t$$

and

$$\frac{d^2y}{dx^2} = \frac{d[dy/dx]/dt}{dx/dt} = \frac{0}{2} = 0 \quad \text{for all values of } t.$$

7. For the parametric equations $x = 2 + \sec\theta$ and $y = 1 + 2\tan\theta$, find dy/dx and d^2y/dx^2 and evaluate these two derivatives when $\theta = \pi/6$.

Solution:

Since $x = 2 + \sec\theta$ and $y = 1 + 2\tan\theta$, you have

$$\frac{dy}{dx} = \frac{dy/d\theta}{dx/d\theta} = \frac{2\sec^2\theta}{\sec\theta\tan\theta} = \frac{2\sec\theta}{\tan\theta} = 2\csc\theta$$

$$\frac{d^2y}{dx^2} = \frac{d[dy/dx]/d\theta}{dx/d\theta} = \frac{-2\csc\theta\cot\theta}{\sec\theta\tan\theta} = -2\cot^3\theta$$

At $\theta = \pi/6$,

$$\frac{dy}{dx} = 2\csc\frac{\pi}{6} = 2(2) = 4$$

$$\frac{d^2y}{dx^2} = -2\cot^3\frac{\pi}{6} = -2(\sqrt{3})^3 = -6\sqrt{3}$$

11. Find equations of the tangent lines shown on the graph of the parametric equations $x = 2\cot\theta$ and $y = 2\sin^2\theta$.

Solution:

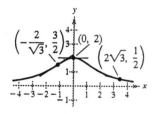

$$\frac{dy}{dx} = \frac{dy/d\theta}{dx/d\theta} = \frac{4\sin\theta\cos\theta}{-2\csc^2\theta} = -2\sin^3\theta\cos\theta$$

At the point $(-2/\sqrt{3},\ 3/2)$, $\theta = 2\pi/3$. When $\theta = \dfrac{2\pi}{3}$,

$$\frac{dy}{dx} = -2\sin^3\frac{2\pi}{3}\cos\frac{2\pi}{3} = -2\left(\frac{\sqrt{3}}{2}\right)^3\left(-\frac{1}{2}\right) = \frac{3\sqrt{3}}{8}.$$

Therefore, the equation of the tangent line is

$$y - \frac{3}{2} = \frac{3\sqrt{3}}{8}\left(x + \frac{2}{\sqrt{3}}\right)$$

$$3\sqrt{3}x - 8y + 18 = 0.$$

At the point $(0, 2)$, $\theta = \pi/2$. When $\theta = \dfrac{\pi}{2}$,

$$\frac{dy}{dx} = -2\sin^3\frac{\pi}{3}\cos\frac{\pi}{2} = -2(1)^3(0) = 0.$$

Therefore, the equation of the tangent line is

$$y - 2 = 0(x - 0)$$
$$y - 2 = 0.$$

At the point $(2\sqrt{3},\ 1/2)$, $\theta = \pi/6$. When $\theta = \dfrac{\pi}{6}$,

$$\frac{dy}{dx} = -2\sin^3\frac{\pi}{6}\cos\frac{\pi}{6} = -2\left(\frac{1}{2}\right)^3\left(\frac{\sqrt{3}}{2}\right) = -\frac{\sqrt{3}}{8}.$$

Therefore, the equation of the tangent line is

$$y - \frac{1}{2} = -\frac{\sqrt{3}}{8}(x - 2\sqrt{3})$$

$$\sqrt{3}x + 8y - 10 = 0.$$

21. Find all points (if any) of horizontal and vertical tangency on the graph of $x = 1 - t$, $y = t^3 - 3t$.

Solution:

Since $\dfrac{dy}{dx} = \dfrac{dy/dt}{dx/dt} = \dfrac{3t^2 - 3}{-1} = 3 - 3t^2$,

the horizontal tangents occur when

$$3 - 3t^2 = 0 \quad \text{or} \quad t = \pm 1$$

The corresponding points are $(0, -2)$, and $(2, 2)$. Since dy/dx is never undefined, there are no points of vertical tangency.

29. Find the length of the arc on the graph of $x = e^{-t} \cos t$, $y = e^{-t} \sin t$ for $0 \le t \le \pi/2$.

Solution:

$$x = e^{-t} \cos t$$
$$\frac{dx}{dt} = -e^{-t}(\sin t + \cos t)$$
$$\left(\frac{dx}{dt}\right)^2 = e^{-2t}(\sin^2 t + 2 \sin t \cos t + \cos^2 t)$$
$$= e^{-2t}(1 + \sin 2t)$$
$$y = e^{-t} \sin t$$
$$\frac{dy}{dt} = e^{-t}(\cos t - \sin t)$$
$$\left(\frac{dy}{dt}\right)^2 = e^{-2t}(\cos^2 t - 2 \sin t \cos t + \sin^2 t)$$
$$= e^{-2t}(1 - \sin 2t)$$
$$\left(\frac{dx}{dt}\right)^2 + \left(\frac{dy}{dt}\right)^2 = 2e^{-2t}$$

Therefore,

$$s = \int_0^{\pi/2} \sqrt{\left(\frac{dx}{dt}\right)^2 + \left(\frac{dy}{dt}\right)^2}\, dt$$

$$= \int_0^{\pi/2} \sqrt{2e^{-2t}}\, dt$$

$$= -\sqrt{2} \int_0^{\pi/2} e^{-t}(-1)\, dt = -\sqrt{2}\left[e^{-t}\right]_0^{\pi/2} = \sqrt{2}(1 - e^{-\pi/2}) \approx 1.12.$$

35. Find the perimeter of the hypocycloid given by
$x = a \cos^3 \theta$, $y = a \sin^3 \theta$.

Solution:

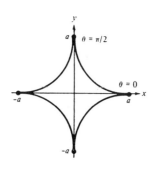

Using $dx/d\theta = -3a \cos^2 \theta \sin \theta$, $dy/d\theta = 3a \sin^2 \theta \cos \theta$, and the symmetry of the graph, the perimeter is given by

$$s = 4 \int_0^{\pi/2} \sqrt{\left(\frac{dx}{d\theta}\right)^2 + \left(\frac{dy}{d\theta}\right)^2}\, d\theta$$

$$= 4 \int_0^{\pi/2} \sqrt{9a^2 \cos^4 \theta \sin^2 \theta + 9a^2 \sin^4 \theta \cos^2 \theta}\, d\theta$$

$$= 4 \int_0^{\pi/2} \sqrt{9a^2 \sin^2 \theta \cos^2 \theta (\cos^2 \theta + \sin^2 \theta)}\, d\theta$$

$$= 4 \int_0^{\pi/2} 3a \sin \theta \cos \theta\, d\theta = 12a \left[\frac{\sin^2 \theta}{2}\right]_0^{\pi/2} = 6a.$$

49. A portion of a sphere or radius r is removed by a circular cone with its vertex at the center of the sphere. Find the surface area removed from the sphere if the vertex of the cone forms an angle of 2θ.

Solution:

The sphere is formed by revolving a circle of radius r about the x-axis. We represent the circle by

$$x = f(\phi) = r \cos \phi \qquad\qquad y = g(\phi) = r \sin \phi$$
$$\frac{dx}{d\phi} = -r \sin \phi \qquad\qquad \frac{dy}{d\phi} = r \cos \phi$$

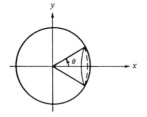

From the integrals of Theorem 10.3 and the accompanying figure, you have

$$S = 2\pi \int_0^\theta g(\phi) \sqrt{\left(\frac{dx}{d\phi}\right)^2 + \left(\frac{dy}{d\phi}\right)^2}\, d\phi$$

$$= 2\pi \int_0^\theta r \sin \phi \sqrt{r^2 \sin^2 \phi + r^2 \cos^2 \phi}\, d\phi$$

$$= 2\pi r^2 \int_0^\theta \sin \phi\, d\phi = -2\pi r^2 \left[\cos \phi\right]_0^\theta = 2\pi r^2 (1 - \cos \theta).$$

53. Find the area of the region between the cissoid
$x = 2\sin^2\theta$, $y = 2\sin^2\theta\tan\theta$ and its asymptote in the first
quadrant (see figure).

Solution:

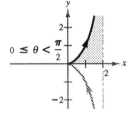

From Exercise 50, you have the following convergent improper
integral

$$A = \int_0^2 y\,dx = \int_0^{\pi/2} y\frac{dx}{d\theta}\,d\theta$$

$$= \int_0^{\pi/2} 2\sin^2\theta\tan\theta(4\sin\theta\cos\theta)\,d\theta$$

$$= 8\int_0^{\pi/2} \sin^4\theta\,d\theta$$

$$= 8\int_0^{\pi/2} \left(\frac{1-\cos 2\theta}{2}\right)^2\,d\theta$$

$$= 2\int_0^{\pi/2} (1 - 2\cos 2\theta + \cos^2 2\theta)\,d\theta$$

$$= 2\int_0^{\pi/2} \left(1 - 2\cos 2\theta + \frac{1+\cos 4\theta}{2}\right)\,d\theta$$

$$= 2\int_0^{\pi/2} \left(\frac{3}{2} - 2\cos 2\theta + \frac{1}{2}\cos 4\theta\right)\,d\theta$$

$$= 2\left[\frac{3}{2}\theta - \sin 2\theta + \frac{1}{8}\sin 4\theta\right]_0^{\pi/2} = 2\left(\frac{3\pi}{4}\right) = \frac{3\pi}{2}$$

10.3   Polar Coordinates and Polar Graphs

5. Plot the point $(\sqrt{2}, 2.36)$ in polar coordinates and find its
corresponding rectangular coordinates.

Solution:

Using the conversion from polar to rectangular coordinates, you
have

$$x = r\cos\theta = \sqrt{2}\cos(2.36) \approx -1.004$$

$$y = r\sin\theta = \sqrt{2}\sin(2.36) \approx 0.996$$

Therefore the rectangular coordinates are $(-1.004, 0.996)$.

9. Find two sets of polar coordinates for the point $(-3, 4)$ given in rectangular coordinates, using $0 \leq \theta < 2\pi$.

Solution:

First,

$$r = \pm\sqrt{x^2 + y^2} = \pm\sqrt{(-3)^2 + 4^2} = \pm 5$$

$$\tan \theta = -\frac{4}{3} \quad \text{and} \quad \arctan\left(-\frac{4}{3}\right) \approx -0.9273$$

Since $(-3, 4)$ lies in the second quadrant, let $\theta = \pi - 0.9273 = 2.214$. Thus one polar representation is

$$(5,\ 2.214) \quad (r > 0,\ 0 \leq \theta < 2\pi)$$

To obtain the second representation, change the sign of r and increase θ by π radians to obtain

$$(-5,\ 5.356) \quad (r < 0,\ 0 \leq \theta < 2\pi).$$

15. Convert the rectangular equation $3x - y + 2 = 0$ to polar form.

Solution:

Since $x = r \cos \theta$ and $y = r \sin \theta$, you have

$$3x - y + 2 = 0$$
$$3(r \cos \theta) - r \sin \theta + 2 = 0$$
$$r(3 \cos \theta - \sin \theta) = -2$$
$$r = \frac{-2}{3 \cos \theta - \sin \theta}.$$

21. Convert the polar equation $r = \sin \theta$ to rectangular form and sketch its graph.

Solution:

Since $r^2 = x^2 + y^2$ and $y = r \sin \theta$, we begin by multiplying both members of the polar equation by r to obtain

$$r = \sin \theta$$
$$r^2 = r \sin \theta$$
$$x^2 + y^2 = y$$
$$x^2 + y^2 - y = 0$$
$$x^2 + \left(y^2 - y + \frac{1}{4} \right) = \frac{1}{4}$$
$$x^2 + \left(y - \frac{1}{2} \right)^2 = \left(\frac{1}{2} \right)^2.$$

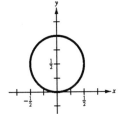

Therefore, the graph is a circle of radius $1/2$ centered at $(0, 1/2)$.

39. Find the slope of the graph of $r = 3(1 - \cos \theta)$ at $\theta = \pi/2$.

Solution:

Since $f(\theta) = 3(1 - \cos \theta)$, you have $f'(\theta) = 3 \sin \theta$ and

$$
\begin{aligned}
\frac{dy}{dx} &= \frac{f'(\theta) \sin \theta + f(\theta) \cos \theta}{f'(\theta) \cos(\theta) - f(\theta) \sin(\theta)} \\
&= \frac{(3 \sin \theta) \sin \theta + 3(1 - \cos \theta) \cos \theta}{(3 \sin \theta) \cos \theta - 3(1 - \cos \theta) \sin \theta} \\
&= \frac{\sin^2 \theta + \cos \theta - \cos^2 \theta}{\sin \theta (2 \cos \theta - 1)} \\
&= \frac{(1 + 2 \cos \theta)(1 - \cos \theta)}{\sin \theta (2 \cos \theta - 1)}.
\end{aligned}
$$

When $\theta = \pi/2$, the slope is

$$\frac{dy}{dx} = \frac{(1 + 0)(1 - 0)}{1(0 - 1)} = -1.$$

43. Find the horizontal and vertical tangents to the polar curve $r = 1 + \sin\theta$.

Solution:

Since $f(\theta) = 1 + \sin\theta$ and $f'(\theta) = \cos\theta$, you have

$$\frac{dy}{dx} = \frac{f'(\theta)\sin\theta + f(\theta)\cos\theta}{f'(\theta)\cos\theta - f(\theta)\sin\theta}$$

$$= \frac{\cos\theta\sin\theta + (1 + \sin\theta)\cos\theta}{\cos^2\theta - (1 + \sin\theta)\sin\theta}$$

$$= \frac{\cos\theta(1 + 2\sin\theta)}{1 - \sin\theta - 2\sin^2\theta}$$

$$= \frac{\cos\theta(1 + 2\sin\theta)}{(1 - 2\sin\theta)(1 + \sin\theta)}.$$

Since $dy/dx = 0$ when $\cos\theta(1 + 2\sin\theta) = 0$ or $\theta = \pi/2,\ 7\pi/6,\ 11\pi/6$, there are horizontal tangents at the points

$$\left(2, \frac{\pi}{2}\right),\ \left(\frac{1}{2}, \frac{7\pi}{6}\right),\ \text{and}\ \left(\frac{1}{2}, \frac{11\pi}{6}\right).$$

Since dy/dx is undefined when $(1 - 2\sin\theta)(1 + \sin\theta) = 0$ or $\theta = \pi/6,\ 5\pi/6,\ 3\pi/2$, there are vertical tangents at the points

$$\left(\frac{3}{2}, \frac{\pi}{6}\right),\ \left(\frac{3}{2}, \frac{5\pi}{6}\right),\ \text{and}\ \left(0, \frac{3\pi}{2}\right).$$

(Note that L'Hopital's Rule must be used to show that dy/dx is undefined at $\theta = 3\pi/2$.)

53. Sketch and identify the graph of $r = 2\cos 3\theta$. Find the tangents at the pole.

Solution:

The equation has the form $r = a\cos(n\theta) = 2\cos 3\theta$ where n is odd. This means the graph is a *rose curve* with $n = 3$ petals. The curve has polar axis symmetry. The relative extrema of r are $(2, 0)$, $(-2, \pi/3)$, and $(2, 2\pi/3)$. Since $r = 0$ and $dr/d\theta \neq 0$ when $\theta = \pi/6$, $\theta = \pi/2$, and $\theta = 5\pi/6$, you have tangents at the pole for these values of θ.

θ	0	$\dfrac{\pi}{12}$	$\dfrac{\pi}{6}$	$\dfrac{\pi}{3}$	$\dfrac{\pi}{2}$	$\dfrac{2\pi}{3}$
r	2	$\sqrt{2}$	0	-2	0	2

71. Use a graphing utility to graph the polar equation $r^2 = 4\sin 2\theta$. Find an interval of θ over which the graph is traced only once.

Solution:

Before using the graphing utility to obtain the graph of the polar equation, observe that the equation has the form

$$r^2 = a^2 \sin 2\theta = 4 \sin 2\theta$$

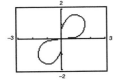 and is a *lemniscate* with symmetry with respect to the pole. (If your graphing utility doesn't have a polar mode, but does have a parametric mode, you can sketch the graph of $r = f(\theta)$ by writing the equation as $x = f(\theta)\cos\theta$ and $y = f(\theta)\sin\theta$.) The entire graph will be traced for $0 \le \theta \le \pi$. Also, the tangents at the pole are $\theta = 0$ and $\theta = \pi/2$ since $r = 0$ and $dr/d\theta \ne 0$ at these values.

 73. Use a graphing utility to sketch the graph of $r = 2 - \sec\theta$ and show that $x = -1$ is a vertical asymptote of the graph.

Solution:

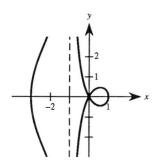 The graph of the polar equation obtained by a graphing utility is shown in the accompanying figure. To locate the vertical asymptote, note that

$$r \to -\infty \quad \text{as} \quad \theta \to \frac{\pi^-}{2}$$

and

$$r \to \infty \quad \text{as} \quad \theta \to \frac{\pi^+}{2}$$

To see that the vertical asymptote is located at $x = -1$, write

$$r = 2 - \frac{1}{\cos\theta} = 2 - \frac{r}{r\cos\theta} = 2 - \frac{r}{x}$$
$$rx = 2x - r$$
$$r(1 + x) = 2x$$
$$r = \frac{2x}{1 + x}$$

Thus $r \to \pm\infty$ as $x \to -1$.

10.4   Area and Arc Length in Polar Coordinates

9. Find the area of the region within the inner loop of the graph of $r = 1 + 2\cos\theta$.

Solution:

Solving the equation $r = 0$ to find the tangents at the pole you have

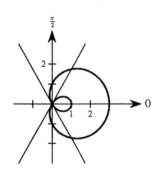

$$1 + 2\cos\theta = 0$$
$$\cos\theta = -\frac{1}{2}$$
$$\theta = \frac{2\pi}{3} \quad \text{or} \quad \theta = \frac{4\pi}{3}.$$

Therefore, the lower half of the inner loop is generated when θ is in the interval $2\pi/3 \leq \theta \leq \pi$. From the symmetry of the graph in the accompanying figure, the area of the inner loop is given by

$$2\int_{2\pi/3}^{\pi} \frac{1}{2}r^2\,d\theta = \int_{2\pi/3}^{\pi} (1 + 2\cos\theta)^2\,d\theta$$
$$= \int_{2\pi/3}^{\pi} (1 + 4\cos\theta + 4\cos^2\theta)\,d\theta$$
$$= \int_{2\pi/3}^{\pi} \left(1 + 4\cos\theta + 4\,\frac{1+\cos 2\theta}{2}\right)d\theta$$
$$= \int_{2\pi/3}^{\pi} (3 + 4\cos\theta + 2\cos 2\theta)\,d\theta$$
$$= \left[3\theta + 4\sin\theta + \sin 2\theta\right]_{2\pi/3}^{\pi}$$
$$= \pi - \frac{3\sqrt{3}}{2} = \frac{2\pi - 3\sqrt{3}}{2}.$$

11. Find the area of the region between the loops of the graph of $r = 1 + 2\cos\theta$.

Solution:

From the symmetry of the graph given in the accompanying figure, the area of the region inside the outer loop is

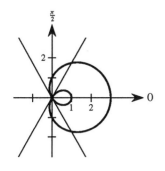

$$2\int_0^{2\pi/3} \frac{1}{2}r^2 d\theta = \int_0^{2\pi/3} (1 + 2\cos\theta)^2 d\theta$$

$$= \Big[3\theta + 4\sin\theta + \sin 2\theta\Big]_0^{2\pi/3} \qquad \text{From Exercise 9}$$

$$= 2\pi + 2\sqrt{3} - \frac{\sqrt{3}}{2} = 2\pi + \frac{3\sqrt{3}}{2}.$$

From Exercise 9, we see that the area of the inner loop is given by $\pi - (3\sqrt{3}/2)$. Finally, the area of the region between the two loops is

$$A = \left(2\pi + \frac{3\sqrt{3}}{2}\right) - \left(\pi - \frac{3\sqrt{3}}{2}\right) = \pi + 3\sqrt{3}.$$

17. Find the points of intersection of the graphs of $r = 4 - 5\sin\theta$ and $r = 3\sin\theta$.

Solution:

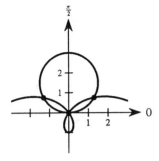

From Section 10.3, you know the graph of $r = 4 - 5\sin\theta$ is a limacon and the graph of $r = 3\sin\theta$ is a circle (see figure). Solving the two equations simultaneously, yields

$$4 - 5\sin\theta = 3\sin\theta$$
$$8\sin\theta = 4$$
$$\sin\theta = \frac{1}{2}$$
$$\theta = \frac{\pi}{6}, \frac{5\pi}{6} \qquad (0 \le \theta < 2\pi).$$

From these values you obtain the points $(3/2, \pi/6)$ and $(3/2, 5\pi/6)$. To test for additional points of intersection, replace r by $-r$ and θ by $\pi + \theta$ in $r = 4 - 5\sin\theta$ to obtain

$$-r = 4 - 5\sin(\pi + \theta) = 4 + 5\sin\theta.$$

Solving this equation simultaneously with $r = 3\sin\theta$, yields

$$-4 - 5\sin\theta = 3\sin\theta$$
$$8\sin\theta = -4$$
$$\sin\theta = -\frac{1}{2}$$
$$\theta = \frac{7\pi}{6}, \frac{11\pi}{6}.$$

The corresponding points are $(-3/2, 7\pi/6/)$ and $(-3/2, 11\pi/6)$. However, these two points coincide with the previous two points. Finally, observe that both curves pass through the pole. Hence there are three points of intersection, $(3/2, \pi/6)$, $(3/2, 5\pi/6)$, and $(0,0)$ as seen in the accompanying figure.

21. Find the points of intersection of the graphs of $r = 4\sin 2\theta$ and $r = 2$.

Solution:

From section 10.3, you know the graph of $r = 4\sin 2\theta$ is a rose curve with 4 petals and is symmetric to the polar axis, the vertical axis and the pole. Also, the graph of $r = 2$ is a circle of radius 2 centered at the pole. Solving the two equations simultaneously, yields

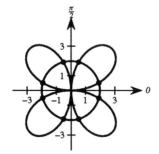

$$4\sin 2\theta = 2$$
$$\sin 2\theta = \frac{1}{2}$$
$$2\theta = \frac{\pi}{6}, \frac{5\pi}{6}$$
$$\theta = \frac{\pi}{12}, \frac{5\pi}{12}.$$

Therefore, the points of intersection for one petal are $(2, \pi/12)$ and $(2, 5\pi/12)$. By symmetry, the other points of intersection are $(2, 7\pi/12)$, $(2, 11\pi/12)$, $(2, 13\pi/12)$, $(2, 17\pi/12)$, $(2, 19\pi/12)$, and $(2, 23\pi/12)$.

27. Find the area of the region common to the interiors of the graphs of $r = 4\sin 2\theta$ and $r = 2$.

Solution:

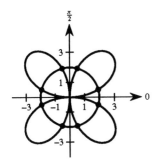

From the accompanying sketch you see that you need only consider the region common to both curves in one petal of the rose curve and multiply the result by four. From Exercise 21 the points of intersection on the first petal occur when $\theta = \pi/12$ and $5\pi/12$. There are three subregions within one petal with the following bounds.

(a) for $0 \le \theta \le \pi/12$, $\quad r = 4\sin 2\theta$

(b) for $\pi/12 \le \theta \le 5\pi/12$, $\quad r = 2$

(c) for $5\pi/12 \le \theta \le \pi/2$, $\quad r = 4\sin 2\theta$

Therefore, the area within one petal is

$$A = \int_0^{\pi/12} \frac{1}{2}(4\sin 2\theta)^2\, d\theta + \int_{\pi/12}^{5\pi/12} \frac{1}{2}(2)^2\, d\theta$$
$$+ \int_{5\pi/12}^{\pi/2} \frac{1}{2}(4\sin 2\theta)^2\, d\theta.$$

By the symmetry of the petal, the first and third integrals are equal. Thus,

$$A = \int_0^{\pi/12} \frac{1}{2}(4\sin 2\theta)^2\, d\theta + \int_{\pi/12}^{5\pi/12} \frac{1}{2}(2)^2\, d\theta$$
$$= 16 \int_0^{\pi/12} \sin^2 2\theta\, d\theta + 2\int_{\pi/12}^{5\pi/12} d\theta$$
$$= 8 \int_0^{\pi/12} (1 - \cos 4\theta)\, d\theta + 2\int_{\pi/12}^{5\pi/12} d\theta$$
$$= 8\left[\theta - \frac{1}{4}\sin 4\theta\right]_0^{\pi/12} + \left[2\theta\right]_{\pi/12}^{5\pi/12}$$
$$= \frac{2\pi}{3} - \frac{2\sqrt{3}}{2} + \frac{2\pi}{3} = \frac{4\pi}{3} - \sqrt{3}.$$

Finally, multiplying by 4, you obtain the total area of

$$\frac{4}{3}(4\pi - 3\sqrt{3}).$$

45. Find the length of the graph of $r = 1/\theta$ over the interval $\pi \le \theta \le 2\pi$.

Solution:

$$s = \int_\alpha^\beta \sqrt{[f(\theta)]^2 + [f'(\theta)]^2}\, d\theta$$

$$= \int_\pi^{2\pi} \sqrt{\left(\frac{1}{\theta}\right)^2 + \left(\frac{-1}{\theta^2}\right)^2}\, d\theta$$

$$= \int_\pi^{2\pi} \frac{1}{\theta^2} \sqrt{\theta^2 + 1}\, d\theta$$

$$= \left[-\frac{\sqrt{\theta^2 + 1}}{\theta} + \ln\left|\theta + \sqrt{\theta^2 + 1}\right| \right]_\pi^{2\pi}$$

$$= \frac{2\sqrt{\pi^2 + 1} - \sqrt{4\pi^2 + 1}}{2\pi} + \ln\left|\frac{2\pi + \sqrt{4\pi^2 + 1}}{\pi + \sqrt{\pi^2 + 1}}\right| \approx 0.7112$$

51. Find the area of the surface formed by revolving the curve $r = e^{a\theta}$ over the interval $0 \le \theta \le \pi/2$ about the line $\theta = \pi/2$.

Solution:

$$S = 2\pi \int_\alpha^\beta f(\theta) \cos\theta \sqrt{[f(\theta)]^2 + [f'(\theta)]^2}\, d\theta$$

$$= 2\pi \int_0^{\pi/2} e^{a\theta}(\cos\theta)\sqrt{(e^{a\theta})^2 + (ae^{a\theta})^2}\, d\theta$$

$$= 2\pi \int_0^{\pi/2} \cos\theta e^{a\theta} \sqrt{e^{2a\theta}(1 + a^2)}\, d\theta$$

$$= 2\pi\sqrt{1 + a^2} \int_0^{\pi/2} \cos\theta e^{2a\theta}\, d\theta$$

$$= 2\pi\sqrt{1 + a^2} \left[\frac{e^{2a\theta}}{4a^2 + 1}(2a\cos\theta + \sin\theta) \right]_0^{\pi/2} \qquad \text{(Integration by Parts)}$$

$$= \frac{2\pi\sqrt{1 + a^2}}{4a^2 + 1}(e^{\pi a} - 2a)$$

10.5 Polar Equations for Conics and Kepler's Laws

11. Sketch the graph of the equation $r = -1/(1 - \sin\theta)$ and identify the curve.

Solution:

From the form of the equation you have

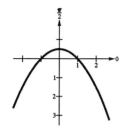

$$r = \frac{-1}{1 - \sin\theta} = \frac{ed}{1 - e\sin\theta}$$

You can conclude that the graph of the equation is a parabola ($e = 1$). Sketch the left half of the parabola by plotting points in the accompanying table. Then using symmetry with respect to $\theta = \pi/2$, sketch the right half.

θ	$-\dfrac{\pi}{2}$	$-\dfrac{\pi}{3}$	$-\dfrac{\pi}{4}$	$-\dfrac{\pi}{6}$	0	$\dfrac{\pi}{6}$	$\dfrac{\pi}{4}$	$\dfrac{\pi}{3}$	$\dfrac{\pi}{2}$
r	-0.50	-0.54	-0.59	-0.67	-1	-2	-3.41	-7.46	Undefined

19. Sketch the graph of the equation $r = 3/(2 + 6\sin\theta)$ and identify the curve.

Solution:

To determine the type of conic, rewrite the equation as

$$r = \frac{3}{2 + 6\sin\theta} = \frac{3/2}{1 + 3\sin\theta} = \frac{ed}{1 - e\sin\theta}.$$

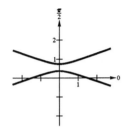

From this form you can conclude that the graph is a hyperbola with $e = 3$. Since $r = f(\sin\theta)$, its graph is symmetric to $\theta = \pi/2$. Therefore, the entries of the following table are solution points of the equation for the right half of the hyperbola. Plot these points and sketch the right half and then sketch the other half by symmetry.

θ	0	$\dfrac{\pi}{6}$	$\dfrac{\pi}{3}$	$\dfrac{\pi}{2}$	$\dfrac{7\pi}{6}$	$\dfrac{4\pi}{3}$	$\dfrac{3\pi}{2}$
r	1.500	0.600	0.417	0.125	-3.000	-0.417	-0.750

27. Find a polar equation for the ellipse with focus at the pole, eccentricity $e = 1/2$ and directrix $y = 1$.

Solution:

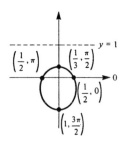

Since the directrix is horizontal and above the pole (see the accompanying figure), choose an equation of the form

$$r = \frac{ed}{1 + e \sin \theta}.$$

Moreover, since the eccentricity of the ellipse is $\frac{1}{2}$ and the directed distance from the focus to the directrix is $d = 1$, you have the equation

$$r = \frac{\frac{1}{2}}{1 + \frac{1}{2} \sin \theta} = \frac{1}{2 + \sin \theta}.$$

31. Find a polar equation of the parabola with focus at the pole and vertex at $(1, -\pi/2)$.

Solution:

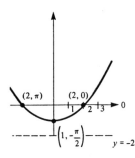

Since the directrix is horizontal and below the pole (see the accompanying figure), choose an equation of the form

$$r = \frac{ed}{1 - e \sin \theta}$$

Moreover, since the eccentricity of a parabola is $e = 1$ and the distance from the focus to the directrix is $d = 2$, you have the equation

$$r = \frac{2}{1 - \sin \theta}.$$

36. Find an equation of the hyperbola with focus at the pole and vertices $(2, 0)$ and $(10, 0)$.

Solution:

Since the vertices of the hyperbola are on the polar axis, its graph is symmetric to the polar axis and we use the equation of the form

$$r = \frac{ed}{1 \pm e \cos \theta}$$

The distance between the vertices is $2a = 10 - 2$ and therefore, $a = 4$. The distance between the center $(6, 0)$ and a focus is $c = 6$. Thus, the eccentricity of the hyperbola is $e = 6/4 = 3/2$ yielding the equation

$$r = \frac{3d/2}{1 \pm (3/2) \cos \theta} = \frac{3d}{2 \pm 3 \cos \theta}.$$

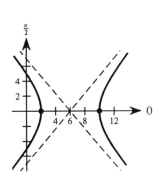

One set of polar coordinates for the vertices of the hyperbola are $(2, 0)$ and $(-10, \pi)$. Substituting these solution points into the form of the equation of the hyperbola yields

$$r = \frac{10}{2 + 3 \cos \theta}.$$

A second set of polar coordinates for the vertices are $(10, 0)$ and $(-2, \pi)$. Substituting these solution points into the form of the equation of the hyperbola yields

$$r = \frac{-10}{2 - 3 \cos \theta}.$$

51. Find the angle ψ for the graph of $r = 6/(1 - \cos \theta)$ at $\theta = 2\pi/3$.

Solution:

Since

$$f(\theta) = r = \frac{6}{1 - \cos \theta}$$

you have

$$f'(\theta) = \frac{-6 \sin \theta}{(1 - \cos \theta)^2}.$$

Now from the definition of the angle between the radial line and the tangent line (see Exercise 46), you have

$$\tan \psi = \left| \frac{f(\theta)}{f'(\theta)} \right| = \left| \frac{6/(1 - \cos \theta)}{-6 \sin \theta / (1 - \cos \theta)^2} \right| = \left| \frac{1 - \cos \theta}{-\sin \theta} \right|.$$

At $\theta = 2\pi/3$,

$$\tan \psi = \left| \frac{1 - \left(-\frac{1}{2}\right)}{-\sqrt{3}/2} \right| = \left| \frac{\frac{3}{2}}{-\sqrt{3}/2} \right| = \sqrt{3}.$$

Therefore, it follows that $\psi = \pi/3$.

Review Exercises for Chapter 10

7. (a) Find dy/dx and all points of horizontal tangency, (b) eliminate the parameter if possible, and (c) sketch the curve represented by the parametric equations $x = 3 + 2\cos\theta$ and $y = 2 + 5\sin\theta$.

Solution:

(a) Since $x = 3 + 2\cos\theta$ and $y = 2 + 5\sin\theta$, you have

$$\frac{dy}{dx} = \frac{dy/d\theta}{dx/d\theta} = \frac{5\cos\theta}{-2\sin\theta} = -\frac{5}{2}\cot\theta$$

The points of horizontal tangency occur at $\theta = \pi/2$ and $\theta = 3\pi/2$. Substituting these values of θ into the set of parametric equations yields the points $(3, 7)$ and $(3, -3)$.

(b) To eliminate the parameter, consider the identity $\sin^2\theta + \cos^2\theta = 1$ and write the parametric equations in the form

$$\frac{x - 3}{2} = \cos\theta, \quad \frac{y - 2}{5} = \sin\theta.$$

By squaring and adding these equations, you obtain

$$\frac{(x - 3)^2}{4} + \frac{(y - 2)^2}{25} = \sin^2\theta + \cos^2\theta = 1$$

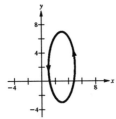

which is an equation for the ellipse centered at $(3, 2)$ with vertices at $(3, -3)$ and $(3, 7)$.

19. Find the length of the involute $x = r(\cos\theta + \theta\sin\theta)$, $y = r(\sin\theta - \theta\cos\theta)$, over the interval $0 \le \theta \le \pi$. (See Exercise 18.)

Solution:

$$x = r(\cos\theta + \theta\sin\theta) \qquad y = r(\sin\theta - \theta\cos\theta)$$
$$\frac{dx}{d\theta} = r\theta\cos\theta \qquad\qquad \frac{dy}{d\theta} = r\theta\sin\theta$$

$$s = \int_0^\pi \sqrt{\left(\frac{dx}{d\theta}\right)^2 + \left(\frac{dy}{d\theta}\right)^2}\, d\theta$$
$$= \int_0^\pi \sqrt{(r\theta\cos\theta)^2 + (r\theta\sin\theta)^2}\, d\theta$$
$$= r\int_0^\pi \theta\, d\theta = r\left[\frac{\theta^2}{2}\right]_0^\pi = \frac{1}{2}\pi^2 r$$

31. Sketch the graph of $r^2 = 4\sin^2 2\theta$.

Solution:

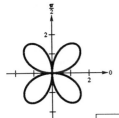

Given the polar equation $r^2 = 4\sin^2 2\theta$, you have $r = \pm 2\sin 2\theta$.

Type of curve: Rose curve with four petals

Symmetry: With respect to the pole, polar axis, and $\theta = \pi/2$

Extrema of r: $(\pm 2, \pi/4)$, $(\pm 2, 3\pi/4)$

Tangents at the pole: $\theta = 0$, $\pi/2$

θ	0	$\dfrac{\pi}{12}$	$\dfrac{\pi}{8}$	$\dfrac{\pi}{6}$	$\dfrac{\pi}{4}$	$\dfrac{\pi}{3}$	$\dfrac{3\pi}{8}$	$\dfrac{5\pi}{12}$	$\dfrac{\pi}{2}$
r	0	± 1	$\pm\sqrt{2}$	$\pm\sqrt{3}$	± 2	$\pm\sqrt{3}$	$\pm\sqrt{2}$	± 1	0

45. Find a rectangular equation that has the same graph as $r = 4\cos 2\theta \sec\theta$.

Solution:

First, replace $\cos 2\theta$ by $2\cos^2\theta - 1$ and $\sec\theta$ by $1/\cos\theta$ to obtain

$$r = 4\cos 2\theta \sec\theta = 4(2\cos^2\theta - 1)\frac{1}{\cos\theta}$$
$$r\cos\theta = 8\cos^2\theta - 4$$

Now since $x = r\cos\theta$ and $r^2 = x^2 + y^2$, you have

$$x = 8\left(\frac{x^2}{r^2}\right) - 4$$
$$= \frac{8x^2 - 4(x^2 + y^2)}{x^2 + y^2}$$
$$x^3 + xy^2 = 4x^2 - 4y^2$$
$$(4 + x)y^2 = (x^2)(4 - x)$$
$$y^2 = x^2\left(\frac{4 - x}{4 + x}\right).$$

Note that you used a graphing utility to obtain the graph of this polar equation in Exercise 37. It follows from the rectangular form of the equation that the vertical asymptote is at $x = -4$.

49. Find a polar equation for the rectangular equation

$$x^2 + y^2 = a^2\left(\arctan\frac{y}{x}\right)^2.$$

Solution:

Since $r^2 = x^2 + y^2$ and $\theta = \arctan\frac{y}{x}$, you have

$$x^2 + y^2 = a^2\left(\arctan\frac{y}{x}\right)^2$$
$$r^2 = a^2\theta^2.$$

57. Find the tangent lines at the pole and all points of vertical and horizontal tangency for the graph of $r = 1 - 2\cos\theta$. Sketch the graph of the equation.

Solution:

The graph has polar axis symmetry and the tangents at the pole are $\theta = \pi/3$ and $\theta = -\pi/3$. To find the points of vertical or horizontal tangency, note that $f'(\theta) = 2\sin\theta$ and find dy/dx as follows:

$$\frac{dy}{dx} = \frac{f'(\theta)\sin\theta + f(\theta)\cos\theta}{f'(\theta)\cos\theta - f(\theta)\sin\theta}$$

$$= \frac{2\sin^2\theta + (1 - 2\cos\theta)\cos\theta}{2\sin\theta\cos\theta - (1 - 2\cos\theta)\sin\theta}$$

$$= \frac{2\sin^2\theta + \cos\theta - 2\cos^2\theta}{4\sin\theta\cos\theta - \sin\theta}$$

$$= \frac{2(1 - \cos^2\theta) + \cos\theta - 2\cos^2\theta}{\sin\theta(4\cos\theta - 1)}$$

$$= \frac{2 + \cos\theta - 4\cos^2\theta}{\sin\theta(4\cos\theta - 1)}$$

The graph has horizontal tangents when $dy/dx = 0$, and this occurs when

$$-4\cos^2\theta + \cos\theta + 2 = 0$$

$$\cos\theta = \frac{-1 \pm \sqrt{1 + 32}}{-8} = \frac{1 \mp \sqrt{33}}{8}.$$

When $\cos\theta = (1 \mp \sqrt{33})/8$,

$$r = 1 - 2\left(\frac{1 \mp \sqrt{33}}{8}\right) = \frac{3 \pm \sqrt{33}}{4}.$$

Therefore, the points of horizontal tangency are

$$\left(\frac{3 - \sqrt{33}}{4}, \arccos\left[\frac{1 + \sqrt{33}}{8}\right]\right) \approx (-0.686, 0.568)$$

$$\left(\frac{3 - \sqrt{33}}{4}, -\arccos\left[\frac{1 + \sqrt{33}}{8}\right]\right) \approx (-0.686, -0.568)$$

$$\left(\frac{3 + \sqrt{33}}{4}, \arccos\left[\frac{1 - \sqrt{33}}{8}\right]\right) \approx (2.186, 2.206)$$

$$\left(\frac{3 + \sqrt{33}}{4}, -\arccos\left[\frac{1 - \sqrt{33}}{8}\right]\right) \approx (2.186, -2.206)$$

The graph has vertical tangents when

$$\sin\theta(4\cos\theta - 1) = 0$$

$$\theta = 0, \ \pi, \ \text{or} \ \pm\arccos\frac{1}{4}.$$

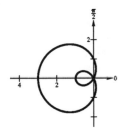

When $\cos\theta = \frac{1}{4}$, $r = 1 - 2(\frac{1}{4}) = \frac{1}{2}$. Thus, the points of vertical tangency are

$$(-1,0), \ (3,\pi), \ \text{and} \ \left(\frac{1}{2}, \pm\arccos\frac{1}{4}\right) \approx (0.5, \pm 1.318)$$

as shown in the accompanying figure.

67. Find the area of the region common to the interiors of $r = 4\cos\theta$ and $r = 2$.

Solution:

To find the point of intersection of the two graphs solve the two equations simultaneously to obtain

$$4\cos\theta = 2$$
$$\cos\theta = \frac{1}{2}$$
$$\theta = \frac{\pi}{3}$$

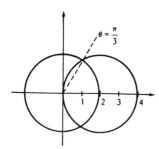

From the accompanying figure it follows that the area is given by

$$A = 2\left[\frac{1}{2}\int_0^{\pi/3} 2^2 \, d\theta + \frac{1}{2}\int_{\pi/3}^{\pi/2} (4\cos\theta)^2 \, d\theta\right]$$

$$= \int_0^{\pi/3} 4 \, d\theta + 16\int_{\pi/3}^{\pi/2} \frac{1 + \cos 2\theta}{2} \, d\theta$$

$$= \left[4\theta\right]_0^{\pi/3} + 8\left[\theta + \frac{1}{2}\sin 2\theta\right]_{\pi/3}^{\pi/2}$$

$$= \frac{8\pi - 6\sqrt{3}}{3}.$$

69. Find the perimeter of a cardiod $r = a(1 - \cos\theta)$.

Solution:

$$r = a(1 - \cos\theta) \quad \text{and} \quad \frac{dr}{d\theta} = a\sin\theta$$

Due to the symmetry with respect to the polar axis, you have

$$s = 2\int_0^\pi \sqrt{r^2 + \left(\frac{dr}{d\theta}\right)^2}\, d\theta$$

$$= 2\int_0^\pi \sqrt{a^2(1 - \cos\theta)^2 + a^2\sin^2\theta}\, d\theta$$

$$= 2\sqrt{2}a \int_0^\pi \sqrt{1 - \cos\theta}\, d\theta$$

$$= 2\sqrt{2}a \int_0^\pi \sqrt{1 - \cos\theta}\frac{\sqrt{1 + \cos\theta}}{\sqrt{1 + \cos\theta}}\, d\theta$$

$$= 2\sqrt{2}a \int_0^\pi \frac{\sin\theta}{\sqrt{1 + \cos\theta}}\, d\theta$$

$$= -4\sqrt{2}a\left[(1 + \cos\theta)^{1/2}\right]_0^\pi = 8a.$$

11 VECTORS AND THE GEOMETRY OF SPACE

11.1 Vectors in the Plane

5. The initial point of a vector **v** is $(1, 2)$ and its terminal point is $(5, 5)$. (a) Sketch the given directed line segment, (b) write the vector in component form, and (c) sketch the vector with its initial point at the origin.

Solution:

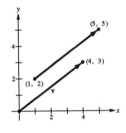

(b) We let $P = (1, 2) = (p_1, p_2)$ and $Q = (5, 5) = (q_1, q_2)$. Then the components of $\mathbf{v} = \langle v_1, v_2 \rangle$ are given by

$$v_1 = q_1 - p_1 = 5 - 1 = 4$$
$$v_2 = q_2 - p_2 = 5 - 2 = 3$$

Thus, $\mathbf{v} = \langle 4, 3 \rangle$.

17. Find the component form of $\mathbf{v} = \mathbf{u} + 2\mathbf{w}$, where $\mathbf{u} = 2\mathbf{i} - \mathbf{j}$ and $\mathbf{w} = \mathbf{i} + 2\mathbf{j}$. Illustrate the indicated vector operations geometrically.

Solution:

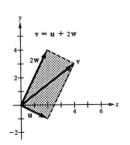

Since $\mathbf{u} = 2\mathbf{i} - \mathbf{j}$ and $2\mathbf{w} = 2(\mathbf{i} + 2\mathbf{j}) = 2\mathbf{i} + 4\mathbf{j}$, you have

$$\mathbf{v} = \mathbf{u} + 2\mathbf{w}$$
$$= (2\mathbf{i} - \mathbf{j}) + (2\mathbf{i} + 4\mathbf{j})$$
$$= (2 + 2)\mathbf{i} + (4 - 1)\mathbf{j} = 4\mathbf{i} + 3\mathbf{j} = \langle 4, 3 \rangle.$$

Geometrically, this sum is illustrated in the accompanying figure.

379

23. Find a and b such that $\mathbf{v} = a\,\mathbf{u} + b\,\mathbf{w}$ where $\mathbf{v} = \langle 3, 0 \rangle$, $\mathbf{u} = \langle 1, 2 \rangle$, and $\mathbf{w} = \langle 1, -1 \rangle$.

Solution:

$$\mathbf{v} = a\,\mathbf{u} + b\,\mathbf{w}$$
$$3\mathbf{i} = a(\mathbf{i} + 2\mathbf{j}) + b(\mathbf{i} - \mathbf{j}) = (a+b)\mathbf{i} + (2a-b)\mathbf{j}$$

Equating coefficients yields

$$a + b = 3$$

and

$$2a - b = 0$$

Solving the system of equations simultaneously you obtain $a = 1$ and $b = 2$.

39. Demonstrate the triangle inequality using the vectors $\mathbf{u} = \langle 2, 1 \rangle$ and $\mathbf{v} = \langle 5, 4 \rangle$.

Solution:
$$\|\mathbf{u}\| = \sqrt{2^2 + 1^2} = \sqrt{5}$$
$$\|\mathbf{v}\| = \sqrt{5^2 + 4^2} = \sqrt{41}$$

Since $\mathbf{u} + \mathbf{v} = \langle 2, 1 \rangle + \langle 5, 4 \rangle = \langle 7, 5 \rangle$ you have

$$\|\mathbf{u} + \mathbf{v}\| = \sqrt{7^2 + 5^2} = \sqrt{74}$$

Therefore,

$$\|\mathbf{u} + \mathbf{v}\| = \sqrt{74} \approx 8.602 \leq 2.236 + 6.403$$
$$\approx \sqrt{5} + \sqrt{41} = \|\mathbf{u}\| + \|\mathbf{v}\|$$

41. Find the vector $\mathbf{v}$ with magnitude $\|\mathbf{v}\| = 4$ and in the same direction as $\mathbf{u} = \langle 1, 1 \rangle$.

Solution:

$$\mathbf{v} = (\text{magnitude of } \mathbf{v})(\text{unit vector in the direction of } \mathbf{v})$$
$$= 4\left(\frac{\mathbf{u}}{\|\mathbf{u}\|} \right)$$
$$= 4\left(\frac{\mathbf{u}}{\sqrt{2}} \right) = 2\sqrt{2}\,\mathbf{u} = \langle 2\sqrt{2},\ 2\sqrt{2} \rangle$$

45. Find a unit vector (a) parallel to and (b) normal to the graph of $f(x) = x^3$ at $(1, 1)$.

Solution:

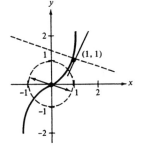

(a) At $(1, 1)$ the slope of the tangent line is $f'(1) = 3$. Therefore a vector $\mathbf{v} = x\,\mathbf{i} + y\,\mathbf{j}$ parallel to the tangent line must have a slope of 3 (see the accompanying figure) and $y = 3x$. Since $\mathbf{v}$ is a unit vector, $x^2 + y^2 = 1$. Thus, you have

$$x^2 + (3x)^2 = 1$$
$$10x^2 = 1$$
$$x = \pm\frac{1}{\sqrt{10}} \quad \text{and} \quad y = \pm\frac{3}{\sqrt{10}}.$$

Finally, you can conclude that

$$\mathbf{v} = \frac{1}{\sqrt{10}}\,\mathbf{i} + \frac{3}{\sqrt{10}}\,\mathbf{j} = \left\langle \frac{1}{\sqrt{10}}, \frac{3}{\sqrt{10}} \right\rangle$$

or

$$\mathbf{v} = -\frac{1}{\sqrt{10}}\,\mathbf{i} - \frac{3}{\sqrt{10}}\,\mathbf{j} = \left\langle -\frac{1}{\sqrt{10}}, -\frac{3}{\sqrt{10}} \right\rangle$$

(b) Similarly, a vector $\mathbf{v} = x\,\mathbf{i} + y\,\mathbf{j}$ normal to the tangent line must have a slope of $-\frac{1}{3}$ (see accompanying figure), and $x = -3y$. Since $\mathbf{v}$ is a unit vector, $x^2 + y^2 = 1$. Thus, you have

$$(-3y)^2 + y^2 = 1$$
$$10y^2 = 1$$
$$y = \pm\frac{1}{\sqrt{10}} \quad \text{and} \quad x = \pm\frac{3}{\sqrt{10}}.$$

Finally, you can conclude that

$$\mathbf{v} = \frac{3}{\sqrt{10}}\,\mathbf{i} - \frac{1}{\sqrt{10}}\,\mathbf{j} = \left\langle \frac{3}{\sqrt{10}}, -\frac{1}{\sqrt{10}} \right\rangle$$

or

$$\mathbf{v} = \frac{-3}{\sqrt{10}}\,\mathbf{i} + \frac{1}{\sqrt{10}}\,\mathbf{j} = \left\langle -\frac{3}{\sqrt{10}}, \frac{1}{\sqrt{10}} \right\rangle$$

51. Find the component form of the vector **v** if it makes an angle of 150° with the positive x-axis and has magnitude $\|\mathbf{v}\| = 2$.

Solution:

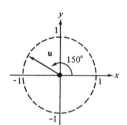

Begin by finding a unit vector **u** making an angle of 150° with the positive x-axis. Since **u** is a unit vector, consider it as the radius of a unit circle as shown in the accompanying figure. Therefore, $x = \cos\theta$ and $y = \sin\theta$, and the component form for **u** is

$$\mathbf{u} = x\mathbf{i} + y\mathbf{j} = \cos(150°)\mathbf{i} + \sin(150°)\mathbf{j} = \left\langle -\frac{\sqrt{3}}{2}, \frac{1}{2} \right\rangle$$

Since $\mathbf{v} = 2\,\mathbf{u}$, you have $\mathbf{v} = \langle -\sqrt{3}, 1 \rangle$.

67. To carry a 100-lb cylindrical weight, two men lift on the ends of short ropes that are tied to an eyelet on the top center of the cylinder. If one rope makes a 20° angle away from the vertical and the other a 30° angle, find (a) the tension in each rope if the resultant force is vertical, and (b) the vertical component of each man's force.

Solution:

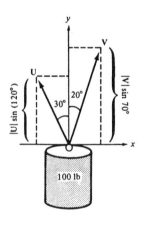

For the weight to be lifted vertically, the sum of the vertical components of **u** and **v** must be 100 and the sum of the horizontal components must be zero. (See the accompanying sketch.) Thus,

$$\|\mathbf{u}\|\sin 120° + \|\mathbf{v}\|\sin 70° = 100 \qquad \text{Vertical}$$
$$\|\mathbf{u}\|\cos 120° + \|\mathbf{v}\|\cos 70° = 0 \qquad \text{Horizontal}$$

and, equivalently

$$\|\mathbf{u}\|\left(\frac{\sqrt{3}}{2}\right) + \|\mathbf{v}\|\sin 70° = 100$$
$$\|\mathbf{u}\|\left(-\frac{1}{2}\right) + \|\mathbf{v}\|\cos 70° = 0.$$

If you multiply the second equation by $\sqrt{3}$ and then add the two equations, you have

$$\|\mathbf{v}\|[\sqrt{3}\cos 70° + \sin 70°] = 100$$
$$\|\mathbf{v}\| = \frac{100}{\sqrt{3}\cos 70° + \sin 70°} \approx 65.27 \text{ lb.}$$

Substituting 65.27 for $\|\mathbf{v}\|$ into the second equation, yields

$$65.27(\cos 70^\circ) = \frac{\|\mathbf{u}\|}{2}$$

or

$$\|\mathbf{u}\| = 130.54(\cos 70^\circ) \approx 44.65 \text{ lb.}$$

(a)   The tension in each rope is

$$\|\mathbf{u}\| = 44.65 \text{ lb,} \qquad \|\mathbf{v}\| = 65.27 \text{ lb.}$$

(b)   The vertical component of each man's force is

$$\|\mathbf{u}\| \sin 120^\circ \approx (44.65)(0.8660) = 38.67 \text{ lb.}$$
$$\|\mathbf{v}\| \sin 70^\circ \approx (65.27)(0.9397) = 61.33 \text{ lb.}$$

(Note that the sum of the two vertical components is 100 lb.)

11.2   Space Coordinates and Vectors in Space

7. Find the lengths of the sides of the triangle with vertices $(1, -3, -2)$, $(5, -1, 2)$, and $(-1, 1, 2)$ and determine whether the triangle is a right triangle, an isosceles triangle, or neither.

Solution:

Let us denote the three given points by A, B, and C, respectively. Then

$$|AB| = \sqrt{(5-1)^2 + (-1+3)^2 + (2+2)^2}$$
$$= \sqrt{16 + 4 + 16} = \sqrt{36} = 6$$
$$|AC| = \sqrt{(-1-1)^2 + (1+3)^2 + (2+2)^2}$$
$$= \sqrt{4 + 16 + 16} = \sqrt{36} = 6$$
$$|BC| = \sqrt{(-1-5)^2 + (1+1)^2 + (2-2)^2}$$
$$= \sqrt{36 + 4} = \sqrt{40} = 2\sqrt{10}$$

Since two sides have equal lengths, the triangle is isosceles.

15. Find the center and radius of the sphere

$$x^2 + y^2 + z^2 - 2x + 6y + 8z + 1 = 0.$$

Solution:

Writing the equation in standard form, yields

$$x^2 + y^2 + z^2 - 2x + 6y + 8z + 1 = 0$$
$$(x^2 - 2x) + (y^2 + 6y) + (z^2 + 8z) = -1$$
$$(x^2 - 2x + 1) + (y^2 + 6y + 9) + (z^2 + 8z + 16) = -1 + 1 + 9 + 16$$
$$(x - 1)^2 + (y + 3)^2 + (z + 4)^2 = 25$$

Thus, the sphere is centered at $(1, -3, -4)$ with radius 5.

31. Find $\mathbf{z}$, where $\mathbf{u} = \langle 1, 2, 3 \rangle$, $\mathbf{w} = \langle 4, 0, -4 \rangle$, and $2\mathbf{z} - 3\mathbf{u} = \mathbf{w}$.

Solution:

$$2\mathbf{z} - 3\mathbf{u} = \mathbf{w}$$
$$2\mathbf{z} = 3\mathbf{u} + \mathbf{w}$$
$$= 3(\mathbf{i} + 2\mathbf{j} + 3\mathbf{k}) + (4\mathbf{i} - 4\mathbf{k})$$
$$= 3\mathbf{i} + 6\mathbf{j} + 9\mathbf{k} + 4\mathbf{i} - 4\mathbf{k}$$
$$= 7\mathbf{i} + 6\mathbf{j} + 5\mathbf{k}$$
$$\mathbf{z} = \frac{7}{2}\mathbf{i} + 3\mathbf{j} + \frac{5}{2}\mathbf{k} = \left\langle \frac{7}{2}, 3, \frac{5}{2} \right\rangle$$

37. Use vectors to determine whether the points $(0, -2, -5)$, $(3, 4, 4)$, and $(2, 2, 1)$ are collinear.

Solution:

If you denote the three points by A, B, and C, respectively, then

$$\overrightarrow{AB} = (3 - 0)\mathbf{i} + (4 + 2)\mathbf{j} + (4 + 5)\mathbf{k} = 3\mathbf{i} + 6\mathbf{j} + 9\mathbf{k}$$
$$\overrightarrow{AC} = (2 - 0)\mathbf{i} + (2 + 2)\mathbf{j} + (1 + 5)\mathbf{k}$$
$$= 2\mathbf{i} + 4\mathbf{j} + 6\mathbf{k}$$
$$= \frac{2}{3}(3\mathbf{i} + 6\mathbf{j} + 9\mathbf{k}) = \frac{2}{3}\overrightarrow{AB}$$

Since $\overrightarrow{AC}$ is a scaler multiple of $\overrightarrow{AB}$, the three points are collinear.

51. Find a unit vector (a) in the direction of $\mathbf{u} = \langle 2, -1, 2 \rangle$ and (b) in the opposite direction of $\mathbf{u}$.

Solution:

(a) Since the magnitude of $\mathbf{u}$ is

$$\|\mathbf{u}\| = \sqrt{2^2 + (-1)^2 + 2^2} = \sqrt{9} = 3,$$

the unit vector in the direction of $\mathbf{u}$ is

$$\frac{\mathbf{u}}{\|\mathbf{u}\|} = \frac{1}{3}\langle 2, -1, 2 \rangle.$$

(b) Since the unit vector in the opposite direction is obtained by multiplying by the scalar -1, you have

$$(-1)\frac{\mathbf{u}}{\|\mathbf{u}\|} = -\frac{1}{3}\langle 2, -1, 2 \rangle.$$

61. Use vectors to find the point that lies two-thirds of the way from $P = (4, 3, 0)$ to $Q = (1, -3, 3)$.

Solution:

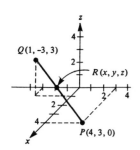

By finding the component form of the vector from P to Q, you have

$$\overrightarrow{PQ} = \langle 1 - 4, \ -3 - 3, \ 3 - 0 \rangle = \langle -3, -6, 3 \rangle.$$

Let $R = (x, y, z)$ be a point on the line segment two-thirds of the way from P to Q. (See the accompanying figure.) Then

$$\overrightarrow{PR} = \langle x - 4, \ y - 3, \ z - 0 \rangle = \langle x - 4, y - 3, z \rangle$$

and

$$\frac{2}{3}\overrightarrow{PQ} = \frac{2}{3}\langle -3, -6, 3 \rangle$$

$$= \langle -2, -4, 2 \rangle = \langle x - 4, y - 3, z \rangle = \overrightarrow{PR}$$

Therefore,

$$x - 4 = -2 \quad \Longrightarrow \quad x = 2$$
$$y - 3 = -4 \quad \Longrightarrow \quad y = -1$$
$$z = 2,$$

and the required point is $(2, -1, 2)$.

65. Let $\mathbf{u} = \mathbf{i} + \mathbf{j}$, $\mathbf{v} = \mathbf{j} + \mathbf{k}$, and $\mathbf{w} = a\,\mathbf{u} + b\,\mathbf{v}$.

(a) Sketch $\mathbf{u}$ and $\mathbf{v}$.

(b) If $\mathbf{w} = \mathbf{0}$, show that a and b are both zero.

(c) Find a and b such that $\mathbf{w} = \mathbf{i} + 2\mathbf{j} + \mathbf{k}$.

(d) Show that no choice of a and b yields $\mathbf{w} = \mathbf{i} + 2\mathbf{j} + 3\mathbf{k}$.

Solution:

(a) See the accompanying figure.

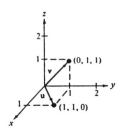

(b)
$$\mathbf{w} = a(\mathbf{i} + \mathbf{j}) + b(\mathbf{j} + \mathbf{k}) = \mathbf{0}$$
$$a\,\mathbf{i} + (a + b)\mathbf{j} + b\,\mathbf{k} = 0\mathbf{i} + 0\mathbf{j} + 0\mathbf{k}$$

Therefore, $a = b = 0$.

(c)
$$\mathbf{w} = a(\mathbf{i} + \mathbf{j}) + b(\mathbf{j} + \mathbf{k}) = \mathbf{i} + 2\mathbf{j} + \mathbf{k}$$
$$a\,\mathbf{i} + (a + b)\mathbf{j} + b\,\mathbf{k} = \mathbf{i} + 2\mathbf{j} + \mathbf{k}$$

Therefore, $a = b = 1$.

(d)
$$\mathbf{w} = a(\mathbf{i} + \mathbf{j}) + b(\mathbf{j} + \mathbf{k} = \mathbf{i} + 2\mathbf{j} + 3\mathbf{k}$$
$$a\,\mathbf{i} + (a + b)\mathbf{j} + b\,\mathbf{k} = \mathbf{i} + 2\mathbf{j} + 3\mathbf{k}$$

Therefore, $a = 1$, $b = 3$, and $a + b = 2$, which is a contradiction.

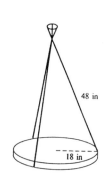

69. The lights in an auditorium are 25-pound disks of radius 18 inches. Each disk is supported by three equally spaced 48-inch wires to the ceiling (see figure). Find the tension in each wire.

Solution:

Because of the way the support system is designed, each wire has the same tension and supports one-third the weight of the light. Consider the wire from P to Q as shown in the accompanying figure. The vector from P to Q is

$$\overrightarrow{PQ} = \langle 0, -18, 6\sqrt{55}\rangle.$$

and the vector representing the force in the wire is a scalar multiple of $\overrightarrow{PQ}$ given by

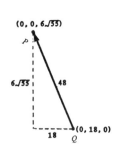

$$\mathbf{F} = c\overrightarrow{PQ} = c\langle 0, -18, 6\sqrt{55}\rangle.$$

The vertical component of $\mathbf{F}$ is $c(6)\sqrt{55}$ and represents one-third the weight of the light. Thus,

$$c(6)\sqrt{55} = \frac{25}{3} \quad \Longrightarrow \quad c = \frac{5\sqrt{55}}{198}.$$

Therefore,

$$\mathbf{F} = \left\langle 0, -\frac{5\sqrt{55}}{11}, \frac{25}{3}\right\rangle$$

and the tension in each of the wires is $\|\mathbf{F}\| = 8.99$ lb.

11.3   The Dot Product of Two Vectors

3. Given the vector $\mathbf{u} = \langle 2, -3, 4\rangle$ and $\mathbf{v} = \langle 0, 6, 5\rangle$ find (a) $\mathbf{u} \cdot \mathbf{v}$, (b) $\mathbf{u} \cdot \mathbf{u}$, (c) $\|\mathbf{u}\|^2$, (d) $(\mathbf{u} \cdot \mathbf{v})\mathbf{v}$, and (e) $\mathbf{u} \cdot (2\mathbf{v})$.

Solution:

(a) For the vectors $\mathbf{u} = \langle u_1, u_2, u_3\rangle$ and $\mathbf{v} = \langle v_1, v_2, v_3\rangle$

$$\mathbf{u} \cdot \mathbf{v} = u_1 v_1 + u_2 v_2 + u_3 v_3$$
$$= 2(0) + (-3)(6) + 4(5) = 2.$$

(b) $\mathbf{u} \cdot \mathbf{u} = 2(2) + (-3)(-3) + 4(4) = 29$

(c) $\|\mathbf{u}\|^2 = 2^2 + (-3)^2 + 4^2 = 29 = \mathbf{u} \cdot \mathbf{u}$

(d) From part (a) you have $\mathbf{u} \cdot \mathbf{v} = 2$. Therefore,

$$(\mathbf{u} \cdot \mathbf{v})\mathbf{v} = 2\mathbf{v} = \langle 0, 12, 10\rangle.$$

(e) From part (a) you have

$$\mathbf{u} \cdot (2\mathbf{v}) = 2(\mathbf{u} \cdot \mathbf{v}) = 2(2) = 4.$$

17. Find the angle θ between $\mathbf{u} = 3\mathbf{i} + 4\mathbf{j}$ and $\mathbf{v} = -2\mathbf{j} + 3\mathbf{k}$.

Solution:

Since

$$\cos\theta = \frac{\mathbf{u} \cdot \mathbf{v}}{\|\mathbf{u}\| \, \|\mathbf{v}\|},$$

you have

$$\cos\theta = \frac{0 - 8 + 0}{\sqrt{25}\sqrt{13}} = \frac{-8\sqrt{13}}{65}$$

and

$$\theta = \arccos\left(\frac{-8\sqrt{13}}{65}\right) \approx 116.3^\circ.$$

21. Determine whether $\mathbf{u} = \langle 4, 3 \rangle$ and $\mathbf{v} = \langle \frac{1}{2}, -\frac{2}{3} \rangle$ are orthogonal, parallel, or neither.

Solution:

$$\mathbf{u} \cdot \mathbf{v} = 4\left(\frac{1}{2}\right) + 3\left(-\frac{2}{3}\right) = 2 - 2 = 0$$

Therefore, $\mathbf{u}$ and $\mathbf{v}$ are orthogonal.

29. What is known about θ, the angle between vectors $\mathbf{u}$ and $\mathbf{v}$, if (a) $\mathbf{u} \cdot \mathbf{v} = 0$? (b) $\mathbf{u} \cdot \mathbf{v} > 0$? (c) $\mathbf{u} \cdot \mathbf{v} < 0$?

Solution:

Assuming that $\mathbf{u}$ and $\mathbf{v}$ are nonzero vectors and $\mathbf{u} \cdot \mathbf{v} = \|\mathbf{u}\| \, \|\mathbf{v}\| \cos\theta$, you have the following:

(a) $\mathbf{u} \cdot \mathbf{v} = 0$ implies that $\cos\theta = 0$, or $\theta = \pi/2$.

(b) $\mathbf{u} \cdot \mathbf{v} > 0$ implies that $\cos\theta > 0$, or $0 \le \theta < \pi/2$.

(c) $\mathbf{u} \cdot \mathbf{v} < 0$ implies that $\cos\theta < 0$, or $\pi/2 < \theta \le \pi$.

31. Find the direction cosines of $\mathbf{u} = \mathbf{i} + 2\mathbf{j} + 2\mathbf{k}$ and demonstrate that the sum of the squares of the direction cosines is 1.

Solution:

Given a vector $\mathbf{u} = u_1\mathbf{i} + u_2\mathbf{j} + u_3\mathbf{k}$ the direction cosines are

$$\cos\alpha = \frac{u_1}{\|\mathbf{u}\|} \qquad \cos\beta = \frac{u_2}{\|\mathbf{u}\|} \qquad \cos\gamma = \frac{u_3}{\|\mathbf{u}\|}.$$

Therefore, for the given vector, you have

$$\cos\alpha = \frac{1}{\sqrt{1^2 + 2^2 + 2^2}} = \frac{1}{\sqrt{9}} = \frac{1}{3}$$

$$\cos\beta = \frac{2}{\sqrt{1^2 + 2^2 + 2^2}} = \frac{2}{\sqrt{9}} = \frac{2}{3}$$

$$\cos\gamma = \frac{2}{\sqrt{1^2 + 2^2 + 2^2}} = \frac{2}{\sqrt{9}} = \frac{2}{3}.$$

The sum of the squares of the direction cosines is

$$\cos^2\alpha + \cos^2\beta + \cos^2\gamma = \left(\frac{1}{3}\right)^2 + \left(\frac{2}{3}\right)^2 + \left(\frac{2}{3}\right)^2 = 1.$$

45. For the vectors $\mathbf{u} = \langle 1, 1, 1 \rangle$ and $\mathbf{v} = \langle -2, -1, 1 \rangle$ find (a) the projection of $\mathbf{u}$ onto $\mathbf{v}$ and (b) the vector component of $\mathbf{u}$ orthogonal to $\mathbf{v}$.

Solution:

(a) The projection of $\mathbf{u}$ onto $\mathbf{v}$ is given by

$$\mathbf{w}_1 = \left(\frac{\mathbf{u}\cdot\mathbf{v}}{\|\mathbf{v}\|^2}\right)\mathbf{v}$$

$$= \left(\frac{-2}{6}\right)\langle -2, -1, 1 \rangle = \left\langle \frac{2}{3}, \frac{1}{3}, -\frac{1}{3} \right\rangle.$$

(b) The vector component of $\mathbf{u}$ orthogonal to $\mathbf{v}$ is given by

$$\mathbf{w}_2 = \mathbf{u} - \mathbf{w}_1$$

$$= \langle 1, 1, 1 \rangle - \left\langle \frac{2}{3}, \frac{1}{3}, -\frac{1}{3} \right\rangle = \left\langle \frac{1}{3}, \frac{2}{3}, \frac{4}{3} \right\rangle.$$

53. An object is dragged 10 feet across a floor, using a force of 85 pounds. Find the work done if the direction of the force is 60° above the horizontal.

Solution:

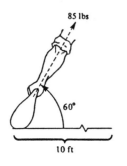

85 lbs

60°

10 ft

The work done by a constant force **F** as its point of application moves along the vector $\overrightarrow{PQ}$ is given by

$$W = \mathbf{F} \cdot \overrightarrow{PQ}.$$

From the accompanying figure it follows that

$$\mathbf{F} = 85[(\cos 60°)\mathbf{i} + (\sin 60°)\mathbf{j}]$$

and

$$\overrightarrow{PQ} = 10\mathbf{i}.$$

Therefore,

$$W = 85[(\cos 60°)(10) + (\sin 60°)(0)] = 425 \text{ ft} \cdot \text{lb}.$$

11.4   The Cross Product of Two Vectors in Space

11. If $\mathbf{u} = \langle 1, 1, 1 \rangle$ and $\mathbf{v} = \langle 2, 1, -1 \rangle$, find $\mathbf{u} \times \mathbf{v}$ and show that it is orthogonal to both $\mathbf{u}$ and $\mathbf{v}$.

Solution:

The cross product is given by

$$\mathbf{u} \times \mathbf{v} = \begin{vmatrix} \mathbf{i} & \mathbf{j} & \mathbf{k} \\ 1 & 1 & 1 \\ 2 & 1 & -1 \end{vmatrix} = \mathbf{i}\begin{vmatrix} 1 & 1 \\ 1 & -1 \end{vmatrix} - \mathbf{j}\begin{vmatrix} 1 & 1 \\ 2 & -1 \end{vmatrix} + \mathbf{k}\begin{vmatrix} 1 & 1 \\ 2 & 1 \end{vmatrix}$$
$$= (-1 - 1)\mathbf{i} - (-1 - 2)\mathbf{j} + (1 - 2)\mathbf{k}$$
$$= -2\,\mathbf{i} + 3\,\mathbf{j} - \mathbf{k} = \langle -2, 3, -1 \rangle.$$

Using the dot product yields

$$\mathbf{u} \cdot (\mathbf{u} \times \mathbf{v}) = (\mathbf{i} + \mathbf{j} + \mathbf{k}) \cdot (-2\,\mathbf{i} + 3\,\mathbf{j} - \mathbf{k}) = -2 + 3 - 1 = 0$$

and

$$\mathbf{v} \cdot (\mathbf{u} \times \mathbf{v}) = (2\,\mathbf{i} + \mathbf{j} - \mathbf{k}) \cdot (-2\,\mathbf{i} + 3\,\mathbf{j} - \mathbf{k}) = -4 + 3 + 1 = 0.$$

Therefore $\mathbf{u} \times \mathbf{v}$ is orthogonal to both $\mathbf{u}$ and $\mathbf{v}$.

21. Verify that the points $(1, 1, 1)$, $(2, 3, 4)$, $(6, 5, 2)$, and $(7, 7, 5)$ are vertices of a parallelogram and find its area.

Solution:

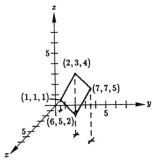

If you denote the four given points by A, B, C, and D, respectively, the vectors representing the sides of the quadrilateral are

$$\overrightarrow{AB} = (2 - 1)\mathbf{i} + (3 - 1)\mathbf{j} + (4 - 1)\mathbf{k} = \mathbf{i} + 2\mathbf{j} + 3\mathbf{k}$$

$$\overrightarrow{AC} = (6 - 1)\mathbf{i} + (5 - 1)\mathbf{j} + (2 - 1)\mathbf{k} = 5\mathbf{i} + 4\mathbf{j} + \mathbf{k}$$

$$\overrightarrow{CD} = (7 - 6)\mathbf{i} + (7 - 5)\mathbf{j} + (5 - 2)\mathbf{k} = \mathbf{i} + 2\mathbf{j} + 3\mathbf{k}$$

$$\overrightarrow{DB} = (7 - 2)\mathbf{i} + (7 - 3)\mathbf{j} + (5 - 4)\mathbf{k} = 5\mathbf{i} + 4\mathbf{j} + \mathbf{k}$$

Since $\overrightarrow{AB} = \overrightarrow{CD}$ and $\overrightarrow{AC} = \overrightarrow{DB}$, the quadrilateral is a parallelogram. The area of the parallelogram is given by $\|\overrightarrow{AB} \times \overrightarrow{AC}\|$.

$$\overrightarrow{AB} \times \overrightarrow{AC} = \begin{vmatrix} \mathbf{i} & \mathbf{j} & \mathbf{k} \\ 1 & 2 & 3 \\ 5 & 4 & 1 \end{vmatrix}$$

$$= \begin{vmatrix} 2 & 3 \\ 4 & 1 \end{vmatrix}\mathbf{i} - \begin{vmatrix} 1 & 3 \\ 5 & 1 \end{vmatrix}\mathbf{j} + \begin{vmatrix} 1 & 2 \\ 5 & 4 \end{vmatrix}\mathbf{k}$$

$$= -10\mathbf{i} + 14\mathbf{j} - 6\mathbf{k} = 2(-5\mathbf{i} + 7\mathbf{j} - 3\mathbf{k})$$

$$\|\overrightarrow{AB} \times \overrightarrow{AC}\| = 2\sqrt{25 + 49 + 9} = 2\sqrt{83}.$$

29. Find the triple scalar product $\mathbf{u} \cdot (\mathbf{v} \times \mathbf{w})$ if $\mathbf{u} = \langle 2, 0, 1 \rangle$, $\mathbf{v} = \langle 0, 3, 0 \rangle$, and $\mathbf{w} = \langle 0, 0, 1 \rangle$.

Solution:

$$\mathbf{v} \times \mathbf{w} = \begin{vmatrix} \mathbf{i} & \mathbf{j} & \mathbf{k} \\ 0 & 3 & 0 \\ 0 & 0 & 1 \end{vmatrix}$$

$$= \mathbf{i}\begin{vmatrix} 3 & 0 \\ 0 & 1 \end{vmatrix} - \mathbf{j}\begin{vmatrix} 0 & 0 \\ 0 & 1 \end{vmatrix} + \mathbf{k}\begin{vmatrix} 0 & 3 \\ 0 & 0 \end{vmatrix} = 3\mathbf{i}$$

Therefore, the triple scalar product is

$$\mathbf{u} \cdot (\mathbf{v} + \mathbf{w}) = 2(3) + 0(0) + 1(0) = 6$$

33. Find the volume of the parallelopiped with vertices $(0, 0, 0)$, $(3, 0, 0)$, $(0, 5, 1)$, $(3, 5, 1)$, $(2, 0, 5)$, $(5, 0, 5)$, $(2, 5, 6)$, and $(5, 5, 6)$.

Solution:

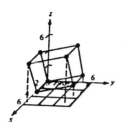

From the accompanying figure, it can be seen that three adjacent sides of the parallelopiped are given by **u**, **v**, and **w** where

$$\mathbf{u} = 3\,\mathbf{i}, \quad \mathbf{v} = 5\,\mathbf{j} + \mathbf{k}, \quad \text{and} \quad \mathbf{w} = 2\,\mathbf{i} + 5\,\mathbf{k}.$$

Since the volume of the parallelopiped is the absolute value of the triple scalar product $\mathbf{u} \cdot (\mathbf{v} \times \mathbf{w})$, you have

$$\mathbf{v} \times \mathbf{w} = \begin{vmatrix} \mathbf{i} & \mathbf{j} & \mathbf{k} \\ 0 & 5 & 1 \\ 2 & 0 & 5 \end{vmatrix} = \begin{vmatrix} 5 & 1 \\ 0 & 5 \end{vmatrix}\mathbf{i} - \begin{vmatrix} 0 & 1 \\ 2 & 5 \end{vmatrix}\mathbf{j} + \begin{vmatrix} 0 & 5 \\ 2 & 0 \end{vmatrix}\mathbf{k}$$

$$= 25\,\mathbf{i} + 2\,\mathbf{j} - 10\,\mathbf{k}$$

$$|\mathbf{u} \cdot (\mathbf{v} \times \mathbf{w})| = |3(25) + 0(2) + 0(-10)| = 75.$$

35. A child applies the brakes on a bicycle by applying a downward force of 20 pounds on the pedal when the crank makes a 40° angle with the horizontal (see figure). Find the torque at P if the crank is 6 inches in length.

Solution:

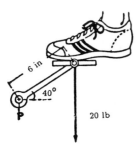

20 lb

If you represent the 20 pound force as $\mathbf{F} = -20\mathbf{k}$ and the lever of length one-half foot as

$$\mathbf{v} = \frac{1}{2}(\cos 40°\mathbf{j} + \sin 40°\mathbf{k}),$$

then the moment of $\mathbf{F}$ about P is given by

$$\mathbf{M} = \mathbf{V} \times \mathbf{F} = \begin{vmatrix} \mathbf{i} & \mathbf{j} & \mathbf{k} \\ 0 & \dfrac{1}{2}\cos 40° & \dfrac{1}{2}\sin 40° \\ 0 & 0 & -20 \end{vmatrix} = -10\cos 40°\,\mathbf{i}.$$

The torque is the magnitude of this moment. Thus,

$$\text{torque} = \|\mathbf{M}\| = 10\cos 40° \approx 7.66 \text{ foot-pounds.}$$

11.5   Lines and Planes in Space

9. Find a set of (a) parametric equation and (b) symmetric equations of the line passing through the points $(5, -3, -2)$ and $(-\frac{2}{3}, \frac{2}{3}, 1)$.

Solution:

(a) If $A = (-\frac{2}{3}, \frac{2}{3}, 1)$ and $B = (5, -3, -2)$, then a direction vector for the line passing through these points is given by

$$\overrightarrow{AB} = \frac{17}{3}\mathbf{i} - \frac{11}{3}\mathbf{j} - 3\mathbf{k} = \frac{1}{3}(17\mathbf{i} - 11\mathbf{j} - 9\mathbf{k})$$

and a set of direction numbers for the line is $a = 17$, $b = -11$, and $c = -9$. Using the form

$$x = x_1 + at, \quad y = y_1 + bt, \quad z = z_1 + ct$$

with $(x_1, y_1, z_1) = (5, -3, -2)$, a set of parametric equation for the line is

$$x = 5 + 17t, \quad y = -3 - 11t, \quad z = -2 - 9t.$$

(b) Solving for t in each equation of part (a) yields

$$t = \frac{x - 5}{17} = \frac{y + 3}{-11} = \frac{z + 2}{-9}.$$

Consequently, the symmetric form

$$\frac{x - 5}{17} = \frac{y + 3}{-11} = \frac{z + 2}{-9}$$

15. Determine if the lines

$$x = 4t + 2 \qquad x = 2s + 2$$
$$y = 3 \qquad y = 2s + 3$$
$$z = -t + 1 \qquad z = s + 1$$

intersect and, if so, find the point of intersection and the cosine of the angle of intersection.

Solution:

At the point of intersection, the coordinates for one line equal the corresponding coordinates for the other line. Thus, you have the three equations.

$$(1) \qquad\qquad 4t + 2 = 2s + 2$$
$$(2) \qquad\qquad 3 = 2s + 3$$
$$(3) \qquad\qquad -t + 1 = s + 1$$

From equation (2) it follows that $s = 0$ and consequently, from equation (3), $t = 0$. Letting $s = t = 0$, equation (1) is satisfied and you can conclude that the two lines intersect. Substituting zero for s or for t, you obtain the point $(2, 3, 1)$.

To find the cosine of the angle of intersection, consider the vectors

$$\mathbf{u} = 4\mathbf{i} - \mathbf{k} \quad \text{and} \quad \mathbf{v} = 2\mathbf{i} + 2\mathbf{j} + \mathbf{k}$$

that have the respective directions of the two given lines. Therefore,

$$\cos \theta = \frac{|\mathbf{u} \cdot \mathbf{v}|}{\|\mathbf{u}\| \, \|\mathbf{v}\|} = \frac{8 - 1}{\sqrt{17}\sqrt{9}} = \frac{7}{3\sqrt{17}} = \frac{7\sqrt{17}}{51}.$$

29. Find an equation of the plane through the points $(0, 0, 0)$, $(1, 2, 3)$, and $(-2, 3, 3)$.

Solution:

To use the form

$$a(x - x_1) + b(y - y_1) + c(z - z_1) = 0$$

you need to know a point in the plane and a vector $\mathbf{n}$ that is normal to the plane. To obtain a normal vector, use the cross product of the vectors $\mathbf{v}_1$ and $\mathbf{v}_2$ from the point $(0, 0, 0)$ to $(1, 2, 3)$ and to $(-2, 3, 3)$, respectively. We have

$$\mathbf{v}_1 = \mathbf{i} + 2\mathbf{j} + 3\mathbf{k} \quad \text{and} \quad \mathbf{v}_2 = -2\mathbf{i} + 3\mathbf{j} + 3\mathbf{k}.$$

Thus, the vector

$$\mathbf{n} = \mathbf{v}_1 \times \mathbf{v}_2 = \begin{vmatrix} \mathbf{i} & \mathbf{j} & \mathbf{k} \\ 1 & 2 & 3 \\ -2 & 3 & 3 \end{vmatrix} = -3\mathbf{i} - 9\mathbf{j} + 7\mathbf{k}$$

is normal to the given plane. Using the direction numbers from $\mathbf{n}$ and the point $(0, 0, 0)$ in the plane, you have

$$-3(x - 0) - 9(y - 0) + 7(z - 0) = 0$$
$$3x + 9y - 7z = 0.$$

35. Find an equation of the plane determined by the two intersecting lines

$$\frac{x - 1}{-2} = y - 4 = z \quad \text{and} \quad \frac{x - 2}{-3} = \frac{y - 1}{4} = \frac{z - 2}{-1}.$$

Solution:

Writing the equations of the lines in parametric form, you have

$$
\begin{array}{ll}
x = 1 - 2t & x = 2 - 3s \\
y = 4 + t & y = 1 + 4s \\
z = t & z = 2 - s
\end{array}
$$

To find the point of intersection of the lines, we equate the expressions for the respective values of x, y, and z, and solve the resulting equations. The solution is $t = s = 1$, and the point of intersection is $(-1, 5, 1)$. Since the direction vectors of the given lines are

$$\mathbf{v}_1 = -2\mathbf{i} + \mathbf{j} + \mathbf{k} \quad \text{and} \quad \mathbf{v}_2 = -3\mathbf{i} + 4\mathbf{j} - \mathbf{k},$$

the vector $\mathbf{n}$ normal to the plane is

$$\mathbf{n} = \mathbf{v}_1 \times \mathbf{v}_2 = \begin{vmatrix} \mathbf{i} & \mathbf{j} & \mathbf{k} \\ -2 & 1 & 1 \\ -3 & 4 & -1 \end{vmatrix} = -5(\mathbf{i} + \mathbf{j} + \mathbf{k}).$$

Therefore, the equation of the plane is

$$1(x + 1) + 1(y - 5) + 1(z - 1) = 0$$
$$x + y + z = 5.$$

37. Find an equation of the plane through the points $(2, 2, 1)$ and $(-1, 1, -1)$ that is perpendicular to the plane $2x - 3y + z = 3$.

Solution:

Let $\mathbf{v}$ be the vector from $(-1, 1, -1)$ to $(2, 2, 1)$, and let $\mathbf{n}$ be a vector normal to the plane $2x - 3y + z = 3$. Then $\mathbf{v}$ and $\mathbf{n}$ both lie in the required plane, where

$$\mathbf{v} = 3\mathbf{i} + \mathbf{j} + 2\mathbf{k} \quad \text{and} \quad \mathbf{n} = 2\mathbf{i} - 3\mathbf{j} + \mathbf{k}.$$

The vector

$$\mathbf{v} \times \mathbf{n} = \begin{vmatrix} \mathbf{i} & \mathbf{j} & \mathbf{k} \\ 3 & 1 & 2 \\ 2 & -3 & 1 \end{vmatrix} = 7\mathbf{i} + \mathbf{j} - 11\mathbf{k}$$

is normal to the required plane. Finally, since the point $(2, 2, 1)$ lies in the plane, an equation for the plane is

$$7(x - 2) + 1(y - 2) - 11(z - 1) = 0$$
$$7x + y - 11z = 5.$$

43. Determine whether the planes $x - 3y + 6z = 4$ and $5x + y - z = 4$ are parallel, orthogonal, or neither. If they are neither parallel or orthogonal, find the angle of intersection.

Solution:

Vectors normal to the planes $x - 3y + 6z = 4$ and $5x + y - z = 4$ are $\mathbf{n}_1 = \langle 1, -3, 6 \rangle$ and $\mathbf{n}_2 = \langle 5, 1, -1 \rangle$, respectively. Since there is no scalar c such that $\mathbf{n}_1 = c\,\mathbf{n}_2$, the vectors, and thus the planes, are not parallel. Also, since $\mathbf{n}_1 \cdot \mathbf{n}_2 \neq 0$, the vectors, and thus the planes, are not orthogonal. The cosine of the angle θ between the two planes is given by

$$\cos \theta = \frac{|\mathbf{n}_1 \cdot \mathbf{n}_2|}{\|\mathbf{n}_1\| \|\mathbf{n}_2\|} = \frac{|5 - 3 - 6|}{\sqrt{46}\sqrt{27}} = \frac{4\sqrt{138}}{414}.$$

Therefore,

$$\theta = \arccos \left(\frac{4\sqrt{138}}{414} \right) \approx 83.5°.$$

57. Find a set of parametric equations for the line of intersection of the planes $3x + 2y - z = 7$ and $x - 4y + 2z = 0$.

Solution:

Let $\mathbf{n}_1 = 3\mathbf{i} + 2\mathbf{j} - \mathbf{k}$ and $\mathbf{n}_2 = \mathbf{i} - 4\mathbf{j} + 2\mathbf{k}$ be the normal vectors to the respective planes. The line of intersection of the two planes will have the same direction as the vector $\mathbf{n}_1 \times \mathbf{n}_2$. Since

$$\mathbf{n}_1 \times \mathbf{n}_2 = \begin{vmatrix} \mathbf{i} & \mathbf{j} & \mathbf{k} \\ 3 & 2 & -1 \\ 1 & -4 & 2 \end{vmatrix}$$

$$= 0\mathbf{i} - 7\mathbf{j} - 14\mathbf{k} = -7(0\mathbf{i} + \mathbf{j} + 2\mathbf{k}),$$

a set of direction numbers for the line of intersection is 0, 1, and 2. By solving the equations for the two planes simultaneously, you can find points on the line of intersection.

$$3x + 2y - z = 7 \implies 6x + 4y - 2z = 14$$

$$x - 4y + 2z = 0 \implies x - 4y + 2z = 0$$

$$7x = 14 \quad \text{or} \quad x = 2$$

By substituting 2 for x, you obtain the equation $2y - z = 1$. If you let $y = 1$, then $z = 1$. Thus, $(2, 1, 1)$ lies on the line of intersection, and you can conclude that a set of parametric equations for the line of intersection is

$$x = 2, \qquad y = 1 + t, \qquad z = 1 + 2t.$$

59. Find the point of intersection (if any) of the line

$$\frac{x - \frac{1}{2}}{1} = \frac{y + \frac{3}{2}}{-1} = \frac{z + 1}{2}$$

and the plane $2x - 2y + z = 12$.

Solution:

The parametric equations for the line are

$$x = \frac{1}{2} + t, \quad y = \frac{-3}{2} - t, \quad z = -1 + 2t.$$

If the line intersects the plane, then the expressions for x, y, and z from the line must satisfy the equation of the plane. Thus,

$$2x - 2y + z = 12$$

$$2\left(\frac{1}{2} + t\right) - 2\left(\frac{-3}{2} - t\right) + (-1 + 2t) = 12$$

$$6t + 3 = 12$$

$$6t = 9 \quad \Longrightarrow \quad t = \frac{3}{2}$$

and you can conclude that the point of intersection occurs when $t = \frac{3}{2}$. This yields the point $(2, -3, 2)$.

69. Find the distance between the two skew lines

$$\frac{x}{1} = \frac{y}{2} = \frac{z}{3} \quad \text{and} \quad \frac{x - 1}{-1} = \frac{y - 4}{1} = \frac{z + 1}{1}.$$

Solution:

Let $\mathbf{n}_1 = \mathbf{i} + 2\mathbf{j} + 3\mathbf{k}$ and $\mathbf{n}_2 = -\mathbf{i} + \mathbf{j} + \mathbf{k}$ be vectors parallel to the given lines, respectively. Then the vector $\mathbf{n}_1 \times \mathbf{n}_2$ will be orthogonal to both lines. By choosing an arbitrary vector $\mathbf{v}$ from one line to the other, you can project $\mathbf{v}$ onto $\mathbf{n}_1 \times \mathbf{n}_2$ to find the distance between the two lines. Since $(0, 0, 0)$ lies on the first line and $(0, 5, 0)$ lies on the second line, you have $\mathbf{v} = 5\mathbf{j}$. Now the absolute value of the component of $\mathbf{v}$ in the direction of $\mathbf{n}_1 \times \mathbf{n}_2$ will be the actual distance between the lines. Since

$$\mathbf{n}_1 \times \mathbf{n}_2 = \begin{vmatrix} \mathbf{i} & \mathbf{j} & \mathbf{k} \\ 1 & 2 & 3 \\ -1 & 1 & 1 \end{vmatrix} = -\mathbf{i} - 4\mathbf{j} + 3\mathbf{k},$$

the distance between the lines is

$$|\text{component of } \mathbf{v} \text{ in direction of } \mathbf{n}_1 \times \mathbf{n}_2| = \frac{|\mathbf{v} \cdot (\mathbf{n}_1 \times \mathbf{n}_2)|}{\|\mathbf{n}_1 \times \mathbf{n}_2\|}$$

$$= \left| \frac{-20}{\sqrt{1 + 16 + 9}} \right|$$

$$= \frac{10\sqrt{26}}{13} \approx 3.92.$$

11.6   Surfaces in Space

11. Describe and sketch the surface defined by the equation $x^2 - y = 0$.

Solution:

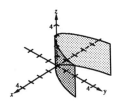

Since the z-coordinate is missing in the equation, the surface is a cylindrical surface with rulings parallel to the z-axis. The generating curve is the parabola $y = x^2$ and the surface is called a *parabolic cylinder*. (See accompanying figure.)

21. Identify and sketch the quadric surface given by $16x^2 - y^2 + 16z^2 = 4$.

Solution:

The equation has the form

$$\frac{x^2}{1/4} - \frac{y^2}{4} + \frac{z^2}{1/4} = 1$$

which is the form for a **hyperboloid of one sheet.** The axis of the hyperboloid is the $y-$axis. The xz-trace $(y = 0)$ is the circle

$$\frac{x^2}{1/4} + \frac{z^2}{1/4} = 1$$

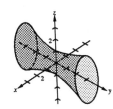

and the xy and yz traces are the hyperbolas

$$\frac{x^2}{1/4} - \frac{y^2}{4} = 1 \quad \text{and} \quad \frac{z^2}{1/4} - \frac{y^2}{4} = 1$$

(See the accompanying figure.)

25. Identify and sketch the quadric surface given by $x^2 - y^2 + z = 0$.

Solution:

The given equation can be written as

$$z = -\frac{x^2}{1} + \frac{y^2}{1}$$

which is the form for a **hyperbolic paraboloid.**

xy-trace $(z = 0)$:	$y = \pm x$	(intersecting lines)
xz-trace $(y = 0)$:	$z = -x^2$	(parabola opening upward)
yz-trace $(x = 0)$:	$z = 13y^2$	(parabola opening downward)

Traces parallel to the xy-coordinate plane are hyperbolas. For example, when $z = 1$, you have $y^2 - x^2 = 1$.

45. Find an equation for the surface of revolution generated by revolving the graph of $z = 2y$ in the yz-plane about the z-axis.

Solution:

Since you are revolving the curve about the $z-$axis, the equation for the surface of revolution has the form $x^2 + y^2 = [a(z)]^2$ where $y = a(z) = z/2$. Therefore, the equation is

$$x^2 + y^2 = \frac{z^2}{4} \quad \text{or} \quad 4x^2 + 4y^2 = z^2.$$

11.7   Cylindrical and Spherical Coordinates

5. Convert the point $(2, -2, -4)$ from rectangular to cylindrical coordinates.

Solution:

Since $x = 2$, $y = -2$, and $z = -4$, you have

$$r = \pm\sqrt{x^2 + y^2} = \pm\sqrt{4+4} = \pm 2\sqrt{2}$$

$$\theta = \arctan\frac{y}{x} = \arctan\frac{-2}{2} = \arctan(-1) = -\frac{\pi}{4} \quad \text{or} \quad \frac{3\pi}{4}$$

$$z = -4.$$

Therefore, one set of corresponding cylindrical coordinates for the given point is $(r, \theta, z) = (2\sqrt{2}, -\pi/4, -4)$. (See the accompanying figure.) Another set of cylindrical coordinates for the given point is $(-2\sqrt{2}, \ 3\pi/4, \ -4)$.

11. Convert the point $(4, \ 7\pi/6, \ 3)$ from cylindrical to rectangular coordinates.

Solution:

Since $\left(4, \dfrac{7\pi}{6}, 3\right) = (r, \theta, z)$, you have

$$x = r\cos\theta = 4\cos\frac{7\pi}{6} = -2\sqrt{3}$$
$$y = r\sin\theta = 4\sin\frac{7\pi}{6} = -2$$
$$z = 3.$$

Therefore, in rectangular coordinates, the point is $(-2\sqrt{3}, -2, 3)$.

17. Find an equation in rectangular coordinates for the equation $r = 2\sin\theta$ in cylindrical coordinates, and sketch its graph.

Solution:

Since $x^2 + y^2 = r^2$ and $y = r\sin\theta$, you have

$$r = 2\sin\theta$$
$$r^2 = 2r\sin\theta$$
$$x^2 + y^2 = 2y$$
$$x^2 + y^2 - 2y + 1 = 1$$
$$x^2 + (y - 1)^2 = 1$$

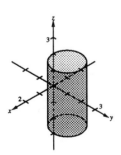

Therefore, the graph of the equation is a circular cylinder with rulings parallel to the z-axis.

402   Chapter 11   Vectors and the geometry of space

23. Convert the point $(-2, 2\sqrt{3}, 4)$ from rectangular to spherical coordinates.

Solution:

Since $(-2, 2\sqrt{3}, 4) = (x, y, z)$, you have

$$\rho = \sqrt{x^2 + y^2 + z^2} = \sqrt{4 + 12 + 16} = 4\sqrt{2}$$

$$\theta = \arctan \frac{y}{x} = \arctan \frac{2\sqrt{3}}{-2} = \frac{2\pi}{3}$$

$$\phi = \arccos \frac{z}{\rho} = \arccos \frac{4}{4\sqrt{2}} = \frac{\pi}{4}.$$

Therefore, in spherical coordinates, the point is $\left(4\sqrt{2}, \frac{2\pi}{3}, \frac{\pi}{4}\right)$.

27. Convert the point $(4, \pi/6, \pi/4)$ from spherical to rectangular coordinates.

Solution:

Since $\left(4, \frac{\pi}{6}, \frac{\pi}{4}\right) = (\rho, \theta, \phi)$, you have

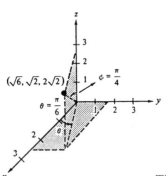

$$x = \rho \sin \phi \cos \theta$$
$$= 4 \sin \frac{\pi}{4} \cos \frac{\pi}{6} = 4\left(\frac{\sqrt{2}}{2}\right)\left(\frac{\sqrt{3}}{2}\right) = \sqrt{6}$$

$$y = \rho \sin \phi \sin \theta$$
$$= 4 \sin \frac{\pi}{4} \sin \frac{\pi}{6} = 4\left(\frac{\sqrt{2}}{2}\right)\left(\frac{1}{2}\right) = \sqrt{2}$$

$$z = \rho \cos \phi = 4 \cos \frac{\pi}{4} = 4\left(\frac{\sqrt{2}}{2}\right) = 2\sqrt{2}.$$

Therefore, in rectangular coordinates, the given point is $(\sqrt{6}, \sqrt{2}, 2\sqrt{2})$.

37. Find an equation in rectangular coordinates for $\rho = 4\cos\phi$.

Solution:

Since

$$\cos\phi = \frac{z}{\sqrt{x^2 + y^2 + z^2}},$$

write

$$\frac{\rho}{4} = \cos\phi = \frac{z}{\sqrt{x^2 + y^2 + z^2}}.$$

Furthermore, since

$$\rho = \sqrt{x^2 + y^2 + z^2},$$

it follows that

$$\frac{\sqrt{x^2 + y^2 + z^2}}{4} = \frac{z}{\sqrt{x^2 + y^2 + z^2}}$$

$$x^2 + y^2 + z^2 = 4z$$

$$x^2 + y^2 + z^2 - 4z = 0.$$

Completing the square on variable z, yields the equation

$$x^2 + y^2 + (z - 2)^2 = 4$$

which represents a sphere of radius 2, centered at $(0, 0, 2)$.

43. Convert the point $(4, -\pi/6, 6)$ from cylindrical to spherical coordinates.

Solution:

Since $\left(4, -\dfrac{\pi}{6}, 6\right) = (r, \theta, z)$, you have

$$\rho^2 = x^2 + y^2 + z^2 = r^2 + z^2 = 16 + 36 = 52$$

or

$$\rho = 2\sqrt{13} > 0.$$

Furthermore,

$$\cos\phi = \frac{z}{\sqrt{x^2 + y^2 + z^2}} = \frac{z}{\sqrt{r^2 + z^2}} = \frac{6}{\sqrt{52}} = \frac{3}{\sqrt{13}}$$

or

$$\phi = \arccos\frac{3}{\sqrt{13}}.$$

Therefore, in spherical coordinates, the given point is

$$\left(2\sqrt{13}, \ -\frac{\pi}{6}, \ \arccos\frac{3}{\sqrt{13}}\right).$$

69. Find an equation in (a) cylindrical coordinates and (b) spherical coordinates for the equation $x^2 + y^2 = 4y$.

Solution:

(a) Since $x^2 + y^2 = r^2$ and $y = 4\sin\theta$, the equation in cylindrical coordinates is

$$x^2 + y^2 = 4y$$
$$r^2 = 4r\sin\theta \implies r = 4\sin\theta.$$

(b) In spherical coordinates, since $x^2 + y^2 = r^2 = \rho^2 \sin^2\phi$ and $y = \rho\sin\phi\sin\theta$, you have

$$x^2 + y^2 = 4y$$
$$\rho^2 \sin^2\phi = 4\rho\sin\phi\sin\theta$$
$$\rho\sin\phi = 4\sin\theta$$
$$\rho = 4\sin\theta\csc\phi.$$

77. Sketch the solid that has the following description in spherical coordinates: $0 \le \theta \le 2\pi$, $0 \le \phi \le \pi/6$, and $0 \le \rho \le a\sec\phi$.

Solution:

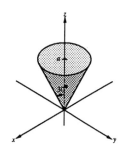

From the constraint $0 \le \rho \le a\sec\phi$, you have

$$0 \le \rho \le a\left(\frac{1}{\cos\phi}\right)$$
$$0 \le \rho\cos\phi \le a$$
$$0 \le z \le a$$

Review Exercises for Chapter 11

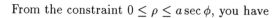

3. Given the points $P = (5, 0, 0)$, $Q = (4, 4, 0)$ and $R = (2, 0, 6)$, let $\mathbf{u} = \overrightarrow{PQ}$ and $\mathbf{v} = \overrightarrow{PR}$. Find (a) the component forms of $\mathbf{u}$ and $\mathbf{v}$, (b) $\mathbf{u} \cdot \mathbf{v}$, (c) $\mathbf{u} \times \mathbf{v}$, (d) an equation of the plane containing P, Q, and R, and (e) a set of parametric equations of the line through P and Q.

Solution:

(a) $\mathbf{u} = \overrightarrow{PQ} = \langle 4 - 5,\ 4 - 0,\ 0 - 0 \rangle = \langle -1, 4, 0 \rangle$

   $\mathbf{v} = \overrightarrow{PR} = \langle 2 - 5,\ 0 - 0,\ 6 - 0 \rangle = \langle -3, 0, 6 \rangle$

(b) $\mathbf{u} \cdot \mathbf{v} = (-1)(-3) + (4)(0) + (0)(6) = 3$

(c)

$$\mathbf{u} \times \mathbf{v} = \begin{vmatrix} \mathbf{i} & \mathbf{j} & \mathbf{k} \\ -1 & 4 & 0 \\ -3 & 0 & 6 \end{vmatrix} = 24\,\mathbf{i} + 6\,\mathbf{j} + 12\,\mathbf{k}$$

$$= 6(4\,\mathbf{i} + \mathbf{j} + 2\,\mathbf{k})$$

(d) A vector normal to the plane is

$$\frac{1}{6}(\mathbf{u} \times \mathbf{v}) = 4\,\mathbf{i} + \mathbf{j} + 2\,\mathbf{k} \quad \text{See part (c)}$$

Therefore, using $(5,\ 0,\ 0)$ as a point in the plane, an equation of the plane is

$$4(x - 5) + 1(y - 0) + 2(z - 0) = 0$$
$$4x + y + 2z = 20$$

(e) Since the direction of the line is determined by $\mathbf{u} = \langle -1, 4, 0 \rangle$ [see part (a)], a set of parametric equations of the line passing through the point $(4, 4, 0)$ is

$$x = 4 - t, \qquad y = 4 + 4t, \qquad z = 0$$

Note that when $t = -1$, $x = 5$, $y = 0$, and $z = 0$. Thus the line passes through the point P.

27. In a manufacturing process an electric hoist lifts 500 pound ingots (see figure). Find the shortest cable connecting points P, O, and Q that can be used if the tension in the cable cannot exceed 750 pounds. (Assume that O is at the midpoint of the cable.)

Solution:

From the accompanying figure, observe that a scalar multiple of the vector $\overrightarrow{OQ}$ is

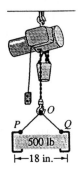

$$\overrightarrow{cOQ} = c(9\,\mathbf{i} - y\,\mathbf{j}).$$

Since the tension in the cable cannot exceed 750 pounds, you have

$$\|\overrightarrow{cOQ}\| = c\sqrt{81 + y^2} \le 750.$$

Since the magnitude of the vertical component of $\overrightarrow{cOQ}$ must equal one-half the weight of the ingot, you have

$$cy = 250 \quad \text{or} \quad c = \frac{250}{y}.$$

Substituting this expression for c into the inequality for the tension yields

$$\frac{250}{y}\sqrt{81 + y^2} \le 750$$

$$\sqrt{81 + y^2} \le 3y$$

$$81 + y^2 \le 9y^2$$

$$81 \le 8y^2 \quad \Longrightarrow \quad y^2 \ge \frac{81}{8}.$$

Therefore, the minimum length of the cable is

$$l = 2\sqrt{81 + \frac{81}{8}} = \frac{27\sqrt{2}}{2} \approx 19.1 \text{ in.}$$

29. Find (a) a set of parametric equations and (b) a set of symmetric equations for the line perpendicular to the xz-coordinate plane passing through the point $(1, 2, 3)$.

Solution:

(a) Any line perpendicular to the xz-plane must have the direction of the vector $\mathbf{v} = 0\,\mathbf{i} + \mathbf{j} + 0\,\mathbf{k}$. Thus, direction numbers for the required line are 0, 1, 0. Since the line passes through the point $(1, 2, 3)$, the parametric equations are

$$x = 1 + (0)t = 1$$
$$y = 2 + t$$
$$z = 3 + (0)t = 3.$$

(b) Since two of the direction numbers are zero, there is no symmetric form.

35. Find an equation of the plane containing the lines

$$\frac{x-1}{-2} = y = z+1 \quad \text{and} \quad \frac{x+1}{-2} = y-1 = x-2.$$

Solution:

We first observe that the lines are parallel since they have the same direction numbers, -2, 1, 1. Therefore, a vector parallel to the plane is $\mathbf{u} = \langle -2, 1, 1 \rangle$. A point on the first line is $(1, 0, -1)$ and a point on the second line is $(-1, 1, 2)$. The vector $\mathbf{v} = \langle 2, -1, -3 \rangle$ connecting these two points is also parallel to the plane. Thus, a normal to the plane is

$$\mathbf{u} \times \mathbf{v} = \begin{vmatrix} \mathbf{i} & \mathbf{j} & \mathbf{k} \\ -2 & 1 & 1 \\ 2 & -1 & -3 \end{vmatrix} = -2(\mathbf{i} + 2\mathbf{j}).$$

Therefore, an equation of the plane is

$$1(x-1) + 2(y-0) + 0(z+1) = 0$$
$$x + 2y = 1.$$

49. Sketch a graph of the surface $16x^2 + 16y^2 - 9z^2 = 0$.

Solution:

The given equation has the form

$$\frac{x^2}{9} + \frac{y^2}{9} - \frac{z^2}{16} = 0$$

which is the form for a cone. The axis of the cone is the z-axis. The xz-trace ($y = 0$) is given by

$$\frac{x^2}{9} - \frac{z^2}{16} = 0 \quad \text{or} \quad z = \pm\frac{4}{3}x.$$

The yz-trace ($x = 0$) is given by

$$\frac{y^2}{9} - \frac{z^2}{16} = 0 \quad \text{or} \quad z = \pm\frac{4}{3}y.$$

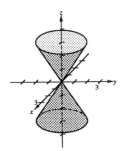

Traces parallel to the xy-plane are circles. For example, when $z = 4$ you have

$$16x^2 + 16y^2 - 9(16) = 0$$
$$x^2 + y^2 = 9.$$

63. Write the equation $r^2(\cos^2\theta - \sin^2\theta) + z^2 = 1$ in rectangular coordinates.

Solution:

We rewrite the equation as

$$r^2\cos^2\theta - r^2\sin^2\theta + z^2 = 1$$

Since $x = r\cos\theta$ and $y = r\sin\theta$, you have

$$x^2 - y^2 + z^2 = 1$$

as the rectangular equation.

12 VECTOR-VALUED FUNCTIONS

12.1 Vector-Valued Functions

---

3. Find the domain and the component functions of the vector-valued function $\mathbf{r}(t) = \ln t\, \mathbf{i} - e^t\, \mathbf{j} - t\, \mathbf{k}$.

Solution:

The component functions of the vector-valued function

$$\mathbf{r}(t) = f(t)\,\mathbf{i} + g(t)\,\mathbf{j} + h(t)\,\mathbf{k}$$

are the real-valued functions f, g, and h. Given the vector-valued function $\mathbf{r}(t) = \ln t\, \mathbf{i} - e^t\, \mathbf{j} - t\, \mathbf{k}$ the component functions and their domains are:

$$\begin{aligned}
f(t) &= \ln t, & 0 < t < \infty \\
g(t) &= -e^t, & -\infty < t < \infty \\
h(t) &= -t, & -\infty < t < \infty
\end{aligned}$$

The intersection of the domains of f, g, and h, is the interval $(0, \infty)$, the domain of $\mathbf{r}$.

21. Sketch the curve represented by the vector-valued function

$$\mathbf{r}(t) = 2\cos t\,\mathbf{i} + 2\sin t\,\mathbf{j} + t\,\mathbf{k}$$

and give the orientation of the curve.

Solution:

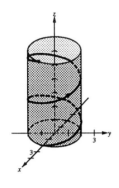

From the first two parametric equations $x = 2\cos t$ and $y = 2\sin t$, you obtain

$$x^2 + y^2 = 4\cos^2 t + 4\sin^2 t = 4.$$

This means that the curve lies on a right circular cylinder of radius 2 centered about the z-axis. To locate the curve on this cylinder, use the third parametric equation $z = t$. Thus the graph of the vector-valued function spirals counterclockwise up the cylinder to produce a circular helix.

39. Sketch the curve represented by the intersection of the surfaces $x^2 + y^2 + z^2 = 4$ and $x + z = 2$. Find a vector-valued function for the curve using the parameter $x = 1 + \sin t$.

Solution:

The equation $x^2 + y^2 + z^2 = 4$ represents a sphere of radius 2 centered at the origin. The equation $x + z = 2$ represents a plane. We let $x = 1 + \sin t$, then

$$z = 2 - x = 1 - \sin t.$$

Substituting into the equation of the sphere, you have

$$x^2 + y^2 + z^2 = 4$$
$$(1 + \sin t)^2 + y^2 + (1 - \sin t)^2 = 4$$
$$2 + 2\sin^2 t + y^2 = 4$$
$$y^2 = 2 - 2\sin^2 t$$
$$y^2 = 2\cos^2 t$$
$$y = \pm\sqrt{2}\cos t.$$

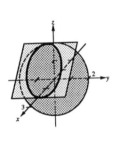

Therefore $x = 1 + \sin t$, $y = \pm\sqrt{2}\cos t$, $z = 1 - \sin t$ and the two vector-valued functions are

$$\mathbf{r}(t) = (1 + \sin t)\,\mathbf{i} + (\sqrt{2}\cos t)\,\mathbf{j} + (1 - \sin t)\,\mathbf{k}$$

and

$$\mathbf{r}(t) = (1 + \sin t)\,\mathbf{i} - (\sqrt{2}\cos t)\,\mathbf{j} + (1 - \sin t)\,\mathbf{k}.$$

t	$-\dfrac{\pi}{2}$	$-\dfrac{\pi}{6}$	0	$\dfrac{\pi}{6}$	$\dfrac{\pi}{2}$
x	0	$\dfrac{1}{2}$	1	$\dfrac{3}{2}$	2
y	0	$\pm\dfrac{\sqrt{6}}{2}$	$\pm\sqrt{2}$	$\pm\dfrac{\sqrt{6}}{2}$	0
z	2	$\dfrac{3}{2}$	1	$\dfrac{1}{2}$	0

41. Sketch the curve (first octant portion) represented by the intersection of the surfaces $x^2 + z^2 = 4$ and $y^2 + z^2 = 4$. Find a vector-valued function for the curve using the parameter $x = t$.

Solution:

The equation $x^2 + z^2 = 4$ represents a cylinder of radius 2 with its axis along the y-axis, while the equation $y^2 + z^2 = 4$ represents a cylinder of radius 2 with its axis along the x-axis. Thus the curve of intersection lies along both cylinders, as shown in the accompanying sketch. Two points on the curve are $(2, 2, 0)$ and $(0, 0, 2)$. To find a set of parametric equations for the curve of intersection, subtract the second equation from the first to obtain

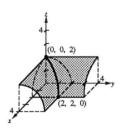

$$
\begin{aligned}
x^2 + z^2 &= 4 \\
-(y^2 + z^2 &= 4) \\
\hline
x^2 - y^2 &= 0 \quad \text{or} \quad y = \pm x.
\end{aligned}
$$

Therefore, in the first octant, if you let $x = t \; (t > 0)$, then parametric equations for the curve are

$$x = t, \quad y = t, \quad z = \sqrt{4 - t^2}$$

and the vector-valued function is

$$\mathbf{r}(t) = t\,\mathbf{i} + t\,\mathbf{j} + \sqrt{4 - t^2}\,\mathbf{k}.$$

45. Evaluate the limit

$$\lim_{t \to 0} \left(t^2 \mathbf{i} + 3t \mathbf{j} + \frac{1 - \cos t}{t} \mathbf{k} \right).$$

Solution:

$$\lim_{t \to 0} \left(t^2 \mathbf{i} + 3t \mathbf{j} + \frac{1 - \cos t}{t} \mathbf{k} \right)$$

$$= \left[\lim_{t \to 0} t^2 \right] \mathbf{i} + \left[\lim_{t \to 0} 3t \right] \mathbf{j} + \left[\lim_{t \to 0} \frac{1 - \cos t}{t} \right] \mathbf{k}$$

$$= 0 \mathbf{i} + 0 \mathbf{j} + 0 \mathbf{k} = \mathbf{0}$$

12.2   Differentiation and Integration of Vector-Valued Functions

3. (a) Sketch the graph of the vector-valued function
$\mathbf{r}(t) = \cos t \, \mathbf{i} + \sin t \, \mathbf{j}$, and (b) sketch the vectors $\mathbf{r}(\pi/2)$ and
$\mathbf{r}'(\pi/2)$. Position the vectors so that the initial point of $\mathbf{r}(\pi/2)$
is at the origin and the initial point of $\mathbf{r}'(\pi/2)$ is at the terminal
point of $\mathbf{r}(\pi/2)$.

Solution:

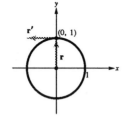

(a) Eliminating the parameter from the parametric equations
$x = \cos t$ and $y = \sin t$, you have

$$x^2 + y^2 = \cos^2 t + \sin^2 t = 1.$$

Therefore, the vector-valued function represents a circle
of radius 1 center at the origin. (See the accompanying
figure.)

(b) $\mathbf{r}(t) = \cos t \, \mathbf{i} + \sin t \, \mathbf{j}$ $\implies$ $\mathbf{r}\left(\dfrac{\pi}{2} \right) = \mathbf{j}$

$\mathbf{r}'(t) = -\sin t \, \mathbf{i} + \cos t \, \mathbf{j}$ $\implies$ $\mathbf{r}'\left(\dfrac{\pi}{2} \right) = -\mathbf{i}$

17. Find $\mathbf{r}'(t)$ for $\mathbf{r}(t) = \langle t \sin t,\ t \cos t,\ t \rangle$.

Solution:

$$\mathbf{r}(t) = \langle t \sin t,\ t \cos t,\ t \rangle$$

$$\mathbf{r}'(t) = \left\langle \frac{d}{dt}[t \sin t],\ \frac{d}{dt}[t \cos t],\ \frac{d}{dt}[t] \right\rangle$$

$$= \langle t \cos t + \sin t,\ -t \sin t + \cos t,\ 1 \rangle$$

19. For the vector-valued functions $\mathbf{r}(t) = t\,\mathbf{i} + 3t\,\mathbf{j} + t^2\,\mathbf{k}$ and $\mathbf{u}(t) = 4t\,\mathbf{i} + t^2\,\mathbf{j} + t^3\,\mathbf{k}$ find the following:
(a) $\mathbf{r}'(t)$ (b) $\mathbf{r}''(t)$
(c) $D_t[\mathbf{r}(t) \cdot \mathbf{u}(t)]$ (d) $D_t[3\,\mathbf{r}(t) - \mathbf{u}(t)]$
(e) $D_t[\mathbf{r}(t) \times \mathbf{u}(t)]$ (f) $D_t[\|\mathbf{r}(t)\|]$

Solution:

(a) Differentiating in a component-by-component basis produces the following.

$$\mathbf{r}'(t) = \mathbf{i} + 3\,\mathbf{j} + 2t\,\mathbf{k}$$

(b) Differentiating $\mathbf{r}'(t)$ in a component by component basis produces the following.

$$\mathbf{r}''(t) = 2\,\mathbf{k}$$

(c)
$$\mathbf{r}(t) \cdot \mathbf{u}(t) = t(4t) + 3t(t^2) + t^2(t^3)$$
$$= 4t^2 + 3t^3 + t^5$$
$$D_t[\mathbf{r}(t) \cdot \mathbf{u}(t)] = 8t + 9t^2 + 5t^4$$

(d)
$$3\,\mathbf{r}(t) - \mathbf{u}(t) = 3(t\,\mathbf{i} + 3t\,\mathbf{j} + t^2\,\mathbf{k}) - (4t\,\mathbf{i} + t^2\,\mathbf{j} + t^3\,\mathbf{k})$$
$$= -t\,\mathbf{i} + (9t - t^2)\,\mathbf{j} + (3t^2 - t^3)\,\mathbf{k}$$
$$D_t[3\,\mathbf{r}(t) - \mathbf{u}(t)] = -\,\mathbf{i} + (9 - 2t)\,\mathbf{j} + (6t - 3t^2)\,\mathbf{k}$$

(e)
$$\mathbf{r}(t) \times \mathbf{u}(t) = \begin{vmatrix} \mathbf{i} & \mathbf{j} & \mathbf{k} \\ t & 3t & t^2 \\ 4t & t^2 & t^3 \end{vmatrix}$$

$$= \begin{vmatrix} 3t & t^2 \\ t^2 & t^3 \end{vmatrix}\mathbf{i} - \begin{vmatrix} t & t^2 \\ 4t & t^3 \end{vmatrix}\mathbf{j} + \begin{vmatrix} t & 3t \\ 4t & t^2 \end{vmatrix}\mathbf{k}$$

$$= 2t^4\,\mathbf{i} - (t^4 - 4t^3)\,\mathbf{j} + (t^3 - 12t^2)\,\mathbf{k}$$

$$D_t[\mathbf{r}(t) \times \mathbf{u}(t)] = 8t^3\,\mathbf{i} + (12t^2 - 4t^3)\,\mathbf{j} + (3t^2 - 24t)\,\mathbf{k}$$

(f)
$$\|\mathbf{r}(t)\| = \sqrt{t^2 + (3t)^2 + (t^2)^2}$$
$$= \sqrt{10t^2 + t^4}$$
$$D_t[\|\mathbf{r}(t)\|] = \frac{1}{2}(10t^2 + t^4)^{-1/2}(20t + 4t^3)$$
$$= \frac{10t + 2t^3}{\sqrt{10t^2 + t^4}} = \frac{10 + 2t^2}{\sqrt{10 + t^2}}$$

29. Evaluate the indefinite integral $\int \left(\frac{1}{t}\mathbf{i} + \mathbf{j} - t^{3/2}\mathbf{k} \right) dt$.

Solution:

$$\int \left(\frac{1}{t}\mathbf{i} + \mathbf{j} - t^{3/2}\mathbf{k} \right) dt$$
$$= \left[\int \frac{1}{t} dt \right] \mathbf{i} + \left[\int dt \right] \mathbf{j} + \left[\int -t^{3/2} dt \right] \mathbf{k}$$
$$= \ln|t|\mathbf{i} + t\mathbf{j} - \frac{2}{5}t^{5/2}\mathbf{k} + \mathbf{C}$$

35. Find the position vector $\mathbf{r}$ if $\mathbf{r}'(t) = 4e^{2t}\mathbf{i} + 3e^t\mathbf{j}$ and $\mathbf{r}(0) = 2\mathbf{i}$.

Solution:

$$\mathbf{r}(t) = \mathbf{i}\int 4e^{2t}\,dt + \mathbf{j}\int 3e^t\,dt = \mathbf{i}[2e^{2t} + c_1] + \mathbf{j}[3e^t + c_2]$$
$$\mathbf{r}(0) = \mathbf{i}[2 + c_1] + \mathbf{j}[3 + c_2] = 2\mathbf{i}$$

Therefore, $2 + c_1 = 2$, or $c_1 = 0$. Also, $3 + c_2 = 0$, or $c_2 = -3$.
Thus,

$$\mathbf{r}(t) = 2e^{2t}\mathbf{i} + (3e^t - 3)\mathbf{j} = 2e^{2t}\mathbf{i} + 3(e^t - 1)\mathbf{j}.$$

41. Evaluate the definite integral $\int_0^1 (8t\mathbf{i} + t\mathbf{j} - \mathbf{k})\,dt$.

Solution:

$$\int_0^1 (8t\mathbf{i} + t\mathbf{j} - \mathbf{k})\,dt = \mathbf{i}\int_0^1 8t\,dt + \mathbf{j}\int_0^1 t\,dt - \mathbf{k}\int_0^1 dt$$
$$= \left[4t^2\right]_0^1 \mathbf{i} + \left[\frac{1}{2}t^2\right]_0^1 \mathbf{j} - \left[t\right]_0^1 \mathbf{k}$$
$$= 4\mathbf{i} + \frac{1}{2}\mathbf{j} - \mathbf{k}$$

12.3 Velocity and Acceleration

3. The motion of an object in the xy-plane is described by the vector- valued function $\mathbf{r}(t) = t^2\,\mathbf{i} + t\,\mathbf{j}$. Sketch a graph of the path and sketch the velocity and acceleration vectors at the point $(4, 2)$.

Solution:

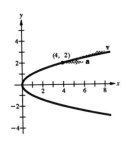

Eliminating the parameter from the parametric equations $x = t^2$ and $y = t$, you obtain the rectangular equation $x = y^2$. Therefore, the object is moving in a parabolic path. (See the accompanying figure.) The velocity is given by

$$\mathbf{v}(t) = \mathbf{r}'(t) = 2t\,\mathbf{i} + \mathbf{j}$$

and the acceleration is given by

$$\mathbf{a}(t) = \mathbf{r}''(t) = 2\,\mathbf{i}.$$

At the point $(4, 2)$, $t = 2$. Thus,

$$\mathbf{v}(2) = 4\,\mathbf{i} + \mathbf{j} \quad \text{and} \quad \mathbf{a}(2) = 2\,\mathbf{i}.$$

15. The position function

$$\mathbf{r}(t) = \langle 4t,\ 3\cos t,\ 3\sin t \rangle$$

describes the path of an object moving in space. Find the velocity, speed, and acceleration of the object

Solution:

Since
$$\mathbf{r}(t) = \langle 4t,\ 3\cos t,\ 3\sin t \rangle,$$

you have

$$\mathbf{v}(t) = \mathbf{r}'(t) = \langle 4,\ -3\sin t,\ 3\cos t \rangle$$
$$\text{speed} = \|\mathbf{v}(t)\| = \sqrt{16 + 9(\sin^2 t + \cos^2 t)} = \sqrt{25} = 5$$

and
$$\mathbf{a}(t) = \mathbf{r}''(t) = \langle 0,\ -3\cos t,\ -3\sin t \rangle.$$

21. Use the acceleration function $\mathbf{a}(t) = t\mathbf{j} + t\mathbf{k}$ to find the velocity and position functions if $\mathbf{v}(1) = 5\mathbf{j}$ and $\mathbf{r}(1) = \mathbf{0}$. Find the position at time $t = 2$.

Solution:

$$\mathbf{v}(t) = \int \mathbf{a}(t)\,dt + \mathbf{C}$$

$$= \int (t\mathbf{j} + t\mathbf{k})\,dt + \mathbf{C}$$

$$= \frac{1}{2}t^2\mathbf{j} + \frac{1}{2}t^2\mathbf{k} + C_1\mathbf{i} + C_2\mathbf{j} + C_3\mathbf{k}$$

$$\mathbf{v}(1) = C_1\mathbf{i} + (\frac{1}{2} + C_2)\mathbf{j} + (\frac{1}{2} + C_3)\mathbf{k} = 5\mathbf{j}$$

Therefore,

$$C_1 = 0$$

$$\frac{1}{2} + C_2 = 5 \qquad \Longrightarrow \qquad C_2 = \frac{9}{2}$$

$$\frac{1}{2} + C_3 = 0 \qquad \Longrightarrow \qquad C_3 = -\frac{1}{2}$$

Thus, the velocity vector is

$$\mathbf{v}(t) = \left(\frac{t^2}{2} + \frac{9}{2}\right)\mathbf{j} + \left(\frac{t^2}{2} - \frac{1}{2}\right)\mathbf{k}$$

$$\mathbf{r}(t) = \int \mathbf{v}(t)\,dt + \mathbf{C}$$

$$= \int \left(\frac{t^2}{2} + \frac{9}{2}\right)\,dt\,\mathbf{j} + \int \left(\frac{t^2}{2} - \frac{1}{2}\right)\,dt\,\mathbf{k} + \mathbf{C}$$

$$= \left(\frac{t^3}{6} + \frac{9}{2}t\right)\mathbf{j} + \left(\frac{t^3}{6} - \frac{1}{2}t\right)\mathbf{k} + C_4\mathbf{i} + C_5\mathbf{j} + C_6\mathbf{k}$$

$$\mathbf{r}(1) = C_4\mathbf{i} + \left(\frac{1}{6} + \frac{9}{2} + C_5\right)\mathbf{j} + \left(\frac{1}{6} - \frac{1}{2} + C_6\right)\mathbf{k}$$

$$= C_4\mathbf{i} + \left(\frac{14}{3} + C_5\right)\mathbf{j} + \left(-\frac{1}{3} + C_6\right)\mathbf{k} = \mathbf{0}$$

This implies that

$$C_4 = 0,\ C_5 = \frac{-14}{3},\ \text{and}\ C_6 = \frac{1}{3}.$$

Thus, the position vector is

$$\mathbf{r}(t) = \left(\frac{t^3}{6} + \frac{9}{2}t - \frac{14}{3}\right)\mathbf{j} + \left(\frac{t^3}{6} - \frac{1}{2}t + \frac{1}{3}\right)\mathbf{k}$$

and

$$\mathbf{r}(2) = \frac{17}{3}\mathbf{j} + \frac{2}{3}\mathbf{k}.$$

25. A baseball, hit 3 feet above the ground, leaves the bat at an angle of 45° and is caught by an outfielder 300 feet from the plate. What was the initial velocity of the ball and how high did it rise if it was caught 3 feet above the ground?

Solution:

Using Theorem 12.3 with $h = 3$, the path of the ball is given by

$$\mathbf{r}(t) = (v_0 \cos 45°)t\,\mathbf{i} + [3 + (v_0 \sin 45° t - 16t^2]\mathbf{j}$$

$$= \left(\frac{tv_0}{\sqrt{2}}\right)\mathbf{i} + \left(3 + \frac{tv_0}{\sqrt{2}} - 16t^2\right)\mathbf{j}$$

We know that the horizontal component is 300 when the vertical component is three. Thus,

$$\frac{tv_0}{\sqrt{2}} = 300 \quad \text{and} \quad 3 + \frac{tv_0}{\sqrt{2}} - 16t^2 = 3.$$

From the first equation we obtain $t = 300\sqrt{2}/v_0$. Substituting this expression into the second equation yields

$$\frac{300\sqrt{2}}{v_0}\left(\frac{v_0}{\sqrt{2}}\right) - 16\left(\frac{300\sqrt{2}}{v_0}\right)^2 = 0$$

$$300 = \frac{16(300^2)(2)}{v_0{}^2}$$

$$v_0{}^2 = 32(300)$$

$$v_0 = \sqrt{9600} = 40\sqrt{6} \approx 97.98 \text{ ft/sec.}$$

The maximum height is reached when the derivative of the vertical component is zero. Thus,

$$y(t) = 3 + \frac{tv_0}{\sqrt{2}} - 16t^2$$

$$= 3 + \frac{t(40\sqrt{6})}{\sqrt{2}} - 16t^2 = 3 + 40\sqrt{3}t - 16t^2$$

$$y'(t) = 40\sqrt{3} - 32t = 0$$

$$t = \frac{40\sqrt{3}}{32} = \frac{5\sqrt{3}}{4}.$$

Finally, the maximum height is

$$y\left(\frac{5\sqrt{3}}{4}\right) = 3 + 40\sqrt{3}\left(\frac{5\sqrt{3}}{4}\right) - 16\left(\frac{5\sqrt{3}}{4}\right)^2$$

$$= 3 + 150 - 75 = 78 \text{ ft.}$$

37. Consider the motion of a point on the circumference of a rolling circle. As the circle rolls, it generates the cycloid

$$\mathbf{r}(t) = b(\omega t - \sin \omega t)\mathbf{i} + b(1 - \cos \omega t)\mathbf{j}.$$

where ω is the constant angular velocity of the circle of radius b. Determine the times when the speed of the point will be (a) zero and (b) maximum.

Solution:

$$\mathbf{r}(t) = b(\omega t - \sin \omega t)\mathbf{i} + b(1 - \cos \omega t)\mathbf{j}$$
$$\mathbf{r}'(t) = \mathbf{v}(t) = b(\omega - \omega \cos \omega t)\mathbf{i} + b\omega \sin \omega t \, \mathbf{j}$$
$$\text{speed} = \|\mathbf{v}(t)\|$$
$$= \sqrt{[b(\omega - \omega \cos \omega t)]^2 + [b\omega \sin \omega t]^2}$$
$$= b\omega \sqrt{1 - 2\cos \omega t + \cos^2 \omega t + \sin^2 \omega t}$$
$$= b\omega \sqrt{2 - 2\cos \omega t} = \sqrt{2} b\omega \sqrt{1 - \cos \omega t}$$

(a) When $\omega t = 0$, 2π, $4\pi, \cdots$, $1 - \cos \omega t = 0$, and therefore, $\|\mathbf{v}(t)\| = 0$. Hence, the speed is zero when the point contacts the surface on which the circle is rolling.

(b) When $\omega t = \pi$, 3π, $\cdots$, $1 - \cos \omega t = 2$, and therefore, $\|\mathbf{v}(t)\| = 2b\omega$, its maximum value. Hence, the speed is maximum when the point is at the top of cycloidal arch.

39. Find the velocity vector and show that it is orthogonal to $\mathbf{r}(t)$ if $\mathbf{r}(t) = b\cos \omega t \, \mathbf{i} + b\sin \omega t \, \mathbf{j}$.

Solution:

The velocity vector is

$$\mathbf{v}(t) = \mathbf{r}'(t) = -b\omega \sin \omega t \, \mathbf{i} + b\omega \cos \omega t \, \mathbf{j}$$

and since

$$\mathbf{r}(t) \cdot \mathbf{v}(t) = -b^2 \omega \sin \omega t \cos \omega t + b^2 \omega \sin \omega t \cos \omega t = 0,$$

it follows that $\mathbf{r}(t)$ and $\mathbf{v}(t)$ are orthogonal.

40. Show that the speed of the particle is $b\omega$ if
$\mathbf{r}(t) = b\cos\omega t\,\mathbf{i} + b\sin\omega t\,\mathbf{j}$.

Solution:

Using the vector-valued function for the velocity found in
Exercise 39, you have

$$\text{speed} = \|\mathbf{v}\| = \sqrt{b^2\omega^2\sin^2\omega t + b^2\omega^2\cos^2\omega t}$$
$$= \sqrt{b^2\omega^2[\sin^2\omega t + \cos^2\omega t]} = b\omega.$$

41. Find the acceleration vector and show that its
direction is always toward the center of the circle if
$\mathbf{r}(t) = b\cos\omega t\,\mathbf{i} + b\sin\omega t\,\mathbf{j}$.

Solution:

Since

$$\mathbf{a}(t) = \mathbf{r}''(t) = [-b\omega^2\cos\omega t]\,\mathbf{i} - [b\omega^2\sin\omega t]\,\mathbf{j}$$
$$= -b\omega^2[\cos\omega t\,\mathbf{i} + \sin\omega t\,\mathbf{j}]$$
$$= -\omega^2\,\mathbf{r}(t),$$

it follows that $\mathbf{a}(t)$ is a negative multiple of a unit vector from
$(0, 0)$ to $(\cos\omega t,\ \sin\omega t)$, and thus $\mathbf{a}(t)$ is directed toward the
origin.

42. Show that the magnitude of the acceleration vector is $b\omega^2$ if
$\mathbf{r}(t) = b\cos\omega t\,\mathbf{i} + b\sin\omega t\,\mathbf{j}$.

Solution:

Using the vector-valued function for the acceleration found in
Exercise 41, you have

$$\|\mathbf{a}(t)\| = b\omega^2\|\cos\omega t\,\mathbf{i} + \sin\omega t\,\mathbf{j}\| = b\omega^2.$$

12.4   Tangent Vectors and Normal Vectors

---

3. Find the unit tangent vector $\mathbf{T}(t)$ and find a set of parametric equations for the line tangent to the helix $\mathbf{r}(t) = 2\cos t\,\mathbf{i} + 2\sin t\,\mathbf{j} + t\,\mathbf{k}$ at the point $(2,\,0,\,0)$.

Solution:

$$\mathbf{r}(t) = 2\cos t\,\mathbf{i} + 2\sin t\,\mathbf{j} + t\,\mathbf{k}$$
$$\mathbf{r}'(t) = -2\sin t\,\mathbf{i} + 2\cos t\,\mathbf{j} + \mathbf{k}$$

The unit tangent vector is

$$\mathbf{T}(t) = \frac{\mathbf{r}'(t)}{\|\mathbf{r}'(t)\|} = \frac{-2\sin t\,\mathbf{i} + 2\cos t\,\mathbf{j} + \mathbf{k}}{\sqrt{(-2\sin t)^2 + (2\cos t)^2 + 1}}$$
$$= \frac{1}{\sqrt{5}}(-2\sin t\,\mathbf{i} + 2\cos t\,\mathbf{j} + \mathbf{k}).$$

Since $t = 0$ at the point $(2,0,0)$, the direction vector for the line is given by $\mathbf{r}'(0) = 2\mathbf{j} + \mathbf{k}$, and the parametric representation of the line is

$$x = 2, \quad y = 2s, \quad z = s.$$

13. If $\mathbf{r} = t\,\mathbf{i} + (1/t)\,\mathbf{j}$, find $\mathbf{T}(t)$, $\mathbf{N}(t)$, $a_{\mathbf{T}}$, and $a_{\mathbf{N}}$ when $t = 1$.

Solution:

$$\mathbf{r}(t) = t\,\mathbf{i} + \frac{1}{t}\,\mathbf{j}$$
$$\mathbf{v}(t) = \mathbf{r}'(t) = \mathbf{i} - \frac{1}{t^2}\,\mathbf{j}$$
$$\mathbf{a}(t) = \mathbf{r}''(t) = \frac{2}{t^3}\,\mathbf{j}$$

At $t = 1$, you have

$$\mathbf{v}(1) = \mathbf{i} - \mathbf{j}, \quad \|\mathbf{v}(1)\| = \sqrt{2}, \quad \text{and} \quad \mathbf{a}(1) = 2\mathbf{j}.$$

Therefore, when $t = 1$,

$$\mathbf{T}(1) = \frac{\mathbf{v}(1)}{\|\mathbf{v}(1)\|} = \frac{\mathbf{i}}{\sqrt{2}} - \frac{\mathbf{j}}{\sqrt{2}} = \frac{\sqrt{2}}{2}(\mathbf{i} - \mathbf{j}).$$

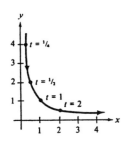

Since $\mathbf{N}(1)$ points toward the concave side of the curve (see the accompanying figure),

$$\mathbf{N}(1) = \frac{\mathbf{i}}{\sqrt{2}} + \frac{\mathbf{j}}{\sqrt{2}} = \frac{\sqrt{2}}{2}(\mathbf{i} + \mathbf{j}).$$

It follows that

$$a_{\mathbf{T}} = \mathbf{a}(1) \cdot \mathbf{T}(1) = (2\mathbf{j}) \cdot \left(\frac{\mathbf{i}}{\sqrt{2}} - \frac{\mathbf{j}}{\sqrt{2}}\right) = \frac{-2}{\sqrt{2}} = -\sqrt{2}$$

and

$$a_{\mathbf{N}} = \mathbf{a}(1) \cdot \mathbf{N}(1) = (2\mathbf{j}) \cdot \left(\frac{\mathbf{i}}{\sqrt{2}} + \frac{\mathbf{j}}{\sqrt{2}}\right) = \frac{2}{\sqrt{2}} = \sqrt{2}.$$

15. If $\mathbf{r}(t) = e^t \cos t\, \mathbf{i} + e^t \sin t\, \mathbf{j}$, find $\mathbf{T}(t)$, $\mathbf{N}(t)$, $a_{\mathbf{T}}$, and $a_{\mathbf{N}}$ when $t = \pi/2$.

Solution:

$$\mathbf{r}(t) = e^t \cos t\, \mathbf{i} + e^t \sin t\, \mathbf{j}$$
$$\mathbf{v}(t) = \mathbf{r}'(t) = e^t(\cos t - \sin t)\mathbf{i} + e^t(\cos t + \sin t)\mathbf{j}$$
$$\mathbf{a}(t) = \mathbf{r}''(t)$$
$$= e^t(-\sin t - \cos t + \cos t - \sin t)\mathbf{i}$$
$$\qquad + e^t(-\sin t + \cos t + \cos t + \sin t)\mathbf{j}$$
$$= e^t(-2\sin t)\mathbf{i} + e^t(2\cos t)\mathbf{j}$$

At $t = \pi/2$, you have

$$\mathbf{v}(\pi/2) = -e^{\pi/2}\mathbf{i} + e^{\pi/2}\mathbf{j}$$
$$\|\mathbf{v}(\pi/2)\| = e^{\pi/2}\sqrt{2}$$
$$\mathbf{a}(\pi/2) = -2e^{\pi/2}\mathbf{i}.$$

Therefore, at $t = \pi/2$,

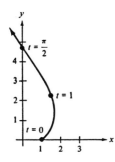

$$\mathbf{T}(\pi/2) = \frac{\mathbf{v}(\pi/2)}{\|\mathbf{v}(\pi/2)\|} = \frac{-\mathbf{i}}{\sqrt{2}} + \frac{\mathbf{j}}{\sqrt{2}} = \frac{\sqrt{2}}{2}(-\mathbf{i} + \mathbf{j}),$$

and since $\mathbf{N}(\pi/2)$ points toward the concave side of the curve, (see figure),

$$\mathbf{N}(\pi/2) = -\frac{\sqrt{2}}{2}(\mathbf{i} + \mathbf{j}).$$

It follows that

$$a_{\mathbf{T}} = \mathbf{a}(\pi/2) \cdot \mathbf{T}(\pi/2) = (-2e^{\pi/2}\mathbf{i}) \cdot \frac{(-\mathbf{i} + \mathbf{j})}{\sqrt{2}} = \sqrt{2}e^{\pi/2}$$

$$a_{\mathbf{N}} = \mathbf{a}(\pi/2) \cdot \mathbf{N}(\pi/2) = (-2e^{\pi/2}\mathbf{i}) \cdot \frac{(-\mathbf{i} - \mathbf{j})}{\sqrt{2}} = \sqrt{2}e^{\pi/2}.$$

31. Use a graphing utility to graph the space curve
$\mathbf{r}(t) = 4t\,\mathbf{i} + 3\cos t\,\mathbf{j} + 3\sin t\,\mathbf{k}$. Also, find $\mathbf{T}(t)$, $\mathbf{N}(t)$, $a_{\mathbf{T}}$, and $a_{\mathbf{N}}$ when $t = \pi/2$.

Solution:

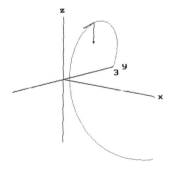

$$\mathbf{r}(t) = 4t\,\mathbf{i} + 3\cos t\,\mathbf{j} + 3\sin t\,\mathbf{k}$$
$$\mathbf{v}(t) = \mathbf{r}'(t) = 4\,\mathbf{i} - 3\sin t\,\mathbf{j} + 3\cos t\,\mathbf{k}$$
$$\|\mathbf{v}(t)\| = \sqrt{16 + 9(\sin^2 t + \cos^2 t)} = \sqrt{25} = 5$$
$$\mathbf{a}(t) = \mathbf{r}''(t) = -3\cos t\,\mathbf{j} - 3\sin t\,\mathbf{k}$$
$$\mathbf{T}(t) = \frac{\mathbf{v}(t)}{\|\mathbf{v}(t)\|} = \frac{1}{5}[4\,\mathbf{i} - 3\sin t\,\mathbf{j} + 3\cos t\,\mathbf{k}]$$
$$\mathbf{N}(t) = \frac{\mathbf{T}'(t)}{\|\mathbf{T}'(t)\|} = \frac{(\frac{1}{5})[-3\cos t\,\mathbf{j} - 3\sin t\,\mathbf{k}]}{(\frac{1}{5})\sqrt{9(\cos^2 t + \sin^2 t)}}$$
$$= \frac{(-\frac{3}{5})[(\cos t)\,\mathbf{j} + (\sin t)\,\mathbf{k}]}{\frac{3}{5}}$$
$$= -\cos t\,\mathbf{j} - \sin t\,\mathbf{k}$$

Therefore,

$$\mathbf{a}(\pi/2) = -3\,\mathbf{k}, \quad \mathbf{T}(\pi/2) = \frac{1}{5}[4\,\mathbf{i} - 3\,\mathbf{j}], \quad \text{and} \quad \mathbf{N}(\pi/2) = \mathbf{k}.$$

Thus,

$$a_{\mathbf{T}} = \mathbf{a}\left(\frac{\pi}{2}\right) \cdot \mathbf{T}\left(\frac{\pi}{2}\right) = 0 \text{ and } a_{\mathbf{N}} = \mathbf{a}\left(\frac{\pi}{2}\right) \cdot \mathbf{N}\left(\frac{\pi}{2}\right) = 3.$$

35. Find the tangential and normal components of acceleration for a projectile fired at an angle θ with the horizontal and with an initial speed of v_0. What are the components when the projectile is at its maximum height?

Solution:

From Theorem 12.3, you have

$$\mathbf{r}(t) = (v_0 t \cos \theta)\,\mathbf{i} + (h + v_0 t \sin \theta - 16t^2)\,\mathbf{j}$$
$$\mathbf{v}(t) = (v_0 \cos \theta)\,\mathbf{i} + (v_0 \sin \theta - 32t)\,\mathbf{j}$$
$$\mathbf{a}(t) = -32\,\mathbf{j}$$
$$\mathbf{T}(t) = \frac{\mathbf{v}(t)}{\|\mathbf{v}(t)\|} = \frac{(v_0 \cos \theta)\mathbf{i} + (v_0 \sin \theta - 32t)\,\mathbf{j}}{\sqrt{v_0{}^2 \cos^2 \theta + (v_0 \sin \theta - 32t)^2}}$$

Since the path of a projectile is concave downward,

$$\mathbf{N}(t) = \frac{(v_0 \sin\theta - 32t)\,\mathbf{i} + (-v_0 \cos\theta)\,\mathbf{j}}{\sqrt{v_0{}^2 \cos^2\theta + (v_0 \sin\theta - 32t)^2}}.$$

Therefore,

$$a_{\mathbf{T}} = \frac{-32(v_0 \sin\theta - 32t)}{\sqrt{v_0{}^2 \cos^2\theta + (v_0 \sin\theta - 32t)^2}}$$

$$a_{\mathbf{N}} = \frac{32v_0 \cos\theta}{\sqrt{v_0{}^2 \cos^2\theta + (v_0 \sin\theta - 32t)^2}}.$$

The projectile will reach its maximum height when the vertical component of velocity is zero, or

$$v_0 \sin\theta - 32t = 0$$

Hence, at the maximum height $a_{\mathbf{T}} = 0$ and $a_{\mathbf{N}} = 32$. Thus, at the maximum height of the projectile, all the acceleration is normal to the path.

38. An object of mass m moves at a constant speed v in a circular path of radius r. The force required to produce the centripetal component of acceleration is called the **centripetal force**. Show that this force is $F = mv^2/r$. Newton's **Law of Universal Gravitation** is given by $F = GMm/d^2$, where d is the distance between the centers of the two bodies of mass M and m. Use this to show that the speed required for circular motion is $v = \sqrt{GM/r}$.

Solution:

Since the motion is circular, the path is described by

$$\mathbf{r}(t) = r\cos\omega t\,\mathbf{i} + r\sin\omega t\,\mathbf{j}.$$

Furthermore,

$$\mathbf{v}(t) = -r\omega \sin\omega t\,\mathbf{i} + r\omega \cos\omega t\,\mathbf{j}$$
$$\|\mathbf{v}(t)\| = r\omega\sqrt{1} = r\omega$$
$$\mathbf{a}(t) = (-r\omega^2 \cos\omega t)\,\mathbf{i} - (r\omega^2 \sin\omega t)\,\mathbf{j}$$
$$\|\mathbf{a}(t)\| = r\omega^2\sqrt{1} = r\omega^2.$$

Since force equals mass times acceleration, you have

$$F = m[\mathbf{a}(t)] = m(r\omega^2) = \frac{m}{r}(r^2\omega^2) = \frac{mv^2}{r}.$$

If you consider the moving object to have a point mass m, then in Newton's Law let $d = r$ and you obtain

$$\frac{mv^2}{r} = \frac{GMm}{r^2}$$

$$v^2 = \frac{GM}{r} \implies v = \sqrt{\frac{GM}{r}}.$$

42. Use the result of Exercise 38 to find the speed necessary for the circular orbit of a syncom satellite in a geosynchronous orbit r miles above the surface of the earth. Let $GM = 9.56 \times 10^4$ mi^3/s^2 and assume the radius of the earth is 4000 miles. [The satellite completes one orbit per sidereal day (23 hours, 56 minutes) and thus appears to remain stationary above a point on the earth.]

Solution:

We let r be the radius of the orbit measured from the center of the earth. Then the distance the satellite travels in one day is $d = 2\pi r$ and the speed is $v = 2\pi r/t = 2\pi r/[(24)(3600)]$. However, from Exercise 38 the speed is also given by $v = \sqrt{9.56(10^4)/r}$. Solving these two equations simultaneously, yields

$$\frac{2\pi r}{(24)(3600)} = \sqrt{\frac{9.56(10^4)}{r}}$$

$$\frac{4\pi^2 r^2}{(24^2)(3600^2)} = \frac{9.56(10^4)}{r}$$

$$r^3 = \frac{9.56(10^4)(24^2)(3600^2)}{4\pi^2}$$

$$r = \sqrt[3]{\frac{9.56(10^4)(24^2)(3600^2)}{4\pi^2}} \approx 26,245 \text{ mi.}$$

Therefore, the altitude above the earth is $26,245 - 4,000 = 22,245$ mi, and the speed is

$$v = \frac{\text{distance}}{\text{time}} = \frac{2\pi(26,245)}{(24)(3,600)}$$

$$\approx 1.91 \text{ mi/sec} \approx 6,871 \text{ mi/hr.}$$

12.5   Arc Length and Curvature

7. Sketch the space curve

$$\mathbf{r}(t) = a\cos t\,\mathbf{i} + a\sin t\,\mathbf{j} + bt\,\mathbf{k}$$

and find the length over the interval $[0, 2\pi]$.

Solution:

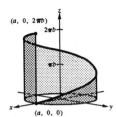

The graph is a circular helix of radius a. (See the accompanying figure.) Since

$$\mathbf{r}'(t) = -a\sin t\,\mathbf{i} + a\cos t\,\mathbf{j} + b\,\mathbf{k},$$

the arc length on the interval $[0, 2\pi]$ is

$$
\begin{aligned}
s &= \int_0^{2\pi} \|\mathbf{r}'(t)\|\,dt \\
&= \int_0^{2\pi} \sqrt{(-a\sin t)^2 + (a\cos t)^2 + b^2}\,dt \\
&= \int_0^{2\pi} \sqrt{a^2(\sin^2 t + \cos^2 t) + b^2}\,dt \\
&= \int_0^{2\pi} \sqrt{a^2 + b^2}\,dt = \left[\sqrt{a^2 + b^2}\,t\right]_0^{2\pi} = 2\pi\sqrt{a^2 + b^2}.
\end{aligned}
$$

12. Consider the space curve represented by the vector-valued function

$$\mathbf{r}(t) = \left\langle 4(\sin t - t\cos t),\ 4(\cos t + t\sin t),\ \tfrac{3}{2}t^2 \right\rangle.$$

(a) Express the length of the arc s on the curve as a function of t by evaluating the integral

$$s = \int_0^t \sqrt{[x'(\tau)]^2 + [y'(\tau)]^2 + [z'(\tau)]^2}\,d\tau.$$

(b) Solve for t in the relationship derived in part (a), and substitute the result into the original set of parametric equations. This yields a parametrization of the curve in terms of the arc length parameter s.

(c) Find the coordinates of the point on the space curve when the length of the arc is $s = \sqrt{5}$ and when $s = 4$.

(d) Verify that $\|\mathbf{r}'(s)\| = 1$.

Solution:

(a) Since $x'(t) = 4t \sin t$, $y'(t) = 4t \cos t$, and $z'(t) = 3t$, you have

$$s = \int_0^t \sqrt{[x'(\tau)]^2 + [y'(\tau)]^2 + [z'(\tau)]^2}\, d\tau$$

$$= \int_0^t \sqrt{[4\tau \sin \tau]^2 + [4\tau \cos \tau]^2 + [3\tau]^2}\, d\tau$$

$$= 5 \int_0^t \tau\, d\tau = \frac{5}{2}\left[\tau^2\right]_0^t = \frac{5t^2}{2}.$$

(b) Since $s = 5t^2/2$, you have

$$t = \sqrt{\frac{2s}{5}} = \frac{\sqrt{10s}}{5}.$$

Therefore, the parametrization of the curve in terms of s is

$$\mathbf{r}(s) = \left\langle 4\left(\sin \frac{\sqrt{10s}}{5} - \frac{\sqrt{10s}}{5} \cos \frac{\sqrt{10s}}{5}\right),\right.$$

$$\left. 4\left(\cos \frac{\sqrt{10s}}{5} - \frac{\sqrt{10s}}{5} \sin \frac{\sqrt{10s}}{5}\right), \frac{3s}{5}\right\rangle.$$

(c) $\mathbf{r}(\sqrt{5}) \approx \langle 1.030, 5.408, 1.342 \rangle$

and

$\mathbf{r}(4) \approx \langle 2.291, 6.029, 2.400 \rangle$.

(d) $\mathbf{r}'(s) = \left\langle \frac{4}{5} \sin \frac{\sqrt{10s}}{5}, \frac{4}{5} \cos \frac{\sqrt{10s}}{5}, \frac{3}{5} \right\rangle$

$$\|\mathbf{r}'(s)\| = \sqrt{\left(\frac{4}{5} \sin \frac{\sqrt{10s}}{5}\right)^2 + \left(\frac{4}{5} \cos \frac{\sqrt{10s}}{5}\right)^2 + \left(\frac{3}{5}\right)^2}$$

$$= \sqrt{\frac{16}{25} + \frac{9}{25}} = 1$$

19. Find the curvature K of the curve $\mathbf{r}(t) = t\,\mathbf{i} + \frac{1}{t}\,\mathbf{j}$ at $t = 1$.

Solution:

From Exercise 13, Section 12.4, you have

$$\mathbf{a}(1) \cdot \mathbf{N}(1) = \sqrt{2} \quad \text{and} \quad \|\mathbf{v}(1)\|^2 = 2.$$

Therefore, the curvature is

$$K = \frac{\mathbf{a}(1) \cdot \mathbf{N}(1)}{\|\mathbf{v}(1)\|^2} = \frac{\sqrt{2}}{2} \approx 0.707.$$

25. Find the curvature K of the curve $\mathbf{r}(t) = e^t \cos t\,\mathbf{i} + e^t \sin t\,\mathbf{j}$.

Solution:

$$\mathbf{r}(t) = e^t \cos t\,\mathbf{i} + e^t \sin t\,\mathbf{j}$$
$$\mathbf{r}'(t) = e^t(\cos t - \sin t)\,\mathbf{i} + e^t(\cos t + \sin t)\,\mathbf{j}$$
$$\|\mathbf{r}'(t)\| = e^t\sqrt{(\cos t - \sin t)^2 + (\cos t + \sin t)^2} = \sqrt{2}e^t$$
$$\mathbf{T}(t) = \frac{\mathbf{r}'(t)}{\|\mathbf{r}'(t)\|} = \frac{1}{\sqrt{2}}[(\cos t - \sin t)\,\mathbf{i} + (\cos t + \sin t)\,\mathbf{j}]$$
$$\mathbf{T}'(t) = \frac{1}{\sqrt{2}}[(-\sin t - \cos t)\,\mathbf{i} + (-\sin t + \cos t)\,\mathbf{j}]$$
$$\|\mathbf{T}'(t)\| = \frac{1}{\sqrt{2}}\sqrt{(-\sin t - \cos t)^2 + (-\sin t + \cos t)^2} = 1$$
$$K = \frac{\|\mathbf{T}'(t)\|}{\|\mathbf{r}'(t)\|} = \frac{1}{\sqrt{2}e^t} = \frac{\sqrt{2}}{2}e^{-t}$$

31. Find the curvature K of the curve $\mathbf{r}(t) = 4t\,\mathbf{i} + 3\cos t\,\mathbf{j} + 3\sin t\,\mathbf{k}$.

Solution:

From Exercise 31, Section 12.4, you have

$$\|\mathbf{T}'(t)\| = \frac{3}{5} \quad \text{and} \quad \|\mathbf{r}'(t)\| = 5.$$

Therefore, the curvature is

$$K = \frac{\|\mathbf{T}'(t)\|}{\|\mathbf{r}'(t)\|} = \frac{3/5}{5} = \frac{3}{25}.$$

35. Find the curvature and radius of curvature of the plane curve $y = 2x^2 + 3$ at $x = -1$.

Solution:

$$y = 2x^2 + 3 \quad y' = 4x \quad y'' = 4$$

$$K = \left| \frac{y''}{[1 + (y')^2]^{3/2}} \right| = \frac{4}{[1 + (4x)^2]^{3/2}}$$

When $x = -1$, the curvature is

$$K = \frac{4}{(1 + 16)^{3/2}} = \frac{4}{17^{3/2}} \approx 0.057$$

and the radius of curvature when $x = -1$ is

$$r = \frac{1}{K} = \frac{17^{3/2}}{4} \approx 17.523.$$

41. Sketch the graph of the function and sketch the circle of curvature to the graph of $y = x + (1/x)$ at the point $(1, 2)$.

Solution:

$$f(x) = x + \frac{1}{x}, \quad f'(x) = 1 - \frac{1}{x^2} = \frac{x^2 - 1}{x^2}, \quad f''(x) = \frac{2}{x^3}$$

At the point $(1, 2)$, $f'(1) = 0$ and $f''(1) = 2$. Thus, at $(1, 2)$ the curvature is

$$K = \left| \frac{2}{(1 + 0^2)^{3/2}} \right| = 2$$

and the radius of curvature is

$$r = \frac{1}{K} = \frac{1}{2}.$$

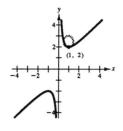

Since the slope of the tangent line to the graph of the function is 0 at the point $(1, 2)$, the normal line is vertical and the center of the closest circular approximation is $(1, \frac{5}{2})$. (See the accompanying figure.) Finally, the equation of the closest circular approximation is

$$(x - 1)^2 + \left(y - \frac{5}{2} \right)^2 = \left(\frac{1}{2} \right)^2.$$

51. Show that the curvature is greatest at the endpoints of the major axis and least at the endpoints of the minor axis for the ellipse given by $x^2 + 4y^2 = 4$.

Solution:

The endpoints of the major axis are $(\pm 2, 0)$ and the endpoints of the minor axis are $(0, \pm 1)$.

$$x^2 + 4y^2 = 4$$
$$2x + 8yy' = 0$$
$$y' = \frac{-x}{4y}$$
$$y'' = \frac{(4y)(-1) - (-x)(4y')}{16y^2} = \frac{-4y - (x^2/y)}{16y^2}$$
$$= \frac{-(4y^2 + x^2)}{16y^3} = \frac{-4}{16y^3} = \frac{-1}{4y^3}$$

The curvature is given by

$$K = \left| \frac{-1/4y^3}{[1 + (-x/4y)^2]^{3/2}} \right| = \left| \frac{-1}{4y^3[(16y^2 + x^2)/16y^2]^{3/2}} \right|$$
$$= \left| \frac{-16}{(16y^2 + x^2)^{3/2}} \right| = \frac{16}{(12y^2 + 4y^2 + x^2)^{3/2}} = \frac{16}{(12y^2 + 4)^{3/2}}.$$

Since $-1 \le y \le 1$, K is greatest when $y = 0$ and smallest when $y = \pm 1$.

Review Exercises for Chapter 12

19. Find a set of parametric equations for the line tangent to the space curve given by $\mathbf{r}(t) = 2\cos t\,\mathbf{i} + 2\sin t\,\mathbf{j} + t\,\mathbf{k}$ at the point $t = 3\pi/4$.

Solution:

$$\mathbf{r}(t) = 2\cos t\,\mathbf{i} + 2\sin t\,\mathbf{j} + t\,\mathbf{k}$$
$$\mathbf{r}(3\pi/4) = -\sqrt{2}\,\mathbf{i} + \sqrt{2}\,\mathbf{j} + (3\pi/4)\,\mathbf{k}$$
$$\mathbf{r}'(t) = -2\sin t\,\mathbf{i} + 2\cos t\,\mathbf{j} + \mathbf{k}$$
$$\mathbf{r}'(3\pi/4) = -\sqrt{2}\,\mathbf{i} - \sqrt{2}\,\mathbf{j} + \mathbf{k}$$

Therefore, the tangent line must pass through the point $(-\sqrt{2},\ \sqrt{2},\ 3\pi/4)$ and have direction numbers $a = -\sqrt{2}$, $b = -\sqrt{2}$, and $c = 1$. Thus, the parametric equations of the line are given by

$$x = -\sqrt{2} - \sqrt{2}t, \quad y = \sqrt{2} - \sqrt{2}t, \quad z = \frac{3\pi}{4} + t.$$

23. Find the indefinite integral $\int \| \cos t\,\mathbf{i} + \sin t\,\mathbf{j} + t\,\mathbf{k}\|\,dt$.

Solution:

$$\int \| \cos t\,\mathbf{i} + \sin t\,\mathbf{j} + t\,\mathbf{k}\|\,dt$$

$$= \int \sqrt{\cos^2 t + \sin^2 t + t^2}\,dt$$

$$= \int \sqrt{1+t^2}\,dt = \frac{1}{2}(t\sqrt{1+t^2} + \ln|t + \sqrt{1+t^2}|) + C$$

39. If the position function of an object is given by $\mathbf{r}(t) = t\,\mathbf{i} + t^2\,\mathbf{j} + \frac{1}{2}t^2\,\mathbf{k}$, find $\mathbf{v}(t)$, $\|\mathbf{v}(t)\|$, $\mathbf{a}(t)$, $\mathbf{a}\cdot\mathbf{T}$, $\mathbf{a}\cdot\mathbf{N}$, and the curvature of the path at time t.

Solution:

$$\mathbf{v}(t) = \mathbf{r}'(t) = \mathbf{i} + 2t\,\mathbf{j} + t\,\mathbf{k}$$

$$\text{speed} = \|\mathbf{v}(t)\| = \sqrt{i^2 + (2t)^2 + t^2} = \sqrt{1+5t^2}$$

$$\mathbf{a}(t) = \mathbf{r}''(t) = 2\,\mathbf{j} + \mathbf{k}$$

$$\mathbf{T}(t) = \frac{\mathbf{v}(t)}{\|\mathbf{v}(t)\|} = \frac{\mathbf{i} + 2t\,\mathbf{j} + t\,\mathbf{k}}{\sqrt{1+5t^2}}$$

$$\mathbf{T}'(t) = \frac{5t}{(1+5t^2)^{3/2}}\mathbf{i} + \frac{2}{(1+5t^2)^{3/2}}\mathbf{j} + \frac{1}{(1+5t^2)^{3/2}}\mathbf{k}$$

$$= \frac{5t\,\mathbf{i} + 2\,\mathbf{j} + \mathbf{k}}{(1+5t^2)^{3/2}}$$

$$\|\mathbf{T}'(t)\| = \frac{\sqrt{(5t)^2 + 2^2 + 1^2}}{(1+5t^2)^{3/2}} = \frac{\sqrt{5}\sqrt{1+5t^2}}{(1+5t^2)^{3/2}} = \frac{\sqrt{5}}{1+5t^2}$$

$$\mathbf{N}(t) = \frac{\mathbf{T}'(t)}{\|\mathbf{T}'(t)\|} = \frac{5t\,\mathbf{i} + 2\,\mathbf{j} + \mathbf{k}}{\sqrt{5}\sqrt{1+5t^2}}$$

$$a_T = \mathbf{a} \cdot \mathbf{T}(t)$$
$$= \frac{2(2t)+1(t)}{\sqrt{1+5t^2}} = \frac{5t}{\sqrt{1+5t^2}}$$
$$a_N = \mathbf{a}(t) \cdot \mathbf{N}(t)$$
$$= \frac{2(2)+1(1)}{\sqrt{5}\sqrt{1+5t^2}} = \frac{5}{\sqrt{5}\sqrt{1+5t^2}} = \frac{\sqrt{5}}{\sqrt{1+5t^2}}$$
$$K = \frac{\mathbf{a}(t) \cdot \mathbf{N}(t)}{\|\mathbf{v}(t)\|} = \frac{\frac{\sqrt{5}}{\sqrt{1+5t^2}}}{1+5t^2} = \frac{\sqrt{5}}{(1+5t^2)^{3/2}}$$

41. Find the length of the space curve given by vector-valued function
$$\mathbf{r}(t) = \frac{1}{2}t\,\mathbf{i} + \sin t\,\mathbf{j} + \cos t\,\mathbf{k}$$
over the interval $0 \le t \le \pi$.

Solution:

$$s = \int_0^\pi \|\mathbf{r}'(t)\|\,dt$$
$$= \int_0^\pi \sqrt{(\tfrac{1}{2})^2 + \cos^2 t + (-\sin t)^2}\,dt$$
$$= \frac{\sqrt{5}}{2}\int_0^\pi dt = \frac{\sqrt{5}}{2}\Big[t\Big]_0^\pi = \frac{\sqrt{5}\pi}{2}$$

13 FUNCTIONS OF
SEVERAL VARIABLES

13.1   Introduction to Functions of Several Variables

9. Find the functional values (a) $f(0,4)$ and (b) $f(1,4)$ if

$$f(x,y) = \int_x^y (2t-3)\, dt$$

Solution:

$$f(x,y) = \int_x^y (2t-3)\, dt = \left[t^2 - 3t\right]_x^y = (y^2 - 3y) - (x^2 - 3x)$$

(a) $f(0,4) = (16-12) - (0-0) = 4$

(b) $f(1,4) = (16-12) - (1-3) = 6$

13. Describe the region R, in the xy–coordinate plane, that corresponds to the domain of $f(x,y) = \sqrt{4 - x^2 - y^2}$. Find the range of $f(x,y)$.

Solution:

Since $f(x,y) = \sqrt{4 - x^2 - y^2}$, you have

$$4 - x^2 - y^2 \geq 0$$
$$4 \geq x^2 + y^2$$

Therefore, the region R is the set of all points inside and on the boundary of the circle $x^2 + y^2 = 4$. The range of f is the set of all real numbers in the interval $[0, 2]$.

15. Describe the region R, in the xy-coordinate plane, that corresponds to the domain of $f(x,y) = \arcsin(x+y)$. Find the range of $f(x,y)$.

Solution:

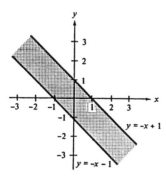

Since $z = \arcsin(x+y)$ implies that $\sin z = x+y$, you can conclude that $|x+y| \leq 1$. Therefore, the region R is such that

$$-1 \leq x+y \leq 1$$
$$-1-x \leq y \leq -x+1.$$

This means that R lies on and between the parallel lines

$$y = -1-x \quad \text{and} \quad y = -x+1$$

as shown in the accompanying figure. The range of the arcsine function is the set of all reals in the interval $[-\pi/2, \ \pi/2]$.

43. (a) Sketch the graph of the surface given by $f(x,y) = x^2 + y^2$.
(b) On the surface of part (a), sketch the graphs of $z = f(1,y)$ and $z = f(x,1)$.

Solution:

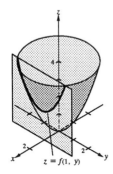

(a) The surface is a paraboloid and its axis is the z−axis. Some conventional traces are

$$
\begin{array}{lll}
yz\text{-trace } (x=0): & z = y^2 & \text{Parabola} \\
xz\text{-trace } (y=0): & z = x^2 & \text{Parabola} \\
\text{Parallel to } xy\text{-plane } (z=4): & 4 = x^2 + y^2 & \text{Circle.}
\end{array}
$$

The domain of f is the set of all points (x,y) in the xy-coordinate plane and the range is the set of all non-negative real numbers. The surface is shown in the accompanying figure.

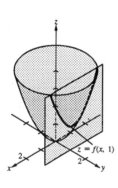

(b) The trace parallel to the yz-coordinate plane when $x = 1$ is the parabola given by

$$z = f(1,y) = 1 + y^2.$$

The trace parallel to the xz-coordinate plane when $y = 1$ is the parabola given by

$$z = f(x,1) = x^2 + 1.$$

The traces are shown in the accompanying figures.

51. Describe the level curves for the function $f(x,y) = x/(x^2+y^2)$. Sketch the level curves for $c = \pm\frac{1}{2}, \pm 1, \pm\frac{3}{2}, \pm 2$.

Solution:

If $f(x,y) = c$, then the level curves are of the form

$$c = \frac{x}{x^2+y^2}$$

$$x^2+y^2 = \frac{x}{c}$$

$$x^2 - \frac{x}{c} + y^2 = 0$$

$$\left(x^2 - \frac{x}{c} + \frac{1}{4c^2}\right) + y^2 = \frac{1}{4c^2}$$

$$\left(x - \frac{1}{2c}\right)^2 + y^2 = \left(\frac{1}{2c}\right)^2.$$

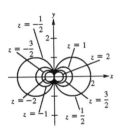

Therefore, each level curve is a circle centered at $(1/2c, 0)$ with radius equal to $1/2c$. For example, if $c = 1$, the level curve has the equation

$$\left(x - \frac{1}{2}\right)^2 + y^2 = \frac{1}{4}.$$

The required level curves are shown in the accompanying figure.

13.2   Limits and Continuity

9. Find

$$\lim_{(x,y)\to(0,1)} \frac{\arcsin(x/y)}{1+xy}$$

and discuss the continuity of the function.

Solution:

Since the limit of a quotient is the quotient of the limits, you have

$$\lim_{(x,y)\to(0,1)} \frac{\arcsin(x/y)}{1+xy} = \frac{\arcsin 0}{1+0} = \frac{0}{1} = 0.$$

A rational function is continuous at every point in its domain. Therefore, the given function is continuous for all points (x, y) in the xy-plane such that $1 + xy \neq 0$, $y \neq 0$, and $|x/y| \leq 1$.

21. Find the limit (if it exists)

$$\lim_{(x,y)\to(0,0)} \frac{-xy^2}{x^2+y^4}$$

by examining the behavior of the function along the paths $x = y^2$ and $x = -y^2$.

Solution:

Along the path $x = y^2$ you have

$$\lim_{(x,y)\to(0,0)} \frac{-xy^2}{x^2+y^4} = \lim_{(y^2,y)\to(0,0)} \frac{-y^2(y^2)}{(y^2)^2+y^4}$$

$$= \lim_{(y^2,y)\to(0,0)} \frac{-y^4}{y^4+y^4} = -\frac{1}{2}.$$

Along the path $x = -y^2$ you have

$$\lim_{(x,y)\to(0,0)} \frac{-xy^2}{x^2+y^4} = \lim_{(-y^2,y)\to(0,0)} \frac{-(-y^2)(y^2)}{(-y^2)^2+y^4}$$

$$= \lim_{(-y^2,y)\to(0,0)} \frac{y^4}{y^4+y^4} = \frac{1}{2}.$$

Since the limits are not the same along different paths, the limit does not exist. The function is discontinuous at $(0, 0)$.

27. Use polar coordinates to find the limit

$$\lim_{(x,y)\to(0,0)} \frac{\sin(x^2+y^2)}{x^2+y^2}.$$

Solution:

We first observe that direct substitution yields the indeterminate from $0/0$. Letting $x = r\cos\theta$, $y = r\sin\theta$, and $r^2 = x^2 + y^2$, yields

$$\lim_{(x,y)\to(0,0)} \frac{\sin(x^2+y^2)}{x^2+y^2} = \lim_{r\to 0} \frac{\sin r^2}{r^2} = 1.$$

37. Discuss the continuity of the composite function $f \circ g$ if $f(t) = 1/t$ and $g(x, y) = 3x - 2y$.

Solution:

$$(f \circ g)(x, y) = f[g(x, y)]$$

$$= \frac{1}{g(x, y)} = \frac{1}{3x - 2y}$$

The composite function is continuous for $y \neq 3x/2$.

39. Given the function $f(x, y) = x^2 - 4y$, find

(a) $\lim\limits_{\Delta x \to 0} \dfrac{f(x + \Delta x, y) - f(x, y)}{\Delta x}$

(b) $\lim\limits_{\Delta y \to 0} \dfrac{f(x, y + \Delta y) - f(x, y)}{\Delta y}$.

Solution:

(a) $\lim\limits_{\Delta x \to 0} \dfrac{f(x + \Delta x, y) - f(x, y)}{\Delta x}$

$$= \lim_{\Delta x \to 0} \frac{[(x + \Delta x)^2 - 4y] - (x^2 - 4y)}{\Delta x}$$

$$= \lim_{\Delta x \to 0} \frac{x^2 + 2x\Delta x + (\Delta x)^2 - 4y - x^2 + 4y}{\Delta x}$$

$$= \lim_{\Delta x \to 0} (2x + \Delta x) = 2x$$

(b) $\lim\limits_{\Delta y \to 0} \dfrac{f(x, y + \Delta y) - f(x, y)}{\Delta y}$

$$= \lim_{\Delta y \to 0} \frac{x^2 - 4(y + \Delta y) - (x^2 - 4y)}{\Delta y}$$

$$= \lim_{\Delta y \to 0} \frac{x^2 - 4y - 4\Delta y - x^2 + 4y}{\Delta y}$$

$$= \lim_{\Delta y \to 0} (-4) = -4$$

13.3   Partial Derivatives

9. Find the first partial derivatives with respect to x and with respect to y for

$$z = \ln \frac{x+y}{x-y}.$$

Solution:

Using the properties of the logarithm function, rewrite the function and obtain

$$z = \ln(x+y) - \ln(x-y).$$

Considering y to be a constant and differentiating with respect to x you have

$$\frac{\partial z}{\partial x} = \frac{1}{x+y}(1) - \frac{1}{x-y}(1) = \frac{-2y}{x^2 - y^2}.$$

Now considering x to be a constant and differentiating with respect to y you have

$$\frac{\partial z}{\partial y} = \frac{1}{x+y}(1) - \frac{1}{x-y}(-1) = \frac{2x}{x^2 - y^2}.$$

17. Find the first partial derivatives with respect to x and with respect to y for

$$z = e^y \sin xy.$$

Solution:

First, considering y to be constant, you have

$$\frac{\partial z}{\partial x} = e^y(\cos xy)(y) = ye^y \cos xy.$$

Now considering x to be constant and using the Product Rule, you have

$$\frac{\partial z}{\partial y} = e^y(\cos xy)(x) + (\sin xy)(e^y)(1) = e^y(x \cos xy + \sin xy).$$

29. Given the function $f(x,y) = \arctan(y/x)$, find $f_x(2,-2)$ and $f_y(2,-2)$.

Solution:

First, considering y to be constant, you have

$$\frac{\partial z}{\partial x} = \frac{1}{1+(y^2/x^2)}\left(\frac{-y}{x^2}\right) = \frac{-y}{x^2+y^2}.$$

Now considering x to be constant, you have

$$\frac{\partial z}{\partial y} = \frac{1}{1+(y^2/x^2)}\left(\frac{1}{x}\right) = \frac{x}{x^2+y^2}.$$

Therefore,

$$f_x(2,-2) = \frac{1}{4} \quad \text{and} \quad f_y(2,-2) = \frac{1}{4}.$$

35. Sketch the curve formed by the intersection of the paraboloid $z = 9x^2 - y^2$ and the plane $y = 3$. Find the slope of the curve at the point $(1, 3, 0)$.

Solution:

The graph of the equation $z = 9x^2 - y^2$ is a hyperbolic paraboloid. The xy-trace ($z = 0$), consists of the intersecting lines $y = \pm 3x$. The yz-trace ($x = 0$) is the parabola $z = -y^2$ opening downward, and the xz-trace ($y = 0$) is the parabola $z = 9x^2$ opening upward. The curve of intersection of the paraboloid and plane $y = 3$ is given by $z = 9x^2 - 9$. It is a parabola opening upward (see figure). Since y is a constant on the curve of intersection, differentiate with respect to x to obtain

$$\frac{\partial z}{\partial x} = 18x$$

At the point $(1, 3, 0)$ the slope is

$$\frac{\partial z}{\partial x} = 18(1) = 18.$$

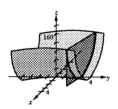

37. Find all the second partial derivatives for $z = x^2 - 2xy + 3y^2$.

Solution:

The first partials are

$$\frac{\partial z}{\partial x} = 2x - 2y \quad \text{and} \quad \frac{\partial z}{\partial y} = -2x + 6y$$

The second partials are

$$\frac{\partial^2 z}{\partial x^2} = \frac{\partial}{\partial x}\left[\frac{\partial z}{\partial x}\right] = 2 \qquad \frac{\partial^2 z}{\partial y\,\partial x} = \frac{\partial}{\partial y}\left[\frac{\partial z}{\partial x}\right] = -2$$

$$\frac{\partial^2 z}{\partial y^2} = \frac{\partial}{\partial y}\left[\frac{\partial z}{\partial y}\right] = 6 \qquad \frac{\partial^2 z}{\partial x\,\partial y} = \frac{\partial}{\partial x}\left[\frac{\partial z}{\partial y}\right] = -2.$$

(Note that the mixed partial derivatives are equals.)

43. Find all the second derivatives for

$$z = \arctan \frac{y}{x}.$$

Solution:

The first partial derivatives were found in Exercise 29.

$$\frac{\partial z}{\partial x} = \frac{-y}{x^2 + y^2} \quad \text{and} \quad \frac{\partial z}{\partial y} = \frac{x}{x^2 + y^2}$$

The second partials are

$$\frac{\partial^2 z}{\partial x^2} = \frac{\partial}{\partial x}\left[\frac{\partial z}{\partial x}\right] = \frac{(x^2+y^2)(0) - (-y)(2x)}{(x^2+y^2)^2} = \frac{2xy}{(x^2+y^2)^2}$$

$$\frac{\partial^2 z}{\partial y\,\partial x} = \frac{\partial}{\partial y}\left[\frac{\partial z}{\partial x}\right] = \frac{(x^2+y^2)(-1) - (-y)(2y)}{(x^2+y^2)^2} = \frac{y^2 - x^2}{(x^2+y^2)^2}$$

$$\frac{\partial^2 z}{\partial y^2} = \frac{\partial}{\partial y}\left[\frac{\partial z}{\partial y}\right] = \frac{(x^2+y^2)(0) - x(2y)}{(x^2+y^2)^2} = \frac{-2xy}{(x^2+y^2)^2}$$

$$\frac{\partial^2 z}{\partial x\,\partial y} = \frac{\partial}{\partial x}\left[\frac{\partial z}{\partial y}\right] = \frac{(x^2+y^2)(1) - x(2x)}{(x^2+y^2)^2} = \frac{y^2 - x^2}{(x^2+y^2)^2}.$$

(Note that the mixed partial derivatives are equal.)

57. Show that the mixed partials f_{xyy}, f_{yxy}, and f_{yyx} are equal if
$f(x, y, z) = e^{-x} \sin yz$.

Solution:

$$f_x(x, y, z) = -e^{-x} \sin yz$$
$$f_{xy}(x, y, z) = -e^{-x}(\cos yz)(z) = -ze^{-x} \cos yz$$
$$f_{xyy}(x, y, z) = -ze^{-x}(-\sin yz)(z) = z^2 e^{-x} \sin yz$$
$$f_y(x, y, z) = e^{-x}(\cos yz)(z) = ze^{-x} \cos yz$$
$$f_{yx}(x, y, z) = -ze^{-x} \cos yz$$
$$f_{yxy}(x, y, z) = -ze^{-x}(-\sin yz)(z) = z^2 e^{-x} \sin yz$$
$$f_y(x, y, z) = e^{-x}(\cos yz)(z) = ze^{-x} \cos yz$$
$$f_{yy}(x, y, z) = ze^{-x}(-\sin yz)(z) = -z^2 e^{-x} \sin yz$$
$$f_{yyx}(x, y, z) = -z^2 e^{-x}(-1) \sin yz = z^2 e^{-x} \sin yx$$

Therefore,

$$f_{xyy}(x, y, z) = f_{yxy}(x, y, z) = f_{yyx}(x, y, z) = z^2 e^{-x} \sin yx.$$

63. Show that $z = \sin(x - ct)$ is a solution of the equation
$\partial^2 z / \partial t^2 = c^2 (\partial^2 z / \partial x^2)$.

Solution:

$$\frac{\partial z}{\partial x} = \cos(x - ct) \qquad \text{and} \qquad \frac{\partial^2 z}{\partial x^2} = -\sin(x - ct)$$

$$\frac{\partial z}{\partial t} = -c\cos(x - ct) \qquad \text{and} \qquad \frac{\partial^2 z}{\partial t^2} = -c^2 \sin(x - ct)$$

Therefore,

$$\frac{\partial^2 z}{\partial t^2} = -c^2 \sin(x - ct) = c^2 \frac{\partial^2 z}{\partial x^2}.$$

13.4   Differentials

5. Find the total differential for $z = x \cos y - y \cos x$.

Solution:

$$dz = \frac{\partial z}{\partial x} dx + \frac{\partial z}{\partial y} dy$$
$$= (\cos y + y \sin x) \, dx + (-x \sin y - \cos x) \, dy$$
$$= (\cos y + y \sin x) \, dx - (x \sin y + \cos x) \, dy$$

9. Find the total differential for $u = (x + y)/(z - 2y)$.

Solution:

Since $u = (x + y)/(z - 2y)$, you have

$$du = \frac{\partial u}{\partial x}\, dx + \frac{\partial u}{\partial y}\, dy + \frac{\partial u}{\partial z}\, dz$$

$$= \frac{1}{z - 2y}\, dx + \frac{(z - 2y)(1) - (x + y)(-2)}{(z - 2y)^2}\, dy + \frac{0 - (x + y)(1)}{(z - 2y)^2}\, dz$$

$$= \frac{1}{z - 2y}\, dx + \frac{2x + z}{(z - 2y)^2}\, dy - \frac{x + y}{(z - 2y)^2}\, dz.$$

11. Given the function $f(x, y) = 9 - x^2 - y^2$, (a) evaluate $f(1, 2)$ and $f(1.05, 2.1)$ and calculate Δz, and (b) use the total differential dz to approximate Δz.

Solution:

(a)
$$f(1, 2) = 9 - 1^2 - 2^2 = 4$$
$$f(1.05, 2.1) = 9 - (1.05)^2 - (2.1)^2 = 3.4875$$
$$\Delta z = f(1.05, 2.1) - f(1, 2) = -0.5125$$

(b)
$$dz = f_x(x, y)\, dx + f_y(x, y)\, dy$$
$$= -2x\, dx - 2y\, dy$$

Letting $x = 1$, $y = 2$, $dx = 0.05$, and $dy = 0.1$, yields

$$dz = -2(1)(0.05) - 2(2)(0.1) = -0.5.$$

23. The radius r and height h of a right circular cylinder are measured with a possible error of 4% and 2%, respectively. Approximate the maximum possible percentage error in measuring the volume.

Solution:

First consider the relative errors in r and h as

$$\frac{dr}{r} = \pm 4\% = \pm 0.04 \quad \text{and} \quad \frac{dh}{h} = \pm 2\% = \pm 0.02.$$

Since $V = \pi r^2 h$, you have

$$dV = 2\pi rh\, dr + \pi r^2\, dh$$

or the relative error in V is

$$\frac{dV}{V} = \frac{2\pi rh\, dr}{\pi r^2 h} + \frac{\pi r^2\, dh}{\pi r^2 h}$$

$$= 2\frac{dr}{r} + \frac{dh}{h} = 2(\pm 0.04) \pm 0.02 = \pm 0.10 = \pm 10\%.$$

31. The inductance L (in micro-henrys) of a straight non-magnetic wire in free space is given by

$$L = 0.00021 \left[\ln \frac{2h}{r} - 0.75 \right]$$

where h is the length of the wire in millimeters and r is the radius of a circular cross section. Approximate L when $r = 2 \pm \frac{1}{16}$ millimeters and $h = 100 \pm \frac{1}{100}$ millimeters.

Solution:

Using the properties of logarithms, rewrite the function as

$$L = 0.00021[\ln(2h) - \ln r - 0.75].$$

Letting $h = 100$, $r = 2$, $dh = \pm\frac{1}{100}$, and $dr = \pm\frac{1}{16}$, the total differential is given by

$$dL = 0.00021 \left(\frac{1}{h}\, dh - \frac{1}{r}\, dr \right)$$

$$= 0.00021 \left[\frac{1}{100} \left(\pm\frac{1}{100} \right) - \frac{1}{2} \left(\pm\frac{1}{16} \right) \right]$$

$$\approx \pm 6.6 \times 10^{-6}.$$

Approximating the inductance, you have

$$L = 0.00021(\ln 200 - \ln 2 - 0.75) \pm dL$$

$$= 8.096 \times 10^{-4} \pm 6.6 \times 10^{-6}.$$

33. Find the values of ϵ_1 and ϵ_2 (see the definition of differentiability) for $f(x,y) = x^2 - 2x + y$ and verify that $\epsilon_1 \to 0$ and $\epsilon_2 \to 0$ as $(\Delta x, \Delta y) \to (0,0)$.

Solution:

$$\Delta z = f(x + \Delta x, y + \Delta y) - f(x,y)$$
$$= (x + \Delta x)^2 - 2(x + \Delta x) + (y + \Delta y) - (x^2 - 2x + y)$$
$$= x^2 + 2x(\Delta x) + (\Delta x)^2 - 2x - 2\Delta x + y + \Delta y - x^2 + 2x - y$$
$$= (2x - 2)\Delta x + (1)\Delta y + \Delta x(\Delta x) + 0(\Delta y)$$
$$= f_x(x,y)\Delta x + f_y(x,y)\Delta y + \epsilon_1(\Delta x) + \epsilon_2(\Delta y)$$

Therefore, $\epsilon_1 = \Delta x$ and $\epsilon_2 = 0$. As $(\Delta x, \Delta y) \to (0,0)$, $\epsilon_1 \to 0$ and $\epsilon_2 \to 0$.

13.5   Chain Rules for Functions of Several Variables

7. If $w = x^2 - y^2$, $x = s\cos t$, and $y = s\sin t$, find $\partial w/\partial s$ and $\partial w/\partial t$ by using the Chain Rule. Evaluate each partial derivative when $s = 3$ and $t = \pi/4$.

Solution:

By the Chain Rule,

$$\frac{\partial w}{\partial s} = \frac{\partial w}{\partial x}\frac{\partial x}{\partial s} + \frac{\partial w}{\partial y}\frac{\partial y}{\partial s} \quad \text{and} \quad \frac{\partial w}{\partial t} = \frac{\partial w}{\partial x}\frac{\partial x}{\partial t} + \frac{\partial w}{\partial y}\frac{\partial y}{\partial t}$$

Therefore,

$$\frac{\partial w}{\partial s} = 2x(\cos t) + (-2y)(\sin t)$$
$$= (2s\cos t)(\cos t) - (2s\sin t)(\sin t)$$
$$= 2s(\cos^2 t - \sin^2 t) = 2s\cos 2t.$$

When $s = 3$ and $t = \pi/4$, $\partial w/\partial s = 2(3)\cos(\pi/2) = 0$. Similarly,

$$\frac{\partial w}{\partial t} = 2x(-s\sin t) - 2y(s\cos t)$$
$$= 2s\cos t(-s\sin t) - 2s\sin t(s\cos t)$$
$$= -4s^2\sin t\cos t = -2s^2\sin 2t.$$

When $s = 3$ and $t = \pi/4$, $\partial w/\partial t = -2(9)\sin(\pi/2) = -18$.

13. If $w = xy + xz + yz$, $x = t - 1$, $y = t^2 - 1$, and $z = t$, find dw/dt (a) by the Chain Rule and (b) by converting w to a function of t before differentiating.

Solution:

(a) By the Chain Rule you have

$$\frac{dw}{dt} = \frac{\partial w}{\partial x}\frac{dx}{dt} + \frac{\partial w}{\partial y}\frac{dy}{dt} + \frac{\partial w}{\partial z}\frac{dz}{dt}$$
$$= (y + z)(1) + (x + z)(2t) + (x + y)(1)$$
$$= (t^2 - 1 + t)(1) + (t - 1 + t)(2t) + (t - 1 + t^2 - 1)(1)$$
$$= 6t^2 - 3 = 3(2t^2 - 1).$$

(b) By writing w as a function of t before differentiating, you have

$$w = (t - 1)(t^2 - 1) + (t - 1)t + (t^2 - 1)t$$
$$= 2t^3 - 3t + 1$$
$$\frac{dw}{dt} = 6t^2 - 3 = 3(2t^2 - 1).$$

17. If $w = \arctan(y/x)$, $x = r\cos\theta$, and $y = r\sin\theta$, find $\partial w/\partial r$ and $\partial w/\partial\theta$ (a) by the Chain Rule and (b) by converting w to a function of r and θ before differentiating.

Solution:

(a) First calculate $\partial w/\partial x$ and $\partial w/\partial y$.

$$w = \arctan\frac{y}{x}$$
$$\frac{\partial w}{\partial x} = \frac{-y/x^2}{1 + (y^2/x^2)} = \frac{-y/x^2}{(x^2 + y^2)/x^2} = \frac{-y}{x^2 + y^2}$$
$$\frac{\partial w}{\partial y} = \frac{1/x}{1 + (y^2/x^2)} = \frac{1/x}{(x^2 + y^2)/x^2} = \frac{x}{x^2 + y^2}$$

Using the Chain Rule yields

$$\frac{\partial w}{\partial r} = \frac{\partial w}{\partial x}\frac{\partial x}{\partial r} + \frac{\partial w}{\partial y}\frac{\partial y}{\partial r}$$
$$= \frac{-y}{x^2 + y^2}(\cos\theta) + \frac{x}{x^2 + y^2}(\sin\theta)$$
$$= \frac{x\sin\theta - y\cos\theta}{x^2 + y^2} = \frac{r\cos\theta\sin\theta - r\sin\theta\cos\theta}{r^2} = 0.$$

Furthermore,

$$\frac{\partial w}{\partial \theta} = \frac{\partial w}{\partial x}\frac{\partial x}{\partial \theta} + \frac{\partial w}{\partial y}\frac{\partial y}{\partial \theta}$$

$$= \frac{-y}{x^2 + y^2}(-r\sin\theta) + \frac{x}{x^2 + y^2}(r\cos\theta)$$

$$= \frac{-r\sin\theta(-r\sin\theta) + r\cos\theta(r\cos\theta)}{r^2} = \frac{r^2}{r^2} = 1.$$

(b) Since

$$w = \arctan\frac{y}{x} = \arctan\left(\frac{r\sin\theta}{r\cos\theta}\right) = \arctan(\tan\theta) = \theta + n\pi,$$

you have

$$\frac{\partial w}{\partial r} = 0 \quad \text{and} \quad \frac{\partial w}{\partial \theta} = 1$$

19. Differentiate implicitly to find the first partial derivatives of z if $x^2 + y^2 + z^2 = 25$.

Solution:

Let

$$F(x, y, z) = x^2 + y^2 + z^2 - 25.$$

Then,

$$F_x(x, y, z) = 2x$$
$$F_y(x, y, z) = 2y$$
$$F_z(x, y, z) = 2z.$$

Now by Theorem 13.8 you have

$$\frac{\partial z}{\partial x} = -\frac{F_x(x, y, z)}{F_z(x, y, z)} = -\frac{2x}{2z} = -\frac{x}{z}$$

$$\frac{\partial z}{\partial y} = -\frac{F_y(x, y, z)}{F_z(x, y, z)} = -\frac{2y}{2z} = -\frac{y}{z}.$$

27. Let θ be the angle between the equal sides of an isosceles triangle, and let x be the length of the sides of equal length. If x is increasing at $\frac{1}{2}$m/hr and θ is increasing at $\pi/90$ rad/hr, find the rate of increase of the area when $x = 6$ and $\theta = \pi/4$.

Solution:

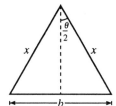

From the accompanying figure you have

$$\sin\frac{\theta}{2} = \frac{b/2}{x} \quad\Longrightarrow\quad b = 2x\sin\frac{\theta}{2}$$

$$\cos\frac{\theta}{2} = \frac{h}{2} \quad\Longrightarrow\quad h = x\cos\frac{\theta}{2}.$$

Therefore, the area of the triangle is

$$A = \frac{1}{2}bh = \frac{1}{2}\left(2x\sin\frac{\theta}{2}\right)\left(x\cos\frac{\theta}{2}\right) = \frac{1}{2}x^2\sin\theta.$$

Differentiating the area function with respect to t and substituting the given information about the triangle yields

$$\frac{dA}{dt} = \frac{\partial A}{\partial x}\frac{dx}{dt} + \frac{\partial A}{\partial\theta}\frac{d\theta}{dt}$$

$$= (x\sin\theta)\frac{dx}{dt} + \left(\frac{1}{2}x^2\cos\theta\right)\frac{d\theta}{dt}$$

$$= 6\sin\frac{\pi}{4}\left(\frac{1}{2}\right) + \left(\frac{1}{2}(6^2)\cos\frac{\pi}{4}\right)\left(\frac{\pi}{90}\right)$$

$$= 5\left(\frac{\sqrt{2}}{2}\right)\left(\frac{1}{2}\right) + \frac{1}{2}(36)\left(\frac{\sqrt{2}}{2}\right)\left(\frac{\pi}{90}\right) = \frac{\sqrt{2}}{10}(15 + \pi) \text{ m}^2/\text{hr}.$$

35. A function $f(x,y)$ is **homogeneous of degree n** if

$$f(tx, ty) = t^n f(x,y)$$

Find the degree of the homogeneous function

$$f(x,y) = x^3 - 3xy^2 + y^3$$

and show that

$$x\,f_x(x,y) + y\,f_y(x,y) = nf(x,y).$$

Solution:

$$f(tx, ty) = (tx)^3 - 3(tx)(ty)^2 + (ty)^3$$
$$= t^3x^3 - 3t^3xy^2 + t^3y^3$$
$$= t^3(x^2 - 3xy^2 + y^3) = t^3 f(x, y)$$

Therefore, the function is homogeneous of degree 3.

$$x\, f_x(x, y) + y\, f_y(x, y) = x(3x^2 - 3y^2) + y(-6xy + 3y^2)$$
$$= 3x^3 - 3xy^2 - 6xy^2 + 3y^3$$
$$= 3x^3 - 9xy^2 + 3y^3$$
$$= 3(x^3 - 3xy^2 + y^3) = 3f(x, y)$$

13.6   Directional Derivatives and Gradients

5. Find the directional derivative of the function
$g(x, y) = \sqrt{x^2 + y^2}$ at the point $P = (3, 4)$ in the direction of
$\mathbf{v} = 3\mathbf{i} - 4\mathbf{j}$.

Solution:

The unit vector, $\mathbf{u}$, in the direction of $\mathbf{v}$ is

$$\mathbf{u} = \frac{\mathbf{v}}{\|\mathbf{v}\|} = \frac{3}{5}\mathbf{i} - \frac{4}{5}\mathbf{j} = \cos\theta\,\mathbf{i} + \sin\theta\,\mathbf{j}.$$

Thus,

$$\cos\theta = \frac{3}{5} \quad \text{and} \quad \sin\theta = -\frac{4}{5}.$$

By Theorem 13.9 you have

$$D_\mathbf{u}f(x, y) = f_x(x, y)\cos\theta + f_y(x, y)\sin\theta$$
$$= \frac{x}{\sqrt{x^2 + y^2}}\left(\frac{3}{5}\right) + \frac{y}{\sqrt{x^2 + y^2}}\left(-\frac{4}{5}\right)$$
$$= \frac{1}{5\sqrt{x^2 + y^2}}(3x - 4y).$$

Therefore,

$$D_\mathbf{u}f(3, 4) = \frac{-7}{25}.$$

Note that $D_\mathbf{u}f(x, y) = \nabla f(x, y) \cdot \mathbf{u}$ where

$$\nabla f(x, y) = \frac{x}{\sqrt{x^2 + y^2}}\mathbf{i} + \frac{y}{\sqrt{x^2 + y^2}}\mathbf{j}.$$

9. Find the directional derivative of $f(x,y,z) = xy + yz + xz$ at the point $P = (1,1,1)$ in the direction of $\mathbf{v} = 2\mathbf{i} + \mathbf{j} - \mathbf{k}$.

Solution:

We begin by finding $\nabla f(x,y,z)$ and a unit vector, $\mathbf{u}$, in the direction of $\mathbf{v}$.

$$\nabla f(x,y,z) = f_x(x,y,z)\mathbf{i} + f_y(x,y,z)\mathbf{j} + f_z(x,y,z)\mathbf{k}$$
$$= (y+z)\mathbf{i} + (x+z)\mathbf{j} + (x+y)\mathbf{k}$$

and

$$\mathbf{u} = \frac{\mathbf{v}}{\|\mathbf{v}\|} = \frac{\sqrt{6}}{6}(2\mathbf{i} + \mathbf{j} - \mathbf{k})$$

Therefore,

$$D_\mathbf{u} f(x,y,z) = \nabla f(x,y,z) \cdot \mathbf{u}$$
$$= \frac{\sqrt{6}}{6}[2(y+z) + (x+z) - (x+y)]$$
$$= \frac{\sqrt{6}}{6}(y + 3z)$$

and

$$D_\mathbf{u} f(1,1,1) = \frac{4\sqrt{6}}{6} = \frac{2\sqrt{6}}{3}$$

17. Find the directional derivative of the function $f(x,y) = x^2 + 4y^2$ at the point $P = (3,1)$ in the direction of $Q = (1,-1)$.

Solution:

A vector in the specified direction is

$$\overrightarrow{PQ} = \mathbf{v} = (1-3)\mathbf{i} + (-1-1)\mathbf{j} = -2\mathbf{i} - 2\mathbf{j}$$

and a unit vector in this direction is

$$\mathbf{u} = \frac{\mathbf{v}}{\|\mathbf{v}\|} = \frac{-2}{\sqrt{8}}\mathbf{i} - \frac{2}{\sqrt{8}}\mathbf{j} = -\frac{1}{\sqrt{2}}\mathbf{i} - \frac{1}{\sqrt{2}}\mathbf{j}.$$

Since $\nabla f(x,y) = f_x(x,y)\mathbf{i} + f_y(x,y)\mathbf{j} = 2x\mathbf{i} + 8y\mathbf{j}$, the gradient at $(3,1)$ is

$$\nabla f(3,1) = 6\mathbf{i} + 8\mathbf{j}.$$

Consequently, at $(3,1)$ the directional derivative is

$$D_\mathbf{u} f(3,1) = \nabla f(3,1) \cdot \mathbf{u}$$
$$= (6\mathbf{i} + 8\mathbf{j}) \cdot \left(-\frac{\sqrt{2}}{2}\mathbf{i} - \frac{\sqrt{2}}{2}\mathbf{j}\right)$$
$$= -3\sqrt{2} - 4\sqrt{2} = -7\sqrt{2}.$$

23. Find the gradient of the function $h(x, y) = x \tan y$ and the maximum value of the directional derivative at the point $P = (2, \pi/4)$.

Solution:

The gradient vector is given by

$$\nabla f(x, y) = f_x(x, y)\,\mathbf{i} + f_y(x, y)\,\mathbf{j} = \tan y\,\mathbf{i} + x \sec^2 y\,\mathbf{j}.$$

At the point $P = (2, \pi/4)$ the gradient is

$$\nabla f(2, \pi/4) = \tan \frac{\pi}{4}\,\mathbf{i} + 2 \sec^2 \frac{\pi}{4}\,\mathbf{j} = \mathbf{i} + 4\,\mathbf{j}.$$

Hence, it follows that the maximum value of the directional derivative at the point $P = (2, \pi/4)$ is

$$\|\nabla f(2, \pi/4)\| = \sqrt{17}.$$

27. Find the gradient of the function $f(x, y, z) = \sqrt{x^2 + y^2 + z^2}$ and the maximum value of the directional derivative at the point $(1, 4, 2)$.

Solution:

The gradient vector is given by

$$\begin{aligned}
\nabla f(x, y, z) \\
&= f_x(x, y, z)\,\mathbf{i} + f_y(x, y, z)\,\mathbf{j} + f_z(x, y, z)\,\mathbf{k} \\
&= \frac{x}{\sqrt{x^2 + y^2 + z^2}}\,\mathbf{i} + \frac{y}{\sqrt{x^2 + y^2 + z^2}}\,\mathbf{j} + \frac{z}{\sqrt{x^2 + y^2 + z^2}}\,\mathbf{k} \\
&= \frac{x\,\mathbf{i} + y\,\mathbf{j} + z\,\mathbf{k}}{\sqrt{x^2 + y^2 + z^2}}.
\end{aligned}$$

At the point $(1, 4, 2)$ the gradient is

$$\nabla f(1, 4, 2,) = \frac{1}{\sqrt{21}}(\mathbf{i} + 4\,\mathbf{j} + 2\,\mathbf{k}).$$

Hence, the maximum value of the directional derivative at the point $(1, 4, 2)$ is

$$\|\nabla f(1, 4, 2)\| = \frac{1}{\sqrt{21}}\sqrt{1 + 16 + 4} = 1.$$

33. If $f(x, y) = 3 - (x/3) - (y/2)$, find $D_{\mathbf{u}} f(3, 2)$ if (a) $\theta = 4\pi/3$ and (b) $\theta = -\pi/6$

Solution:

(a) By Theorem 13.9 the directional derivative is

$$D_{\mathbf{u}} f(x, y) = f_x(x, y) \cos \theta + f_y(x, y) \sin \theta$$
$$= -\frac{1}{3} \cos \theta - \frac{1}{2} \sin \theta.$$

For $\theta = 4\pi/3$, $x = 3$, and $y = 2$, you have

$$D_{\mathbf{u}} f(3, 2) = -\frac{1}{3} \cos \frac{4\pi}{3} - \frac{1}{2} \sin \frac{4\pi}{3}$$
$$= -\frac{1}{3} \left(-\frac{1}{2} \right) - \frac{1}{2} \left(-\frac{\sqrt{3}}{2} \right) = \frac{2 + 3\sqrt{3}}{12}.$$

(b) For $\theta = -\frac{\pi}{6}$, $x = 3$, and $y = 2$, you have

$$D_{\mathbf{u}} f(3, 2) = -\frac{1}{3} \cos \left(-\frac{\pi}{6} \right) - \frac{1}{2} \sin \left(-\frac{\pi}{6} \right)$$
$$= -\frac{1}{3} \left(\frac{\sqrt{3}}{2} \right) - \frac{1}{2} \left(-\frac{1}{2} \right) = \frac{3 - 2\sqrt{3}}{12}.$$

35. If $f(x, y) = 3 - (x/3) - (y/2)$, find $D_{\mathbf{v}} f(3, 2)$ if (a) $\mathbf{v}$ is the vector from $(1, 2)$ to $(-2, 6)$ and (b) $\mathbf{v}$ is the vector from $(3, 2)$ to $(4, 5)$.

Solution:

(a) Let $\mathbf{u}$ be a unit vector in the direction of $\mathbf{v}$. Then

$$\mathbf{v} = (-2 - 1)\mathbf{i} + (6 - 2)\mathbf{j} = -3\mathbf{i} + 4\mathbf{j}$$

and

$$\mathbf{u} = \frac{\mathbf{v}}{\|\mathbf{v}\|} = \frac{-3\mathbf{i} + 4\mathbf{j}}{\sqrt{25}} = -\frac{3}{5}\mathbf{i} + \frac{4}{5}\mathbf{j}.$$

At $(3, 2)$,

$$\nabla f(3, 2) = f_x(3, 2)\mathbf{i} + f_y(3, 2)\mathbf{j} = -\frac{1}{3}\mathbf{i} - \frac{1}{2}\mathbf{j}.$$

Therefore, the directional derivative in the direction of **v** is

$$\nabla f(3,2) \cdot \mathbf{u} = \left(-\frac{1}{3}\right)\left(-\frac{3}{5}\right) + \left(-\frac{1}{2}\right)\left(\frac{4}{5}\right) = \frac{1}{5} - \frac{2}{5} = -\frac{1}{5}.$$

(b) Let **u** be a unit vector in the direction of **v**. Then

$$\mathbf{v} = (4-3)\mathbf{i} + (5-2)\mathbf{j} = \mathbf{i} + 3\mathbf{j}$$

and

$$\mathbf{u} = \frac{\mathbf{v}}{\|\mathbf{v}\|} = \frac{\mathbf{i} + 3\mathbf{j}}{\sqrt{10}} = \frac{\sqrt{10}}{10}\mathbf{i} + \frac{3\sqrt{10}}{10}\mathbf{j}.$$

Therefore, the directional derivative in the direction of **v** is

$$\nabla f(3,2) \cdot \mathbf{u} = \left(-\frac{1}{3}\right)\left(\frac{\sqrt{10}}{10}\right) + \left(-\frac{1}{2}\right)\left(\frac{3\sqrt{10}}{10}\right) = -\frac{11\sqrt{10}}{60}.$$

37. If $f(x,y) = 3 - (x/3) - (y/2)$, find the maximum value of the directional derivative at $(3, 2)$.

Solution:

By Theorem 13.11 the maximum value of the directional derivative is $\|\nabla f(3,2)\|$.

$$f(x,y) = 3 - \frac{x}{3} - \frac{y}{2}$$

$$\nabla f(x,y) = f_x(x,y)\mathbf{i} + f_y(x,y)\mathbf{j} = -\frac{1}{3}\mathbf{i} - \frac{1}{2}\mathbf{j}$$

Therefore, the maximum value of the directional derivative at $(3, 2)$ is

$$\|\nabla f(3,2)\| = \sqrt{\frac{1}{9} + \frac{1}{4}} = \frac{\sqrt{13}}{6}.$$

49. Use the gradient to find a unit normal vector to the graph of $9x^2 + 4y^2 = 40$ at $P = (2, -1)$. Sketch the results.

Solution:

The ellipse given by the equation $9x^2 + 4y^2 = 40$ corresponds to the level curve with $c = 0$ to the function

$$f(x, y) = 9x^2 + 4y^2 - 40.$$

By Theorem 13.12, $\nabla f(x_0, y_0)$ yields a normal vector to the level curve at the point (x_0, y_0). Therefore,

$$\nabla f(x, y) = 18x\,\mathbf{i} + 8y\,\mathbf{j}$$

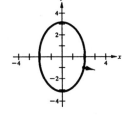

and at $(2, -1)$, a normal vector is

$$\nabla f(2, -1) = 36\,\mathbf{i} - 8\,\mathbf{j} = 4(9\,\mathbf{i} - 2\,\mathbf{j})$$

Since, $\|\nabla f(2, -1)\| = 4\sqrt{9^2 + (-2)^2} = 4\sqrt{85}$, a unit normal vector is

$$\frac{\sqrt{85}}{85}(9\,\mathbf{i} - 2\,\mathbf{j}).$$

51. The temperature field at any point on a metal plate is given by $T = x/(x^2 + y^2)$. Find the direction of greatest increase in heat at the point $(3, 4)$.

Solution:

The direction of greatest increase in temperature at $(3, 4)$ will be the direction of the gradient $\nabla T(x, y)$ at that point. Since

$$T_x(x, y) = \frac{(x^2 + y^2)(1) - x(2x)}{(x^2 + y^2)^2} = \frac{y^2 - x^2}{(x^2 + y^2)^2}$$

$$T_y(x, y) = \frac{(x^2 + y^2)(0) - x(2y)}{(x^2 + y^2)^2} = \frac{-2xy}{(x^2 + y^2)^2},$$

the gradient at $(3, 4)$ is

$$\nabla T(3, 4) = T_x(3, 4)\,\mathbf{i} + T_y(3, 4)\,\mathbf{j}$$

$$= \frac{7}{(25)^2}\,\mathbf{i} - \frac{24}{(25)^2}\,\mathbf{j} = \frac{1}{625}(7\,\mathbf{i} - 24\,\mathbf{j}).$$

13.7   Tangent Planes and Normal Lines

---

7. Find a unit normal vector to the surface $z - x \sin y = 4$ at the point $(6, \pi/6, 7)$.

Solution:

Writing the equation for the surface as a function of three variables yields

$$F(x, y, z) = z - x \sin y - 4.$$

By Theorem 13.13, a normal vector to the surface, $F(x, y, z) = 0$, at (x_0, y_0, z_0) is given by $\nabla F(x_0, y_0, z_0)$.

$$\nabla F(x, y, z) = F_x(x, y, z)\,\mathbf{i} + F_y(x, y, z)\,\mathbf{j} + F_z(x, y, z)\,\mathbf{k}$$
$$= -\sin y\,\mathbf{i} - x \cos y\,\mathbf{j} + \mathbf{k}$$
$$\nabla F(6, \pi/6, 7) = -\frac{1}{2}\mathbf{i} - 3\sqrt{3}\,\mathbf{j} + \mathbf{k}$$

Now, the unit normal vector to the surface is

$$\frac{\nabla F(6, \pi/6, 7)}{\|\nabla F(6, \pi/6, 7)\|} = \frac{\sqrt{113}}{113}(-\mathbf{i} - 6\sqrt{3}\,\mathbf{j} + 2\,\mathbf{k}).$$

---

13. Find the equation of the tangent plane for the function $f(x, y) = y/x$ at the point $(1, 2, 2)$.

Solution:

Begin by writing the equation of the surface as

$$\frac{y}{x} - z = 0.$$

Then considering

$$F(x, y, z) = \frac{y}{x} - z$$

you have

$$F_x(x, y, z) = \frac{-y}{x^2}, \quad F_y(x, y, z) = \frac{1}{x}, \quad \text{and} \quad F_z(x, y, z) = -1.$$

At the point $(1, 2, 2)$, the partial derivatives are

$$F_x(1,2,2) = -2, \quad F_y(1,2,2) = 1, \quad \text{and} \quad F_z(1,2,2) = -1$$

Therefore, by Theorem 13.13, the equation of the tangent plane at $(1, 2, 2)$ is

$$F_x(1,2,2)(x-1) + F_y(1,2,2)(y-2) + F_z(1,2,2)(z-2) = 0$$
$$-2(x-1) + 1(y-2) - (z-2) = 0$$
$$-2x + y - z + 2 = 0.$$

23. Find the equation of the tangent plane to $xy^2 + 3x - z^2 = 4$ at the point $(2, 1, -2)$.

Solution:

Let $F(x, y, z) = xy^2 + 3x - z^2 - 4$. Then $\nabla F(2, 1, -2)$ is normal to the tangent plane at $(2, 1, -2)$.

$$\nabla F(x, y, z) = F_x(x, y, z)\mathbf{i} + F_y(x, y, z)\mathbf{j} + F_z(x, y, z)\mathbf{k}$$
$$= (y^2 + 3)\mathbf{i} + 2xy\mathbf{j} - 2z\mathbf{k}$$
$$\nabla F(2, 1, -2) = 4\mathbf{i} + 4\mathbf{j} + 4\mathbf{k}$$

Therefore, the equation of the tangent plane is

$$4(x-2) + 4(y-1) + 4(z+2) = 0$$
$$x + y + z = 1.$$

25. Find an equation for the tangent plane and find symmetric equations of the normal line to the surface $x^2 + y^2 + z = 9$ at the point $(1, 2, 4)$.

Solution:

Let $F(x, y, z) = x^2 + y^2 + z - 9$. Then $\nabla F(1, 2, 4)$ is normal to the surface at $(1, 2, 4)$.

$$\nabla F(x, y, z) = F_x(x, y, z)\mathbf{i} + F_y(x, y, z)\mathbf{j} + F_z(x, y, z)\mathbf{k}$$
$$= 2x\mathbf{i} + 2y\mathbf{j} + \mathbf{k}$$
$$\nabla F(1, 2, 4) = 2\mathbf{i} + 4\mathbf{j} + \mathbf{k}$$

Therefore, the equation of the tangent plane is

$$2(x-1)+4(y-2)+1(z-4)=0$$
$$2x+4y+z=14.$$

Since the normal line to the surface at $(1, 2, 4)$ is parallel to $\nabla F(1,2,4)$, the direction numbers for the line are 2, 4, and 1. Therefore, symmetric equations for a normal line at $(1,2,4)$ are

$$\frac{x-1}{2}=\frac{y-2}{4}=\frac{z-4}{1}.$$

29. Find an equation for the tangent plane and find symmetric equations of the normal line to the surface $z = \arctan(y/x)$ at the point $(1,1,\pi/4)$.

Solution:

Let $F(x,y,z) = \arctan(y/x) - z$. Then

$$\nabla F(x,y,z) = F_x(x,y,z)\,\mathbf{i} + F_y(x,y,z)\,\mathbf{j} + F_z(x,y,z)\,\mathbf{k}$$
$$= \frac{-y}{x^2+y^2}\,\mathbf{i} + \frac{x}{x^2+y^2}\,\mathbf{j} - \mathbf{k}$$
$$\nabla F(1,1,\pi/4) = -\frac{1}{2}\,\mathbf{i} + \frac{1}{2}\,\mathbf{j} - \mathbf{k} = -\frac{1}{2}(\mathbf{i}-\mathbf{j}+2\,\mathbf{k}).$$

Since $\nabla F(1,1,\pi/4)$ is normal to the surface at the point $(1,1,\pi/4)$, an equation of the tangent plane is

$$(x-1)-(y-1)+2\left(z-\frac{\pi}{4}\right)=0$$
$$x-y+2z=\frac{\pi}{2}$$

and symmetric equations for the normal line are

$$\frac{x-1}{1}=\frac{y-1}{-1}=\frac{z-(\pi/4)}{2}.$$

31. For the surfaces described by $x^2 + y^2 = 5$ and $z = x$, (a) find symmetric equations of the tangent line to their curve of intersection at the point (2, 1, 2), and (b) find the cosine of the angle between the gradients of the surfaces at the same point. State whether the surfaces are orthogonal at the point of intersection.

Solution:

Let $f(x, y, z) = x^2 + y^2 - 5$ and $g(x, y, z) = x - z$. Then

$$\nabla f(x, y, z) = 2x\,\mathbf{i} + 2y\,\mathbf{j} \implies \nabla f(2, 1, 2) = 4\mathbf{i} + 2\mathbf{j}$$

and

$$\nabla g(x, y, z) = \mathbf{i} - \mathbf{k} \implies \nabla g(2, 1, 2) = \mathbf{i} - \mathbf{k}.$$

(a) Since ∇f and ∇g are each normal to their respective surfaces, the vector $\nabla f \times \nabla g$ will be tangent to both surfaces at the point (2, 1, 2) on the curve of intersection. Therefore, from

$$\nabla f \times \nabla g = \begin{vmatrix} \mathbf{i} & \mathbf{j} & \mathbf{k} \\ 4 & 2 & 0 \\ 1 & 0 & -1 \end{vmatrix} = -2\mathbf{i} + 4\mathbf{j} - 2\mathbf{k} = -2(\mathbf{i} - 2\mathbf{j} + \mathbf{k})$$

it follows that direction numbers for the tangent line are 1, -2, and 1. Hence symmetric equations for the tangent line at (2, 1, 2) are

$$\frac{x-2}{1} = \frac{y-1}{-2} = \frac{z-2}{1}.$$

(b) The angle between ∇f and ∇g at (2, 1, 2) is such that

$$\cos\theta = \frac{\nabla f \cdot \nabla g}{\|\nabla f\|\,\|\nabla g\|} = \frac{4 + 0 - 0}{\sqrt{20}\sqrt{2}} = \frac{4}{\sqrt{40}} = \frac{4}{2\sqrt{10}} = \frac{\sqrt{10}}{5}.$$

Therefore, the surfaces are **not** orthogonal at the point of intersection.

45. Show that the tangent plane to the ellipsoid

$$\frac{x^2}{a^2} + \frac{y^2}{b^2} + \frac{z^2}{c^2} = 1$$

at the point (x_0, y_0, z_0) can be written in the form

$$\frac{x_0 x}{a^2} + \frac{y_0 y}{b^2} + \frac{z_0 z}{c^2} = 1.$$

Solution:

We let $F(x, y, z) = \dfrac{x^2}{a^2} + \dfrac{y^2}{b^2} + \dfrac{z^2}{c^2} - 1$. Then

$$\nabla F(x, y, z) = F_x(x, y, z)\,\mathbf{i} + F_y(x, y, z)\,\mathbf{j} + f_z(x, y, z)\,\mathbf{k}$$

$$= \frac{2x}{a^2}\,\mathbf{i} + \frac{2y}{b^2}\,\mathbf{j} + \frac{2z}{c^2}\,\mathbf{k}$$

$$\nabla F(x_0, y_0, z_0) = 2\left[\frac{x_0}{a^2}\,\mathbf{i} + \frac{y_0}{b^2}\,\mathbf{j} + \frac{z_0}{c^2}\,\mathbf{k}\right]$$

Now since $\nabla F(x_0, y_0, z_0)$ is normal to the surface at the point (x_0, y_0, z_0), an equation of the tangent plane is

$$\frac{x_0}{a^2}(x - x_0) + \frac{y_0}{b^2}(y - y_0) + \frac{z_0}{c^2}(z - z_0) = 0$$

$$\left[\frac{x_0 x}{a^2} + \frac{y_0 y}{b^2} + \frac{z_0 z}{c^2}\right] - \left[\frac{x_0{}^2}{a^2} + \frac{y_0{}^2}{b^2} + \frac{z_0{}^2}{c^2}\right] = 0$$

$$\left[\frac{x_0 x}{a^2} + \frac{y_0 y}{b^2} + \frac{z_0 z}{c^2}\right] - 1 = 0$$

$$\frac{x_0 x}{a^2} + \frac{y_0 y}{b^2} + \frac{z_0 z}{c^2} = 1.$$

13.8   Extrema of Functions of Two Variables

13. Examine the function $h(x, y) = x^2 - y^2 - 2x - 4y - 4$ for relative extrema and saddle points.

Solution:

Since $h_x(x, y) = 2x - 2 = 2(x - 1) = 0$ when $x = 1$ and
$h_y(x, y) = -2y - 4 = -2(y + 2) = 0$ when $y = -2$, there is one
critical point, $(1, -2)$. Since

$$h_{xx}(x, y) = 2, \; h_{yy}(x, y) = -2, \text{ and } h_{xy}(x, y) = 0,$$

you have

$$d = h_{xx}(1, -2)h_{yy}(1, -2) - [h_{xy}(1, -2)]^2 = -4 - 0 = -4 < 0$$

Therefore, by part 3 of Theorem 13.17, you conclude that there
is a saddle point at $(1, -2, -1)$.

17. Examine the function $f(x, y) = x^3 - 3xy + y^3$ for relative
extrema and saddle points.

Solution:

Since $f_x(x, y) = 3x^2 - 3y$ and $f_y(x, y) = -3x + 3y^2$, any critical
points must be the simultaneous solutions of the system of
equations
$$3x^2 - 3y = 0 \quad \text{and} \quad 3y^2 - 3x = 0.$$

From the first equation it follows that $y = x^2$. Making this
substitution for y in the second equation yields

$$3(x^2)^2 - 3x = 0$$
$$x^4 - x = 0$$
$$x(x^3 - 1) = 0$$

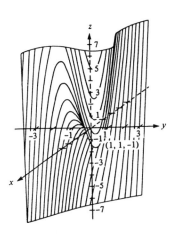

Therefore, the critical points are $(0, 0)$ and $(1, 1)$. Since

$$f_{xx}(x, y) = 6x, \quad f_{yy}(x, y) = 6y, \quad f_{xy}(x, y) = -3,$$

you have $f_{xx}(0, 0) = 0$ and

$$d = f_{xx}(0, 0)f_{yy}(0, 0) - [f_{xy}(0, 0)]^2 = 0 - 9 < 0.$$

Therefore, by Theorem 13.17, you can conclude that
$(0, 0, 0)$ is a saddle point of f. At $(1, 1)$ you have
$f(1, 1) = -1$, $f_{xx}(1, 1) = 6 > 0$ and

$$d = f_{xx}(1, 1)f_{yy}(1, 1) - [f_{xy}(1, 1)]^2 = (6)(6) - 9 > 0.$$

By Theorem 13.17, the point $(1, 1, -1)$ is a relative minimum.

35. Find the absolute extrema of $f(x, y) = x^2 + 2xy + y^2$ over the region $R = \{(x, y) : x^2 + y^2 \leq 8\}$.

Solution:

We first observe that f is a perfect square trinomial and can be written as

$$f(x, y) = (x + y)^2.$$

Therefore, the minimum value of f is zero and occurs at all points in R such that $y = -x$. From the partial derivatives, you have

$$f_x(x, y) = 2(x + y) = f_y(x, y)$$

Each point lying on the line $y = -x$ in the xy-coordinate plane is a critical point. However, from the discussion above, you know these points always yield the minimum value of zero. Thus, by Theorem 13.15, the maximum must occur at point(s) on the boundary of R. There the function f will be maximum when the absolute value of $x + y$ is maximum. Thus, you have

$$f(-2, -2) = f(2, 2) = 16$$

as the absolute maxima of f over the region R.

41. For the function $f(x, y) = x^{2/3} + y^{2/3}$, find the critical points and text for relative extrema. List the critical points for which the Second-Partials Test fails.

Solution:

From the first partial derivatives you have

$$f_x(x, y) = \frac{2}{3}x^{-1/3} = \frac{2}{3\sqrt[3]{x}}$$

and

$$f_y(x, y) = \frac{2}{3}y^{-1/3} = \frac{2}{3\sqrt[3]{y}}.$$

Since neither first partial derivative exists at $(0, 0)$, it is a critical point. Moreover,

$$f_{xx}(x, y) = -\frac{2}{9x\sqrt[3]{x}} \text{ and } f_{yy}(x, y) = -\frac{2}{9x\sqrt[3]{x}}$$

do not exist at $(0, 0)$ and thus the Second-Partials Test fails. Since $x^{2/3} + y^{2/3} \geq 0$ for all x and y, it follows that the absolute minimum of f is $f(0, 0) = 0$.

13.9   Applications of Extrema of Functions of Two Variables

7. Find three positive numbers whose sum is 30 and the sum of whose squares is minimum.

Solution:

Let x, y, and z be the numbers and let $s = x^2 + y^2 + z^2$. Since $x + y + z = 30$, it is necessary to minimize

$$s = x^2 + y^2 + (30 - x - y)^2$$

Setting the first partial derivatives equal to zero yields

$$s_x = 2x - 2(30 - x - y) = 0 \quad \Longrightarrow \quad 2x + y = 30$$
$$s_y = 2y - 2(30 - x - y) = 0 \quad \Longrightarrow \quad 2x + 4y = 60$$

Subtracting the first equation from the second you obtain the equation $3y = 30$ or $y = 10$. Substituting this value into the previous equations and solving yields the critical values $x = y = z = 10$. These values give us the desired minimum, since $s_{xx}(10, 10) = 4 > 0$ and

$$s_{xx}(10, 10)s_{yy}(10, 10) - [s_{xy}(10, 10)]^2 = (4)(4) - 2^2 > 0.$$

11. A waterline is built from point P to point S and must pass through regions where construction costs differ (see figure). Find x and y so the total cost C will be minimum if the cost per mile in dollars is $3k$ from P to Q, $2k$ from Q to R, and k from R to S.

Solution:

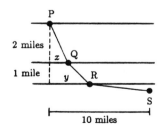

$$C = (\text{cost per mile})(\text{distance from } P \text{ to } Q)$$
$$+ (\text{cost per mile})(\text{distance from } Q \text{ to } R)$$
$$+ (\text{cost per mile})(\text{distance from } R \text{ to } S)$$
$$= 3k\sqrt{x^2 + 4} + 2k\sqrt{(y - x)^2 + 1} + k(10 - y)$$

Setting the first partials equal to zero yields the system

$$\frac{\partial C}{\partial x} = k\left[\frac{3x}{\sqrt{x^2+4}} + \frac{-2(y-x)}{\sqrt{(y-x)^2+1}}\right] = 0$$

$$\frac{\partial C}{\partial y} = k\left[\frac{2(y-x)}{\sqrt{(y-x)^2+1}} - 1\right] = 0$$

From the equation $\partial C/\partial y = 0$ you have,

$$\frac{2(y-x)}{\sqrt{(y-x)^2+1}} = 1$$

$$2(y-x) = \sqrt{(y-x)^2+1}$$
$$4(y-x)^2 = (y-x)^2+1$$
$$3(y-x)^2 = 1$$
$$y-x = \pm\frac{1}{\sqrt{3}} \implies y = x \pm \frac{1}{\sqrt{3}}.$$

Substituting the result with the positive root into the equation $\partial C/\partial x = 0$, you obtain

$$\frac{3x}{\sqrt{x^2+4}} = \frac{2(y-x)}{\sqrt{(y-x)^2+1}}$$

$$\frac{3x}{\sqrt{x^2+4}} = \frac{2/\sqrt{3}}{\sqrt{4/3}}$$

$$3x = \sqrt{x^2+4}$$
$$9x^2 = x^2+4$$
$$8x^2 = 4.$$

Therefore, you have

$$x = \frac{1}{\sqrt{2}} \approx 0.707 \text{ mi}$$

$$y = x + \frac{\sqrt{3}}{3} = \frac{3\sqrt{2}+2\sqrt{3}}{6} \approx 1.284 \text{ miles.}$$

17. An eaves trough with trapezoidal cross sections is formed by turning up the edges of a sheet of aluminum which is w inches wide. Find the cross section of maximum area.

Solution:

From the accompanying figure observe that the area of a trapezoidal cross section is given by

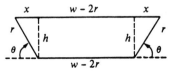

$$A = h\left[\frac{(w - 2r) + [(w - 2r) + 2x]}{2}\right]$$

$$= (w - 2r + x)h$$

where $x = r\cos\theta$ and $h = r\sin\theta$. Substituting these expressions for x and h, you have

$$A(r, \theta) = (w - 2r + r\cos\theta)(r\sin\theta)$$
$$= wr\sin\theta - 2r^2\sin\theta + r^2\sin\theta\cos\theta$$
$$A_r(r, \theta) = w\sin\theta - 4r\sin\theta + 2r\sin\theta\cos\theta$$
$$= \sin\theta(w - 4r + 2r\cos\theta) = 0 \implies w = r(4 - 2\cos\theta)$$

and

$$A_\theta(r, \theta) = wr\cos\theta - 2r^2\cos\theta + r^2\cos 2\theta = 0.$$

Substituting the expression for w from $A_r(r, \theta) = 0$ into the equation $A_\theta(r, \theta) = 0$, you have

$$r^2(4 - 2\cos\theta)\cos\theta - 2r^2\cos\theta + r^2(2\cos^2\theta - 1) = 0$$
$$r^2(2\cos\theta - 1) = 0 \text{ or } \cos\theta = \frac{1}{2}.$$

The first partial derivatives are zero when $\theta = \pi/3$ and $r = w/3$. (Ignore the solution $r = \theta = 0$.) Thus, the trapezoid of maximum area occurs when each edge of width $w/3$ is turned up 60° from the horizontal.

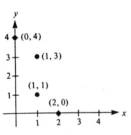

25. For the points in the accompanying figure, (a) find the least squares regression line, and (b) calculate S, the sum of squared errors.

Solution:

(a)

x	y	xy	x^2
0	4	0	0
1	3	3	1
1	1	1	1
2	0	0	4
$\sum x_i = 4$	$\sum y_i = 8$	$\sum x_i y_i = 4$	$\sum x_i^2 = 6$

By Theorem 13.18, you have

$$a = \frac{n \sum x_i y_i - \sum x_i \sum y_i}{n \sum x_i^2 - \left(\sum x_i \right)^2} = \frac{4(4) - 4(8)}{4(6) - 4^2} = -2$$

$$b = \frac{1}{n}\left(\sum y_i - a \sum x_i \right) = \frac{1}{4}[8 + 2(4)] = 4.$$

Therefore, the least squares regression line is

$$f(x) = -2x + 4.$$

(b) $S = \sum [f(x_i) - y_i]^2$
$$= (4 - 4)^2 + (2 - 3)^2 + (2 - 1)^2 + (0 - 0)^2 = 2.$$

39. Use the result of Exercise 35 to find the least squares regression quadratic for the points $(0, 0)$, $(2, 2)$, $(3, 6)$, and $(4, 12)$.

Solution:

From Exercise 35, you have that the least squares regression quadratic for the points $(x_1, y_1), (x_2, y_2), \ldots, (x_n, y_n)$ is

$$y = ax^2 + bx + c$$

where a, b, and c are the solutions to the system

$$a \sum_{i=1}^{n} x_i^{\,4} + b \sum_{i=1}^{n} x_i^{\,3} + c \sum_{i=1}^{n} x_i^{\,2} = \sum_{i=1}^{n} x_i^{\,2} y_i$$

$$a \sum_{i=1}^{n} x_i^{\,3} + b \sum_{i=1}^{n} x_i^{\,2} + c \sum_{i=1}^{n} x_i = \sum_{i=1}^{n} x_i y_i$$

$$a \sum_{i=1}^{n} x_i^{\,2} + b \sum_{i=1}^{n} x_i + cn = \sum_{i=1}^{n} y_i.$$

For the given points, you have

$$\sum x_i = 9, \quad \sum x_i^{\,2} = 29, \quad \sum x_i^{\,3} = 99, \quad \sum x_i^{\,4} = 353$$

$$\sum y_i = 20, \quad \sum x_i y_i = 70, \quad \sum x_i^{\,2} y_i = 192.$$

The resulting system of equations is

$$353a + 99b + 29c = 192$$
$$99a + 29b + 9c = 70$$
$$29a + 9b + 4c = 20$$

Solving this system yields $a = 1$, $b = -1$, and $c = 0$. Therefore, the least squares regression quadratic is

$$y = x^2 - x.$$

13.10   Lagrange Multipliers

5. Use Lagrange multipliers to find the minimum of
$f(x, y) = x^2 - y^2$ subject to the constraint $x - 2y + 6 = 0$.
Assume x and y are positive.

Solution:

Begin by letting
$$g(x, y) = x - 2y + 6.$$

Then, since $\nabla f(x, y) = 2x\,\mathbf{i} - 2y\,\mathbf{j}$ and $\lambda \nabla g(x, y) = \lambda(\mathbf{i} - 2\mathbf{j})$, you have the following system of equations.

$$2x = \lambda \qquad f_x(x,y) = \lambda g_x(x,y)$$
$$-2y = -2\lambda \qquad f_y(x,y) = \lambda g_y(x,y)$$
$$x - 2y + 6 = 0 \qquad Constraint$$

From the first equation you have $\lambda = 2x$. Substituting this result into the second equation, you have $y = 2x$. Substituting this result into the constraint yields

$$x - 2(2x) + 6 = 0$$
$$-3x = -6 \quad \Longrightarrow \quad x = 2.$$

Since $y = 2x$, you have $y = 4$ and the required minimum is

$$f(2,4) = 2^2 - 4^2 = -12.$$

17. Use Lagrange multipliers to find the maximum of $f(x,y,z) = xyz$, subject to the constraints $x + y + z = 32$ and $x - y + z = 0$. Assume x, y, and z are positive.

Solution:

In this case you have two constraints which you can denote by g and h.

$$g(x,y,z) = x + y + z - 32$$
$$h(x,y,z) = x - y + z$$

Since $\nabla f(x,y,z) = yz\,\mathbf{i} + xz\,\mathbf{j} + xy\,\mathbf{j}$, $\lambda \nabla g(x,y,z) = \lambda\mathbf{i} + \lambda\mathbf{j} + \lambda\mathbf{k}$, and $\mu \nabla h(x,y,z) = \mu\mathbf{i} - \mu\mathbf{j} + \mu\mathbf{k}$, you have the following system of equations.

$$yz = \lambda + \mu \qquad f_x(x,y,z) = \lambda g_x(x,y,z) + \mu h_x(x,y,z)$$
$$xz = \lambda - \mu \qquad f_y(x,y,z) = \lambda g_y(x,y,z) + \mu h_y(x,y,z)$$
$$xy = \lambda + \mu \qquad f_z(x,y,z) = \lambda g_z(x,y,z) + \mu h_z(x,y,z)$$
$$x + y + z = 32 \qquad Constraint\ 1$$
$$x - y + z = 0 \qquad Constraint\ 2$$

From the first and third equations you have

$$yz = xy \quad \text{or} \quad z = x.$$

Substituting this result into the second constraint, you have

$$y = 2x.$$

Finally, substituting these two results into the first constraint yields

$$4x = 32 \implies x = 8$$
$$y = 2x = 16$$
$$z = x = 8.$$

Therefore, the maximum value of f, subject to the given constraints is

$$f(8, 16, 8) = 8(16)(8) = 1024.$$

20. Use Lagrange multipliers to find the maximum of $f(x, y, z) = xyz$ subject to the constraints $x^2 + z^2 = 5$ and $x - 2y = 0$. Assume x, y, and z are positive.

Solution:

In this case you have two constraints given by f and g.

$$g(x, y, z) = x^2 + z^2 - 5$$
$$h(x, y, z) = x - 2y$$

Since $\nabla f(x, y, z) = yz\,\mathbf{i} + xz\,\mathbf{j} + xy\,\mathbf{k}$, $\lambda\nabla g(x, y, z) = 2\lambda x\,\mathbf{i} + 2\lambda z\,\mathbf{k}$, and $\mu\nabla h(x, y, z) = \mu\mathbf{i} - 2\mu\mathbf{j}$, you have the following system of equations.

$$yz = 2\lambda x + \mu \qquad f_x(x, y, z) = \lambda g_x(x, y, z) + \mu h_x(x, y, z)$$
$$xz = -2\mu \qquad f_y(x, y, z) = \lambda g_y(x, y, z) + \mu h_y(x, y, z)$$
$$xy = 2\lambda z \qquad f_z(x, y, z) = \lambda g_z(x, y, z) + \mu h_z(x, y, z)$$
$$x^2 + z^2 = 5 \qquad Constraint\ 1$$
$$x - 2y = 0 \qquad Constraint\ 2$$

From the second constraint it follows that $x = 2y$. Substituting this into second, third and fourth equations above, you obtain

$$2yz = -2\mu \implies \mu = -yz$$
$$2y^2 = 2\lambda z \implies \lambda = \frac{y^2}{z}$$
$$4y^2 + z^2 = 5 \implies z^2 = 5 - 4y^2$$

Now substitute these results into the first equation above to obtain

$$yz = 2\left(\frac{y^2}{z}\right)(2y) + (-yz)$$

$$2yz^2 - 4y^3 = 0$$
$$2y(z^2 - 2y^2) = 0$$
$$2y(5 - 4y^2 - 2y^2) = 0$$
$$2y(5 - 6y^2) = 0$$

$$y = 0 \quad \text{or} \quad y = \pm\sqrt{\frac{5}{6}}$$

Since x, y, and z are postive, you have $y = \sqrt{\frac{5}{6}}$, $x = 2\sqrt{\frac{5}{6}}$ and $z = \sqrt{\frac{10}{6}} = \sqrt{\frac{5}{3}}$. Therefore, the maximum value of f is

$$f\left(2\sqrt{\frac{5}{6}}, \sqrt{\frac{5}{6}}, \sqrt{\frac{5}{3}}\right) = \left(2\sqrt{\frac{5}{6}}\right)\left(\sqrt{\frac{5}{6}}\right)\sqrt{\frac{5}{3}} = \frac{5}{3}\sqrt{\frac{5}{3}}.$$

25. Use Lagrange multipliers to find the minimum distance from the point $(2, 1, 1)$ to the plane $x + y + z = 1$.

Solution:

Let (x, y, z) be an arbitrary point in the given plane. Then

$$s = \sqrt{(x-2)^2 + (y-1)^2 + (z-1)^2}$$

represents the distance between $(2, 1, 1)$ and a point in the plane. To simplify our calculations, minimize s^2 rather than s. With $g(x, y, z) = x + y + z - 1$ as the constraint, you have $\nabla s^2 = 2(x-2)\mathbf{i} + 2(y-1)\mathbf{j} + 2(z-1)\mathbf{k}$ and $\lambda\nabla g(x, y, z) = \lambda\mathbf{i} + \lambda\mathbf{j} + \lambda\mathbf{k}$. Therefore,

$$2(x-2) = \lambda \qquad s_x{}^2(x,y,z) = \lambda g_x(x,y,z)$$
$$2(y-1) = \lambda \qquad s_y{}^2(x,y,z) = \lambda g_y(x,y,z)$$
$$2(z-1) = \lambda \qquad s_z{}^2(x,y,z) = \lambda g_z(x,y,z)$$
$$x + y + z - 1 = 0 \qquad Constraint.$$

From the first three equations, you can conclude that

$$\lambda = 2(x-2) = 2(y-1) = 2(z-1)$$

or that $x = y + 1$ and $z = y$. Therefore, from the constraint, you have

$$(y + 1) + y + y - 1 = 3y = 0 \quad \text{or} \quad y = 0.$$

Thus $x = 1$ and $z = 0$ and the point $(1, 0, 0)$ in the plane $x + y + z = 1$, is closest to the given point $(2, 1, 1)$. The minimum distance is

$$s = \sqrt{(1 - 2)^2 + (0 - 1)^2 + (0 - 1)^2} = \sqrt{3}.$$

31. A cargo container (in the shape of a rectangular solid) must have a volume of 480 cubic feet. Use Lagrange multipliers to find the dimensions of the container of this size which has minimum cost if the bottom will cost \$5 per square foot to construct and the sides and top will cost \$3 per square foot to construct.

Solution:

Letting x, y, and z be the length, width and height of the solid, respectively, it is necessary to minimize the cost function

$$C(x, y, z) = 5xy + 3(2yz + xy + 2xz) = 8xy + 6yz + 6xz$$

subject to the constraint $xyz = 480$. First, write the constraint as

$$g(x, y, z) = xyz - 480.$$

Then, since $\nabla C(x, y, z) = (8y + 6z)\mathbf{i} + (8x + 6z)\mathbf{j} + (6y + 6x)\mathbf{k}$ and $\lambda \nabla g(x, y, z) = \lambda(yz\,\mathbf{i} + xz\,\mathbf{j} + xy\,\mathbf{k})$, you obtain the following system of equations.

$$8y + 6z = \lambda yz \qquad C_x(x, y, z) = \lambda g_x(x, y, z)$$
$$8x + 6z = \lambda xz \qquad C_y(x, y, z) = \lambda g_y(x, y, z)$$
$$6y + 6x = \lambda xy \qquad C_z(x, y, z) = \lambda g_z(x, y, z)$$
$$xyz - 480 = 0 \qquad \textit{Constraint}$$

We now multiply the first equation by x, the second by $-y$ and add to obtain

$$6xy - 6yz = 0 \quad \Longrightarrow \quad y = x.$$

Next, multiply the first equation by x, the third by $-z$ and add to obtain

$$8xy - 6yz = 0 \quad \Longrightarrow \quad z = \frac{4}{3}x.$$

Finally, substitute these results into constraint to obtain

$$x(x)\left(\frac{4}{3}x\right) = 480$$

$$x^3 = 360$$

$$x = y = \sqrt[3]{360} \quad \text{and} \quad z = \frac{4}{3}\sqrt[3]{360}.$$

Review Exercises for Chapter 13

---

15. Find the first partial derivatives of

$$g(x, y) = \frac{xy}{x^2 + y^2}.$$

Solution:

Using the Quotient Rule you have

$$g_x(x, y) = \frac{(x^2 + y^2)y - xy(2x)}{(x^2 + y^2)^2} = \frac{y(y^2 - x^2)}{(x^2 + y^2)^2}$$

$$g_y(x, y) = \frac{(x^2 + y^2)x - xy(2y)}{(x^2 + y^2)^2} = \frac{x(x^2 - y^2)}{(x^2 + y^2)^2}.$$

25. Find all second partial derivatives for $h(x, y) = x \sin y + y \cos x$ and verify that the second mixed partials are equal.

Solution:

Since $h(x, y) = x \sin y + y \cos x$ you have

$$h_x(x, y) = \sin y - y \sin x \qquad h_{xx}(x, y) = -y \cos x$$
$$h_y(x, y) = x \cos y + \cos x \qquad h_{yy}(x, y) = -x \sin y.$$

Furthermore,

$$h_{xy}(x, y) = \cos y - \sin x \quad \text{and} \quad h_{yx}(x, y) = \cos y - \sin x.$$

29. Show that the function $z = y/(x^2 + y^2)$ satisfies the **Laplace Equation**

$$\frac{\partial^2 z}{\partial x^2} + \frac{\partial^2 z}{\partial y^2} = 0.$$

Solution:

$$\frac{\partial z}{\partial x} = \frac{-2xy}{(x^2 + y^2)^2} = -2y\left[\frac{x}{(x^2 + y^2)^2}\right]$$

$$\frac{\partial^2 z}{\partial x^2} = -2y\left[\frac{(x^2 + y^2)^2 - x(2)(x^2 + y^2)(2x)}{(x^2 + y^2)^4}\right]$$

$$= \frac{2y(3x^2 - y^2)}{(x^2 + y^2)^3}$$

$$\frac{\partial z}{\partial y} = \frac{(x^2 + y^2) - y(2y)}{(x^2 + y^2)^2} = \frac{x^2 - y^2}{(x^2 + y^2)^2}$$

$$\frac{\partial^2 z}{\partial y^2} = \frac{(x^2 + y^2)^2(-2y) - (x^2 - y^2)(2)(x^2 + y^2)(2y)}{(x^2 + y^2)^4}$$

$$= \frac{-2y(3x^2 - y^2)}{(x^2 + y^2)^3}$$

Therefore,

$$\frac{\partial^2 z}{\partial x^2} + \frac{\partial^2 z}{\partial y^2} = \frac{2y(3x^2 - y^2)}{(x^2 + y^2)^3} + \frac{(-2y)(3x^2 - y^2)}{(x^2 + y^2)^3} = 0.$$

35. The volume of a right circular cone is $V = \frac{1}{3}\pi r^2 h$. Find the maximum approximate error in the volume due to possible errors of $\frac{1}{8}$ inch in the measured values of r and h, if these values are found to be 2 and 5 inches, respectively.

Solution:

Using the total differential you have

$$V = \frac{1}{3}\pi r^2 h$$

$$dV = V_r \, dr + V_h \, dh$$

$$= \frac{2}{3}\pi r h \, dr + \frac{1}{3}\pi r^2 \, dh.$$

Now, letting $r = 2$, $h = 5$, and $dr = dh = \pm\frac{1}{8}$ you obtain the maximum approximate error.

$$dV = \frac{2}{3}\pi(2)(5)\left(\pm\frac{1}{8}\right) + \frac{1}{3}\pi(2)^2\left(\pm\frac{1}{8}\right)$$

$$= \pm\frac{5}{6}\pi \pm \frac{1}{6}\pi = \pm\pi \text{ in}^3.$$

37. Find $\partial u/\partial r$ and $\partial u/\partial t$ if $u = x^2+y^2+z^2$, $x = r\cos t$, $y = r\sin t$, and $z = t$. (a) Use the Chain Rule and (b) check the result by substituting the expressions for x, y, and z before differentiating.

Solution:

(a) By the Chain Rule,

$$\frac{\partial u}{\partial r} = \frac{\partial u}{\partial x}\frac{\partial x}{\partial r} + \frac{\partial u}{\partial y}\frac{\partial y}{\partial r} + \frac{\partial u}{\partial z}\frac{\partial z}{\partial r}$$
$$= 2x(\cos t) + 2y(\sin t) + 2z(0)$$
$$= 2r\cos t(\cos t) + 2r\sin t(\sin t)$$
$$= 2r(\cos^2 t + \sin^2 t) = 2r$$

and

$$\frac{\partial u}{\partial t} = \frac{\partial u}{\partial x}\frac{\partial x}{\partial t} + \frac{\partial u}{\partial y}\frac{\partial y}{\partial t} + \frac{\partial u}{\partial z}\frac{\partial z}{\partial t}$$
$$= 2x(-r\sin t) + 2y(r\cos t) + 2z(1)$$
$$= 2r\cos t(-r\sin t) + 2r\sin t(r\cos t) + 2t$$
$$= -2r^2\sin t\cos t + 2r^2\sin t\cos t + 2t = 2t.$$

(b) By first substituting the expressions for x, y, and z, you have
$$u = r^2\cos^2 t + r^2\sin^2 t + t^2$$
$$= r^2(\cos^2 t + \sin^2 t) + t^2 = r^2 + t^2.$$

Therefore,
$$\frac{\partial u}{\partial r} = 2r \quad \text{and} \quad \frac{\partial u}{\partial t} = 2t.$$

43. Find the gradient and the maximum value of the directional derivative of the function

$$f(x,y) = \frac{y}{x^2 + y^2}$$

at the point $(1,1)$.

Solution:

The gradient is given by

$$\nabla f(x,y) = f_x(x,y)\,\mathbf{i} + f_y(x,y)\,\mathbf{j} = \frac{-2xy}{(x^2+y^2)^2}\,\mathbf{i} + \frac{x^2 - y^2}{(x^2+y^2)^2}\,\mathbf{j}.$$

At the point $(1, 1)$ the gradient is

$$\nabla f(1,1) = \frac{-2(1)(1)}{(1^2+1^2)^2}\mathbf{i} + \frac{1^2-1^2}{(1^2+1^2)^2}\mathbf{j} = -\frac{1}{2}\mathbf{i}.$$

The maximum value of the directional derivative at the point $(1, 1)$ is given by

$$\|\nabla f(1,1)\| = \frac{1}{2}.$$

49. For the surface defined by $f(x,y) = -9 + 4x - 6y - x^2 - y^2$, find an equation of the tangent plane and parametric equations for the normal line at the point $(2, -3, 4)$.

Solution:

We start by defining a function F and finding its first partial derivatives.

$$F(x,y,z) = z - f(x,y) = z + 9 - 4x + 6y + x^2 + y^2$$
$$F_x(x,y,z) = -4 + 2x$$
$$F_y(x,y,z) = 6 + 2y$$
$$F_z(x,y,z) = 1$$

Then at $(2, -3, 4)$ you have $F_x(2, -3, 4) = 0$, $F_y(2, -3, 4) = 0$, and $F_z(2, -3, 4) = 1$. Therefore, by Theorem 13.13 an equation of the tangent plane at $(2, -3, 4)$ is

$$0(x - 2) + 0(y + 3) + 1(z - 4) = 0 \quad \text{or} \quad z = 4.$$

Furthermore, a normal line at $(2, -3, 4)$ has direction numbers $0, 0, 1$, and its parametric equations are

$$x = 2, \quad y = -3, \quad z = 4 + t.$$

55. Locate and classify any extrema of the function

$$f(x,y) = xy + \frac{1}{x} + \frac{1}{y}.$$

Solution:

We begin by setting the first partials of f equal to zero.

$$f(x,y) = xy + \frac{1}{x} + \frac{1}{y}$$

$$f_x(x,y) = y - \frac{1}{x^2} = 0 \quad \Longrightarrow \quad x^2 y = 1$$

$$f_y(x,y) = x - \frac{1}{y^2} = 0 \quad \Longrightarrow \quad xy^2 = 1$$

Thus, $x^2 y = xy^2$ or $x = y$. Substituting this result into $f_x(x,y) = 0$ yields

$$f_x(x,y) = y - \frac{1}{x^2}$$

$$= x - \frac{1}{x^2}$$

$$= \frac{x^3 - 1}{x^2} = 0 \quad \Longrightarrow \quad x = 1.$$

Therefore, the critical point is $(1, 1)$. We now use the Second Derivative Test and obtain

$$f_{xx} = \frac{2}{x^3}, \quad f_{xy} = 1, \quad f_{yy} = \frac{2}{y^3}.$$

At the critical point $(1,1)$, you have $f(1,1) = 3$, $f_{xx}(1,1) = 2 > 0$ and $f_{xx}(1,2)f_{yy}(1,1) - (f_{xy}(1,1))^2 = 3 > 0$. Thus, $(1, 1, 3)$ is a relative minimum.

59. Use Lagrange multipliers to locate and identify any extrema of the function $f(x, y) = x^2 y$ subject to the constraint $x + 2y = 2$.

Solution:

First, let
$$g(x, y) = x + 2y - 2.$$

Then, since $\nabla f(x, y) = 2xy\,\mathbf{i} + x^2\,\mathbf{j}$ and $\lambda \nabla g(x, y) = \lambda\,\mathbf{i} + 2\lambda\,\mathbf{j}$ you obtain the following system of equations

$$
\begin{aligned}
2xy &= \lambda & f_x(x, y) &= \lambda g_x(x, y) \\
x^2 &= 2\lambda & f_y(x, y) &= \lambda f_y(x, y) \\
x + 2y - 2 &= 0 & &\text{Constraint.}
\end{aligned}
$$

Substitution of the expression for λ from the first equation into the second equation yields

$$x^2 = 2(2xy)$$
$$x^2 - 4xy = 0$$
$$x(x - 4y) = 0 \quad \Longrightarrow \quad x = 0 \text{ or } x = 4y.$$

Now, substitute these values for x into the constraint. When $x = 0$ you have

$$2y - 2 = 0 \quad \Longrightarrow \quad y = 1.$$

Therefore, the relative minimum of f is

$$f(0, 1) = 0.$$

When $x = 4y$ you have

$$4y + 2y - 2 = 0$$
$$6y = 2 \quad \Longrightarrow \quad y = \frac{1}{3} \text{ and } x = \frac{4}{3}.$$

Thus, the relative maximum of f is

$$f\left(\frac{4}{3}, \frac{1}{3}\right) = \frac{16}{27}.$$

14 MULTIPLE INTEGRATION

14.1   Iterated Integrals and Area in the Plane

7. Evaluate $\displaystyle\int_{e^y}^{y} \frac{y \ln x}{x}\, dx.$

Solution:

Considering y to be constant and integrating with respect to x yields

$$\int_{e^y}^{y} y(\ln x)\left(\frac{1}{x}\right) dx = \left[\frac{y(\ln x)^2}{2}\right]_{e^y}^{y}$$

$$= \frac{y}{2}[(\ln y)^2 - (\ln e^y)^2] = \frac{y}{2}[(\ln y)^2 - y^2].$$

13. Evaluate $\displaystyle\int_{1}^{2}\int_{0}^{4} (x^2 - 2y^2 + 1)\, dx\, dy.$

Solution:

$$\int_{1}^{2}\int_{0}^{4} (x^2 - 2y^2 + 1)\, dx\, dy = \int_{1}^{2}\left[\frac{x^3}{3} - 2xy^2 + x\right]_{0}^{4} dy$$

$$= \int_{1}^{2}\left(\frac{64}{3} - 8y^2 + 4\right) dy$$

$$= \int_{1}^{2}\left(\frac{76}{3} - 8y^2\right) dy$$

$$= \frac{1}{3}\left[76y - 8y^3\right]_{1}^{2}$$

$$= \frac{1}{3}[152 - 64 - 76 + 8] = \frac{20}{3}$$

19. Evaluate $\displaystyle\int_0^{\pi/2}\int_0^{\sin\theta}\theta r\,dr\,d\theta$.

Solution:

$$\int_0^{\pi/2}\int_0^{\sin\theta}\theta r\,dr\,d\theta = \int_0^{\pi/2}\left[\frac{\theta r^2}{2}\right]_0^{\sin\theta}d\theta$$

$$= \frac{1}{2}\int_0^{\pi/2}\theta\sin^2\theta\,d\theta$$

$$= \frac{1}{2}\int_0^{\pi/2}\theta\left(\frac{1-\cos 2\theta}{2}\right)d\theta$$

$$= \frac{1}{4}\int_0^{\pi/2}(\theta-\theta\cos 2\theta)\,d\theta$$

Using integration by parts or a table of integrals yields

$$\int_0^{\pi/2}\int_0^{\sin\theta}\theta r\,dr\,d\theta = \frac{1}{4}\left[\frac{\theta^2}{2}-\left(\frac{1}{4}\cos 2\theta+\frac{\theta}{2}\sin 2\theta\right)\right]_0^{\pi/2}$$

$$= \frac{1}{4}\left(\frac{\pi^2}{8}+\frac{1}{4}-0-0+\frac{1}{4}+0\right)$$

$$= \frac{1}{4}\left(\frac{\pi^2}{8}+\frac{1}{2}\right)=\frac{\pi^2}{32}+\frac{1}{8}.$$

23. Evaluate the improper iterated integral

$$\int_0^\infty\int_0^\infty xye^{-(x^2+y^2)}dx\,dy.$$

Solution:

$$\int_0^\infty\int_0^\infty xye^{-(x^2+y^2)}\,dx\,dy = -\frac{1}{2}\int_0^\infty\int_0^\infty ye^{-(x^2+y^2)}(-2x)dx\,dy$$

$$= -\frac{1}{2}\int_0^\infty ye^{-(x^2+y^2)}\Big]_0^\infty dy$$

$$= -\frac{1}{2}\int_0^\infty -ye^{-y^2}\,dy$$

$$= -\frac{1}{4}\int_0^\infty e^{-y^2}(-2y)\,dy$$

$$= -\frac{1}{4}\left[e^{-y^2}\right]_0^\infty = \frac{1}{4}.$$

35. Sketch the region R whose area is given by the iterated integral

$$\int_0^1 \int_{y^2}^{\sqrt[3]{y}} dx\, dy.$$

Switch the order of integration, and show that both orders yield the same area.

Solution:

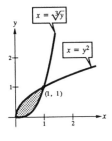

From the limits of integration, you know that when $0 \le y \le 1$, then $y^2 \le x \le \sqrt[3]{y}$. This implies that the region R is bounded by the curves $x = y^2$ and $x = \sqrt[3]{y}$. Therefore, the region R can be sketched as in the accompanying figure. If you interchange the order of integration so that x is the outer variable then $0 \le x \le 1$. Solving for y in the equations $x = y^2$ and $x = \sqrt[3]{y}$, yields $y = \sqrt{x}$ and $y = x^3$. Thus $x^3 \le y \le \sqrt{x}$, and the area of R is given by the iterated integral

$$\int_0^1 \int_{x^3}^{\sqrt{x}} dy\, dx$$

Evaluating each iterated integral, you have

$$\int_0^1 \int_{y^2}^{\sqrt[3]{y}} dx\, dy = \int_0^1 (\sqrt[3]{y} - y^2)\, dy = \left[\frac{3}{4} y^{4/3} - \frac{1}{3} y^3\right]_0^1 = \frac{5}{12}$$

$$\int_0^1 \int_{x^3}^{\sqrt{x}} dy\, dx = \int_0^1 (\sqrt{x} - x^3)\, dx = \left[\frac{2}{3} x^{3/2} - \frac{1}{4} x^4\right]_0^1 = \frac{5}{12}.$$

51. Evaluate the iterated integral $\displaystyle\int_0^1 \int_y^1 \sin x^2 dx\, dy.$

Solution:

Since it is not possible to perform the inner integration, it is necessary to switch the order of integration. From the given limits of integration, it follows that

$$y \le x \le 1 \qquad \text{(Inner limits of integration)}$$

which means that the region R is bounded on the left by the line $x = y$ and on the right by $x = 1$. Furthermore, since

$$0 \le y \le 1, \qquad \text{(Outer limits of integration)}$$

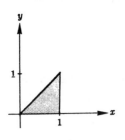

it follows that R is bounded below by the x-axis as shown in the accompanying figure. Now to change the order of integration to $dy\,dx$ observe that the outer limits have the constant bounds $0 \le x \le 1$ and the inner limits have bounds $0 \le y \le x$. Therefore,

$$\int_0^1 \int_y^1 \sin x^2\, dx\, dy = \int_0^1 \int_0^x \sin x^2\, dy\, dx$$

$$= \int_0^1 \left[y \sin x^2 \right]_0^x dx$$

$$= \int_0^1 x \sin x^2\, dx$$

$$= \frac{1}{2}\left[-\cos x^2 \right]_0^1 = \frac{1}{2}(1 - \cos 1).$$

14.2   Double Integrals and Volume

9. Sketch the region R and evaluate the integral

$$\int_0^6 \int_{y/2}^3 (x + y)dx\, dy.$$

Solution:

From the given limits of integration, it follows that

$$\frac{y}{2} \le x \le 3 \qquad \text{(Inner limits of integration)}$$

which means that the region R is bounded on the left by $x = y/2$ and on the right by $x = 3$. Furthermore, since

$$0 \le y \le 6, \qquad \text{(Outer limits of integration)}$$

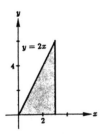

it follows that R is bounded by the x-axis as shown in the accompanying figure.

$$\int_0^6 \int_{y/2}^3 (x+y)dx\,dy = \int_0^6 \left(\frac{1}{2}x^2 + xy\right)\Big]_{y/2}^3 dy$$

$$= \int_0^6 \left[\left(\frac{9}{2} + 3y\right) - \left(\frac{y^2}{8} + \frac{y^2}{2}\right)\right] dy$$

$$= \int_0^6 \left(\frac{9}{2} + 3y - \frac{5y^2}{8}\right) dy$$

$$= \left[\frac{9}{2}y + \frac{3}{2}y^2 - \frac{5}{24}y^3\right]_0^6 = 36$$

15. Given

$$\int_R \int \frac{y}{x^2+y^2} dA$$

where R is the triangle bounded by $y = x$, $y = 2x$, and $x = 2$, set up the integrals for $dx\,dy$ and for $dy\,dx$. Use the most convenient order to evaluate the integral over the region R.

Solution:

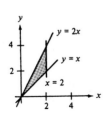

$$\int_R \int \frac{y}{x^2+y^2}\,dx\,dy$$

$$= \int_0^2 \int_{y/2}^y \frac{y}{x^2+y^2}\,dx\,dy + \int_2^4 \int_{y/2}^2 \frac{y}{x^2+y^2}\,dx\,dy$$

and

$$\int_R \int \frac{y}{x^2+y^2}\,dy\,dx = \int_0^2 \int_x^{2x} \frac{y}{x^2+y^2}\,dy\,dx$$

$$= \frac{1}{2}\int_0^2 \left[\ln(x^2+y^2)\right]_x^{2x} dx$$

$$= \frac{1}{2}\int_0^2 [\ln(5x^2) - \ln(2x^2)]\,dx$$

$$= \frac{1}{2}\int_0^2 \ln\frac{5}{2}\,dx$$

$$= \frac{1}{2}\left[x\ln\frac{5}{2}\right]_0^2 = \ln\frac{5}{2}$$

23. Use a double integral to find the volume of the solid in the accompanying figure.

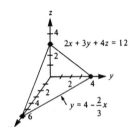

Solution:

By letting $z = 0$, it follows that the base of the solid is the triangle in the xy-plane bounded by the graphs of $2x + 3y = 12$, $x = 0$, and $y = 0$. This plane region is both vertically and horizontally simple. If the order of integration is $dy\, dx$, then the bounds of the region are

$$\text{variable bounds for } y : 0 \le y \le 4 - \frac{2}{3}x$$
$$\text{constant bounds for } x : 0 \le x \le 6.$$

Therefore, the volume is

$$V = \int_0^6 \int_0^{4-(2/3)x} \left(3 - \frac{1}{2}x - \frac{3}{4}y \right) dy\, dx$$

$$= \int_0^6 \left(3y - \frac{1}{2}xy - \frac{3}{8}y^2 \right) \Bigg]_0^{4-(2/3)x} dx$$

$$= \int_0^6 \left(6 - 2x + \frac{1}{6}x^2 \right) dx$$

$$= \left[6x - x^2 + \frac{1}{18}x^3 \right]_0^6 = 12.$$

35. Use a double integral to find the volume of the first octant portion of the solid of intersection of the cylinders $x^2 + z^2 = 1$ and $y^2 + z^2 = 1$.

Solution:

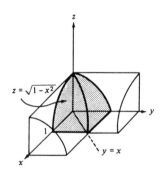

The accompanying figure shows the solid in the first octant. Divide this solid in two equal parts by the plane $y = x$ and find $\frac{1}{2}$ of the total volume. Therefore, integrate the function $z = \sqrt{1 - x^2}$ over the triangle bounded by $y = 0$, $y = x$, and $x = 1$.

$$\text{Constant bounds for } x : 0 \le x \le 1$$
$$\text{Variable bounds for } y : 0 \le y \le x$$

$$V = 2 \int_0^1 \int_0^x \sqrt{1 - x^2}\, dy\, dx$$

$$= 2 \int_0^1 x\sqrt{1 - x^2}\, dx$$

$$= \left[-\frac{2}{3}(1 - x^2)^{3/2} \right]_0^1 = \frac{2}{3}.$$

(Thus, the volume is twice that of Exercise 30.)

41. Use a symbolic integration utility to find the volume of the solid bounded by the paraboloid $z = 4 - x^2 - y^2$ and the xy-plane.

Solution:

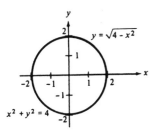

Because of the symmetry of the paraboloid, find the volume of the solid in the first octant (one-fourth the total volume). Thus, you integrate over the first-quadrant portion of the circle in the accompanying figure.

Constant bounds for $x : 0 \le x \le 2$

Variable bounds for $y : 0 \le y \le \sqrt{4 - x^2}$

Using this information and a symbolic integration utility you have

$$V = 4 \int_0^2 \int_0^{\sqrt{4-x^2}} (4 - x^2 - y^2)\, dy\, dx = 8\pi.$$

Note that without the aid of a computer or calculator you would have the following.

$$V = 4 \int_0^2 \int_0^{\sqrt{4-x^2}} (4 - x^2 - y^2)\, dy\, dx$$

$$= 4 \int_0^2 \left[4y - x^2 y - \frac{1}{3}y^3 \right]_0^{\sqrt{4-x^2}} dx$$

$$= \frac{8}{3} \int_0^2 (4 - x^2)^{3/2}\, dx$$

$$= \frac{8}{3} \int_0^{\pi/2} (2 \cos \theta)^3 (2 \cos \theta)\, d\theta \qquad (\text{Let } x = 2 \sin \theta)$$

$$= \frac{128}{3} \int_0^{\pi/2} \cos^4 \theta\, d\theta = 8\pi.$$

49. Evaluate the iterated integral $\displaystyle\int_0^{\ln 10} \int_{e^x}^{10} \frac{1}{\ln y}\, dy\, dx.$

Solution:

Since it is not possible to perform the inner integration, it is necessary to switch the order of integration. From the given limits of integration, you know that

$$e^x \leq y \leq 10 \qquad \text{(Inner limits of integration)}$$

and

$$0 \leq x \leq \ln 10 \qquad \text{(Outer limits of integration).}$$

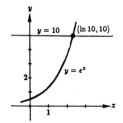

The region R is shown in the accompanying figure. Observe from the figure that if the order of integration is reversed, the outer limits of integration are $1 \leq y \leq 10$ and the inner limits of integration are $0 \leq x \leq \ln y$. Therefore,

$$\int_0^{\ln 10} \int_{e^x}^{10} \frac{1}{\ln y}\, dy\, dx = \int_1^{10} \int_0^{\ln y} \frac{1}{\ln y}\, dx\, dy$$

$$= \int_1^{10} \left[\frac{x}{\ln y}\right]_0^{\ln y} dy$$

$$= \int_1^{10} 1\, dy = \left[y\right]_1^{10} = 9.$$

55. For a particular company, the Cobb-Douglas production function is

$$f(x, y) = 100x^{0.6}y^{0.4}.$$

Estimate the average production level if the number of units of labor varies between 200 and 250 and the number of units of capital varies between 300 and 325.

Solution:

The average value of $f(x, y)$ over the region R is

$$\text{average} = \frac{1}{A} \int_R \int f(x, y)\, dA.$$

The plane region R is a rectangle bounded by $200 \leq x \leq 250$ and $300 \leq y \leq 325$. Therefore, its area is

$$A = (250 - 200)(325 - 300) = 1250.$$

and the average value of f over the region is

$$\text{average} = \frac{1}{1250} \int_{300}^{325} \int_{200}^{250} 100x^{0.6}y^{0.4}\, dx\, dy$$

$$= \frac{1}{1250} \int_{300}^{325} (100y^{0.4})\frac{x^{1.6}}{1.6}\Big]_{200}^{250}\, dy$$

$$= \frac{128,844.1}{1250} \int_{300}^{325} y^{0.4}\, dy$$

$$= 103.075\left[\frac{y^{1.4}}{1.4}\right]_{300}^{325} \approx 25,645.$$

14.3   Change of Variables: Polar Coordinates

5. Evaluate the double integral

$$\int_0^{\pi/2} \int_0^{1+\sin\theta} \theta\, dr\, d\theta$$

and sketch the region R.

Solution:

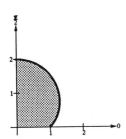

From the given limits of integration, you know that the inner limits are $0 \le r \le 1+\sin\theta$ and the outer limits are $0 \le \theta \le \pi/2$. Hence, the region R is the first quadrant portion of the cardiod $r = 1+\sin\theta$ as shown in the accompanying figure.

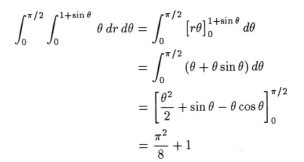

$$\int_0^{\pi/2} \int_0^{1+\sin\theta} \theta\, dr\, d\theta = \int_0^{\pi/2} \left[r\theta\right]_0^{1+\sin\theta} d\theta$$

$$= \int_0^{\pi/2} (\theta + \theta\sin\theta)\, d\theta$$

$$= \left[\frac{\theta^2}{2} + \sin\theta - \theta\cos\theta\right]_0^{\pi/2}$$

$$= \frac{\pi^2}{8} + 1$$

(*Note:* To evaluate $\int \theta\sin\theta\, d\theta$ use integration by parts with $u = \theta$ and $dv = \sin\theta\, d\theta$.)

17. Evaluate the double integral

$$\int_0^2 \int_0^{\sqrt{2x-x^2}} xy \, dy \, dx$$

by changing to polar coordinates.

Solution:

From the limits of integration it follows that

$$0 \le x \le 2$$
$$0 \le y \le \sqrt{2x - x^2} = \sqrt{1 - (x-1)^2}.$$

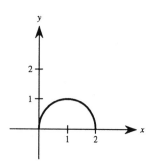

Therefore, the region of integration is bounded by the semicircle in the first quadrant with radius 1 and center $(1, 0)$ as shown in the accompanying figure. In polar coordinates the bounds are

$$0 \le \theta \le \frac{\pi}{2}$$
$$0 \le r \le \cos \theta$$

Consequently, the double integral in polar coordinates is

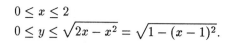

$$\int_0^{\pi/2} \int_0^{2\cos\theta} \overbrace{(r\cos\theta}^{x} \, \overbrace{(r\sin\theta)}^{y} \, \overbrace{r \, dr \, d\theta}^{dA}$$

$$= \int_0^{\pi/2} \int_0^{2\cos\theta} r^3 \sin\theta \cos\theta \, dr \, d\theta$$

$$= \frac{1}{4} \int_0^{\pi/2} \left[r^4 \sin\theta \cos\theta \right]_0^{2\cos\theta} d\theta$$

$$= 4 \int_0^{\pi/2} \cos^5 \theta \sin\theta \, d\theta$$

$$= -\frac{4}{6} \left[\cos^6 \theta \right]_0^{\pi/2} = \frac{2}{3}.$$

19. Combine the sum of the double integrals

$$\int_0^2 \int_0^x \sqrt{x^2 + y^2}\, dy\, dx + \int_2^{2\sqrt{2}} \int_0^{\sqrt{8-x^2}} \sqrt{x^2 + y^2}\, dy\, dx$$

into a single double integral using polar coordinates. Evaluate the resulting double integral.

Solution:

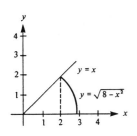

From the accompanying figure we can see that R has the bounds

$$0 \le y \le x \qquad\qquad 0 \le x \le 2$$
$$0 \le y \le \sqrt{8 - x^2} \qquad 2 \le x \le 2\sqrt{2}.$$

and these bounds form a sector of a circle. In polar coordinates the bounds are

$$0 \le r \le 2\sqrt{2} \quad \text{and} \quad 0 \le \theta \le \frac{\pi}{4}$$

where the integrand is $\sqrt{x^2 + y^2} = r$. Consequently, the double integral in polar coordinates is

$$\int_0^{\pi/4} \int_0^{2\sqrt{2}} (r)\, \overbrace{r\, dr\, d\theta}^{dA} = \int_0^{\pi/4} \left[\frac{1}{3}r^3\right]_0^{2\sqrt{2}} d\theta$$

$$= \frac{16\sqrt{2}}{3} \int_0^{\pi/4} d\theta = \frac{4\sqrt{2}\pi}{3}.$$

23. Use polar coordinates to evaluate $\int_R \int f(x,y)\, dA$, where $f(x,y) = \arctan(y/x)$ and R is such that $x^2 + y^2 \le 1$, $0 \le x$, and $0 \le y$.

Solution:

Since $x^2 + y^2 = r^2 \le 1$, $\tan\theta = y/x$, and $dA = r\, dr\, d\theta$, you have

$$\int_0^1 \int_0^{\sqrt{1-x^2}} \arctan\left(\frac{y}{x}\right) dy\, dx = \int_0^{\pi/2} \int_0^1 \theta r\, dr\, d\theta$$

$$= \frac{1}{2} \int_0^{\pi/2} \left[\theta r^2\right]_0^1 d\theta$$

$$= \frac{1}{2} \int_0^{\pi/2} \theta\, d\theta = \frac{1}{4} \left[\theta^2\right]_0^{\pi/2} = \frac{\pi^2}{16}.$$

29. Use a double integral in polar coordinates to find the volume of the solid inside the hemisphere $z = \sqrt{16 - x^2 - y^2}$ and the cylinder $x^2 + y^2 - 4x = 0$.

Solution:

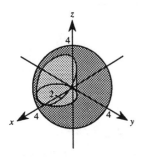

Writing the equation for the cylinder in polar form you have $r = 4\cos\theta$. You can see from the accompanying figure that in polar coordinates R has the bounds

$$0 \le r \le 4\cos\theta \quad \text{and} \quad -\frac{\pi}{2} \le \theta \le \frac{\pi}{2}$$

and $z = \sqrt{16 - x^2 - y^2} = \sqrt{16 - r^2}$. Since the solid is symmetric with respect to the xz-plane, the volume V is given by

$$V = 2 \int_0^{\pi/2} \int_0^{4\cos\theta} \sqrt{16 - r^2}\, r\, dr\, d\theta$$

$$= -\frac{2}{3} \int_0^{\pi/2} [(16 - 16\cos^2\theta)^{3/2} - 64]\, d\theta$$

$$= \frac{128}{3} \int_0^{\pi/2} (1 - \sin^3\theta)\, d\theta = \frac{64}{9}(3\pi - 4).$$

14.4 Center of Mass and Moments of Inertia

1. Find the mass and center of mass for the rectangular lamina with vertices $(0,0)$, $(a,0)$, $(0,b)$, and (a,b) if (a) $\rho = k$ and (b) $\rho = ky$.

Solution:

(a) $\quad m = \int_0^a \int_0^b k\, dy\, dx = kab$

$$M_x = \int_0^a \int_0^b ky\, dy\, dx = \frac{kab^2}{2}$$

$$M_y = \int_0^a \int_0^b kx\, dy\, dx = \frac{ka^2b}{2}$$

$$\bar{x} = \frac{M_y}{m} = \frac{ka^2b/2}{kab} = \frac{a}{2}, \qquad \bar{y} = \frac{M_x}{m} = \frac{kab^2/2}{kab} = \frac{b}{2}$$

(b) $m = \displaystyle\int_0^a \int_0^b ky\,dy\,dx = \dfrac{kab^2}{2}$

$M_x = \displaystyle\int_0^a \int_0^b ky^2\,dy\,dx = \dfrac{kab^3}{3}$

$M_y = \displaystyle\int_0^a \int_0^b kxy\,dy\,dx = \dfrac{ka^2b^2}{4}$

$\bar{x} = \dfrac{M_y}{m} = \dfrac{ka^2b^2/4}{kab^2/2} = \dfrac{a}{2}, \qquad \bar{y} = \dfrac{M_y}{m} = \dfrac{kab^3/3}{kab^2/2} = \dfrac{2}{3}b$

13. Find the mass and center of mass of the lamina of density $\rho = ky$ and bounded by $y = \sin(\pi x/L)$, $y = 0$, $x = 0$, and $x = L$.

Solution:

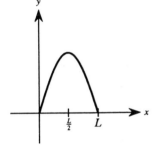

By symmetry, you have $\bar{x} = \dfrac{L}{2}$.

$m = \displaystyle\int_0^L \int_0^{\sin(\pi x/L)} ky\,dy\,dx$

$\quad = \dfrac{k}{2}\displaystyle\int_0^L \sin^2\left(\dfrac{\pi x}{L}\right)dx$

$\quad = \dfrac{k}{4}\displaystyle\int_0^L \left[1 - \cos\left(\dfrac{2\pi x}{L}\right)\right]dx$

$\quad = \dfrac{k}{4}\left[x - \dfrac{L}{2\pi}\sin\left(\dfrac{2\pi x}{L}\right)\right]_0^L = \dfrac{kL}{4}$

$M_x = \displaystyle\int_0^L \int_0^{\sin(\pi x/L)} ky^2\,dy\,dx$

$\quad = \dfrac{k}{3}\displaystyle\int_0^L \sin^3\left(\dfrac{\pi x}{L}\right)dx$

$\quad = \dfrac{k}{3}\displaystyle\int_0^L \left[\sin\left(\dfrac{\pi x}{L}\right) - \cos^2\left(\dfrac{\pi x}{L}\right)\sin\left(\dfrac{\pi x}{L}\right)\right]dx$

$\quad = \dfrac{kL}{3\pi}\left[-\cos\left(\dfrac{\pi x}{L}\right) + \dfrac{1}{3}\cos^3\left(\dfrac{\pi x}{L}\right)\right]_0^L = \dfrac{4kL}{9\pi}$

$\bar{y} = \dfrac{M_x}{m} = \dfrac{4kL/9\pi}{kL/4} = \dfrac{16}{9\pi}$

19. Find the mass and center of mass of one petal $\left(-\frac{\pi}{6} \leq \theta \leq \frac{\pi}{6}\right)$ of the rose curve $r = 2 \cos \theta$. Use a symbolic integration utility to perform the required integration.

Solution:

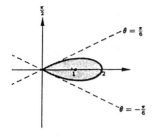

Since the lamina is of uniform density and symmetric to the polar axis (see accompanying figure) you have $\bar{y} = 0$.

$$m = \int_R \int k \, dA$$
$$= \int_{-\pi/6}^{\pi/6} \int_0^{2 \cos 3\theta} kr \, dr \, d\theta \qquad \text{(polar coordinates)}$$
$$= \frac{k\pi}{3}$$

$$M_y = \int_R \int kx \, dA$$
$$= k \int_{-\pi/6}^{\pi/6} \int_0^{2 \cos 3\theta} (r \cos \theta) \, r \, dr \, d\theta \qquad \text{(polar coordinates)}$$
$$= \frac{27\sqrt{3}}{40}$$

Therefore, $\bar{x} = \dfrac{M_y}{m} = \dfrac{27\sqrt{3}k/40}{k\pi/3} = \dfrac{81\sqrt{3}}{40\pi} \approx 1.12.$

29. Find I_x, I_y, I_0, $\bar{\bar{x}}$, and $\bar{\bar{y}}$ for the lamina with density $\rho = kx$ and bounded by the graphs of $y = 4 - x^2$, $y = 0$, and $x > 0$.

Solution:

Since $\rho = kx$, you have

$$m = \int_0^2 \int_0^{4-x^2} kx \, dy \, dx$$
$$= k \int_0^2 xy \Big]_0^{4-x^2} dx$$
$$= k \int_0^2 x(4 - x^2) \, dx = \left[-\frac{k}{4}(4 - x^2)^2 \right]_0^2 = 4k.$$

Furthermore,

$$I_x = \int_R \int y^2 \rho \, dA = \int_0^2 \int_0^{4-x^2} kxy^2 \, dy \, dx$$

$$= \frac{k}{3} \int_0^2 xy^3 \Big]_0^{4-x^2} dx$$

$$= \frac{k}{3} \int_0^2 x(4-x^2)^3 \, dx$$

$$= \left[-\frac{k}{24}(4-x^2)^4 \right]_0^2 = \frac{32k}{3}$$

and

$$I_y = \int_R \int x^2 \rho \, dA = \int_0^2 \int_0^{4-x^2} kx^3 \, dy \, dx$$

$$= k \int_0^2 [x^3 y]_0^{4-x^2} dx$$

$$= k \int_0^2 (4x^3 - x^5) \, dx$$

$$= k \left[x^4 - \frac{1}{6}x^6 \right]_0^2 = \frac{16k}{3}.$$

Therefore, $I_0 = I_x + I_y = 16k$. Finally,

$$\overline{\overline{x}} = \sqrt{\frac{I_y}{m}} = \sqrt{\frac{16k/3}{4k}} = \frac{2}{\sqrt{3}} = \frac{2\sqrt{3}}{3}$$

$$\overline{\overline{y}} = \sqrt{\frac{I_x}{m}} = \sqrt{\frac{32k/3}{4k}} = \frac{4}{\sqrt{6}} = \frac{2\sqrt{6}}{3}.$$

 35. Find the moment of inertia of a uniform circular disk $x^2 + y^2 = b^2$ about the line $x = a$ $(a > b)$. Use a symbolic integration utility to perform the required integrations.

Solution:

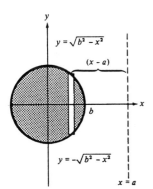

$$I = \int_R \int (\text{distance})^2 \text{mass}$$

$$= \int_{-r}^r \int_{-\sqrt{b^2-x^2}}^{\sqrt{b^2-x^2}} (x-a)^2(k) \, dy \, dx$$

$$= 2k \int_0^\pi \int_0^b (r\cos\theta - a)^2 \, r \, dr \, d\theta \quad \text{(Polar coordinates)}$$

$$= \frac{kb^2\pi}{4}(b^2 + 4a^2)$$

43. Determine the location of the horizontal axis y_a at which the vertical gate in a dam shown in the accompanying figure is required to be hinged so that there is no moment causing rotation under the specified loading. The model for y_a is given by

$$y_a = \overline{y} - \frac{I_{\overline{y}}}{hA}$$

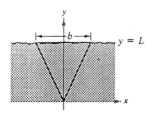

where $\overline{y}$ is the y-coordinate of the centroid of the gate, $I_{\overline{y}}$ is the moment of inertia of the gate about the line $y = \overline{y}$, h is the depth of the centroid below the surface of the water, and A is the area of the gate.

Solution:

The equations of the lines bounding the triangular gate are $y = -2Lx/b$, $y = 2Lx/b$, and $y = L$. Therefore,

$$A = \frac{1}{2}bL$$

$$M_x = 2\left[\frac{1}{2}\int_0^{b/2}\left(L + \frac{2Lx}{b}\right)\left(L - \frac{2Lx}{b}\right)dx\right]$$

$$= \int_0^{b/2}\left(L^2 - \frac{4L^2}{b^2}x^2\right)dx$$

$$= \left[L^2x - \frac{4L^2}{3b^2}x^3\right]_0^{b/2} = \frac{bL^2}{3}$$

$$\overline{y} = \frac{bL^2/3}{bL/2} = \frac{2}{3}L$$

$$I_{\overline{y}} = 2\int_0^L\int_0^{by/2L}\left(\frac{2L}{3} - y\right)^2 dx\,dy$$

$$= 2\int_0^L\left(\frac{2L}{3} - y\right)^2\left(\frac{by}{2L}\right)dy$$

$$= \frac{b}{L}\int_0^L\left(\frac{4L^2}{9}y - \frac{4L}{3}y^2 + y^3\right)dy = \frac{bL^3}{36}$$

$$y_a = \overline{y} - \frac{I_{\overline{y}}}{hA}$$

$$= \frac{2L}{3} - \frac{bL^3/36}{(b/3)(bL/2)} = \frac{L}{2}.$$

14.5 Surface Area

3. Find the area of the surface $f(x, y) = 8 + 2x + 2y$ over the region $R = \{(x, y) : x^2 + y^2 \leq 4\}$.

Solution:

The first partial derivatives of f are

$$f_x(x, y) = 2 \quad \text{and} \quad f_y(x, y) = 2$$

and from the formula for surface area it follows that

$$\sqrt{1 + [f_x(x, y)]^2 + [f_y(x, y)]^2} = 3$$

Therefore, the surface area is given by

$$S = \int_R \int 3 \, dA = 3 \int_R \int dA = 3(\text{area of } R) = 3(4\pi) = 12\pi.$$

11. Find the area of the surface $f(x, y) = \sqrt{x^2 + y^2}$ over the region $R = \{(x, y) : 0 \leq f(x, y) \leq 1\}$.

Solution:

Observe that you are to find the surface area of that part of the cone $f(x, y) = \sqrt{x^2 + y^2}$ inside the cylinder $x^2 + y^2 = 1$. Therefore, the region R in the xy-plane is a circle of radius 1 centered at the origin. Furthermore, the first partial derivatives of f are

$$f_x(x, y) = \frac{x}{\sqrt{x^2 + y^2}} \quad \text{and} \quad f_y(x, y) = \frac{y}{\sqrt{x^2 + y^2}}$$

and from the formula for surface area, you have

$$\sqrt{1 + [f_x(x, y)]^2 + [f_y(x, y)]^2} = \sqrt{1 + \frac{x^2 + y^2}{x^2 + y^2}} = \sqrt{2}.$$

Therefore, the surface area is given by

$$S = \int_R \int \sqrt{2} \, dA = \sqrt{2} \int_R \int dA = \sqrt{2}(\text{area of } R) = \sqrt{2}\pi.$$

29. Set up the double integral which gives the area of the surface on the graph of $f(x, y) = e^{-x} \sin y$ over the region $R = \{(x, y) : x^2 + y^2 \leq 4\}$.

Solution:

The first partial derivatives of f are
$$f_x(x, y) = -e^{-x} \sin y \quad \text{and} \quad f_y(x, y) = e^{-x} \cos y$$
and from the formula for surface area, you have
$$\sqrt{1 + [f_x(x, y)]^2 + [f_y(x, y)]^2} = \sqrt{1 + e^{-2x} \sin^2 y + e^{-2x} \cos^2 y}$$
$$= \sqrt{1 + e^{-2x}}.$$

We integrate over the circle $x^2 + y^2 = 4$.

Constant bounds for x : $-2 \leq x \leq 2$

Variable bounds for y : $-\sqrt{4 - x^2} \leq y \leq \sqrt{4 - x^2}$

Therefore,
$$S = \int_{-2}^{2} \int_{-\sqrt{4-x^2}}^{\sqrt{4-x^2}} \sqrt{1 + e^{-2x}} \, dy \, dx.$$

33. Find the surface area of the solid of intersection of the cylinders $x^2 + z^2 = 1$ and $y^2 + z^2 = 1$.

Solution:

The accompanying figure shows the surface in the first octant. We divide this surface into two equal parts by the plane $y = x$ and thus find $\frac{1}{16}$ of the total surface area. Therefore, find the area of the surface $z = \sqrt{1 - x^2}$ over the triangle bounded by $y = 0$, $y = x$, and $x = 1$.

$z = \sqrt{1 - x^2}$

$x^2 + z^2 = 1$

$y^2 + z^2 = 1$

$$\frac{\partial z}{\partial x} = \frac{-x}{\sqrt{1 - x^2}} \quad \text{and} \quad \frac{\partial z}{\partial y} = 0$$

Therefore,

$$S = 16 \int_0^1 \int_0^x \sqrt{1 + \left(\frac{\partial z}{\partial x}\right)^2 + \left(\frac{\partial z}{\partial y}\right)^2} \, dy \, dx$$

$$= 16 \int_0^1 \int_0^x \sqrt{1 + \frac{x^2}{1 - x^2}} \, dy \, dx$$

$$= 16 \int_0^1 \int_0^x \frac{1}{\sqrt{1^2 - x^2}} \, dy \, dx$$

$$= 16 \int_0^1 \frac{x}{\sqrt{1 - x^2}} \, dx = \left[-16\sqrt{1 - x^2} \right]_0^1 = 16.$$

35. A new auditorium is built with a foundation in the shape of $\frac{1}{4}$ of a circle of radius 50 feet. Therefore, it forms a region R bounded by the graph of $x^2 + y^2 = 50^2$ with $x \geq 0$ and $y \geq 0$. The equations for the floor and ceiling are given by the following.

$$z = \frac{x+y}{5} \qquad \text{Floor: incline plane}$$

$$z = 20 + \frac{xy}{100} \qquad \text{Ceiling}$$

(a) Calculate the volume of the room. This is needed to determine the heating and cooling requirements.

(b) Find the surface area of the ceiling.

Solution:

(a) $\displaystyle V = \int_0^{50} \int_0^{\sqrt{50^2 - x^2}} \left[\left(20 + \frac{xy}{100} \right) - \left(\frac{x+y}{5} \right) \right] dy\, dx$

$\displaystyle = \int_0^{\pi/2} \int_0^{50} \left[20 + \frac{1}{100} r^2 \sin\theta \cos\theta \right.$

$\displaystyle \left. - \frac{1}{5}(r\sin\theta + r\cos\theta) \right] r\, dr\, d\theta$

$\approx 30,416 \text{ ft}^3$

(b) From the equation for the ceiling you have

$$\frac{\partial z}{\partial x} = \frac{y}{100} \quad \text{and} \quad \frac{\partial z}{\partial y} = \frac{x}{100}.$$

$$S = \int_0^{50} \int_0^{\sqrt{50^2 - x^2}} \sqrt{1 + \left(\frac{y}{100}\right)^2 + \left(\frac{x}{100}\right)^2}\, dy\, dx$$

$$= \int_0^{\pi/2} \int_0^{50} \sqrt{1 + \frac{r^2}{100}}\, r\, dr\, d\theta \approx 2082 \text{ ft}^2$$

14.6   Triple Integrals and Applications

7. Evaluate $\displaystyle\int_0^9 \int_0^{y/3} \int_0^{\sqrt{y^2-9x^2}} z \, dz \, dx \, dy.$

Solution:

$$\int_0^9 \int_0^{y/3} \int_0^{\sqrt{y^2-9x^2}} z \, dz \, dx \, dy = \int_0^9 \int_0^{y/3} \left[\frac{1}{2}z^2\right]_0^{\sqrt{y^2-9x^2}} dx \, dy$$

$$= \frac{1}{2} \int_0^9 \int_0^{y/3} (y^2 - 9x^2) \, dx \, dy$$

$$= \frac{1}{2} \int_0^9 \left[xy^2 - 3x^3\right]_0^{y/3} dy$$

$$= \frac{1}{9} \int_0^9 y^3 \, dy = \frac{1}{36}\left[y^4\right]_0^9 = \frac{729}{4}$$

15. Sketch the solid region whose volume is given by the triple integral

$$\int_0^1 \int_y^1 \int_0^{\sqrt{1-y^2}} dz \, dx \, dy$$

and rewrite the integral using $dz \, dy \, dx$ as the order of integration.

Solution:

We have

$$\begin{array}{ll}
\text{Constant bounds on } y: & 0 \le y \le 1 \\
\text{Variable bounds on } x: & y \le x \le 1 \\
\text{Variable bounds on } z: & 0 \le z \le \sqrt{1-y^2}.
\end{array}$$

From the upper bound on z, you have

$$z = \sqrt{1-y^2} \quad \text{or} \quad y^2 + z^2 = 1$$

a cylinder of radius 1 with the x-axis as its axis. Therefore, the triple integral gives the volume of the solid in the first octant bounded by the graphs of $z = \sqrt{1-y^2}$, $z = 0$, $x = y$, and

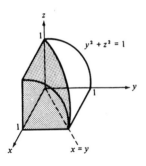

$x = 1$. From the accompanying sketch of the solid we observe the following bounds when the order of integration is $dz\,dy\,dx$

Constant bounds on x : $0 \le x \le 1$

Variable bounds on y : $0 \le y \le x$

Variable bounds on z : $0 \le z \le \sqrt{1 - y^2}$.

Therefore, the integral is $\displaystyle \int_0^1 \int_0^x \int_0^{\sqrt{1-y^2}} dz\,dy\,dx$.

23. Use a triple integral to find the volume of the solid bounded by the cylinders $z = 4 - x^2$ and $y = 4 - x^2$ in the first octant.

Solution:

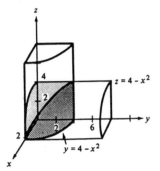

In the first octant you have $0 \le z \le 4 - x^2$, $0 \le y \le 4 - x^2$, and $0 \le x \le 2$. Therefore, the volume is

$$V = \int_0^2 \int_0^{4-x^2} \int_0^{4-x^2} dz\,dy\,dx$$

$$= \int_0^2 \int_0^{4-x^2} (4 - x^2)\,dy\,dx$$

$$= \int_0^2 (4 - x^2)^2\,dx$$

$$= \int_0^2 (16 - 8x^2 + x^4)\,dx$$

$$= \left[16x - \frac{8x^3}{3} + \frac{x^5}{5}\right]_0^2 = 32 - \frac{64}{3} + \frac{32}{5} = \frac{256}{15}.$$

29. Find the mass and center of mass of the solid with density $\rho(x, y, z) = kxy$ and bounded by the graphs of $x = 0$, $x = b$, $y = 0$, $y = b$, $z = 0$, and $z = b$.

Solution:

The mass of the cube is

$$m = \int_0^b \int_0^b \int_0^b kxy\,dz\,dy\,dx = k\int_0^b \int_0^b bxy\,dy\,dx$$

$$= kb\int_0^b \left[\frac{xy^2}{2}\right]_0^b dx = \frac{kb^3}{2} \int_0^b x\,dx = \frac{kb^3}{4}\left[x^2\right]_0^b = \frac{kb^5}{4}.$$

Furthermore,

$$M_{yz} = \int_0^b \int_0^b \int_0^b x(kxy)\, dz\, dy\, dx = k \int_0^b \int_0^b x^2 y(b)\, dy\, dx$$

$$= kb \int_0^b \left[\frac{x^2 y^2}{2} \right]_0^b dx = \frac{kb^3}{2} \int_0^b x^2\, dx = \frac{kb^3}{6} \left[x^3 \right]_0^b = \frac{kb^6}{6}.$$

By the symmetry of the cube and of $\rho = kxy$, you have $M_{xz} = M_{yz} = kb^6/6$. Moreover,

$$M_{xy} = \int_0^b \int_0^b \int_0^b z(kxy)\, dz\, dy\, dx = k \int_0^b \int_0^b \left[\frac{xyz^2}{2} \right]_0^b dy\, dx$$

$$= \frac{kb^2}{2} \int_0^b \int_0^b xy\, dy\, dx = \frac{kb^2}{2} \int_0^b \left[\frac{xy^2}{2} \right]_0^b dx$$

$$= \frac{kb^4}{4} \int_0^b x\, dx = \frac{kb^4}{4} \left[\frac{x^2}{2} \right]_0^b = \frac{kb^6}{8}$$

Finally,

$$\bar{x} = \frac{M_{yz}}{m} = \frac{kb^6/6}{kb^5/4} = \frac{2b}{3}$$

$$\bar{y} = \bar{x} = \frac{2b}{3}$$

$$\bar{z} = \frac{M_{xy}}{m} = \frac{kb^6/8}{kb^5/4} = \frac{b}{2}.$$

41. Verify the moments of inertia

$$I_x = I_z = \frac{1}{12} m(3a^2 + L^2) \quad \text{and} \quad I_y = \frac{1}{2} ma^2$$

for the cylinder of uniform density in the accompanying figure.

Solution:

The solid is a right circular of radius a, length L, and uniform density $\rho(x, y, z) = k$. Thus the mass of the cylinder is

$$m = k(\text{volume}) = k\pi a^2 L$$

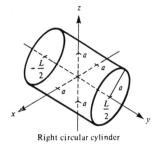

Right circular cylinder

Now

$$I_x = \int_{-a}^{a} \int_{-\sqrt{a^2-x^2}}^{\sqrt{a^2-x^2}} \int_{-L/2}^{L/2} (y^2 + z^2)k \, dy \, dz \, dx$$

$$= 8k \int_{0}^{a} \int_{0}^{\sqrt{a^2-x^2}} \int_{0}^{L/2} (y^2 + z^2) \, dy \, dz \, dx \qquad \text{(By symmetry)}$$

$$= \frac{kL}{3} \int_{0}^{a} \int_{0}^{\sqrt{a^2-x^2}} (L^2 + 12z^2) \, dz \, dx$$

$$= \frac{kL}{3} \int_{0}^{a} [L^2 \sqrt{a^2 - x^2} + 4(a^2 - x^2)^{3/2}] \, dx$$

Now let $x = a \sin \theta$. Then $\sqrt{a^2 - x^2} = a \cos \theta$, $dx = a \cos \theta \, d\theta$, and

$$I_x = \frac{ka^2 L^3}{3} \int_{0}^{\pi/2} \cos^2 \theta \, d\theta + \frac{4ka^4 L}{3} \int_{0}^{\pi/2} \cos^4 \theta \, d\theta$$

By Wallis's Formula you have

$$I_x = \frac{ka^2 L^3}{3} \left(\frac{1}{2}\right)\left(\frac{\pi}{2}\right) + \frac{4ka^4 L}{3} \left(\frac{1}{2}\right)\left(\frac{3}{4}\right)\left(\frac{\pi}{2}\right)$$

$$= \frac{k\pi a^2 L}{12}(L^2 + 3a^2) = \frac{1}{12} m(3a^2 + L^2).$$

By symmetry $I_x = I_z$. To find I_y, change only the integrand in the triple integral given above and obtain

$$I_y = 8k \int_{0}^{a} \int_{0}^{\sqrt{a^2-x^2}} \int_{0}^{L/2} (x^2 + z^2) \, dy \, dz \, dx.$$

Proceeding with integrations similar to those given above yields

$$I_y = \frac{k\pi a^4 L}{2} = k\pi a^2 L \left(\frac{a^2}{2}\right) = \frac{1}{2} ma^2.$$

14.7   Triple Integrals in Cylindrical and Spherical Coordinates

3. Evaluate the triple integral $\int_{0}^{\pi/2} \int_{0}^{2\cos^2\theta} \int_{0}^{4-r^2} r \sin\theta \, dz \, dr \, d\theta.$

Solution:

$$\int_0^{\pi/2} \int_0^{2\cos^2\theta} \int_0^{4-r^2} r\sin\theta \, dz \, dr \, d\theta$$

$$= \int_0^{\pi/2} \int_0^{2\cos^2\theta} \left[rz\sin\theta \right]_0^{4-r^2} dr \, d\theta$$

$$= \int_0^{\pi/2} \int_0^{2\cos^2\theta} r(4-r^2)\sin\theta \, dr \, d\theta$$

$$= \int_0^{\pi/2} \left[-\frac{1}{4}(4-r^2)^2 \sin\theta \right]_0^{2\cos^2\theta} d\theta$$

$$= \int_0^{\pi/2} (8\cos^4\theta - 4\cos^8\theta)\sin\theta \, d\theta$$

$$= \left[-\frac{8}{5}\cos^5\theta + \frac{4}{9}\cos^9\theta \right]_0^{\pi/2} = \frac{52}{45}$$

9. Sketch the solid region whose volume is given by the integral

$$\int_0^{\pi/2} \int_0^3 \int_0^{e^{-r^2}} r \, dz \, dr \, d\theta$$

and evaluate the integral.

Solution:

We first observe that the triple integral is written in terms of cylindrical coordinates. From the limits of integration you have

Constant bounds on θ : $0 \leq \theta \leq \pi/2$

Constant bounds on r : $0 \leq r \leq 3$

Variable bounds on z : $0 \leq z \leq e^{-r^2}$

The limits on r determine a circular cylinder of radius 3 having the z-axis as its axis. The limits on θ and z restrict us to the first-octant portion of the cylinder bounded by the surface

$$z = e^{-r^2} = e^{-(x^2+y^2)}.$$

The solid is shown in the accompanying figure. The value of the integral is given by

$$\int_0^{\pi/2} \int_0^3 \int_0^{e^{-r^2}} r \, dz \, dr \, d\theta = \int_0^{\pi/2} \int_0^3 re^{-r^2} dr \, d\theta$$

$$= -\frac{1}{2}(e^{-9} - 1) \int_0^{\pi/2} d\theta = \frac{\pi}{4}(1 - e^{-9}).$$

15. Convert the integral

$$\int_{-a}^{a} \int_{-\sqrt{a^2-x^2}}^{\sqrt{a^2-x^2}} \int_{a}^{a+\sqrt{a^2-x^2-y^2}} x \, dz \, dy \, dx$$

from rectangular coordinates to both cylindrical and spherical coordinates and evaluate the simplest integral.

Solution:

Observe that the solid S over which you are integrating is the top half of the sphere of radius a and center $(0,0,a)$. Projecting the solid onto the xy-plane forms a circle of radius a. Thus have

Constant bounds on θ : $0 \le \theta \le 2\pi$

Constant bounds on r : $0 \le r \le a$

Variable bounds on z : $a \le z \le a + \sqrt{a^2 - x^2 - y^2}$
$$= a + \sqrt{a^2 - r^2}.$$

Finally, since $x = r \cos \theta$, you obtain the integral in cylindrical coordinates:

$$\int_{0}^{2\pi} \int_{0}^{a} \int_{a}^{a+\sqrt{a^2-r^2}} (r \cos \theta) r \, dz \, dr \, d\theta$$

$$= \int_{0}^{2\pi} \int_{0}^{a} \int_{a}^{a+\sqrt{a^2-r^2}} r^2 \cos \theta \, dz \, dr \, d\theta.$$

To write the integral in spherical coordinates, first write the equation of the sphere in spherical coordinates:

$$z = a + \sqrt{a^2 - x^2 - y^2}$$
$$x^2 + y^2 + (z - a)^2 = a^2$$
$$x^2 + y^2 + z^2 - 2az + a^2 = a^2$$
$$\rho^2 - 2a(\rho \cos \phi) = 0 \quad (\text{Since } z = \rho \cos \phi)$$
$$\rho = 2a \cos \phi.$$

We next write the equation of the plane $z = a$ in spherical coordinates and obtain

$$z = a$$
$$\rho \cos \phi = a$$
$$\rho = a \sec \phi.$$

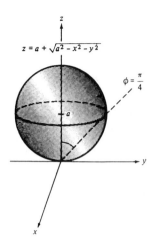

From the accompanying figure it follows that the bounds on ϕ are $0 \le \phi \le \pi/4$. Since $x = \rho \sin \phi \cos \theta$, you can write the integral in spherical coordinates:

$$\int_0^{\pi/4} \int_0^{2\pi} \int_{a \sec \phi}^{2a \cos \phi} (\rho \sin \phi \cos \theta)(\rho^2 \sin \phi) \, d\rho \, d\theta \, d\phi$$

$$= \frac{1}{4} \int_0^{\pi/4} \int_0^{2\pi} \sin^2 \phi \cos \theta [(2a \cos \phi)^4 - (a \sec \phi)^4] \, d\theta \, d\phi$$

$$= \frac{a^4}{4} \int_0^{\pi/4} \left[(16 \sin^2 \phi \cos^4 \phi - \sin^2 \phi \sec^4 \phi) \sin \theta \right]_0^{2\pi} \, d\phi$$

$$= \frac{a^4}{4} \int_0^{\pi/4} 0 \, d\phi = 0.$$

25. Use a triple integral in cylindrical coordinates to find the volume of the solid inside both the sphere $x^2 + y^2 + z^2 = a^2$ and the cylinder $[x - (a/2)]^2 + y^2 = (a/2)^2$.

Solution:

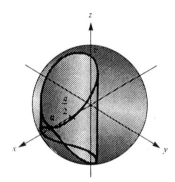

The accompanying figure shows the sphere and cylinder. Using the cylindrical coordinate system, the equation of the cylinder is given by

$$r = a \cos \theta, \quad (-\pi/2 \le \theta \le \pi/2)$$

and the equation of the sphere is given by

$$r^2 + z^2 = a^2 \quad \text{or} \quad z = \pm\sqrt{a^2 - r^2}.$$

Therefore, the volume is

$$V = \int_{-\pi/2}^{\pi/2} \int_0^{a \cos \theta} \int_{-\sqrt{a^2 - r^2}}^{\sqrt{a^2 - r^2}} r \, dz \, dr \, d\theta$$

$$= 4 \int_0^{\pi/2} \int_0^{a \cos \theta} r \sqrt{a^2 - r^2} \, dr \, d\theta$$

$$= -\frac{4}{3} \int_0^{\pi/2} [(a^2 - a^2 \cos^2 \theta)^{3/2} - a^3] \, d\theta$$

$$= \frac{4a^3}{3} \int_0^{\pi/2} (1 - \sin^3 \theta) \, d\theta = \frac{4a^3}{3} \left(\frac{\pi}{2} - \frac{2}{3} \right).$$

34. Use spherical coordinates to find the center of mass of the solid of uniform density lying between two concentric hemispheres of radii r and R, where $r < R$.

Solution:

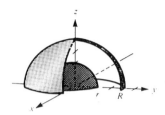

First, without loss of generality, you can position the hemispheres with their centers at the origin and bases on the xy-coordinate plane. Then, by symmetry, $\bar{x} = \bar{y} = 0$. Since the solid has uniform density k, the mass is given by

$$ m = k(\text{volume}) = k\left(\frac{2}{3}\pi R^3 - \frac{2}{3}\pi r^3\right) = \frac{2}{3}k\pi(R^3 - r^3). $$

Using symmetry, you have

$$ M_{xy} = \int\!\!\int_Q\!\!\int z(\text{density})\,dV $$

$$ = 4k\int_0^{\pi/2}\int_0^{\pi/2}\int_r^R \rho^3 \cos\phi \sin\phi\,d\rho\,d\theta\,d\phi $$

$$ = \frac{1}{2}k(R^4 - r^4)\int_0^{\pi/2}\int_0^{\pi/2} \sin 2\phi\,d\theta\,d\phi $$

$$ = \frac{1}{4}k\pi(R^4 - r^4)\int_0^{\pi/2} \sin 2\phi\,d\phi $$

$$ = -\frac{1}{8}k\pi(R^4 - r^4)\big[\cos 2\phi\big]_0^{\pi/2} = \frac{1}{4}k\pi(R^4 - r^4). $$

Therefore,

$$ \bar{z} = \frac{M_{xy}}{m} = \frac{k\pi(R^4 - r^4)/4}{2k\pi(R^3 - r^3)/3} = \frac{3(R^4 - r^4)}{8(R^3 - r^3)}. $$

14.8   Change of Variables: Jacobians

7. Find the Jacobian for the change of variables

$$x = e^u \sin v \qquad y = e^u \cos v.$$

Solution:

Using the definition the Jacobian, you have

$$\frac{\partial(x,y)}{\partial(u,v)} = \begin{vmatrix} \dfrac{\partial x}{\partial u} & \dfrac{\partial y}{\partial u} \\[2mm] \dfrac{\partial x}{\partial v} & \dfrac{\partial y}{\partial v} \end{vmatrix} = \begin{vmatrix} e^u \sin v & e^u \cos v \\ e^u \cos v & -e^u \sin v \end{vmatrix}$$

$$= -e^{2u} \sin^2 v - e^{2u} \cos^2 v = -e^{2u}.$$

13. Evaluate

$$\int_R \int 4(x+y)e^{x-y}\, dy\, dx$$

using the change of variables $x = \tfrac{1}{2}(u+v)$ and $y = \tfrac{1}{2}(u-v)$ and letting R be the region in the accompanying figure.

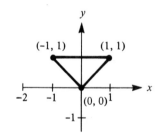

Solution:

Begin by solving for u and v in terms of x and y and obtain

$$u = x + y \qquad \text{and} \qquad v = x - y.$$

Since the region R is bounded by the lines $x + y = 0$, $x - y = 0$ and $y = 1$, the bounds for the region S in the uv-plane are $u = 0, v = 0$, and $v = u - 2$. The Jacobian for the change of variables is

$$\frac{\partial(x,y)}{\partial(u,v)} = \begin{vmatrix} \dfrac{\partial x}{\partial u} & \dfrac{\partial y}{\partial u} \\[2mm] \dfrac{\partial x}{\partial v} & \dfrac{\partial y}{\partial v} \end{vmatrix} = \begin{vmatrix} \dfrac{1}{2} & \dfrac{1}{2} \\[2mm] \dfrac{1}{2} & -\dfrac{1}{2} \end{vmatrix} = -\frac{1}{2}.$$

Therefore, by Theorem 14.6 you have

$$\int_R \int 4(x+y)e^{x-y}\,dy\,dx = \int_S \int 4ue^v \left|\frac{\partial(x,y)}{\partial(u,v)}\right| dv\,du$$

$$= 2\int_0^2 \int_{u-2}^0 ue^v\,dv\,du$$

$$= 2\int_0^2 (u - ue^{u-2})\,du$$

$$= 2\left[\frac{1}{2}u^2 - e^{u-2}(u-1)\right]_0^2 = 2(1 - e^{-2}).$$

19. Use a change of variables to evaluate

$$\int_R \int \sqrt{(x-y)(x+4y)}\,dy\,dx$$

where R is the region bounded by the parallelogram with vertices $(0,0)$, $(1,1)$, $(5,0)$, and $(4,-1)$.

Solution:

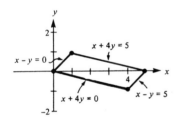

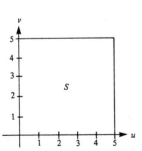

The region R is bounded by the graphs of $x - y = 0$, $x - y = 5$, $x + 4y = 0$ and $x + 4y = 5$. (See the accompanying figure.) By letting $u = x - y$ and $v = x + 4y$, you have

$$x = \frac{1}{5}(4u + v) \qquad \text{and} \qquad y = -\frac{1}{5}(u - v)$$

Thus, the Jacobian is

$$\frac{\partial(x,y)}{\partial(u,v)} = \begin{vmatrix} \dfrac{\partial x}{\partial u} & \dfrac{\partial y}{\partial u} \\[2mm] \dfrac{\partial x}{\partial v} & \dfrac{\partial y}{\partial v} \end{vmatrix} = \begin{vmatrix} \dfrac{4}{5} & -\dfrac{1}{5} \\[2mm] \dfrac{1}{5} & \dfrac{1}{5} \end{vmatrix} = \frac{1}{5}$$

Therefore, by Theorem 14.6 you obtain

$$\int_R \int \sqrt{(x-y)(x+4y)}\,dy\,dx = \int_S \int \sqrt{uv}\left|\frac{\partial(x,y)}{\partial(u,v)}\right| dv\,du$$

$$= \frac{1}{5}\int_0^5 \int_0^5 \sqrt{uv}\,dv\,du$$

$$= \frac{2\sqrt{5}}{3}\int_0^5 \sqrt{u}\,du = \frac{100}{9}.$$

Review Exercises for Chapter 14

7. Evaluate

$$\int_0^h \int_0^x \sqrt{x^2 + y^2}\, dy\, dx$$

by using the coordinate system that makes the integration easiest.

Solution:

We choose to use polar coordinates since $r = \sqrt{x^2 + y^2}$. To rewrite the limits, you first find the equation of the line $x = h$ in polar coordinates.

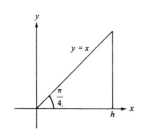

$$x = h$$
$$r \cos \theta = h$$
$$r = h\left(\frac{1}{\cos \theta}\right) = h \sec \theta$$

Therefore,

$$\int_0^h \int_0^x \sqrt{x^2 + y^2}\, dy\, dx = \int_0^{\pi/4} \int_0^{h \sec \theta} (r)r\, dr\, d\theta$$

$$= \frac{h^3}{3} \int_0^{\pi/4} \sec^3 \theta\, d\theta$$

$$= \frac{h^3}{3} \left[\frac{\sec \theta \tan \theta}{2} + \frac{1}{2} \ln |\sec \theta + \tan \theta|\right]_0^{\pi/4}$$

$$= \frac{h^3}{6}[\sqrt{2} + \ln (\sqrt{2} + 1)].$$

19. If R is the larger region between the circle $x^2 + y^2 = 25$ and the line $x = 3$, write the limits for the double integral $\int_R \int f(x, y)\, dA$ for both orders of integration. Compute the area by letting $f(x, y) = 1$.

Solution:

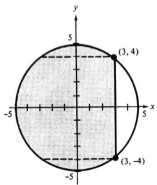

$$\int_R \int f(x,y)\, dA$$

$$= \int_{-5}^{3} \int_{-\sqrt{25-x^2}}^{\sqrt{25-x^2}} f(x,y)\, dy\, dx$$

$$= \int_{-5}^{-4} \int_{-\sqrt{25-y^2}}^{\sqrt{25-y^2}} f(x,y)\, dx\, dy$$

$$+ \int_{-4}^{4} \int_{-\sqrt{25-y^2}}^{3} f(x,y)\, dx\, dy + \int_{4}^{5} \int_{-\sqrt{25-y^2}}^{\sqrt{25-y^2}} f(x,y)\, dx\, dy$$

The area of R is

$$A = 2 \int_{-5}^{3} \int_{0}^{\sqrt{25-x^2}} dy\, dx = 2 \int_{-5}^{3} \sqrt{25-x^2}\, dx$$

$$= 2\left(\frac{1}{2}\right)\left[x\sqrt{25-x^2} + 25\arcsin\left(\frac{x}{5}\right)\right]_{-5}^{3}$$

$$= 3(4) + 25\arcsin\left(\frac{3}{5}\right) - 0 - 25\left(-\frac{\pi}{2}\right)$$

$$= 12 + \frac{25\pi}{2} + 25\arcsin\left(\frac{3}{5}\right) \approx 67.36.$$

(*Note:* The area of the entire circle is $25\pi \approx 78.54$.)

35. Find the area of the surface on the function
$f(x,y) = 16 - x^2 - y^2$ over the region
$R = \{(x,y) : x^2 + y^2 \le 16\}$.

Solution:

Since $z = 16 - x^2 - y^2$,

$$\frac{\partial z}{\partial x} = -2x \quad \text{and} \quad \frac{\partial z}{\partial y} = -2y.$$

$$S = \int_R \int \sqrt{1 + \left(\frac{\partial z}{\partial x}\right)^2 + \left(\frac{\partial z}{\partial y}\right)^2}\, dy\, dx$$

$$= \int_{-4}^{4} \int_{-\sqrt{16-x^2}}^{\sqrt{16-x^2}} \sqrt{1 + 4x^2 + 4y^2}\, dy\, dx$$

$$= 4 \int_{0}^{4} \int_{0}^{\sqrt{16-x^2}} \sqrt{1 + 4(x^2 + y^2)}\, dy\, dx$$

$$= \frac{1}{2} \int_{0}^{\pi/2} \int_{0}^{4} \sqrt{1 + 4r^2}\,(8r)\, dr\, d\theta$$

$$= \frac{1}{3} \int_{0}^{\pi/2} (65^{3/2} - 1)\, d\theta = \frac{\pi}{6}(65\sqrt{65} - 1)$$

39. Find the center of mass of the portion of the solid of constant density in the first octant bounded by $x^2 + y^2 + z^2 = a^2$.

Solution:

The solution S is the first-octant portion of a sphere of radius a and therefore, because of its symmetry, the coordinates of the center of mass are equal. Let k be the constant density; then the mass is

$$m = k(\text{volume}) = k\left(\frac{1}{8}\right)\left(\frac{4}{3}\pi a^3\right) = \frac{k}{6}\pi a^3.$$

Now,

$$M_{xy}$$

$$= \int \int_S \int z(\text{density})\, dS \qquad \text{(rectangular coordinates)}$$

$$= k \int_{0}^{\pi/2} \int_{0}^{\pi/2} \int_{0}^{a} (\rho \cos \phi)(\rho^2 \sin \phi)\, d\rho\, d\theta\, d\phi \qquad \text{(spherical coordinates)}$$

$$= \frac{ka^4}{4} \int_{0}^{\pi/2} \int_{0}^{\pi/2} \cos \phi \sin \phi\, d\theta\, d\phi$$

$$= \frac{k\pi a^4}{8} \int_{0}^{\pi/2} \cos \phi \sin \phi\, d\phi$$

$$= \frac{k\pi a^4}{8} \left[\frac{1}{2} \sin^2 \phi\right]_{0}^{\pi/2} = \frac{k\pi a^4}{16}.$$

Therefore,

$$\bar{x} = \bar{y} = \bar{z} = \frac{M_{xy}}{m} = \frac{k\pi a^4/16}{k\pi a^3/6} = \frac{3a}{8}.$$

15 VECTOR ANALYSIS

15.1   Vector Fields

5. Sketch several representative vectors in the vector field

$$\mathbf{F}(x,y) = x\,\mathbf{i} + y\,\mathbf{j}$$

Solution:

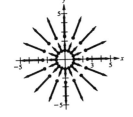

We will plot vectors of equal magnitude and, in this case, they lie along circles given by

$$\|\mathbf{F}(x,y)\| = \sqrt{x^2 + y^2} = c \quad \Longrightarrow \quad x^2 + y^2 = c^2$$

For $c = 1$, sketch several vectors $x\,\mathbf{i} + y\,\mathbf{j}$ of magnitude 1 on the circle given $x^2 + y^2 = 1$. For $c = 4$, sketch several vectors $x\,\mathbf{i} + y\,\mathbf{j}$ of magnitude 2 on the circle given by $x^2 + y^2 = 4$. (See the accompanying figure.)

23. Determine if $\mathbf{F}(x,y) = xe^{x^2 y}(2y\,\mathbf{i} + x\,\mathbf{j})$ is conservative and, if so, find the potential function $f(x,y)$.

Solution:

Since $\mathbf{F}(x,y) = 2xye^{x^2 y}\,\mathbf{i} + x^2 e^{x^2 y}\,\mathbf{j}$, it follows from Theorem 15.1, that $\mathbf{F}$ is conservative since

$$\frac{\partial}{\partial y}[2xye^{x^2 y}] = 2x^3 ye^{x^2 y} + 2xe^{x^2 y} = \frac{\partial}{\partial x}[x^2 e^{x^2 y}].$$

If f is a function such that $\nabla f(x,y) = f_x(x,y)\,\mathbf{i} + f_y(x,y)\,\mathbf{j}$, then you have

$$f_x(x,y) = 2xye^{x^2 y} \quad \text{and} \quad f_y(x,y) = x^2 e^{x^2 y}.$$

509

To reconstruct the function f from these two partial derivatives, integrate $f_x(x, y)$ with respect to x and $f_y(x, y)$ with respect to y as follows.

$$f(x, y) = \int f_x(x, y)\, dx = \int 2xye^{x^2y}\, dx = e^{x^2y} + g(y) + K$$

$$f(x, y) = \int f_y(x, y)\, dy = \int x^2 e^{x^2y}\, dy = e^{x^2y} + h(x) + K$$

These two expressions for $f(x, y)$ are the same if $g(y) = h(x) = 0$. Therefore, you have

$$f(x, y) = e^{x^2y} + K.$$

29. Find the curl of the vector field $\mathbf{F}(x, y, z) = xyz\,\mathbf{i} + y\mathbf{j} + z\,\mathbf{k}$ at the point $(1, 2, 1)$.

Solution:

It follows from the definition of the curl that

$\text{curl } \mathbf{F}(x, y, z)$

$$= \begin{vmatrix} \mathbf{i} & \mathbf{j} & \mathbf{k} \\ \dfrac{\partial}{\partial x} & \dfrac{\partial}{\partial y} & \dfrac{\partial}{\partial z} \\ xyz & y & z \end{vmatrix}$$

$$= \begin{vmatrix} \dfrac{\partial}{\partial y} & \dfrac{\partial}{\partial z} \\ y & z \end{vmatrix} \mathbf{i} - \begin{vmatrix} \dfrac{\partial}{\partial x} & \dfrac{\partial}{\partial z} \\ xyz & z \end{vmatrix} \mathbf{j} + \begin{vmatrix} \dfrac{\partial}{\partial x} & \dfrac{\partial}{\partial y} \\ xyz & y \end{vmatrix} \mathbf{k}$$

$$= (0 - 0)\mathbf{i} - (0 - xy)\mathbf{j} + (0 - xz)\mathbf{k} = xy\,\mathbf{j} - xz\,\mathbf{k}.$$

Therefore, $\text{curl } \mathbf{F}(1, 2, 1) = 2\mathbf{j} - \mathbf{k}$.

33. Find the curl of the vector field

$$\mathbf{F}(x, y, z) = \left(\arctan\frac{x}{y}\right)\mathbf{i} + (\ln\sqrt{x^2 + y^2})\mathbf{j} + \mathbf{k}.$$

Solution:

$$\mathbf{curl}\,\mathbf{F}(x, y, z) = \begin{vmatrix} \mathbf{i} & \mathbf{j} & \mathbf{k} \\[6pt] \dfrac{\partial}{\partial x} & \dfrac{\partial}{\partial y} & \dfrac{\partial}{\partial z} \\[8pt] \arctan\dfrac{x}{y} & \ln\sqrt{x^2 + y^2} & 1 \end{vmatrix}$$

$$= (0 - 0)\mathbf{i} - (0 - 0)\mathbf{j}$$

$$+ \left[\frac{x}{x^2 + y^2} - \frac{-x/y^2}{1 + (x^2/y^2)}\right]\mathbf{k} = \frac{2x}{x^2 + y^2}\,\mathbf{k}$$

37. Determine if $\mathbf{F}(x, y, z) = \sin y\,\mathbf{i} - x\cos y\,\mathbf{j} + \mathbf{k}$ is conservative and, if so, find the potential function $f(x, y, z)$.

Solution:

Using Theorem 15.2, you have

$$\mathbf{F}(x, y, z) = \sin y\,\mathbf{i} - x\cos y\,\mathbf{j} + \mathbf{k} = M\,\mathbf{i} + N\,\mathbf{j} + P\,\mathbf{k}.$$

Since

$$\frac{\partial}{\partial y} = 0 = \frac{\partial N}{\partial z}, \quad \frac{\partial P}{\partial x} = 0 = \frac{\partial M}{\partial z},$$

$$\frac{\partial N}{\partial x} = -\cos y \neq \cos y = \frac{\partial M}{\partial y},$$

it follows that $\mathbf{F}$ is not conservative.

41. Determine if $\mathbf{F}(x, y, z) = (1/y)\mathbf{i} - (x/y^2)\mathbf{j} + (2z - 1)\mathbf{k}$ is conservative and if so find the potential function $f(x, y, z)$.

Solution:

Using Theorem 15.2, you have

$$\mathbf{F}(x, y, z) = \frac{1}{y}\mathbf{i} - \frac{x}{y^2}\mathbf{j} + (2x - 1)\mathbf{k} = M\mathbf{i} + N\mathbf{j} + P\mathbf{k}.$$

Since

$$\frac{\partial P}{\partial y} = 0 = \frac{\partial N}{\partial z}, \quad \frac{\partial P}{\partial x} = 0 = \frac{\partial M}{\partial z}, \quad \frac{\partial N}{\partial x} = -\frac{1}{y^2} = \frac{\partial M}{\partial y},$$

it follows that $\mathbf{F}$ is conservative. Now, if f is a function such that $\mathbf{F}(x, y, z) = \nabla f(x, y, z)$, then

$$f_x(x, y, z) = \frac{1}{y}, \quad f_y(x, y, z) = -\frac{x}{y^2}, \quad f_z(x, y, z) = 2z - 1$$

and by integrating with respect to x, y, and z respectively, you obtain

$$f(x, y, z) = \int M\,dx = \int \frac{1}{y}\,dx = \frac{x}{y} + g(y, z) + K$$

$$f(x, y, z) = \int N\,dy = \int -\frac{x}{y^2}\,dy = \frac{x}{y} + h(x, z) + K$$

$$f(x, y, z) = \int P\,dz = \int (2z - 1)\,dz = z^2 - z + k(x, y) + K$$

By comparing these three versions of f, you can conclude that

$$g(y, z) = z^2 - z, \quad h(x, z) = z^2 - z, \quad \text{and} \quad k(x, y) = \frac{x}{y}.$$

Therefore,

$$f(x, y, z) = \frac{x}{y} + z^2 - z + K.$$

45. Find **curl**(**curl F**) if $F(x, y, z) = xyz\,\mathbf{i} + y\,\mathbf{j} + z\,\mathbf{k}$.

Solution:

From Exercise 29 you have $\mathbf{curl\ F}(x, y, z) = xy\,\mathbf{j} - xz\,\mathbf{k}$. Thus,

$$\mathbf{curl(curl\ F)}(x, y, z)] = \begin{vmatrix} \mathbf{i} & \mathbf{j} & \mathbf{k} \\ \dfrac{\partial}{\partial x} & \dfrac{\partial}{\partial y} & \dfrac{\partial}{\partial z} \\ 0 & xy & -xz \end{vmatrix}$$

$$= (0 - 0)\mathbf{i} - (-z - 0)\mathbf{j} + (y - 0)\mathbf{k}$$
$$= z\,\mathbf{j} + y\,\mathbf{k}.$$

49. Find the divergence of the vector field

$$F(x, y, z) = \sin x\,\mathbf{i} + \cos y\,\mathbf{j} + z^2\,\mathbf{k}.$$

Solution:

By definition,

$$\operatorname{div} F(x, y, z) = \frac{\partial}{\partial x}[\sin x] + \frac{\partial}{\partial y}[\cos y] + \frac{\partial}{\partial z}[z^2]$$
$$= \cos x - \sin y + 2z.$$

57. Find the divergence of the curl of the vector field

$$F(x, y, z) = xyz\,\mathbf{i} + y\,\mathbf{j} + z\,\mathbf{k}.$$

Solution:

From Exercise 29 you have $\mathbf{curl\ F}(x, y, z) = xy\,\mathbf{j} - xz\,\mathbf{k}$. By definition,

$$\operatorname{div}(\mathbf{curl\ F}) = \frac{\partial}{\partial x}[0] + \frac{\partial}{\partial y}[xy] + \frac{\partial}{\partial z}[-xz] = x - x = 0.$$

15.2   Line Integrals

9. Evaluate $\int_C (x^2 + y^2 + z^2)\, ds$ along the path

$C: \mathbf{r}(t) = \sin t\,\mathbf{i} + \cos t\,\mathbf{j} + 8t\,\mathbf{k}$ for $0 \leq t \leq \pi/2$.

Solution:

Since $r'(t) = \cos t\,\mathbf{i} - \sin t\,\mathbf{j} + 8\,\mathbf{k}$, you have

$$ds = \|\mathbf{r}'(t)\|\, dt = \sqrt{[x'(t)]^2 + [y'(t)]^2 + [z'(t)]^2}\, dt$$
$$= \sqrt{\cos^2 t + (-\sin t)^2 + 8^2}\, dt = \sqrt{65}\, dt.$$

It follows that

$$\int_C (x^2 + y^2 + z^2)\, ds = \int_0^{\pi/2} (\sin^2 t + \cos^2 t + 64t^2)\sqrt{65}\, dt$$
$$= \sqrt{65} \int_0^{\pi/2} (1 + 64t^2)\, dt$$
$$= \sqrt{65}\left[t + \frac{64}{3}t^3 \right]_0^{\pi/2} = \frac{\sqrt{65}\,\pi}{6}(3 + 16\pi^2).$$

13. Evaluate $\int_C (x^2 + y^2)\, ds$ along the path $C: x^2 + y^2 = 1$ from $(1, 0)$ counterclockwise to $(0, 1)$.

Solution:

Since the path is one-fourth the unit circle, it can be represented by $\mathbf{r}(t) = \cos t\,\mathbf{i} + \sin t\,\mathbf{j}$ for $0 \leq t \leq \pi/2$. Therefore,

$$\mathbf{r}'(t) = -\sin t\,\mathbf{i} + \cos t\,\mathbf{j}$$
$$ds = \|\mathbf{r}'(t)\|\, dt = \sqrt{(-\sin t)^2 + (\cos t)^2}\, dt = dt.$$

It follows that

$$\int_C (x^2 + y^2)\, ds = \int_0^{\pi/2} (\cos^2 t + \sin^2 t)\, dt$$
$$= \int_0^{\pi/2} dt = \frac{\pi}{2}.$$

17. Evaluate $\displaystyle\int_C (x + 4\sqrt{y})\,ds$ counterclockwise around the triangle with vertices $(0, 0)$, $(1, 0)$, and $(0, 1)$.

Solution:

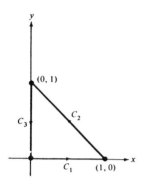

Path C has parts as shown in the accompanying figure.

$$C_1 : \quad x = t, \qquad\qquad\qquad y = 0, \qquad 0 \le t \le 1$$
$$ds = \sqrt{1 + 0}\,dt = dt$$
$$C_2 : \quad x = 1 - t, \qquad\qquad y = t, \qquad 0 \le t \le 1$$
$$ds = \sqrt{1 + 1}\,dt = \sqrt{2}\,dt$$
$$C_3 : \quad x = 0, \qquad\qquad\qquad y = 1 - t, \quad 0 \le t \le 1$$
$$ds = \sqrt{0 + 1}\,dt = dt$$

Therefore,

$$\int_C (x + 4\sqrt{y})\,dx$$
$$= \int_0^1 t\,dt + \int_0^1 [(1 - t) + 4\sqrt{t}]\sqrt{2}\,dt + \int_0^1 4\sqrt{1 - t}\,dt$$
$$= \left[\frac{t^2}{2}\right]_0^1 + \sqrt{2}\left[t - \frac{t^2}{2} + \frac{8}{3}t^{3/2}\right]_0^1 - \left[\frac{8}{3}(1 - t)^{3/2}\right]_0^1$$
$$= \frac{1}{2} + \sqrt{2}\left(\frac{19}{6}\right) + \frac{8}{3} = \frac{19}{6}(1 + \sqrt{2}).$$

23. Evaluate $\displaystyle\int_C \mathbf{F} \cdot d\mathbf{r}$ for $\mathbf{F}(x, y) = 3x\,\mathbf{i} + 4y\,\mathbf{j}$ and $\mathbf{r}(t) = (2\cos t)\mathbf{i} + (2\sin t)\mathbf{j}$ for $0 \le t \le \pi/2$.

Solution:

Since $x(t) = 2\cos t$ and $y(t) = 2\sin t$, the vector field can be written $\mathbf{F}(x, y) = 6\cos t\,\mathbf{i} + 8\sin t\,\mathbf{j}$. Use the fact that $\mathbf{r}'(t) = -2\sin t\,\mathbf{i} + 2\cos t\,\mathbf{j}$ and write the following.

$$\int_C \mathbf{F} \cdot d\mathbf{r} = \int_a^b \mathbf{F}(x(t), y(t)) \cdot \mathbf{r}'(t)\,dt$$
$$= \int_0^{\pi/2} (-12\sin t \cos t + 16\sin t \cos t)\,dt$$
$$= 4\int_0^{\pi/2} \sin t \cos t\,dt = 4\left[\frac{\sin^2 t}{2}\right]_0^{\pi/2} = 2$$

29. Find the work done by the force field $\mathbf{F}(x,y) = -x\,\mathbf{i} - 2y\,\mathbf{j}$ on an object moving along the path given by $y = x^3$ from $(0, 0)$ to $(2, 8)$.

Solution:

Since the path C is given by $x = t$ and $y = t^3$, with $0 \leq t \leq 2$, you have $\mathbf{F}(t) = -t\,\mathbf{i} - 2t^3\,\mathbf{j}$ and $\mathbf{r}'(t) = \mathbf{i} + 3t^2\,\mathbf{j}$. Therefore,

$$W = \int_C \mathbf{F}(t) \cdot \mathbf{r}'(t)\, dt = \int_0^2 (-t - 6t^5)\, dt = \left[-\frac{t^2}{2} - t^6 \right]_0^2 = -66.$$

39. If the tangent vector $\mathbf{r}'(t)$ is orthogonal to the force vector $\mathbf{F}$ then

$$\int_C \mathbf{F} \cdot d\mathbf{r} = 0$$

regardless of the initial and terminal points of C. Demonstrate this for

$$\mathbf{F}(x,y) = (x^3 - 2x^2)\mathbf{i} + \left(x - \frac{y}{2} \right)\mathbf{j} \quad \text{and} \quad \mathbf{r}(t) = t\,\mathbf{i} + t^2\,\mathbf{j}.$$

Solution:

Since the path C is given by $x = t$ and $y = t^2$, you have

$$\mathbf{F}(x,y) = (t^3 - 2t^2)\mathbf{i} + \left(t - \frac{t^2}{2} \right)\mathbf{j} \quad \text{and} \quad \mathbf{r}'(t) = \mathbf{i} + 2t\,\mathbf{j}.$$

Therefore,

$$\int_C \mathbf{F} \cdot d\mathbf{r} = \int_a^b \mathbf{F}(x(t), y(t)) \cdot \mathbf{r}'(t)\, dt$$

$$= \int_a^b (t^3 - 2t^2 + 2t^2 - t^3)\, dt = \int_a^b 0\, dt = 0.$$

49. Evaluate $\displaystyle\int_C (2x - y)\,dx + (x + 3y)\,dy$ along the parabolic path $x = t$ and $y = 2t^2$ from $(0,\,0)$ to $(2,\,8)$.

Solution:

The path C is given by $x = t$, $y = 2t^2$, with $0 \le t \le 2$, $dx = dt$, and $dy = 4t\,dt$. Therefore,

$$
\begin{aligned}
\int_C (2x - y)\,dx + (x + 3y)\,dy &= \int_0^2 (2t - 2t^2)\,dt + (t + 6t^2)4t\,dt \\
&= \int_0^2 (2t + 2t^2 + 24t^3)\,dt \\
&= \left[t^2 + \frac{2t^3}{3} + 6t^4 \right]_0^2 \\
&= 4 + \frac{16}{3} + 96 = \frac{316}{3}.
\end{aligned}
$$

53. Find the area of the lateral surface under the function $f(x, y) = xy$ and over the circle $x^2 + y^2 = 1$ from $(1,\,0)$ to $(0,\,1)$.

Solution:

We represent C parametrically as $x = \cos t$ and $y = \sin t$ for $0 \le t \le \pi/2$. Then $f(x, y) = xy = \cos t \sin t$ and

$$
ds = \sqrt{[x'(t)]^2 + [y'(t)]^2}\,dt = \sqrt{(-\sin t)^2 + (\cos t)^2}\,dt = dt.
$$

Therefore,

$$
\begin{aligned}
\text{area} &= \int_C f(x, y)\,ds \\
&= \int_0^{\pi/2} \cos t \sin t\,dt = \left[\frac{1}{2}\sin^2 t \right]_0^{\pi/2} = \frac{1}{2}.
\end{aligned}
$$

15.3   Conservative Vector Fields and Independence of Path

3. Show that the value of $\displaystyle\int_C \mathbf{F} \cdot d\mathbf{r}$ is the same for each parametric representation of C if $\mathbf{F}(x, y) = y\,\mathbf{i} - x\,\mathbf{j}$.

(a) $\mathbf{r}_1(\theta) = \sec\theta\,\mathbf{i} + \tan\theta\,\mathbf{j}, \qquad 0 \le \theta \le \pi/3$

(b) $\mathbf{r}_2(t) = \sqrt{t+1}\,\mathbf{i} + \sqrt{t}\,\mathbf{j}, \qquad 0 \le t \le 3$

Solution:

(a) For $\mathbf{r}_1(\theta) = \sec\theta\,\mathbf{i} + \tan\theta\,\mathbf{j}$, you have $\mathbf{F}(x,y) = \tan\theta\,\mathbf{i} - \sec\theta\,\mathbf{j}$ and $\mathbf{r}_1{'}(\theta) = \sec\theta\tan\theta\,\mathbf{i} + \sec^2\theta\,\mathbf{j}$. Therefore,

$$
\begin{aligned}
\int_C \mathbf{F} \cdot d\mathbf{r}_1 &= \int_0^{\pi/3} \mathbf{F} \cdot \mathbf{r}_1{'}\,d\theta \\
&= \int_0^{\pi/3} (\sec\theta\tan^2\theta - \sec^3\theta)\,d\theta \\
&= -\int_0^{\pi/3} \sec\theta\,d\theta \\
&= \left[-\ln|\sec\theta + \tan\theta| \right]_0^{\pi/3} = -\ln(2 + \sqrt{3}).
\end{aligned}
$$

(b) For $\mathbf{r}_2(t) = \sqrt{t+1}\,\mathbf{i} + \sqrt{t}\,\mathbf{j}$, you have $\mathbf{F}(t) = \sqrt{t}\,\mathbf{i} - \sqrt{t+1}\,\mathbf{j}$ and $\mathbf{r}_2{'}(t) = \dfrac{1}{2\sqrt{t+1}}\,\mathbf{i} + \dfrac{1}{2\sqrt{t}}\,\mathbf{j}$. Therefore,

$$
\begin{aligned}
\int_C \mathbf{F} \cdot d\mathbf{r}_2 &= \int_0^3 \mathbf{F} \cdot \mathbf{r}_2{\prime}\,dt \\
&= \int_0^3 \left(\frac{\sqrt{t}}{2\sqrt{t+1}} - \frac{\sqrt{t+1}}{2\sqrt{t}} \right) dt = \frac{1}{2}\int_0^3 \frac{-1}{\sqrt{t+1}\sqrt{t}}\,dt.
\end{aligned}
$$

If $u = \sqrt{t+1}$, then $u^2 = t+1$, $u^2 - 1 = t$, $2u\,du = dt$, and $\sqrt{u^2 - 1} = \sqrt{t}$. Thus,

$$
\begin{aligned}
\frac{1}{2}\int_0^3 \frac{-1}{\sqrt{t+1}\sqrt{t}}\,dt &= -\int_1^2 \frac{du}{\sqrt{u^2 - 1}} \\
&= \left[-\ln|u + \sqrt{u^2 + 1}| \right]_1^2 = -\ln(2 + \sqrt{3}).
\end{aligned}
$$

11. Find the value of the line integral

$$\int_C 2xy\,dx + (x^2 + y^2)\,dy$$

along the paths

(a) C : ellipse $(x^2/25) + (y^2/16) = 1$ from $(5, 0)$ to $(0, 4)$.

(b) C : parabola $y = 4 - x^2$ from $(2, 0)$ to $(0, 4)$.

Solution:

(a) Observe that the vector field $\mathbf{F}(x, y) = 2xy\,\mathbf{i} + (x^2 + y^2)\,\mathbf{j}$ is conservative, since

$$\frac{\partial}{\partial y}[2xy] = 2x = \frac{\partial}{\partial x}[x^2 + y^2].$$

Therefore, the line integral is independent of path and you can replace the path along the ellipse from $(5, 0)$ to $(0, 4)$ with a path which will simplify the integration. One possibility is the path along the coordinate axes from $(5, 0)$ to $(0, 0)$ and then from $(0, 0)$ to $(0, 4)$. Along the path from $(5, 0)$ to $(0, 0)$ you have $y = 0$ and $dy = 0$. On the path from $(0, 0)$ to $(0, 4)$ you have $x = 0$ and $dx = 0$. Hence,

$$\int_C 2xy\,dx + (x^2 + y^2)\,dy$$

$$= \int_5^0 0\,dx + (x^2)0 + \int_0^4 (0)(0) + (0 + y^2)\,dy$$

$$= \left[\frac{y^3}{3}\right]_0^4 = \frac{64}{3}.$$

(b) Using the same method as in part (a) replace the path along the parabola by the path along the axes from $(2, 0)$ to $(0, 0)$ and then from $(0, 0)$ to $(0, 4)$. Thus,

$$\int_C 2xy\,dx + (x^2 + y^2)\,dy$$

$$= \int_2^0 0\,dx + (x^2 + 0)(0) + \int_0^4 (0)(0) + (0 + y^2)\,dy$$

$$= \left[\frac{y^3}{3}\right]_0^4 = \frac{64}{3}.$$

(Since the line integral is path-independent, you could have used the Fundamental Theorem.)

15. Find the value of the line integral $\int_C \mathbf{F} \cdot d\mathbf{r}$ where

$$\mathbf{F}(x, y, z) = (2y + x)\,\mathbf{i} + (x^2 - z)\,\mathbf{j} + (2y - 4z)\,\mathbf{k}$$

and

(a) $\mathbf{r}_1(t) = t\,\mathbf{i} + t^2\,\mathbf{j} + \mathbf{k}$ $0 \leq t \leq 1$

(b) $\mathbf{r}_2(t) = t\,\mathbf{i} + t\,\mathbf{j} + (2t - 1)^2\,\mathbf{k}$ $0 \leq t \leq 1.$

Solution:

(a) Along the path $\mathbf{r}_1(t)$, you have

$$\mathbf{F}(x, y, z) = (2t^2 + t)\,\mathbf{i} + (t^2 - 1)\,\mathbf{j} + (2t^2 - 4)\,\mathbf{k}$$

and

$$\mathbf{r}_1' = \mathbf{i} + 2t\,\mathbf{j}.$$

Therefore,

$$\int_C \mathbf{F} \cdot d\mathbf{r}_1 = \int_a^b \mathbf{F}(x(t), y(t), z(t)) \cdot \mathbf{r}_1'(t)\, dt$$

$$= \int_0^1 (2t^3 + 2t^2 - t)\, dt = \frac{2}{3}.$$

(b) Along the path $\mathbf{r}_2(t)$, you have

$$\mathbf{F}(x, y, z) = (2t + t)\,\mathbf{i} + [t^2 - (2t - 1)^2]\,\mathbf{j} + [2t - 4(2t - 1)^2]\,\mathbf{k}$$
$$= 3t\,\mathbf{i} + (-3t^2 + 4t - 1)\,\mathbf{j} + (-16t^2 + 18t - 4)\,\mathbf{k}$$

and

$$\mathbf{r}_2'(t) = \mathbf{i} + \mathbf{j} + 4(2t - 1)\,\mathbf{k}.$$

Therefore,

$$\int_C \mathbf{F} \cdot d\mathbf{r}_2 = \int_a^b \mathbf{F}(x(t), y(t), z(t)) \cdot \mathbf{r}_2'(t)\, dt$$

$$= \int_0^1 (-128t^3 + 205t^2 - 97t + 15)\, dt = \frac{17}{6}$$

23. Use the Fundamental Theorem to evaluate

$$\int_C e^x \sin y \, dx + e^x \cos y \, dy$$

where C is one arch of the cycloid $x = \theta - \sin\theta$, $y = 1 - \cos\theta$ from $(0, 0)$ to $(2\pi, 0)$.

Solution:

Since

$$\frac{\partial}{\partial y}[e^x \sin y] = e^x \cos y = \frac{\partial}{\partial x}[e^x \cos y],$$

the integral is path-independent. Therefore, you can evaluate the line integral by using the Fundamental Theorem. We begin by finding the potential function for the vector field $\mathbf{F}(x, y) = e^x \sin y \, \mathbf{i} + e^x \cos y \, \mathbf{j}$. If f is a potential function of $\mathbf{F}$, then

$$f_x(x, y) = e^x \sin y \quad \text{and} \quad f_y(x, y) = e^x \cos y$$

and you have

$$f(x, y) = \int f_x(x, y) \, dx = \int e^x \sin y \, dx = e^x \sin y + g(y) + K$$

$$f(x, y) = \int f_y(x, y) \, dy = \int e^x \cos y \, dy = e^x \sin y + h(x) + K$$

These two expressions for $f(x, y)$ are the same if $g(y) = h(x) = 0$. Therefore, you have

$$f(x, y) = e^x \sin y + K$$

and

$$\int_C e^x \sin y \, dx + e^x \cos y \, dy = f(2\pi, 0) - f(0, 0)$$

$$= e^{2\pi}(0) - e^0(0) = 0.$$

[Since this line integral is path-independent you could have integrated along the x-axis from $(0, 0)$ to $(2\pi, 0)$ with $y = 0$ and $dy = 0$. This would have given the result of zero immediately.]

15.4   Green's Theorem

3. Verify Green's Theorem by evaluating both integrals

$$\int_C y^2\, dx + x^2\, dy = \int_R\!\!\int \left(\frac{\partial N}{\partial x} - \frac{\partial M}{\partial y} \right) dA$$

where C is the boundary of the region lying between $y = x$ and $y = x^2/4$.

Solution:

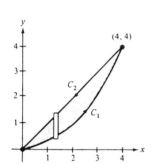

(a) As a line integral define C_1 and C_2 as shown in the accompanying figure.

$$C_1 : \; x = t, \qquad y = \frac{t^2}{4}, \qquad dx = dt, \; dy = \frac{t}{2}\, dt, \; 0 \le t \le 4$$
$$C_2 : \; x = 4 - t, \; y = 4 - t, \; dx = dy = -dt, \qquad 0 \le t \le 4$$

Thus,

$$\int_C y^2\, dx + x^2\, dy$$

$$= \int_{C_1} \left(\frac{t^4}{16} + \frac{t^3}{2} \right) dt + \int_{C_2} 2(4 - t)^2(-dt)$$

$$= \int_0^4 \left(\frac{t^4}{16} + \frac{t^3}{2} \right) dt + 2 \int_0^4 (4 - t)^2(-1)\, dt$$

$$= \left[\frac{t^5}{5(16)} + \frac{t^4}{2(4)} + \frac{2(4 - t)^3}{3} \right]_0^4$$

$$= \frac{64}{5} + 32 - \frac{128}{3} = \frac{32}{15}.$$

(b) By Green's Theorem you have

$$\int_R\!\!\int \left(\frac{\partial N}{\partial x} - \frac{\partial M}{\partial y} \right) dA$$

$$= \int_R\!\!\int (2x - 2y)\, dy\, dx$$

$$= \int_0^4 \int_{x^2/4}^x (2x - 2y)\, dy\, dx$$

$$= \int_0^4 \left[2xy - y^2 \right]_{x^2/4}^x dx$$

$$= \int_0^4 \left(x^2 - \frac{x^3}{2} + \frac{x^4}{16} \right) dx$$

$$= \left[\frac{x^3}{3} - \frac{x^4}{8} + \frac{x^5}{80} \right]_0^4 = \frac{63}{3} - 32 + \frac{64}{5} = \frac{32}{15}.$$

9. Use Green's theorem to evaluate the integral

$$\int_C (y - x)\, dx + (2x - y)\, dy$$

where C is the boundary of the region lying inside the rectangle with vertices $(5, 3)$, $(-5, 3)$, $(-5, -3)$, and $(5, -3)$ and outside the square with vertices $(1, 1)$, $(-1, 1)$, $(-1, -1)$, and $(1, -1)$.

Solution:

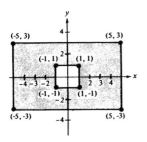

Using $M(x, y) = y - x$, $N(x, y) = 2x - y$, and the accompanying figure, you have

$$\int_C (y - x)\, dx + (2x - y)\, dy = \int_R \int \left(\frac{\partial N}{\partial x} - \frac{\partial M}{\partial y} \right) dA$$

$$= \int_R \int (2 - 1)\, dA$$

$$= \text{area of region}$$

$$= (\text{area of rectangle}) - (\text{area of square})$$

$$= 6(10) - 2(2) = 56.$$

15. Use Green's Theorem to evaluate

$$\int_C 2 \arctan \frac{y}{x}\, dx + \ln (x^2 + y^2)\, dy$$

where C is the boundary of the ellipse $x = 4 + 2 \cos \theta$, $y = 4 + \sin \theta$.

Solution:

By Green's Theorem, you have

$$\int_C M\, dx + N\, dy = \int_C 2 \arctan \frac{y}{x}\, dx + \ln (x^2 + y^2)\, dy$$

$$= \int_R \int \left(\frac{\partial N}{\partial x} - \frac{\partial M}{\partial y} \right) dA$$

$$= \int_R \int \left[\frac{2x}{x^2 + y^2} - \frac{2(1/x)}{1 + (y/x)^2} \right] dA$$

$$= \int_R \int \left[\frac{2x}{x^2 + y^2} - \frac{2x}{x^2 + x^2} \right] dA = 0.$$

27. Use a line integral to find the area of the region R bounded by $y = 2x + 1$ and $y = 4 - x^2$.

Solution:

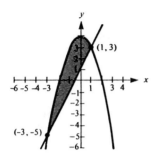

The region R (see the accompanying figure) is enclosed by the path C given by

$$C_1 : \quad x = t, \quad y = 2t + 1, \qquad\qquad -3 \leq t \leq 1$$
$$dx = dt, \quad dy = 2\,dt$$
$$C_2 : \quad x = 1 - t, \quad y = 4 - (1 - t)^2 = 3 + 2t - t^2, \quad 0 \leq t \leq 4$$
$$dx = -\,dt, \quad dy = (2 - 2t)\,dt.$$

Therefore, by Theorem 15.9 the area of R is

$$A = \frac{1}{2} \int_C x\,dy - y\,dx$$

$$= \frac{1}{2} \int_{C_1} t(2\,dt) - (2t + 1)\,dt$$

$$\qquad + \frac{1}{2} \int_{C_2} (1 - t)(2 - 2t)\,dt - (3 + 2t - t^2)(-dt)$$

$$= \frac{1}{2} \int_{-3}^{1} (-1)\,dt + \frac{1}{2} \int_0^4 (5 - 2t + t^2)\,dt$$

$$= \left[-\frac{t}{2} \right]_{-3}^{1} + \frac{1}{2} \left[5t - t^2 + \frac{t^3}{3} \right]_0^4$$

$$= -2 + \frac{1}{2} \left(20 - 16 + \frac{64}{3} \right) = \frac{32}{3}.$$

33. The centroid of the region having area A and bounded by the simple closed path C is

$$\bar{x} = \frac{1}{2A} \int_C x^2\,dy \quad \text{and} \quad \bar{y} = \frac{-1}{2A} \int_C y^2\,dx.$$

Find the centroid of the region R bounded by the graphs of $y = x^3$, $y = x$, $0 \leq x \leq 1$.

Solution:

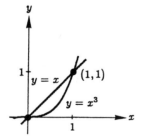

The region R (see the accompanying figure) enclosed by the path C given by

$$C_1 : \quad x = t, \quad y = t^3, \qquad\qquad dx = dt, \quad dy = 3t^2\,dt, \quad 0 \leq t \leq 1$$
$$C_2 : \quad x = 1 - t, \quad y = 1 - t, \quad dx = -\,dt, \quad dy = -\,dt, \quad 0 \leq t \leq 1.$$

Therefore, by Theorem 15.9 the area of R is

$$A = \frac{1}{2} \int_C x\,dy - y\,dx$$

$$= \frac{1}{2} \int_{C_1} t(3t^2)\,dt - t^3\,dt + \frac{1}{2} \int_{C_2} (1-t)(-dt) - (1-t)(-dt)$$

$$= \frac{1}{2} \int_0^1 2t^3\,dt = \left[\frac{1}{4}t^4\right]_0^1 = \frac{1}{4}.$$

Thus,

$$\overline{x} = \frac{1}{2A} \int_C x^2\,dy = 2\left[\int_{C_1} t^2(3t^2)\,dt + \int_{C_2} (1-t)^2(-dt)\right]$$

$$= 2\int_0^1 [3t^4 - (1-t)^2]\,dt$$

$$= 2\left[\frac{3}{5}t^5 + \frac{1}{3}(1-t)^3\right]_0^1$$

$$= 2\left[\frac{3}{5} - \frac{1}{3}\right] = \frac{8}{15}$$

and

$$\overline{y} = -\frac{1}{2A} \int_C y^2\,dx = -2\left[\int_{C_1} t^6\,dt + \int_{C_2} (1-t)^2(-dt)\right]$$

$$= -2\int_0^1 [t^6 - (1-t)^2]\,dt$$

$$= -2\left[\frac{1}{7}t^7 + \frac{1}{3}(1-t)^3\right]_0^1$$

$$= -2\left[\frac{1}{7} - \frac{1}{3}\right] = \frac{8}{21}.$$

37. The area of a plane region in polar coordinates is

$$A = \frac{1}{2} \int_C r^2\,d\theta.$$

Find the area of the region R bounded by the inner loop of the limacon $r = 1 + 2\cos\theta$.

Solution:

The inner loop of $r = 1 + 2\cos\theta$ starts at $\theta = 2\pi/3$ and ends at $\theta = 4\pi/3$ (see figure). Hence the area enclosed by this inner loop is

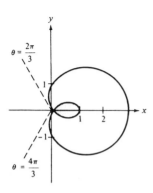

$$A = \frac{1}{2} \int_{2\pi/3}^{4\pi/3} (1 + 2\cos\theta)^2 \, d\theta$$

$$= \frac{1}{2} \int_{2\pi/3}^{4\pi/3} \left[1 + 4\cos\theta + 4\left(\frac{1 + \cos 2\theta}{2}\right)\right] d\theta$$

$$= \frac{1}{2} \int_{2\pi/3}^{4\pi/3} (3 + 4\cos\theta + 2\cos 2\theta) \, d\theta$$

$$= \frac{1}{2} \left[3\theta + 4\sin\theta + \sin 2\theta\right]_{2\pi/3}^{4\pi/3} = \pi - \frac{3\sqrt{3}}{2}.$$

15.5 Parametric Surfaces

7. Find the rectangular equation for the surface by eliminating the parameters from the vector-valued function

$$\mathbf{r}(u, v) = 2\cos u\,\mathbf{i} + v\,\mathbf{j} + 2\sin u\,\mathbf{k}.$$

Identify the surface, and sketch its graph.

Solution:

To identify the surface, you can use the trigonometric identity $\sin^2\theta + \cos^2\theta = 1$ to eliminate the parameter u and obtain

$$x^2 + z^2 = (2\cos u)^2 + (2\sin u)^2 = 4.$$

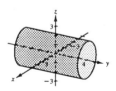

Since $y = v$ can be any real number, the surface is a circular cylinder of radius 2 and rulings parallel to the y-axis.

31. Write a set of parametric equations for the surface of revolution obtained by revolving

$$x = f(z) = \sin z, \quad 0 \le z \le \pi$$

about the z-axis.

Solution:

If you use the parameters u and v and let $x = f(u) \cos v$ and $y = f(u) \sin v$, then

$$x^2 + y^2 = [f(u) \cos v]^2 + [f(u) \sin v]^2 = [f(u)]^2.$$

Therefore, a parametric representation of the surface of revolution is

$$x = \sin u \cos v \quad y = \sin u \sin v \quad \text{and} \quad z = u.$$

where $0 \le u \le \pi$ and $0 \le v \le 2\pi$.

34. Find an equation of the tangent plane to the surface

$$\mathbf{r}(u, v) = u\mathbf{i} + v\mathbf{j} + \sqrt{uv}\,\mathbf{k}$$

at the point $(1, 1, 1)$.

Solution:

The point in the uv-plane that is mapped to the point $(x, y, z) = (1, 1, 1)$ is $(u, v) = (1, 1)$. The partial derivatives of $\mathbf{r}$ are

$$\mathbf{r}_u = \mathbf{i} + \frac{v}{2\sqrt{uv}}\mathbf{k} \quad \text{and} \quad \mathbf{r}_v = \mathbf{j} + \frac{u}{2\sqrt{uv}}\mathbf{k}.$$

Therefore, $\mathbf{r}_u(1, 1) = \mathbf{i} + \frac{1}{2}\mathbf{k}$ and $\mathbf{r}_v(1, 1) = \mathbf{j} + \frac{1}{2}\mathbf{k}$. The normal vector at the point $(1, 1, 1)$ on the surface is

$$\mathbf{v}_u \times \mathbf{v}_v = \begin{vmatrix} \mathbf{i} & \mathbf{j} & \mathbf{k} \\ 1 & 0 & \frac{1}{2} \\ 0 & 1 & \frac{1}{2} \end{vmatrix} = -\frac{1}{2}(\mathbf{i} + \mathbf{j} - 2\mathbf{k}).$$

Thus, an equation of the tangent plane at $(1, 1, 1)$ is

$$(x - 1) + (y - 1) - 2(z - 1) = 0$$
$$x + y - 2z = 0.$$

41. Find the surface area of the part of the cone

$$\mathbf{r}(u,v) = au\cos v\,\mathbf{i} + au\sin v\,\mathbf{j} + u\,\mathbf{k}$$

where $0 \le u \le b$ and $0 \le v \le 2\pi$.

Solution:

Begin by calculating $\mathbf{r}_u$ and $\mathbf{r}_v$.

$$\mathbf{r}_u = a\cos v\,\mathbf{i} + a\sin v\,\mathbf{j} + \mathbf{k}$$

$$\mathbf{r}_v = -au\sin v\,\mathbf{i} + au\cos v\,\mathbf{j}$$

The cross product of these two vectors is

$$\mathbf{v}_u \times \mathbf{v}_v = \begin{vmatrix} \mathbf{i} & \mathbf{j} & \mathbf{k} \\ a\cos v & a\sin v & 1 \\ -au\sin v & au\cos v & 0 \end{vmatrix} = -au\cos v\,\mathbf{i} - au\sin v\,\mathbf{j} + a^2 u\,\mathbf{k}$$

which implies that

$$\begin{aligned} \|\mathbf{r}_u \times \mathbf{u}_v\| &= \sqrt{(-au\cos v)^2 + (-au\sin v)^2 + (a^2 u)^2} \\ &= \sqrt{a^2 u^2 + a^4 u^2} \\ &= au\sqrt{1+a^2} \qquad\qquad (0 < a,\ 0 \le u). \end{aligned}$$

Finally, the surface area of the specified portion of the cone is

$$\begin{aligned} A &= \int_R \int \|\mathbf{r}_u \times \mathbf{r}_v\| = \int_0^{2\pi} \int_0^b au\sqrt{1+a^2}\,du\,dv \\ &= a\sqrt{1+a^2} \int_0^{2\pi} \frac{b^2}{2}\,dv \\ &= \pi a b^2 \sqrt{1+a^2}. \end{aligned}$$

15.6   Surface Integrals

  7. Use a symbolic integration utility to evaluate the surface integral

$$\int_S\int xy\,dS$$

over the surface given by $z = 9 - x^2$ for $0 \leq x \leq 2$, and $0 \leq y \leq x$.

Solution:

Begin by writing the equation for the surface S as $z = g(x,y) = 9 - x^2$ so that $g_x(x,y) = -2x$ and $g_y(x,y) = 0$, to obtain

$$\sqrt{1 + [g_x(x,y)]^2 + [g_y(x,y)]^2} = \sqrt{1 + 4x^2}$$

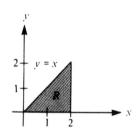

Using Theorem 15.10 and the accompanying figure, you have

$$\int_S\int xy\,dS = \int_R\int f(x,y,g(x,y))\sqrt{1 + [g_x(x,y)]^2 + [g_y(x,y)]^2}\,dA$$

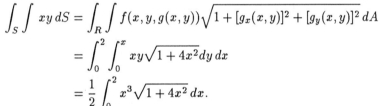

$$= \int_0^2 \int_0^x xy\sqrt{1+4x^2}\,dy\,dx$$

$$= \frac{1}{2}\int_0^2 x^3\sqrt{1+4x^2}\,dx.$$

Using a symbolic integration utility to evaluate the last integral yields

$$\int_S\int xy\,dS = \frac{391\sqrt{17}+1}{240}.$$

If you do not have access to a symbolic integration utility, use trigonometric substitution letting $2x = \tan\theta$ and $2\,dx = \sec^2\theta\,d\theta$ to obtain

$$\int_S\int xy\,dS = \frac{1}{32}\int_0^{\arctan 4} \tan^3\theta\sec^3\theta\,d\theta$$

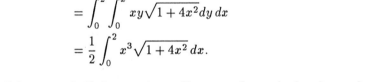

$$= \frac{1}{32}\left[\frac{1}{5}\sec^5\theta - \frac{1}{3}\sec^3\theta\right]_0^{\arctan 4}$$

$$= \frac{391\sqrt{17}+1}{240}.$$

17. Evaluate the surface integral

$$\int_S\int \sqrt{x^2+y^2+z^2}\,dS$$

over the surface $z=\sqrt{x^2+y^2}$ for $x^2+y^2\le 4$.

Solution:

Begin by writing the equation of the surface S as
$z=g(x,y)=\sqrt{x^2+y^2}$, so that $g_x(x,y)=x/\sqrt{x^2+y^2}$ and
$g_y(x,y)=y/\sqrt{x^2+y^2}$ to obtain

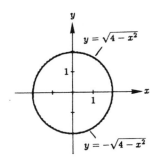

$$\sqrt{1+[g_x(x,y)]^2+[g_y(x,y)]^2}=\sqrt{1+\frac{x^2}{x^2+y^2}+\frac{y^2}{x^2+y^2}}=\sqrt{2}.$$

Using Theorem 15.10 and the accompanying figure, yields

$$\int_S\int \sqrt{x^2+y^2+z^2}\,dS$$

$$=\int_R\int f(x,y,g(x,y))\sqrt{1+[g_x(x,y)]^2+[g_y(x,y)]^2}\,dA$$

$$=\int_{-2}^{2}\int_{-\sqrt{4-x^2}}^{\sqrt{4-x^2}} \sqrt{x^2+y^2+(\sqrt{x^2+y^2})^2}\,\sqrt{2}\,dy\,dx$$

$$=2\int_{-2}^{2}\int_{-\sqrt{4-x^2}}^{\sqrt{4-x^2}} \sqrt{x^2+y^2}\,dy\,dx$$

$$=2\int_{0}^{2\pi}\int_{0}^{2} r(r\,dr\,d\theta) \qquad \text{(polar coordinates)}$$

$$=2\int_{0}^{2\pi}\left[\frac{r^3}{3}\right]_0^2 d\theta=\frac{16}{3}\int_0^{2\pi} d\theta=\frac{32\pi}{3}.$$

23. Find $\int_S\int \mathbf{F}\cdot\mathbf{N}\,dS$ (the flux of $\mathbf{F}$ through S where $\mathbf{N}$ is the
unit upper normal to S) if $\mathbf{F}(x,y,z)=x\,\mathbf{i}+y\,\mathbf{j}+z\,\mathbf{k}$ and S is
the surface given by $z=9-x^2-y^2$ for $0\le z$.

Solution:

The vector field $\mathbf{F}$, over the surface S, is given by

$$\mathbf{F}(x,y,z)=x\,\mathbf{i}+y\,\mathbf{j}+z\,\mathbf{k}=x\,\mathbf{i}+y\,\mathbf{j}+(9-x^2-y^2)\,\mathbf{k}.$$

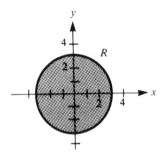

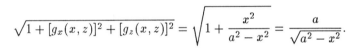

We write the equation for the surface S as
$z = g(x,y) = 9 - x^2 - y^2$ so that $g_x(x,y) = -2x$ and
$g_y(x,y) = -2y$. Then by Theorem 15.11 you have

$$\int_S \int \mathbf{F} \cdot \mathbf{N} \, dS$$

$$= \int_R \int \mathbf{F} \cdot [-g_x(x,y)\mathbf{i} - g_y(x,y)\mathbf{j} + \mathbf{k}] \, dA$$

$$= \int_R \int [x\mathbf{i} + y\mathbf{j} + (9 - x^2 - y^2)\mathbf{k}] \cdot (2x\mathbf{i} + 2y\mathbf{j} + \mathbf{k}) \, dA$$

$$= \int_R \int (9 + x^2 + y^2) \, dA$$

$$= 4 \int_0^{\pi/2} \int_0^3 (9 + r^2) r \, dr \, d\theta \qquad \text{(polar coordinates)}$$

$$= \frac{243\pi}{2}.$$

33. Find I_z for the lamina $x^2 + y^2 = a^2$ $(0 \le z \le h)$ of uniform
density 1.

Solution:

Note that S does not define z as a function of x and y. Hence,
project onto the xz-plane, so that $y = \sqrt{a^2 - x^2} = g(x,z)$ and
obtain

$$\sqrt{1 + [g_x(x,z)]^2 + [g_z(x,z)]^2} = \sqrt{1 + \frac{x^2}{a^2 - x^2}} = \frac{a}{\sqrt{a^2 - x^2}}.$$

Therefore,

$$I_z = \int_S \int (x^2 + y^2)(1) \, dS$$

$$= \int_R \int a^2 \sqrt{1 + [g_x(x,z)]^2 + [g_z(x,z)]^2} \, dA$$

$$= 4a^2 \int_0^a \int_0^h \frac{a}{\sqrt{a^2 - x^2}} \, dz \, dx$$

$$= 4a^3 h \int_0^a \frac{1}{\sqrt{a^2 - x^2}} \, dx$$

$$= 4a^3 h \left[\arcsin \frac{x}{a} \right]_0^a = 2\pi a^3 h.$$

15.7   Divergence Theorem

2. Verify the Divergence Theorem by evaluating $\displaystyle\int_S\int \mathbf{F}\cdot\mathbf{N}\,dS$ as a surface integral and as a triple integral where $\mathbf{F}(x,y,z)=2x\,\mathbf{i}-2y\,\mathbf{j}+z^2\,\mathbf{k}$ and S is the cylinder given by $x^2+y^2=1$ for $0\le z\le h$.

Solution:

As a *surface integral*, you have

$$S_1:\ x^2+y^2\le 1,\qquad z=h$$
$$S_2:\ x^2+y^2=1,\qquad 0\le z\le h$$
$$S_3:\ x^2+y^2\le 1,\qquad z=0$$

For the surface S_1 you have $\mathbf{N}_1(x,y,z)=\mathbf{k}$, $\mathbf{F}(x,y,z)\cdot\mathbf{N}_1(x,y,z)=z^2$, and $dS_1=dA$. Therefore,

$$\int_{S_1}\int \mathbf{F}\cdot\mathbf{N}\,dS=\int_{S_1}\int z^2\,dS=\int_{S_1}\int h^2\,dS=\pi h^2.$$

For the surface S_2 you have

$$\mathbf{N}_2(x,y,z)=\frac{2x\,\mathbf{i}+2y\,\mathbf{j}}{\sqrt{4x^2+4y^2}}=x\,\mathbf{i}+y\,\mathbf{j}$$

and

$$\mathbf{F}(x,y,z)\cdot\mathbf{N}_2(x,y,z)=2x^2-2y^2=2x^2-2(1-x^2)=4x^2-2.$$

Note that S_2 does **not** define z as a function of x and y. Hence, project onto the xz-plane so that $y=g(x,z)=\sqrt{1-x^2}$ and obtain

$$dS_2=\sqrt{1+[g_x(x,z)]^2+[g_z(x,z)]^2}\,dx\,dz=\frac{1}{\sqrt{1-x^2}}\,dx\,dz.$$

Therefore,

$$\begin{aligned}
\int_{S_2}\int \mathbf{F}\cdot\mathbf{N}\,dS&=\int_{S_2}\int\frac{4x^2-2}{\sqrt{1-x^2}}\,dx\,dz\\
&=\int_0^h\int_{-1}^1\left[\frac{4x^2}{\sqrt{1-x^2}}-\frac{2}{\sqrt{1-x^2}}\right]dx\,dz\\
&=\int_0^h\left[2\arcsin x-2x\sqrt{1-x^2}-2\arcsin x\right]_{-1}^1 dz\\
&=\int_0^h 0\,dz=0.
\end{aligned}$$

For the surface S_3 you have $\mathbf{N}_3(x, y, z) = -\mathbf{k}$ and $\mathbf{F}(x, y, z) \cdot \mathbf{N}(x, y, z) = -z^2$. Since $z = 0$ on S_3, you have

$$\int_{S_3} \int \mathbf{F} \cdot \mathbf{N} \, dS = 0.$$

Thus, the surface integral for the cylinder S is

$$\int_S \int \mathbf{F} \cdot \mathbf{N} \, dS = \pi h^2 + 0 + 0 = \pi h^2.$$

Using the *Divergence Theorem*, you have

$$\text{div } \mathbf{F}(x, y, z) = 2 - 2 + 2z = 2z.$$

Thus,

$$\int \int_Q \int 2z \, dV = 4 \int_0^1 \int_0^{\sqrt{1-x^2}} \int_0^h 2z \, dz \, dy \, dx$$

$$= 4 \int_0^1 \int_0^{\sqrt{1-x^2}} [z^2]_0^h \, dy \, dx$$

$$= 4 \int_0^1 \int_0^{\sqrt{1-x^2}} h^2 \, dy \, dx$$

$$= 4h^2 \int_0^1 [y]_0^{\sqrt{1-x^2}} \, dx = 4h^2 \int_0^1 \sqrt{1 - x^2} \, dx$$

$$= 4h^2 (\text{area of quarter circle of radius 1})$$

$$= 4h^2 \left(\frac{\pi}{4} \right) = \pi h^2$$

9. Use the Divergence Theorem to evaluate $\int \int \mathbf{F} \cdot \mathbf{N} \, dS$, where $\mathbf{F}(x, y, z) = x\mathbf{i} + y\mathbf{j} + z\mathbf{k}$ and S is the sphere given by $x^2 + y^2 + z^2 = 4$.

Solution:

Since div $\mathbf{F}(x, y, z) = 1 + 1 + 1 = 3$, it follows that

$$\int \int_Q \int \text{div } \mathbf{F} \, dV = 3 \int \int_Q \int dV$$

$$= 3(\text{volume of sphere of radius 2})$$

$$= 3 \left[\frac{4\pi 2^3}{3} \right] = 32\pi.$$

17. Evaluate $\displaystyle\int_S\int \mathbf{curl\ F}\cdot\mathbf{N}\,dS$ where

$\mathbf{F}(x,y,z)=(4xy+z^2)\mathbf{i}+(2x^2+6yz)\mathbf{j}+2xz\,\mathbf{k}$ and S the closed surface of the solid bounded by the graphs of $x=4$, $z=9-y^2$, and the coordinate planes.

Solution:

Using the Divergence Theorem, you have

$$\int_S\int \mathbf{curl\ F}\cdot\mathbf{N}\,dS=\int\int_Q\int \mathrm{div}(\mathbf{curl\ F})\,dV.$$

$$\mathbf{curl\ F}(x,y,z)=\begin{vmatrix}\mathbf{i}&\mathbf{j}&\mathbf{k}\\[4pt]\dfrac{\partial}{\partial x}&\dfrac{\partial}{\partial y}&\dfrac{\partial}{\partial z}\\[6pt]4xy+z^2&2x^2+6yz&2xz\end{vmatrix}$$

$$=-6y\mathbf{i}-(2z-2z)\mathbf{j}+(4x-4x)\mathbf{k}=-6y\mathbf{i}.$$

Therefore, $\mathrm{div}\,(\mathbf{curl\ F}(x,y,z))=0$ and

$$\int_S\int \mathbf{curl\ F}\cdot\mathbf{N}\,dS=\int\int_Q\int \mathrm{div}(\mathbf{curl\ F})\,dV=0.$$

15.8   Stokes' Theorem

9. Verify Stokes' Theorem by evaluating $\displaystyle\int_C \mathbf{F}\cdot\mathbf{T}\,ds$
as a line integral and as a double integral, where
$\mathbf{F}(x,y,z)=xyz\mathbf{i}+y\mathbf{j}+z\,\mathbf{k}$ and S is the portion of the plane
$3x+4y+2z=12$ lying in the first octant.

Solution:

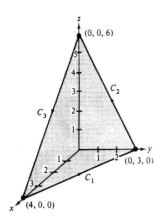

As a *line integral*, integrate along the three paths shown in the accompanying figure and obtain

$$\int_C \mathbf{F}\cdot\mathbf{T}\,ds=\int_C xyz\,dx+y\,dy+z\,dz$$

$$=\int_{C_1}0\,dx+y\,dy+0\,dz+\int_{C_2}0\,dx+y\,dy+z\,dz$$

$$+\int_{C_3}0\,dx+0\,dy+z\,dz$$

$$=\int_0^3 y\,dy+\int_3^0 y\,dy+\int_0^6 z\,dz+\int_6^0 z\,dz=0.$$

As a *double integral*, begin by finding **curl F**.

$$\text{curl } \mathbf{F}(x,y,z) = \begin{vmatrix} \mathbf{i} & \mathbf{j} & \mathbf{k} \\ \dfrac{\partial}{\partial x} & \dfrac{\partial}{\partial y} & \dfrac{\partial}{\partial z} \\ xyz & y & z \end{vmatrix} = xy\mathbf{j} - xz\,\mathbf{k}.$$

The upward normal is $\mathbf{N} = 3\mathbf{i} + 4\mathbf{j} + 2\mathbf{k}$.

$$\int_S \int (\text{curl } \mathbf{F}) \cdot \mathbf{N}\, dS$$

$$= \int_R \int (xy\mathbf{j} - xz\,\mathbf{k}) \cdot (3\mathbf{i} + 4\mathbf{j} + 2\mathbf{k})\, dA$$

$$= \int_R \int (4xy - 2xz)\, dA$$

$$= \int_0^4 \int_0^{3(4-x)/4} \left[4xy - 2x\left(6 - 2y - \frac{3x}{2}\right)\right] dy\, dx$$

$$= \int_0^4 \int_0^{3(4-x)/4} (8xy + 3x^2 - 12x)\, dy\, dx$$

$$= \int_0^4 \left(36x - 18x^2 + \frac{9x^3}{4} + 9x^2 - \frac{9x^3}{4} - 36x + 9x^2\right) dx$$

$$= \int_0^4 (0)\, dx = 0.$$

11. Use Stokes' Theorem to evaluate $\displaystyle\int_C \mathbf{F} \cdot d\mathbf{r}$ for $\mathbf{F}(x,y,z) = 2y\mathbf{i} + 3z\mathbf{j} - x\mathbf{k}$, where C is the triangle whose vertices are $(0, 0, 0)$, $(0, 2, 0)$, and $(1, 1, 1)$. (See the accompanying figure.)

Solution:

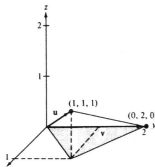

$$\text{curl } \mathbf{F} = \begin{vmatrix} \mathbf{i} & \mathbf{j} & \mathbf{k} \\ \dfrac{\partial}{\partial x} & \dfrac{\partial}{\partial y} & \dfrac{\partial}{\partial z} \\ 2y & 3z & -x \end{vmatrix}$$

$$= (0 - 3)\mathbf{i} - (-1 - 0)\mathbf{j} + (0 - 2)\mathbf{k} = -3\mathbf{i} + \mathbf{j} - 2\mathbf{k}$$

Using the coordinates of the vertices of the triangle you obtain the vectors **u** and **v** forming two of its edges. They are

$u = i + j + k$ and $v = 0i + 2j + k = 2j$. Therefore, a vector normal to the surface is given by

$$u \times v = \begin{vmatrix} i & j & k \\ 1 & 1 & 1 \\ 0 & 2 & 0 \end{vmatrix} = -2i + 2k.$$

and a unit vector normal to the surface is

$$N = \frac{u \times v}{\|u \times v\|} = \frac{-2i + 2k}{2\sqrt{2}} = \frac{-i + k}{\sqrt{2}}.$$

Thus, the surface (plane) is given by $f(x, y, z) = -x + z$, and you have $dS = \sqrt{1 + 1}\, dA = \sqrt{2}\, dA$. We conclude that

$$\int_C F \cdot dr = \int\int_S (\text{curl } F) \cdot N \, dS$$

$$= \int\int_R \frac{(3 - 2)}{\sqrt{2}} \sqrt{2} \, dA$$

$$= \int\int_R dA = \text{area of triangle} = \left(\frac{1}{2}\right)(2)(1) = 1.$$

17. Use Stokes' Theorem to evaluate $\int_C F \cdot dr$ for
$F(x, y, z) = -\ln\sqrt{x^2 + y^2}\, i + \arctan \dfrac{x}{y}\, j + k$, where S is the first octant portion of the plane $z = 9 - 2x - 3y$ over one petal of the rose curve $r = 2\sin 2\theta$.

Solution:

$$\text{curl } F = \begin{vmatrix} i & j & k \\ \dfrac{\partial}{\partial x} & \dfrac{\partial}{\partial y} & \dfrac{\partial}{\partial z} \\ -\ln\sqrt{x^2 + y^2} & \arctan\dfrac{x}{y} & 1 \end{vmatrix}$$

$$= \left[\frac{1/y}{1 + (x^2/y^2)} + \frac{y}{x^2 + y^2} \right] k = \left[\frac{2y}{x^2 + y^2} \right] k.$$

Since S is the first octant portion of the plane $z = 9 - 2x - 3y$ over one petal of $r = 2\sin 2\theta$, you have

$$N = \frac{2i + 3j + k}{\sqrt{14}}$$

and

$$dS = \sqrt{1 + (-2)^2 + (-3)^2}\, dA = \sqrt{14}\, dA.$$

Therefore,

$$\int_S \int \mathbf{curl\ F} \cdot \mathbf{N}\, dS$$

$$= \int_R \int \frac{2y}{x^2 + y^2} \frac{1}{\sqrt{14}} \sqrt{14}\, dA$$

$$= \int_R \int \frac{2y}{x^2 + y^2}\, dA$$

$$= \int_0^{\pi/2} \int_0^{2\sin 2\theta} \frac{2r \sin \theta}{r^2}\, r\, dr\, d\theta \qquad \text{(polar coordinates)}$$

$$= \int_0^{\pi/2} \int_0^{4\sin \theta \cos \theta} 2\sin \theta\, dr\, d\theta$$

$$= \int_0^{\pi/2} 8 \sin^2 \theta \cos \theta\, d\theta = \left[\frac{8 \sin^3 \theta}{3}\right]_0^{\pi/2} = \frac{8}{3}.$$

Review Exercises for Chapter 15

11. Determine if $\mathbf{F}(x, y, z) = (yz\,\mathbf{i} - xz\,\mathbf{j} - xy\,\mathbf{k})/(y^2 z^2)$ is conservative and if it is, find the potential function.

Solution:

$$\mathbf{curl\ F}(x, y, z)$$

$$= \begin{vmatrix} \mathbf{i} & \mathbf{j} & \mathbf{k} \\ \dfrac{\partial}{\partial x} & \dfrac{\partial}{\partial y} & \dfrac{\partial}{\partial z} \\ \dfrac{1}{yz} & \dfrac{-x}{y^2 z} & \dfrac{-x}{yz^2} \end{vmatrix}$$

$$= \left(\frac{x}{y^2 z^2} - \frac{x}{y^2 z^2}\right)\mathbf{i} - \left(\frac{-1}{yz^2} - \frac{-1}{yz^2}\right)\mathbf{j} + \left(\frac{-1}{y^2 z} - \frac{-1}{y^2 z}\right)\mathbf{k} = 0$$

Therefore, $\mathbf{F}$ is conservative. Now, if f is a function such that $\mathbf{F}(x, y, z) = \nabla f(x, y, z)$, then

$$f_x(x, y, z) = \frac{1}{yz}, \quad f_y(x, y, z) = -\frac{x}{y^2 z}, \text{ and } f_z(x, y, z) = -\frac{x}{yz^2}.$$

and by integrating with respect to x, y, and z separately, you obtain

$$f(x, y, z) = \int \frac{1}{yz}\, dx = \frac{x}{yz} + g(y, z) + K$$

$$f(x, y, z) = \int -\frac{x}{y^2 z}\, dy = \frac{x}{yz} + h(x, z) + K$$

$$f(x, y, z) = \int -\frac{x}{yz^2}\, dz = \frac{x}{yz} + k(x, y) + K.$$

By comparing these three versions of $f(x, y, z)$, you can conclude that $g(y, z) = h(x, z) = k(x, y) = 0$, and

$$f(x, y, z) = \frac{x}{yz} + K.$$

15. Find the divergence and the curl of the vector field

$$\mathbf{F}(x, y, z) = (\cos y + y \cos x)\,\mathbf{i} + (\sin x - x \sin y)\,\mathbf{j} + xyz\,\mathbf{k}.$$

Solution:

$$\operatorname{div} \mathbf{F}(x, y, z) = \frac{\partial}{\partial x}[\cos y + y \cos x] + \frac{\partial}{\partial y}[\sin x - x \sin y] + \frac{\partial}{\partial z}[xyz]$$

$$= -y \sin x - x \cos y + xy$$

curl $\mathbf{F}(x, y, z)$

$$= \begin{vmatrix} \mathbf{i} & \mathbf{j} & \mathbf{k} \\ \dfrac{\partial}{\partial x} & \dfrac{\partial}{\partial y} & \dfrac{\partial}{\partial z} \\ \cos y + y \cos x & \sin x - x \sin y & xyz \end{vmatrix}$$

$$= (xz - 0)\,\mathbf{i} - (yz - 0)\,\mathbf{j} + (\cos x - \sin y + \sin y - \cos x)\,\mathbf{k}$$

$$= xz\,\mathbf{i} - yz\,\mathbf{j}$$

25. Evaluate $\displaystyle\int_C (2x - y)\, dx + (x + 3y)\, dy$:

(a) C is the line segment from $(0,0)$ to $(2,-3)$.

(b) C is one counterclockwise revolution on the circle $x = 3\cos t$ and $y = 3\sin t$.

Solution:

(a) $C: x = t,\ dx = dt,\ y = -3t/2,\ dy = (-3/2)\, dt,\ 0 \le t \le 2$. Therefore,

$$\int_C (2x - y)\, dx + (x + 3y)\, dy = \int_0^2 \left[\frac{7t}{2}\, dt + \left(-\frac{7t}{2}\right)\left(-\frac{3}{2}\, dt\right) \right]$$

$$= \int_0^2 \frac{35}{4} t\, dt = \left[\frac{35}{8} t^2\right]_0^2 = \frac{35}{2}.$$

(b) $C: x = 3\cos t,\ dx = -3\sin t\, dt,\ y = 3\sin t$, $dy = 3\cos t\, dt,\ 0 \le t \le 2\pi$. Therefore,

$$\int_C (2x - y)\, dx + (x + 3y)\, dy$$

$$= \int_0^{2\pi} [(6\cos t - 3\sin t)(-3\sin t) + (3\cos t + 9\sin t)(3\cos t)]\, dt$$

$$= \int_0^{2\pi} (9\sin t \cos t + 9)\, dt = \left[\frac{9\sin^2 t}{2} + 9t\right]_0^{2\pi} = 18\pi.$$

33. Evaluate $\displaystyle\int_C \mathbf{F} \cdot d\mathbf{r}$, where $\mathbf{F}(x,y,z) = (y-z)\mathbf{i}+(z-x)\mathbf{j}+(x-y)\mathbf{k}$ and C is the curve of the intersection of the paraboloid $z = x^2 + y^2$ and the plane $x + y = 0$ from the point $(-2,2,8)$ to $(2,-2,8)$.

Solution:

On the curve of intersection $z = x^2 + (-x)^2 = 2x^2$. Hence C is given by $x = t,\ y = -t,\ z = 2t^2,\ -2 \le t \le 2$, and you have

$$\mathbf{r}(t) = t\mathbf{i} - t\mathbf{j} + 2t^2\mathbf{k} \quad \text{and} \quad \mathbf{r}' = \mathbf{i} - \mathbf{j} + 4t\mathbf{k}.$$

Thus,

$$\mathbf{F}(x,y,z) = (y-z)\mathbf{i} + (z-x)\mathbf{j} + (x-y)\mathbf{k}$$
$$= (-t - 2t^2)\mathbf{i} + (2t^2 - t)\mathbf{j} + 2t\mathbf{k}$$

and

$$\int_C \mathbf{F} \cdot d\mathbf{r} = \int_a^b \mathbf{F}(x(t), y(t), z(t)) \cdot \mathbf{r}'(t) \, dt$$

$$= \int_{-2}^{2} (-2t^2 - t - 2t^2 + t + 8t^2) \, dt = \int_{-2}^{2} 4t^2 \, dt = \frac{64}{3}.$$

38. Use the Fundamental Theorem to evaluate

$$\int_C y \, dx + x \, dy + \frac{1}{z} \, dz$$

where C is a smooth curve from $(0, 0, 1)$ to $(4, 4, 4)$.

Solution:

Since

$$\frac{\partial}{\partial y}\left[\frac{1}{z}\right] = 0 = \frac{\partial}{\partial z}[x], \quad \frac{\partial}{\partial x}\left[\frac{1}{z}\right] = 0 = \frac{\partial}{\partial z}[y], \quad \frac{\partial}{\partial x}[x] = 1 = \frac{\partial}{\partial y}[y],$$

the vector field $\mathbf{F}(x, y, z) = y\mathbf{i} + x\mathbf{j} + (1/z)\mathbf{k}$ is conservative. Integrating y, x, and $1/z$ with respect to x, y, and z, respectively, yields the potential function $f(x, y, z) = xy + \ln z + C$.

Therefore, by the Fundamental Theorem (Theorem 15.5), you have

$$\int_C y \, dx + x \, dy + \frac{1}{z} \, dz = f(4, 4, 4) - f(0, 0, 1) = 16 + \ln 4.$$

43. Use Green's Theorem to evaluate $\int_C xy \, dx + x^2 \, dy$ where C is the boundary of the region between the graphs of $y = x^2$ and $y = x$.

Solution:

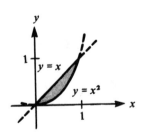

By Green's Theorem and the accompanying figure, you have

$$\int_C M(x, y) \, dx + N(x, y) \, dy = \int_C xy \, dx + x^2 \, dy$$

$$= \int_R \int \left[\frac{\partial N}{\partial x} - \frac{\partial M}{\partial y}\right] dA$$

$$= \int_0^1 \int_{x^2}^{x} x \, dy \, dx$$

$$= \int_0^1 (x^2 - x^3) \, dx = \frac{1}{12}.$$

16 DIFFERENTIAL EQUATIONS

16.1   Definitions and basic concepts

7. Classify the differential equation $(y'')^2 + 3y' - 4y = 0$ according to type, order and degree (when applicable).

Solution:

This is an ordinary differential equation and the term $(y'')^2$ indicates that the order is 2 and the degree is 2.

17. Verify that $u = e^{-t} \sin bx$ is a solution to the differential equation

$$b^2 \frac{\partial u}{\partial t} = \frac{\partial^2 u}{\partial x^2}.$$

Solution:

Since $y = e^{-t} \sin bx$, it follows that

$$\frac{\partial u}{\partial t} = -e^{-t} \sin bx$$

$$\frac{\partial u}{\partial x} = e^{-t}(\cos bx)b = be^{-t} \cos bx$$

$$\frac{\partial^2 u}{\partial x^2} = be^{-t}(-\sin bx)b = -b^2 e^{-t} \sin bx.$$

Therefore,

$$b^2 \frac{\partial u}{\partial t} = b^2(-e^{-t} \sin bx) = -b^2 e^{-t} \sin bx = \frac{\partial^2 u}{\partial x^2}.$$

21. Determine if $y = e^{-2x}$ is a solution to the differential equation $y^{(4)} - 16y = 0$.

Solution:

Since $y = e^{-2x}$, you have

$$y' = -2e^{-2x}$$
$$y'' = 4e^{-2x}$$
$$y''' = -8e^{-2x}$$
$$y^{(4)} = 16e^{-2x}.$$

Therefore,

$$y^{(4)} - 16y = 16e^{-2x} - 16(e^{-2x}) = 0$$

and the function is a solution to the differential equation.

31. It is known that $y = Ce^{kx}$ is a solution to the differential equation

$$\frac{dy}{dx} = 0.07y.$$

Is it possible to determine C or k from the information given? If so, find its value.

Solution:

Since $y = Ce^{kx}$ is a solution to the differential equation, begin by substituting the derivative of the solution into the differential equation.

$$y = Ce^{kx}$$
$$\frac{dy}{dx} = Ce^{kx}(k) = k\left(Ce^{ky}\right) = ky = 0.07y$$

Therefore, $k = 0.07$.

39. Verify that $y = C_1 \sin 3x + C_2 \cos 3x$ is a solution to the differential equation $y'' + 9y = 0$. Then find the particular solution satisfying the initial conditions $y = 2$ and $y' = 1$ when $x = \pi/6$.

Solution:

Since $y = C_1 \sin 3x + C_2 \cos 3x$, you have

$$y' = 3C_1 \cos 3x - 3C_2 \sin 3x$$
$$y'' = -9C_1 \sin 3x - 9C_2 \cos 3x.$$

Therefore,

$$y'' + 9y = -9C_1 \sin 3x - 9C_2 \cos 3x + 9(C_1 \sin 3x + C_2 \cos 3x) = 0$$

and the function is a solution to the given differential equation. Furthermore, since $y = 2$ and $y' = 1$ when $x = \pi/6$, you have

$$2 = C_1 \sin 3\left(\frac{\pi}{6}\right) + C_2 \cos 3\left(\frac{\pi}{6}\right) \quad \Longrightarrow \quad 2 = C_1(1) + C_2(0)$$

$$1 = 3C_1 \cos 3\left(\frac{\pi}{6}\right) - 3C_2 \sin 3\left(\frac{\pi}{6}\right) \quad \Longrightarrow \quad 1 = 3C_1(0) - 3C_2(1)$$

Therefore, $C_1 = 2$, $C_2 = -1/3$, and the particular solution is

$$y = 2 \sin 3x - \frac{1}{3} \cos 3x.$$

16.2   Separation of Variables in First-Order Equations

5. Solve the differential equation $(2 + x)y' = 3y$ by separation of variables.

Solution:

Begin by separating the variables as follows.

$$(2 + x)y' = 3y$$
$$(2 + x)\frac{dy}{dx} = 3y$$
$$\frac{1}{y}\,dy = \frac{3}{2 + x}\,dx.$$

By integration you obtain

$$\int \frac{1}{y} \, dy = 3 \int \frac{1}{2+x} \, dx$$
$$\ln |y| = 3 \ln |2 + x| + \ln C$$
$$\ln y = \ln |C(2+x)^3|$$
$$y = C(2+x)^3.$$

13. Solve the differential equation $y(x+1) + y' = 0$ by separation of variables and find the particular solution satisfying $y(-2) = 1$.

Solution:

Begin by separating the variables to obtain

$$y(x+1) + y' = 0$$
$$\frac{dy}{dx} = -y(x+1)$$
$$\frac{dy}{y} = -(x+1) \, dx.$$

Integration yields

$$\int \frac{dy}{y} = -\int (x+1) \, dx$$
$$\ln |y| = -\frac{(x+1)^2}{2} + C_1$$
$$y = e^{C_1 - (x+1)^2/2} = Ce^{-(x+1)^2/2}.$$

Since $y = 1$ when $x = -2$, it follows that

$$1 = Ce^{-(-2+1)^2/2} = Ce^{-1/2} \quad \text{or} \quad C = e^{1/2}$$

Therefore, the particular solution is

$$y = e^{1/2}e^{-(x+1)^2/2} = e^{[-(x+1)^2/2+1/2]} = e^{-(x^2+2x)/2}.$$

37. Solve the homogeneous differential equation $y' = xy/(x^2 - y^2)$.

Solution:

Letting $y = vx$, yields

$$y' = \frac{xy}{x^2 - y^2}$$

$$v + x\frac{dv}{dx} = \frac{x(vx)}{x^2 - v^2x^2} = \frac{v}{1 - v^2}$$

$$x\frac{dv}{dx} = \frac{v}{1 - v^2} - v = \frac{v^3}{1 - v^2}$$

$$x\,dv = \frac{v^3}{1 - v^2}\,dx.$$

Separating variables, you obtain

$$\frac{1 - v^2}{v^3}\,dv = \frac{dx}{x}$$

$$\int \left(v^{-3} - \frac{1}{v} \right) dv = \int \frac{dx}{x}$$

$$\frac{v^{-2}}{-2} - \ln|v| = \ln|x| + \ln|C_1|$$

$$\frac{1}{-2v^2} = \ln|v| + \ln|x| + \ln|C_1| = \ln|C_1vx|$$

$$\frac{x^2}{-2y^2} = \ln|C_1y|$$

$$e^{-x^2/2y^2} = C_1y \quad \Longrightarrow \quad y = Ce^{-x^2/2y^2}.$$

41. Solve the homogeneous differential equation

$$\left[x\sec\left(\frac{y}{x}\right) + y \right] dx - x\,dy = 0,$$

and find the particular solution satisfying the $y(1) = 0$.

Solution:

Letting $y = vx$, yields

$$\left[x\sec\left(\frac{vx}{x}\right) + vx \right] dx - x(v\,dx + x\,dv) = 0$$

$$x\sec v\,dx + vx\,dx - vx\,dx - x^2\,dx = 0$$

$$x\sec v\,dx = x^2\,dv$$

$$\frac{x}{x^2}\,dx = \frac{dv}{\sec v}.$$

Integration yields

$$\int \frac{dx}{x} = \int \cos v \, dv$$

$$\ln |x| = \sin v + C_1 = \sin \frac{y}{x} + C_1.$$

Since $y = 0$ when $x = 1$, you have

$$0 = \sin (0) + C_1 = C_1$$

and it follows that

$$\ln |x| = \sin \frac{y}{x} \quad \Longrightarrow \quad x = e^{\sin (y/x)}.$$

45. The rate of decomposition of radioactive radium is proportional to the amount present at a given instant. Find the percentage of a present amount that remains after 25 years if the half-life of radioactive radium is 1600 years.

Solution:

Letting y represent the percentage amount of radioactive radium present at time t, you have the differential equation

$$y' = ky$$

and the general solution has the form

$$y = Ce^{kt} = C(e^k)^t.$$

Since $y = 100\% = 1$ when $t = 0$, it follows that

$$1 = Ce^0 = C.$$

Furthermore, $y = \frac{1}{2}$ when $t = 1600$. Hence

$$\frac{1}{2} = e^{k(1600)} = (e^k)^{1600} \quad \text{or} \quad e^k = \left(\frac{1}{2}\right)^{1/1600}$$

and you have

$$y = \left(\frac{1}{2}\right)^{t/1600}.$$

Finally, when $t = 25$, you have

$$y = \left(\frac{1}{2}\right)^{25/1600} = \left(\frac{1}{2}\right)^{1/64}$$

$$\approx 0.989 = 98.9\% \text{ of the original amount.}$$

57. Find the orthogonal trajectories of the family of curves $y^2 = Cx^3$, and sketch several members of each family.

Solution:

First, solve for C in the given equation and obtain

$$C = \frac{y^2}{x^3}.$$

Then, differentiating $y^2 = Cx^3$ implicitly with respect to x and substituting the expression for C given above, you have

$$2yy' = 3Cx^2 = 3\left(\frac{y^2}{x^3}\right)x^2 = \frac{3y^2}{x}.$$

Therefore,

$$\frac{dy}{dx} = \frac{3y}{2x} \qquad \text{Slope of given family}$$

Since dy/dx represents the slope of the given family of curves at (x, y), it follows that the orthogonal family has the negative reciprocal slope, and you write

$$\frac{dy}{dx} = -\frac{2x}{3y} \qquad \text{Slope of orthogonal family}$$

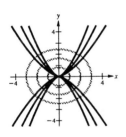

Now, find the orthogonal family by separating variables and integrating to obtain

$$3 \int y \, dy = -2 \int x \, dx$$

$$\frac{3}{2}y^2 = -x^2 + K$$

$$2x^2 + 3y^2 = K.$$

16.3   Exact First-Order Equations

3. Determine whether the differential equation

$$(3y^2 + 10xy^2)\,dx + (6xy - 2 + 10x^2y)\,dy = 0$$

is exact. If it is, find the general solution.

Solution:

Since

$$M(x,y)\,dx + N(x,y)\,dy = (3y^2+10xy^2)\,dx$$
$$+ (6xy - 2 + 10x^2y)\,dy = 0,$$

you have

$$\frac{\partial M}{\partial y} = 6y + 20xy = \frac{\partial N}{\partial x}$$

and the equation is exact. We next find a function f such that $f_x(x,y) = M(x,y)$ and $f_y(x,y) = N(x,y)$. The general solution, $f(x,y) = C$, is given by

$$f(x,y) = \int (3y^2 + 10xy^2)\,dx = 3xy^2 + 5x^2y^2 + g(y).$$

Differentiating with respect to y yields

$$f_y(x,y) = 6xy + 10x^2y + g'(y) = \overbrace{6xy - 2 + 10x^2y}^{N(x,y)}.$$

Therefore,

$$g'(y) = -2$$

and

$$g(y) = \int -2\,dy = -2y + C_1.$$

Thus, the general solution is

$$f(x,y) = C \quad \text{or} \quad 3xy^2 + 5x^2y^2 - 2y = C.$$

9. Determine whether the differential equation

$$\frac{1}{(x-y)^2}(y^2\,dx + x^2\,dy) = 0$$

is exact. If it is, find the general solution.

Solution:

Since

$$M(x,y)\,dx + N(x,y)\,dy = \frac{y^2}{(x-y)^2}\,dx + \frac{x^2}{(x-y)^2}\,dy = 0,$$

you have

$$\frac{\partial M}{\partial y} = \frac{(x-y)^2(2y) - y^2(2)(x-y)(-1)}{(x-y)^4}$$

$$= \frac{2xy}{(x-y)^3}$$

$$\frac{\partial N}{\partial x} = \frac{(x-y)^2(2x) - x^2(2)(x-y)(1)}{(x-y)^4}$$

$$= \frac{-2xy}{(x-y)^3}.$$

Since $\dfrac{\partial M}{\partial y} \neq \dfrac{\partial N}{\partial x}$, the differential equation is not exact.

13. Find the particular solution of the differential equation

$$\frac{1}{x^2+y^2}(x\,dx + y\,dy) = 0$$

that satisfies the boundary condition $y(0) = 4$.

Solution:

Since $M(x,y) = \dfrac{x}{x^2+y^2}$ and $N(x,y) = \dfrac{y}{x^2+y^2}$, you have

$$\frac{\partial M}{\partial y} = \frac{-2xy}{(x^2+y^2)^2} = \frac{\partial N}{\partial x}$$

and the equation is exact. We next find a function f such that $f_x(x,y) = M(x,y)$ and $f_y(x,y) = N(x,y)$.

$$f(x,y) = \int \frac{x}{x^2+y^2}\,dx = \ln\sqrt{x^2+y^2} + g(y)$$

$$f(x,y) = \int \frac{y}{x^2+y^2}\,dy = \ln\sqrt{x^2+y^2} + h(x)$$

Therefore, $g(y) = h(x) = C_1$ and the general solution to the differential equation is

$$f(x, y) = C \quad \text{or} \quad \ln \sqrt{x^2 + y^2} = C.$$

Finally, since $y = 4$ when $x = 0$, you have $\ln 4 = C$, and the particular solution is $\ln \sqrt{x^2 + y^2} = \ln 4$ or $x^2 + y^2 = 16$.

21. Solve the differential equation

$$(x + y) \, dx + \tan x \, dy = 0$$

by using an integrating factor that is a function of x or y alone.

Solution:

Since

$$\frac{(\partial M / \partial y) - (\partial N / \partial x)}{N} = \frac{1 - \sec^2 x}{\tan x} = h(x) \text{ and}$$

$$\int \frac{1 - \sec^2 x}{\tan x} \, dx = -\int \frac{\tan^2 x}{\tan x} \, dx = -\int \tan x \, dx = \ln|\cos x|,$$

it follows that

$$e^{\int h(x) dx} = e^{\ln|\cos x|} = \cos x$$

is an integrating factor. Thus

$$(x + y) \cos x \, dx + \cos x \tan x \, dy = 0$$
$$(x + y) \cos x \, dx + \sin x \, dy = 0$$

is exact. Thus,

$$f(x, y) = \int \sin x \, dy = y \sin x + g(x)$$

and

$$f_x(x, y) = y \cos x + g'(x) = \overbrace{(x + y) \cos x}^{M(x,y)}.$$

Therefore,

$$g'(x) = x \cos x$$

$$g(x) = \int x \cos x \, dx = x \sin x + \cos x + C_1 \qquad \text{Integration by parts}$$

and

$$f(x, y) = y \sin x + x \sin x + \cos x + C_1$$
$$= (x + y) \sin x + \cos x + C_1.$$

Finally, the general solution is

$$f(x, y) = C \quad \text{or} \quad (x + y) \sin x + \cos x = C.$$

29. Use the integrating factor $x^{-2}y^{-3}$ to find the general solution of the differential equation

$$(-y^5 + x^2 y)\,dx + (2xy^4 - 2x^3)\,dy = 0.$$

Solution:

Multiplying the given differential equation by the integrating factor $x^{-2}y^{-3}$ yields

$$\left(-\frac{y^2}{x^2} + \frac{1}{y^2}\right)dx + \left(\frac{2y}{x} - \frac{2x}{y^3}\right)dy = 0.$$

Since

$$\frac{\partial M}{\partial y} = -\frac{2y}{x^2} - \frac{2}{y^3} = \frac{\partial N}{\partial x},$$

the differential equation is exact. We next find a function f such that $f_x(x,y) = M(x,y)$ and $f_y(x,y) = N(x,y)$.

$$f(x,y) = \int \left(-\frac{y^2}{x^2} + \frac{1}{y^2}\right)dx = \frac{y^2}{x} + \frac{x}{y^2} + g(y)$$

$$f(x,y) = \int \left(\frac{2y}{x} - \frac{2x}{y^3}\right)dy = \frac{y^2}{x} + \frac{x}{y^2} + h(x)$$

Therefore, $g(y) = h(x) = C_1$ and the general solution to the differential equation is

$$f(x,y) = C \quad \text{or} \quad \frac{y^2}{x} + \frac{x}{y^2} = C.$$

16.4 First-Order Linear Differential Equations

7. Solve the first-order linear differential equation $y' - y = \cos x$.

Solution:

The equation $y' - y = \cos x$ is linear with $P(x) = -1$. Thus,

$$\int P(x)\,dx = -x \quad \text{and} \quad e^{\int P(x)\,dx} = e^{-x}$$

Multiplying both members of the differential equation in standard form by the integrating factor e^{-x} yields

$$y'e^{-x} - ye^{-x} = e^{-x}\cos x$$

$$\frac{d}{dx}[ye^{-x}] = e^{-x}\cos x$$

$$ye^{-x} = \int e^{-x}\cos x \, dx$$

$$= \frac{1}{2}(e^{-x}\sin x - e^{-x}\cos x) + C$$

$$y = \frac{1}{2}(\sin x - \cos x) + Ce^{x}.$$

11. Solve the first-order linear differential equation

$$(x-1)y' + y = x^2 - 1.$$

Solution:

In standard form the equation is

$$y' + \left(\frac{1}{x-1}\right)y = x + 1.$$

Thus,

$$\int P(x)\,dx = \int \frac{1}{x-1}\,dx = \ln|x-1|$$

and the integrating factor is

$$e^{\int P(x)\,dx} = e^{\ln|x-1|} = x - 1$$

Therefore, multiplying both members of the standard form of the differential equation yields

$$y'(x-1) + y = (x+1)(x-1)$$

$$\frac{d}{dx}[y(x-1)] = (x+1)(x-1)$$

$$y(x-1) = \int (x-1)(x+1)\,dx$$

$$= \int (x^2 - 1)\,dx = \frac{x^3}{3} - x + C_1$$

$$y = \frac{x^3 - 3x + C}{3(x-1)}.$$

15. Solve the differential equation $y' + y \tan x = \sec x + \cos x$. Find the particular solution if $y = 1$ when $x = 0$.

Solution:

The integrating factor is given by

$$e^{\int P\,dx} = e^{\int \tan x\,dx} = e^{\ln|\sec x|} = \sec x$$

Therefore, multiplying both members of the differential equation in standard form yields

$$y'(\sec x) + (\sec x \tan x)y = \sec x(\sec x + \cos x)$$

$$\frac{d}{dx}[y \sec x] = \sec^2 x + 1$$

$$y \sec x = \int (\sec^2 x + 1)\,dx$$

$$= \tan x + x + C$$

$$y = \cos x(\tan x + x + C)$$

$$= \sin x + x \cos x + C \cos x.$$

Since $y = 1$ when $x = 0$, you have

$$1 = \sin 0 + 0 \cos 0 + C \cos 0$$

$$1 = C$$

Therefore, the particular solution is $y = \sin x + (x + 1) \cos x$.

21. Solve the Bernoulli differential equation

$$y' + \left(\frac{1}{x}\right)y = xy^2.$$

Solution:

For the given Bernoulli equation you have $n = 2$, and $1 - n = -1$. If $z = y^{1-n} = y^{-1}$, then $z' = -y^{-2}y'$. Multiplying both members of the differential equation produces

$$y' + \left(\frac{1}{x}\right)y = xy^2$$

$$-y^{-2}y' - \left(\frac{1}{x}\right)y^{-1} = -x$$

$$z' - \left(\frac{1}{x}\right)z = -x.$$

This equation is linear in z. Using $P(x) = -1/x$ produces

$$-\int \frac{1}{x}\, dx = -\ln|x| \quad \Longrightarrow \quad e^{-\ln|x|} = \frac{1}{x}.$$

as the integrating factor. Multiplying the linear equation by this factor produces

$$z'\left(\frac{1}{x}\right) - \left(\frac{1}{x^2}\right) z = -1$$

$$\frac{d}{dx}\left[\frac{z}{x}\right] = -1$$

$$\frac{z}{x} = \int (-1)\, dx = -x + C$$

$$z = -x^2 + Cx$$

$$\frac{1}{y} = -x^2 + Cx$$

$$y = \frac{1}{Cx - x^2}.$$

33. Glucose is added intravenously to the bloodstream at the rate q units per minute, and the body removes glucose from the bloodstream at a rate proportional to the amount present. Assume $Q(t)$ is the amount of glucose in the bloodstream at time t.

   (a) Determine the differential equation describing the rate of change with respect to time of glucose in the bloodstream.

   (b) Solve the differential equation, letting $Q = Q_0$ when $t = 0$.

   (c) Find the limit of $Q(t)$ as $t \to \infty$.

Solution:

   (a) Let $Q(t)$ be the amount of glucose in the blood stream at any time. The rate of change of Q is given by

$$\frac{dQ}{dt} = (\text{rate administered}) - (\text{rate removed})$$

$$\frac{dQ}{dt} = q - kQ$$

$$\frac{dQ}{dt} + kQ = q$$

   where t is time in minutes and k is a constant of proportionality.

(b) To solve this linear equation begin by finding the integrating factor

$$e^{\int P(t)dt} = e^{\int k\,dt} = e^{kt}$$

Therefore, by Theorem 16.4 the general solution has the form

$$Qe^{kt} = \int qe^{kt} = \frac{q}{k}e^{kt} + C$$

$$Q(t) = \frac{q}{k} + Ce^{-kt}$$

Since $Q(0) = Q_0$, you have

$$Q_0 = \frac{q}{k} + C$$

$$Q_0 - \frac{q}{k} = C$$

$$Q(t) = \frac{q}{k} + (Q_0 - \frac{q}{k})e^{-kt}.$$

(c) $$\lim_{t\to\infty} Q(t) = \lim_{t\to\infty}\left[\frac{q}{k} + \left(Q_0 - \frac{q}{k}\right)e^{-kt}\right] = \frac{q}{k}$$

36. A 200-gal tank is full of a solution containing 25 lb of concentrate. Starting at time $t = 0$, distilled water is admitted to the tank at the rate of 10 gal/min, and the well-stirred solution is withdrawn at the same rate.

(a) Find the amount of the concentrate in the solution as a function of t.

(b) Find the time at which concentrate must be added if it is done when the amount of concentrate in the tank reaches 15 lb.

Solution:

(a) Using Exercise 35, you have $r_1 = 10$, $r_2 = 10$, $q_1 = 0$, and $v_0 = 200$. Thus,

$$\frac{dQ}{dt} + \frac{10Q}{200 + (0)t} = 0$$

$$\frac{dQ}{dt} = \frac{-10Q}{200} = -\frac{Q}{20}$$

Separating variables, you have

$$\frac{dQ}{Q} = -\frac{dt}{20}$$

$$\ln |Q| = -\frac{t}{20} + C_1$$

$$Q(t) = e^{C_1-(t/20)} = Ce^{-t/20}.$$

(b) When $Q = 15$, you have

$$15 = 25e^{-t/20}$$

$$\frac{3}{5} = e^{-t/20}$$

$$\ln 3 - \ln 5 = -\frac{t}{20}$$

$$t = 20(\ln 5 - \ln 3) \approx 10.2 \text{ min.}$$

43. Solve the differential equation $\dfrac{dy}{dx} = \dfrac{e^{2x+y}}{e^{x-y}}$ by any appropriate method.

Solution:

Separating variables yields

$$e^{2x+y}\, dx = e^{x-y}\, dy$$

$$e^{2x}e^y\, dx = e^x e^{-y}\, dy$$

$$e^x\, dx = e^{-2y}\, dy$$

$$\int e^x\, dx = \int e^{-2y}\, dy$$

$$e^x = -\frac{1}{2}e^{-2y} + C_1$$

$$2e^x + e^{-2y} = C.$$

49. Solve the differential equation $2xy\,dx + (x^2 + \cos y)\,dy = 0$ by any appropriate method.

Solution:

Since

$$\frac{\partial M}{\partial y} = 2x = \frac{\partial N}{\partial x}$$

the equation is exact. Thus,

$$f(x, y) = \int 2xy\,dx + g(y) = x^2 y + g(y)$$

$$f_y(x, y) = x^2 + g'(y) = \overbrace{x^2 + \cos y}^{N(x,y)}$$

Therefore,

$$g(y) = \int \cos y\,dy = \sin y + C_1$$

and

$$f(x, y) = x^2 y + \sin y + C_1.$$

Hence, the general solution is

$$x^2 y + \sin y = C.$$

55. Solve the differential equation $(x^2 y^4 - 1)\,dx + x^3 y^3\,dy = 0$ by any appropriate method.

Solution:

Rewriting the differential equation you have

$$x^3 y^3 \frac{dy}{dx} + x^2 y^4 = 1$$

$$\frac{dy}{dx} + \left(\frac{1}{x}\right) y = x^{-3} y^{-3}.$$

For this Bernoulli equation, $n = -3$ and use the substitution $z = y^{1-n} = y^4$ and $z' = 4y^3 y'$. Multiplying the equation in standard form by $4y^3$ produces

$$4y^3 y' + 4\left(\frac{1}{x}\right) y^4 = \frac{4}{x^3}$$

$$z' + \left(\frac{4}{x}\right) z = \frac{4}{x^3}.$$

This equation in linear in z. Using $P(x) = 4/x$ produces

$$\int P(x)\, dx = \int \frac{4}{x}\, dx = 4\ln|x| = \ln x^4$$

which implies that $e^{\ln x^4} = x^4$ is an integrating factor. Multiplying the linear equation by this factor produces

$$x^4 z' + 4x^3 z = 4x$$
$$\frac{d}{dx}[x^4 z] = 4x$$
$$x^4 z = \int 4x\, dx = 2x^2 + C$$
$$x^4 y^4 = 2x^2 + C$$
$$x^4 y^4 - 2x^2 = C.$$

16.5   Second-Order Homogeneous Linear Equations

7. Find the general solution of the linear differential equation $y'' - y' - 6y = 0$.

Solution:

The characteristic equation

$$m^2 - m - 6 = 0 \quad \text{or} \quad (m - 3)(m + 2) = 0$$

has two distinct real roots, $m_1 = 3$ and $m_2 = -2$. Thus the general solution is

$$y = C_1 e^{m_1 x} + C_2 e^{m_2 x} = C_1 e^{3x} + C_2 e^{-2x}.$$

11. Find the general solution of the linear differential equation $y'' + 6y' + 9y = 0$.

Solution:

The characteristic equation

$$m^2 + 6m + 9 = 0 \quad \text{or} \quad (m + 3)^2 = 0$$

has two equal roots given by $m = -3$. Thus the general solution is

$$y = C_1 e^{m_1 x} + C_2 x e^{m_1 x} = C_1 e^{-3x} + C_2 x e^{-3x} = (C_1 + C_2 x)e^{-3x}.$$

21. Find the general solution of the linear differential equation $y'' - 3y' + y = 0$.

Solution:

The characteristic equation

$$m^2 - 3m + 1 = 0$$

has the distinct real roots (using the quadratic formula)

$$m = \frac{3 \pm \sqrt{9 - 4}}{2} = \frac{3 \pm \sqrt{5}}{2}.$$

Thus, the general solution is

$$y = C_1 e^{(3+\sqrt{5})x/2} + C_2 e^{(3-\sqrt{5})x/2}.$$

23. Find the general solution of the linear differential equation $9y'' - 12y' + 11y = 0$.

Solution:

The characteristic equation

$$9m^2 - 12m + 11 = 0$$

has complex roots.

$$m = \frac{12 \pm \sqrt{144 - 4(9)(11)}}{2(9)}$$

$$= \frac{12 \pm \sqrt{-252}}{18} = \frac{12 \pm 6i\sqrt{7}}{18} = \frac{2}{3} \pm \frac{\sqrt{7}}{3} i$$

Thus, $\alpha = 2/3$ and $\beta = \sqrt{7}/3$ and the general solution is

$$y = C_1 e^{2x/3} \cos \frac{\sqrt{7}x}{3} + C_2 e^{2x/3} \sin \frac{\sqrt{7}x}{3}$$

$$= e^{2x/3} \left[C_1 \cos \frac{\sqrt{7}x}{3} + C_2 \sin \frac{\sqrt{7}x}{3} \right].$$

39. Describe the motion of a 32-pound weight suspended on a spring. Assume that the weight stretches the spring $\frac{2}{3}$ foot from its natural position, it is pulled $\frac{1}{2}$ foot below the equilibrium and released, and the motion takes place in a medium which furnishes a damping force of magnitude $\frac{1}{8}$ its speed at all times.

Solution:

By Hooke's Law, $32 = k(2/3)$ so that $k = 48$. Moreover, since the weight w is given by mg, it follows that $m = w/g = \frac{32}{32} = 1$. Also the damping force is given by $-\frac{1}{8}(dy/dt)$. Thus the differential equation modeling the oscillations of the weight is

$$\frac{d^2y}{dt^2} = -\frac{1}{8}\left(\frac{dy}{dt}\right) - 48y$$

$$\frac{d^2y}{dt^2} + \frac{1}{8}\left(\frac{dy}{dt}\right) + 48y = 0.$$

The characteristic equation is

$$8m^2 + m + 384 = 0$$

with complex roots

$$m = -\frac{1}{16} \pm \frac{\sqrt{12,287}\,i}{16}.$$

Therefore, the general solution is

$$y(t) = e^{-t/16}\left(C_1 \cos \frac{\sqrt{12,287}\,t}{16} + C_2 \sin \frac{\sqrt{12,287}\,t}{16}\right).$$

Using the initial conditions, you have

$$y(0) = C_1 = \frac{1}{2}$$

$$y'(0) = e^{-t/16}\left[\left(-\frac{\sqrt{12,287}}{16}C_1 - \frac{C_2}{16}\right)\sin \frac{\sqrt{12,287}\,t}{16}\right.$$
$$\left. + \left(\frac{\sqrt{12,287}}{16}C_2 - \frac{C_1}{16}\right)\cos \frac{\sqrt{12,287}\,t}{16}\right]$$

$$y'(0) = \frac{\sqrt{12,287}}{16}C_2 - \frac{C_1}{16} = 0 \implies C_2 = \frac{\sqrt{12,287}}{24,574}.$$

and the particular solution

$$y(t) = \frac{e^{-t/16}}{2}\left[\cos \frac{\sqrt{12,287}\,t}{16} + \frac{\sqrt{12,287}}{12,287}\sin \frac{\sqrt{12,287}\,t}{16}\right].$$

53. Use the Wronskian to verify the linear independence of the functions $y_1 = e^{ax} \sin bx$ and $y_2 = e^{ax} \cos bx$.

Solution:

$$W(y_1, y_2) = \begin{vmatrix} y_1 & y_2 \\ y_1' & y_2' \end{vmatrix}$$

$$= \begin{vmatrix} e^{ax} \sin bx & e^{ax} \cos bx \\ e^{ax}(b \cos bx + a \sin bx) & e^{ax}(a \cos bx - b \sin bx) \end{vmatrix}$$

$$= e^{2ax} \begin{vmatrix} \sin bx & \cos bx \\ b \cos bx + a \sin bx & a \cos bx - b \sin bx \end{vmatrix}$$

$$= e^{2ax}(a \sin bx \cos bx - b \sin^2 bx - b \cos^2 bx - a \sin bx \cos bx)$$

$$= -be^{2ax}$$

Since $W(y_1, y_2) \neq 0$, y_1 and y_2 are linearly independent.

16.6   Second-Order Nonhomogeneous Linear Equations

7. Solve the differential equation $y'' + y = x^3$ by the method of undetermined coefficients. Find the particular solution satisfying the initial conditions $y(0) = 1$ and $y'(0) = 0$.

Solution:

The characteristic equation $m^2 + 1 = 0$ has roots $m = \pm i$. Thus,

$$y_h = C_1 \cos x + C_2 \sin x.$$

Since $F(x) = x^3$, choose y_p to be

$$y_p = A + Bx + Cx^2 + Dx^3.$$

Thus, $y_p' = B + 2Cx + 3Dx^2$ and $y_p'' = 2C + 6Dx$. Substitution into the differential equation yields

$$y'' + y = x^3$$
$$(2C + 6Dx) + (A + Bx + Cx^2 + Dx^3) = x^3$$
$$(2C + A) + (6D + B)x + Cx^2 + Dx^3 = x^3.$$

Therefore,

$$2C + A = 0, \qquad 6D + B = 0, \qquad C = 0, \qquad D = 1$$

from which it follows that $A = 0$ and $B = -6$. Thus, the general solution is

$$y = y_h + y_p = C_1 \cos x + C_2 \sin x - 6x + x^3.$$

Since $y(0) = 1$, $y'(0) = 0$, and $y' = -C_1 \sin x + C_2 \cos x - 6 + 3x^2$, you have

$$1 = C_1(1) + C_2(0) = C_1$$
$$0 = -C_1(0) + C_2(1) - 6 \quad \text{or} \quad C_2 = 6.$$

Finally, the particular solution is

$$y = \cos x + 6 \sin x - 6x + x^3.$$

15. Solve the differential equation $y'' + 9y = \sin 3x$ by the method of undetermined coefficients.

Solution:

The characteristic equation $m^2 + 9 = 0$ has roots $m = \pm 3i$ and you have

$$y_h = C_1 \cos 3x + C_2 \sin 3x$$

Since $F(x) = \sin 3x$, consider

$$y_p = A \cos 3x + B \sin 3x$$

However, the terms of y_p are *not* independent of those of y_h. Thus, multiply both terms by x and write

$$y_p = Ax \cos 3x + Bx \sin 3x$$
$$y_p' = -3Ax \sin 3x + A \cos 3x + 3Bx \cos 3x + B \sin 3x$$
$$y_p'' = -6A \sin 3x + 6B \cos 3x - 9Ax \cos 3x - 9Bx \sin 3x$$

Substitution into the given differential equation and simplifying yields

$$y_p'' + 9y_p = -6A \sin 3x + 6B \cos 3x = \sin 3x.$$

Therefore, $A = -\frac{1}{6}$ and $B = 0$ and the general solution is

$$y = y_h + y_p = C_1 \cos 3x + C_2 \sin 3x - \frac{x}{6} \cos 3x$$
$$= \left(C_1 - \frac{x}{6} \right) \cos 3x + C_2 \sin 3x.$$

23. Solve the differential equation $y'' + 4y = \csc 2x$ by the method of variation of parameters.

Solution:

The characteristic equation, $m^2 + 4 = 0$, has solutions $m = \pm 2i$. Hence,

$$y_h = C_1 \cos 2x + C_2 \sin 2x$$

Replacing C_1 and C_2 by u_1 and u_2 produces

$$y_p = u_1 \cos 2x + u_2 \sin 2x.$$

By the method of variation of parameters you obtain the following system of equations.

$$u_1{}' \cos 2x + u_2{}' \sin 2x = 0$$
$$u_1{}'(-2 \sin 2x) + u_2{}'(2 \cos 2x) = \csc 2x$$

Multiplying the first equation by $2 \sin 2x$ and the second by $\cos 2x$, and then adding the equations yields $u_2{}' = \frac{1}{2} \cot 2x$. Substituting this result into the first equation, yields $u_1{}' = -\frac{1}{2}$. Integration yields

$$u_1 = \int -\frac{1}{2} dx = -\frac{x}{2}$$

$$u_2 = \int \frac{1}{2} \cot 2x \, dx = \frac{1}{4} \ln |\sin 2x|$$

and it follows that

$$y = y_h + y_p$$
$$= C_1 \cos 2x + C_2 \sin 2x - \frac{x}{2} \cos 2x + \frac{1}{4} \sin 2x \ln |\sin 2x|$$
$$= \left(C_1 - \frac{x}{2} \right) \cos 2x + \left(C_2 + \frac{1}{4} \ln |\sin 2x| \right) \sin 2x.$$

29. The oscillating motion of a 24 pound weight suspended by a spring is modeled by

$$\frac{24}{32}y'' + 48y = \frac{24}{32}(48\sin 4t)$$

where t is time in seconds and y is the displacement in feet. Solve the differential equation if the initial displacement is $y(0) = \frac{1}{4}$ and the initial velocity is $y'(0) = 0$.

Solution:

$$\frac{24}{32}y'' + 48y = \frac{24}{32}(48\sin 4t)$$
$$y'' + 64y = 48\sin 4t$$

The characteristic equation, $m^2 + 64 = 0$, has roots $m = \pm 8i$ and you have

$$y_h = C_1\cos 8t + C_2\sin 8t.$$

Since the derivatives of even order of the sine function is a sine function, let $y_p = A\sin 4t$. Therefore, $y_p' = 4A\cos 4t$ and $y_p'' = -16A\sin 4t$. Substituting these results in the differential equation and simplifying yields

$$y_p'' + 64y_p = -16A\sin 4t + 64A\sin 4t = 48A\sin 4t = 48\sin 4t.$$

Thus, $A = 1$ and

$$y = y_h + y_p = C_1\cos 8t + C_2\sin 8t + \sin 4t$$
$$y' = y_h' + y_p' = -8C_1\sin 8t + 8C_2\cos 8t + 4\cos 4t.$$

Since $y = \dfrac{1}{4}$ and $y' = 0$ when $t = 0$, you have

$$\frac{1}{4} = C_1$$

$$0 = 8C_2 + 4 \quad \text{or} \quad C_2 = -\frac{1}{2}.$$

Therefore,

$$y = \frac{1}{4}\cos 8t - \frac{1}{2}\sin 8t + \sin 4t.$$

16.7   Series Solutions of Differential Equations

3. Verify that the power series solution of the differential equation $y'' - 9y = 0$ is equivalent to the solution found using techniques in Sections 16.1–16.5.

Solution:

Assume $y = \displaystyle\sum_{n=0}^{\infty} a_n x^n$ is a solution. Then,

$$y' = \sum_{n=1}^{\infty} n a_n x^{n-1} \quad \text{and} \quad y'' = \sum_{n=2}^{\infty} n(n-1) a_n x^{n-2}.$$

Substituting these series in the differential equation yields

$$y'' - 9y = 0$$

$$\sum_{n=2}^{\infty} n(n-1) a_n x^{n-2} - 9 \sum_{n=0}^{\infty} a_n x^n = 0$$

$$\sum_{n=2}^{\infty} n(n-1) a_n x^{n-2} = \sum_{n=0}^{\infty} 9 a_n x^n$$

The index of summation is changed by replacing n by $n+2$ in the left sum to insure that x^n occurs in both sums. Thus,

$$\sum_{n=0}^{\infty} (n+2)(n+1) a_{n+2} x^n = \sum_{n=0}^{\infty} 9 a_n x^n.$$

Equating coefficients yields

$$(n+2)(n+1) a_{n+2} = 9 a_n$$

from which you obtain the recursion formula

$$a_{n+2} = \frac{9 a_n}{(n+2)(n+1)} \qquad (n \geq 0)$$

Thus the coefficients of the series solution are

$$a_2 = \frac{9}{2}a_0 = \frac{3^2}{2}a_0 \qquad\qquad a_3 = \frac{9}{6}a_1 = \frac{3^2}{3 \cdot 2}a_1$$

$$a_4 = \frac{9}{12}a_2 = \frac{3^4}{4 \cdot 3 \cdot 2}a_0 \qquad\qquad a_5 = \frac{9}{20}a_3 = \frac{3^4}{5 \cdot 4 \cdot 3 \cdot 2}a_1$$

$$\begin{array}{c} \cdot \\ \cdot \\ \cdot \end{array} \qquad\qquad\qquad\qquad \begin{array}{c} \cdot \\ \cdot \\ \cdot \end{array}$$

$$a_{2k} = \frac{(3)^{2k}}{(2k)!}a_0 \qquad\qquad a_{2k+1} = \frac{(3)^{2k}}{(2k+1)!}a_1$$

Thus, you can represent the general solution as the sum of two power series—one for the even-powered terms with coefficients in terms of a_0 and one for the odd-powered terms with coefficients in terms of a_1. Thus, you have

$$y = \left(a_0 x^0 + \frac{3^2}{2!}a_0 x^2 + \frac{3^4}{4!}a_0 x^4 + \cdots \right)$$

$$+ \left(a_1 x + \frac{3^2}{3!}a_1 x^3 + \frac{3^4}{5!}a_1 x^5 + \cdots \right)$$

$$= a_0 \left[1 + \frac{(3x)^2}{2!} + \frac{(3x)^4}{4!} + \cdots \right] + \frac{a_1}{3}\left[3x + \frac{(3x)^3}{3!} + \frac{(3x)^5}{5!} + \cdots \right]$$

$$= a_0 \sum_{k=0}^{\infty} \frac{(3x)^{2k}}{(2k)!} + \frac{a_1}{3}\sum_{k=0}^{\infty} \frac{(3x)^{2k+1}}{(2k+1)!}.$$

Observe that $y'' - 9y = 0$ is a second-order homogeneous differential equation with characteristic equation $m^2 - 9 = 0$. Thus, the general solution is

$$y = C_1 e^{3x} + C_2 e^{-3x}.$$

The following reconciles this form of the solution with the series solution given above.

$$y = C_1 e^{3x} + C_2 e^{-3x}$$

$$= C_1 \sum_{n=0}^{\infty} \frac{(3x)^n}{n!} + C_2 \sum_{n=0}^{\infty} \frac{(-3x)^n}{n!}$$

$$= C_1 \left[1 + (3x) + \frac{(3x)^2}{2!} + \cdots \right]$$

$$+ C_2 \left[1 + (-3x) + \frac{(-3x)^2}{2!} + \cdots \right]$$

$$= C_1 \left[1 + \frac{(3x)^2}{2!} + \frac{(3x)^4}{4!} + \cdots \right]$$

$$+ C_1 \left[(3x) + \frac{(3x)^3}{3!} + \frac{(3x)^5}{5!} + \cdots \right]$$

$$+ C_2 \left[1 + \frac{(-3x)^2}{2!} + \frac{(-3x)^4}{4!} + \cdots \right]$$

$$+ C_2 \left[(-3x) + \frac{(-3x)^3}{3!} + \frac{(-3x)^5}{5!} + \cdots \right]$$

$$= (C_1 + C_2) \left[1 + \frac{(3x)^2}{2!} + \frac{(3x)^4}{4!} + \cdots \right]$$

$$+ (C_1 - C_2) \left[(3x) + \frac{(3x)^3}{3!} + \cdots \right]$$

$$= a_0 \sum_{k=0}^{\infty} \frac{(3x)^{2k}}{(2k)!} + \frac{a_1}{3} \sum_{k=0}^{\infty} \frac{(3x)^{2k+1}}{(2k+1)!}$$

where $C_1 + C_2 = a_0$ and $C_1 - C_2 = a_1/3$.

9. Use power series to solve the differential equation $y'' - xy' = 0$ and find the interval of convergence.

Solution:

Assume $y = \sum_{n=0}^{\infty} a_n x^n$ is a solution. Then

$$y' = \sum_{n=1}^{\infty} n a_n x^{n-1} \quad \text{and} \quad y'' = \sum_{n=2}^{\infty} n(n-1) a_n x^{n-2}.$$

Thus the equation $y'' - xy' = 0$ is written as

$$\sum_{n=2}^{\infty} n(n-1)a_n x^{n-2} - \sum_{n=0}^{\infty} n a_n x^n = 0$$

$$\sum_{n=2}^{\infty} n(n-1)a_n x^{n-2} = \sum_{n=0}^{\infty} n a_n x^n.$$

The index of summation is changed by replacing n by $n+2$ in the left sum to insure that x^n occurs in both sums. Thus,

$$\sum_{n=0}^{\infty} (n+2)(n+1)a_{n+2} x^n = \sum_{n=0}^{\infty} n a_n x^n.$$

Equating coefficients, you have

$$(n+2)(n+1)a_{n+2} = n a_n$$

$$a_{n+2} = \frac{n a_n}{(n+2)(n+1)}. \qquad \text{(Recursion Formula)}$$

Thus, the coefficients of the series solution are

$$a_2 = \frac{0}{2}a_0 = 0 \qquad\qquad a_3 = \frac{1}{3 \cdot 2}a_1$$

$$a_4 = \frac{2}{12}a_2 = 0 \qquad\qquad a_5 = \frac{3}{5 \cdot 4}a_3 = \frac{1 \cdot 3}{2 \cdot 3 \cdot 4 \cdot 5}a_1$$

$$a_7 = \frac{5}{7 \cdot 6}a_5 = \frac{1 \cdot 3 \cdot 5}{2 \cdot 3 \cdot 4 \cdot 5 \cdot 6 \cdot 7}a_1$$

$$\vdots \qquad\qquad\qquad\qquad \vdots$$

$$a_{2k} = 0 \qquad\qquad a_{2k+1} = \frac{1 \cdot 3 \cdot 5 \cdots (2k-1)}{1 \cdot 2 \cdot 3 \cdot 4 \cdot 5 \cdots (2k+1)}a_1$$

$$= \frac{a_1}{2 \cdot 4 \cdot 6 \cdots (2k)(2k+1)}$$

$$= \frac{a_1}{2^k(1 \cdot 2 \cdot 3 \cdots k)(2k+1)}$$

$$= \frac{a_1}{2^k k!(2k+1)}$$

Since the coefficients of the terms with even powers of x are all zero, the solution is

$$y = a_1 \sum_{k=0}^{\infty} \frac{x^{2k+1}}{2^k k!(2k+1)}.$$

You can find the interval of convergence by using the Ratio Test with $u_k = \dfrac{x^{2k+1}}{2^k k!(2k+1)}$.

$$\lim_{k \to \infty} \left| \frac{u_{k+1}}{u_k} \right| = \lim_{k \to \infty} \left| \frac{x^{2k+3}}{2^{k+1}(k+1)!(2k+3)} \cdot \frac{2^k k!(2k+1)}{x^{2k+1}} \right|$$

$$= \lim_{k \to \infty} \left| \frac{(2k+1)x^2}{2(k+1)(2k+3)} \right| = 0$$

for any value of x. Therefore, the interval of convergence is $(-\infty, \infty)$.

15. Use Taylor's Theorem to find the series solution of $y'' - 2xy = 0$ given that $y(0) = 1$ and $y'(0) = -3$. Use the first six terms of the series to approximate y when $x = \frac{1}{4}$.

Solution:

Taylor's Theorem for $c = 0$ is

$$y = y(0) + y'(0)x + \frac{y''(0)}{2!}x^2 + \frac{y'''(0)}{3!}x^3 + \cdots.$$

Since $y'' = 2xy$, $y(0) = 1$ and $y'(0) = -3$, you have

$$y(0) = 1$$
$$y'(0) = -3$$

$$\begin{aligned} y'' &= 2xy & y''(0) &= 0 \\ y''' &= 2xy' + 2y & y'''(0) &= 2 \\ y^{(4)} &= 2xy'' + 4y' & y^{(4)}(0) &= -12 \\ y^{(5)} &= 2xy''' + 6y'' & y^{(5)}(0) &= 0 \\ y^{(6)} &= 2xy^{(4)} + 8y''' & y^{(6)}(0) &= 16 \\ y^{(7)} &= 2xy^{(5)} + 10y^{(4)} & y^{(7)}(0) &= -120. \end{aligned}$$

Therefore, the first six terms of the Taylor series are

$$y = 1 - 3x + 0x^2 + \frac{2}{3!}x^3 - \frac{12}{4!}x^4 + 0x^5 + \frac{16}{6!}x^6 - \frac{120}{7!}x^7 + \cdots$$

$$= 1 - 3x + \frac{1}{3}x^3 - \frac{1}{2}x^4 + \frac{1}{45}x^6 - \frac{1}{42}x^7 + \cdots.$$

Finally, at $x = \frac{1}{4}$ you have

$$y = 1 - \frac{3}{4} + \frac{1}{3 \cdot 4^3} - \frac{1}{2 \cdot 4^4} + \frac{1}{45 \cdot 4^6} + \frac{1}{42 \cdot 4^7} \approx 0.253.$$

17. Use the differential equation $y' - y = 0$ to verify that the series

$$y = \sum_{n=0}^{\infty} \frac{x^n}{n!}$$

converges to the function $y = e^x$ on the interval $(-\infty, \infty)$.

Solution:

Since the solution to a differential equation is unique, you only need to show that both the series and the exponential function are solutions to the given differential equation. Beginning with the series you have

$$y = \sum_{n=0}^{\infty} \frac{x^n}{n!}$$

$$y' = \sum_{n=1}^{\infty} \frac{nx^{n-1}}{n!} = \sum_{n=1}^{\infty} \frac{x^{n-1}}{(n-1)!} = \sum_{n=0}^{\infty} \frac{x^n}{n!} = y.$$

Therefore, the function represented by the series is a solution of the differential equation $y' - y = 0$. Since

$$y = e^x = y',$$

the exponential function is also a solution to the differential equation.

Review Exercises for Chapter 16

15. Find the general solution of the first-order differential equation

$$\frac{dy}{dx} - \frac{y}{x} = \frac{x}{y}.$$

Solution:

The given equation

$$\frac{dy}{dx} - \left(\frac{1}{x}\right)y = xy^{-1}$$

is a Bernoulli equation with $n = -1$, and $1 - n = 2$. Let $z = y^{1-n} = y^2$, then $z' = 2yy'$. Multiplying the original differential equation by $2y$ produces

$$2yy' - 2\left(\frac{1}{x}\right) y^2 = 2x$$

$$z' - \left(\frac{2}{x}\right) z = 2x.$$

This equation is linear in z. Using $P(x) = -2/x$ yields

$$\int P(x)\, dx = \int \frac{-2}{x}\, dx = -2\ln x = \ln x^{-2}$$

which implies that $e^{\ln x^{-2}} = x^{-2}$ is an integrating factor. Multiplying the linear equation by this factor produces

$$\left(\frac{1}{x^2}\right) z' - \left(\frac{2}{x^3}\right) z = \frac{2}{x}$$

$$\frac{d}{dx}\left[\frac{1}{x^2}(z)\right] = \frac{2}{x}$$

$$z\left(\frac{1}{x^2}\right) = \int \frac{2}{x}\, dx = 2\ln|x| + C$$

$$y^2 = 2x^2 \ln|x| + Cx^2 = x^2 \ln x^2 + Cx^2.$$

19. Find the general solution of the first-order differential equation
$$(2x - 2y^3 + y)\, dx + (x - 6xy^2)\, dy = 0.$$

Solution:

Since
$$\frac{\partial M}{\partial y} = 1 - 6y^2 = \frac{\partial N}{\partial x}$$

the equation is exact. Therefore, there exists a function f such that

$$f_x(x, y) = 2x - 2y^3 + y \quad \text{and} \quad f_y(x, y) = x - 6xy^2.$$

Integration yields

$$f(x, y) = \int (2x - 2y^3 + y)\, dx = x^2 - 2xy^3 + xy + g(y)$$

$$f(x, y) = \int (x - 6xy^2)\, dy = xy - 2xy^3 + h(x)$$

Reconciling these two version of f, yields $h(x) = x^2$, $g(y) = 0$ and

$$f(x, y) = x^2 - 2xy^3 + xy.$$

Therefore, the general solution $f(x, y) = C$ is

$$x^2 - 2xy^3 + xy = C.$$

29. Find the general solution of the first-order differential equation $(1 + x^2)\,dy = (1 + y^2)\,dx$.

Solution:

Since the variables are separable, you have

$$\frac{dy}{1 + y^2} = \frac{dx}{1 + x^2}$$

$$\int \frac{dy}{1 + y^2} = \int \frac{dx}{1 + x^2}$$

$$\arctan y = \arctan x + C_1$$

Now taking the tangent of this equation and using the identity for $\tan (A - B)$, yields

$$\tan (\arctan y - \arctan x) = \tan C_1$$

$$\frac{\tan \arctan y - \tan \arctan x}{1 + \tan (\arctan y) \tan (\arctan x)} = C$$

$$\frac{y - x}{1 + xy} = C.$$

as the general solution.

43. Assume that the rate of change in the number of miles s of road cleared per hour by a snowplow is inversely proportional to the height h of snow.

(a) Write and solve the differential equation to find s as a function of h.

(b) Find the particular solution if $s = 25$ miles when $h = 2$ inches and $s = 12$ miles when $h = 10$ inches ($2 \leq h \leq 15$).

Solution:

(a) Since the rate of change of the number of miles s of road cleared per hour is inversely proportional to the height of the snow h, the differential equation is

$$\frac{ds}{dh} = \frac{k}{h}.$$

This is a separable differential equation. Therefore,

$$\frac{ds}{dh} = \frac{k}{h}$$

$$ds = k\frac{1}{h}\,dh$$

$$\int ds = k\int \frac{1}{h}\,dh$$

$$s = k\ln h + C.$$

(b) To find the particular solution use the fact that $s = 25$ when $h = 2$ and $s = 12$ when $h = 10$ to obtain the following system of equations:

$$25 = k\ln 2 + C$$
$$12 = k\ln 10 + C$$

Subtracting the second equation from the first yields

$$13 = k(\ln 2 - \ln 10) = k\ln \frac{1}{5} = -k\ln 5.$$

Thus, $k = -13/\ln 5$. Substituting this expression for k into the first equation of the system produces

$$25 = -\frac{13}{\ln 5}\ln 2 + C \quad \Longrightarrow \quad C = 25 + \frac{13}{\ln 5}\ln 2.$$

Substituting the results for k and C into the general solution of the differential equation yields the particular solution

$$s = -\frac{13}{\ln 5}\ln h + 25 + \frac{13}{\ln 5}\ln 2$$
$$= 25 - \frac{13}{\ln 5}(\ln h - \ln 2)$$
$$= 25 - \frac{13\ln(h/2)}{\ln 5} \quad 2 \le h \le 15.$$

53. Find the general solution of the second-order differential equation $y'' - 2y' + y = 2xe^x$.

Solution:

The characteristic equation $m^2 - 2m + 1 = 0$ has equal roots given by $m = 1$. Thus the general solution is

$$y_h = C_1 e^x + C_2 x e^x.$$

Since $F(x) = 2xe^x$, use the Method of Undetermined Coefficients with

$$y_p = (A + Bx)e^x = Ae^x + Bxe^x.$$

This function is not independent of y_h since it already contains both of these terms. Therefore, multiply by x^2 to obtain

$$y_p = Ax^2 e^x + Bx^3 e^x = e^x(Ax^2 + Bx^3)$$
$$y_p{}' = 2Axe^x + (3B + A)x^2 e^x + Bx^3 e^x$$
$$y_p{}'' = 2Ae^x + (4A + 6B)xe^x + (A + 6B)x^2 e^x + Bx^3 e^x.$$

Substituting into the given differential equation, yields

$$2Ae^x + (4A + 6B)xe^x + (A + 6B)x^2 e^x + Bx^3 e^x - 4Axe^x$$
$$-2(3B + A)x^2 e^x - 2Bx^3 e^x + Ax^2 e^x + Bx^3 e^x = 2xe^x$$

or

$$2Ae^x + 6Bxe^x = 2xe^x.$$

Equating coefficients, yields $A = 0$ and $6B = 2$, or $B = \frac{1}{3}$. Hence, $y_p = \frac{1}{3}x^3 e^x$, and the general solution is

$$y = y_h + y_p$$
$$= C_1 e^x + C_2 x e^x + \frac{1}{3}x^3 e^x$$
$$= \left(C_1 + C_2 x + \frac{1}{3}x^3 \right) e^x.$$